Springer Series in
SOLID-STATE SCIENCES 148

Springer Series in
SOLID-STATE SCIENCES

Series Editors:
M. Cardona   P. Fulde   K. von Klitzing   R. Merlin   H.-J. Queisser   H. Störmer

The Springer Series in Solid-State Sciences consists of fundamental scientific books prepared by leading researchers in the field. They strive to communicate, in a systematic and comprehensive way, the basic principles as well as new developments in theoretical and experimental solid-state physics.

136 **Nanoscale Phase Separation and Colossal Magnetoresistance**
The Physics of Manganites and Related Compounds
By E. Dagotto

137 **Quantum Transport in Submicron Devices**
A Theoretical Introduction
By W. Magnus and W. Schoenmaker

138 **Phase Separation in Soft Matter Physics**
Micellar Solutions, Microemulsions, Critical Phenomena
By P.K. Khabibullaev and A.A. Saidov

139 **Optical Response of Nanostructures**
Microscopic Nonlocal Theory
By K. Cho

140 **Fractal Concepts in Condensed Matter Physics**
By T. Nakayama and K. Yakubo

141 **Excitons in Low-Dimensional Semiconductors**
Theory, Numerical Methods, Applications   By S. Glutsch

142 **Two-Dimensional Coulomb Liquids and Solids**
By Y. Monarkha and K. Kono

143 **X-Ray Multiple-Wave Diffraction**
Theory and Application
By S.-L. Chang

144 **Physics of Transition Metal Oxides**
By S. Maekawa, T. Tohyama, S.E. Barnes, S. Ishihara, W. Koshibae, and G. Khaliullin

145 **Point-Contact Spectroscopy**
By Yu.G. Naidyuk and I.K. Yanson

146 **Optics of Semiconductors and Their Nanostructures**
Editors: H. Kalt and M. Hetterich

147 **Electron Scattering in Solid Matter**
A Theoretical and Computational Treatise
By J. Zabloudil, R. Hammerling, L. Szunyogh, and P. Weinberger

148 **Physical Acoustics in the Solid State**
By B. Lüthi

Volumes 90–135 are listed at the end of the book.

B. Lüthi

Physical Acoustics in the Solid State

With 188 Figures

Springer

Professor Bruno Lüthi
Physikalisches Institut
Universität Frankfurt
D-60054 Frankfurt
E-mail: luethi@physik.uni-frankfurt.de

Series Editors:

Professor Dr., Dres. h. c. Manuel Cardona
Professor Dr., Dres. h. c. Peter Fulde*
Professor Dr., Dres. h. c. Klaus von Klitzing
Professor Dr., Dres. h. c. Hans-Joachim Queisser

Max-Planck-Institut für Festkörperforschung, Heisenbergstrasse 1, 70569 Stuttgart, Germany
* Max-Planck-Institut für Physik komplexer Systeme, Nöthnitzer Strasse 38
 01187 Dresden, Germany

Professor Dr. Roberto Merlin
Department of Physics, 5000 East University, University of Michigan
Ann Arbor, MI 48109-1120, USA

Professor Dr. Horst Störmer
Dept. Phys. and Dept. Appl. Physics, Columbia University, New York, NY 10027 and
Bell Labs., Lucent Technologies, Murray Hill, NJ 07974, USA

ISSN 0171-1873

ISBN 3-540-22910-8 Springer Berlin Heidelberg New York

Library of Congress Control Number: 2004111958

This work is subject to copyright. All rights are reserved, whether the whole or part of the material is concerned, specifically the rights of translation, reprinting, reuse of illustrations, recitation, broadcasting, reproduction on microfilm or in any other way, and storage in data banks. Duplication of this publication or parts thereof is permitted only under the provisions of the German Copyright Law of September 9, 1965, in its current version, and permission for use must always be obtained from Springer. Violations are liable to prosecution under the German Copyright Law.

Springer is a part of Springer Science+Business Media.
springeronline.com

© Springer-Verlag Berlin Heidelberg 2005
Printed in Germany

The use of general descriptive names, registered names, trademarks, etc. in this publication does not imply, even in the absence of a specific statement, that such names are exempt from the relevant protective laws and regulations and therefore free for general use.

Production and Typesetting: PTP-Berlin Protago-T$_E$X-Production GmbH, Germany
Cover concept: eStudio Calamar Steinen
Cover production: *design & production* GmbH, Heidelberg

Printed on acid-free paper   SPIN: 10980897   57/3141/YU - 5 4 3 2 1 0

for Ursula

Preface

This book gives an up to date survey of acoustic effects in the Solid State. After a review of the different experimental techniques and an introduction to the theory of elasticity, emphasizing the symmetry aspects, applications are given for the different fields of condensed matter physics. These applications include metals and semiconductors, superconductivity, unstable magnetic moments including heavy fermion physics, magnetism, structural and magnetic phase transitions, low dimensional systems, amorphous systems and symmetry related experiments. The main emphasis is on more recent developments not covered in books written 30 years ago. Actually, acoustic experiments have been performed in all modern fields of solid state physics.

Acknowledgments

Various chapters of the book were read by different colleagues: Wolfram Brenig, Peter Fulde, Ute Löw, Bernd Wolf and Sergei Zherlitsyn. I would like to thank them for their helpful remarks and constructive criticism. Part of the book was written while the author was staying at the Max Planck Institutes in Dresden. I would like to thank Frank Steglich and Peter Fulde for the hospitality and many discussions. Much of the work presented here has been developed with students, postdoctoral fellows and colleagues. I would like to thank all of them for their contributions, their help and for sharing their insights. Finally I would like to thank Bernd Wolf for his help during all stages of the project, in particular with the figures. I also would like to thank Carolyne and Malcolm Agnew for their help on the final layout in the LATEXversion.

Frankfurt am Main, July 2004 *Bruno Lüthi*

Contents

1 **Introduction** 1

2 **Experimental Techniques** 5
   2.1   Transducer 6
   2.2   Sound Velocity and Attenuation, Experimental Techniques... 7
         2.2.1   Simple Ultrasonic Set-Up 7
         2.2.2   Relative Sound Velocity and Attenuation Changes 9
         2.2.3   Ultrasonics at Very Low Temperatures 11
         2.2.4   Absolute Sound Velocity Measurements 12
         2.2.5   Resonant Ultrasound Spectroscopy, RUS 14
         2.2.6   Vibrating Reed Technique 16
   2.3   Phonon Echoes 16
   2.4   Ultrasonics in Pulsed Magnetic Fields 17
   2.5   Surface Acoustic Wave Generation and Detection 20
   2.6   Microwave Ultrasonics 21
   2.7   Brillouin Scattering 23
   2.8   Thermal Expansion and Magneto-striction,
        Thermal Conductivity 25
         2.8.1   Thermal Expansion and Magneto-striction 25
         2.8.2   Thermal Conductivity 25

3 **Elasticity** 27
   3.1   Strains, Stresses and Elastic Constants 27
   3.2   Symmetry Aspect 32
   3.3   Third Order Elastic Constants 36
   3.4   Elastic Stability, Elastic Isotropy 37
   3.5   Surface Acoustic Waves, SAW 39
   3.6   Lattice Dynamics 44
         3.6.1   Phonon Dispersion 44
         3.6.2   Debye Theory of Lattice Dynamics 45

4 **Thermodynamics and Phase Transitions** 47
   4.1   Thermodynamic Potentials 47
   4.2   Background Elastic Constant and Attenuation 49
         4.2.1   Background Elastic Constant, Thermal Expansion
                and Specific Heat 49

		4.2.2	Sound Dissipation Due to Phonons in Insulators	52

    4.3 Landau Theory, Strain–Order Parameter Coupling 52
          4.3.1 Landau Theory for Second Order Phase Transitions .. 53
          4.3.2 Scaling Relations 54
          4.3.3 Landau Theory for a First Order Phase Transition.... 55
          4.3.4 Strain–Order Parameter Coupling 56
          4.3.5 Fluctuation Effects 59
          4.3.6 Landau–Khalatnikov Theory 60
    4.4 Ginzburg Criterion and Marginal Dimensionality 62
    4.5 Adiabatic and Isothermal Quantities 63

5 Acoustic Waves in the Presence of Magnetic Ions 67
    5.1 Strain–Magnetic Ion Interaction 67
          5.1.1 Magnetic Interactions 68
          5.1.2 Single Ion–Strain Interaction 70
          5.1.3 Exchange Striction 72
    5.2 Thermodynamic Functions for 4f–Rare Earth Ions
         in the Presence of Crystal Fields 73
          5.2.1 Specific Heat and Thermal Expansion 75
          5.2.2 Magnetic Susceptibility and Elastic Constants 77
          5.2.3 Other Experimental Methods
                to Determine Magneto-elastic Coupling Constants 80
          5.2.4 Magneto-elastic Coupling Constants 80
    5.3 Susceptibilities with Interactions 82
    5.4 External and Internal Strains 83
    5.5 Paramagnetic Spin–Phonon Interaction
         for 3d-Transition Metal Ions in Crystals 87
    5.6 Nuclear Acoustic Resonance 91

6 Ultrasonics at Magnetic Phase Transitions 93
    6.1 Magnetic Phase Transition 94
          6.1.1 Spin–Phonon Coupling Mechanism 94
          6.1.2 Critical Attenuation Coefficient 95
          6.1.3 Sound Velocity Effects
                near Magnetic Phase Transitions 101
    6.2 Spin Reorientation Phase Transition 103
    6.3 Sound Propagation
         in the Spin-Density Wave Anti-ferromagnet Chromium 106

7 Ultrasonics at Structural Transitions 109
    7.1 Charge Order 110
    7.2 Cooperative Jahn–Teller Effect
         and Quadrupolar (Orbital) Transition 119
          7.2.1 Case of Transition Metal Compounds 121
          7.2.2 Case of Rare Earth Compounds 125
          7.2.3 Case of Actinide Compounds 139

		7.2.4 Effective Quadrupole–Quadrupole Interaction 144

       7.2.4 Effective Quadrupole–Quadrupole Interaction 144
       7.2.5 Summary... 146
  7.3 Ferro-elastic and Martensitic Phase Transitions............. 147
  7.4 Other Structural Phase Transitions 151
       7.4.1 Piezo-distortive Ferro-electrics 151
       7.4.2 Other Non-electronic Compounds 151
       7.4.3 Strain Is Not Order Parameter 152
       7.4.4 Conclusion 156

8 Metals and Semiconductors................................ 157
  8.1 Deformation Potential Coupling 157
  8.2 Elastic Constants and Ultrasonic Attenuation,
      Case of $ql_e < 1$ 159
       8.2.1 Electronic Redistribution Mechanism 159
       8.2.2 Alpher–Rubin Effect in Magnetic Fields 164
  8.3 Ultrasonic Propagation, Case of $ql_e > 1$ 166
       8.3.1 Ultrasonic Attenuation and Dispersion 166
       8.3.2 Geometric Resonances 167
  8.4 Magneto-acoustic Quantum Oscillations 169
       8.4.1 Theory... 169
       8.4.2 Applications 172
  8.5 Ultrasonics in Semiconductors and Semimetals 177
       8.5.1 Inter-valley Scattering Effect....................... 177
       8.5.2 Acousto-electric Effect 177
       8.5.3 Sound Wave Amplification
             in Piezoelectric Semiconductors 178
       8.5.4 Ultrasonic Amplification in Bismuth 179

9 Unstable Moment Compounds.............................. 181
  9.1 Experimental Characterisation 183
       9.1.1 Mixed Valence Compounds 183
       9.1.2 Kondo Alloys and Kondo Lattices 184
  9.2 Electron–Lattice Coupling in Mixed-Valence Systems 187
  9.3 Electron–Phonon Coupling in Heavy Fermion Systems 196
       9.3.1 Introduction 196
       9.3.2 Kondo Volume Collapse 198
       9.3.3 Special Sound Propagation Effects for Large Ω 199
       9.3.4 Scaling Approach for the Temperature Dependence
             of Thermal Quantities 202
       9.3.5 Sound Wave Effects in Magnetic Fields,
             Meta-magnetic Transition 213
       9.3.6 Non-Fermi Liquid Effects.......................... 219

10 Ultrasonics in Superconductors 223
  10.1 Introduction 223
  10.2 Electron–Strain Coupling in Superconductors 225

	10.2.1 Elastic Constants................................. 226
	10.2.2 Ultrasonic Attenuation............................ 227
10.3	Conventional Superconductors 229
10.4	High Temperature Superconductors...................... 237
10.5	Sound Wave Interaction with the Flux Line Lattice 241
10.6	Ultrasonic Surface Wave Attenuation in Superconductors 246
10.7	Unconventional Superconductivity 247
	10.7.1 Heavy Fermion Superconductivity 249
	10.7.2 Other Unconventional Superconductors 264
	10.7.3 Other Methods to Probe the Energy Gap Structure... 267
	10.7.4 Summary.. 268

11 Coupling to Collective Excitations 271

11.1	Plasmons and Helicons 271
	11.1.1 Dielectric Tensor 271
	11.1.2 Plasma Polariton 272
	11.1.3 Helicons and Alfven Waves 274
	11.1.4 Helicon–Phonon Interaction 275
11.2	Magneto-elastic Waves 277
	11.2.1 Ferromagnetic Spinwaves with Dipolar Interaction 277
	11.2.2 Magneto-static Modes 279
	11.2.3 Spinwave–Phonon Interaction 280
	11.2.4 Magneto-elastic Gap 282
	11.2.5 Experiments with Magneto-elastic Waves in a Ferrimagnet 284
	11.2.6 High Power Level Effects 286
	11.2.7 Sound Wave Experiments in Anti-ferromagnets....... 286

12 Ultrasonics in Low Dimensional Spin and Electronic Peierls-Systems 291

12.1	Magnetic Properties of Low Dimensional Spin Systems 293
	12.1.1 Uniform Chain 293
	12.1.2 Dimerized Chains 295
12.2	Temperature Dependence of Elastic Constants in Low Dimensional Spin Systems 300
	12.2.1 Temperature Dependence of Elastic Constants in Quasi One-Dimensional Spin Systems.............. 300
	12.2.2 Case of Two-Dimensional Dimer Spin Systems 303
12.3	Magnetic Field Effects 309
12.4	Thermal Conductivity in Low Dimensional Spin Systems 319
12.5	Peierls and Spin Peierls Effects 323
12.6	Perovskite-Type Layer-Structure Materials.................. 326
12.7	Bose–Einstein Condensation of Magnons in $TlCuCl_3$ 328
12.8	Conclusion 328

13 Symmetry Effects with Sound Waves 331
13.1 Magnetic Field Induced Symmetry Breaking 332
13.2 Rotationally Invariant Magneto-elastic Effects 335
13.3 Magneto-acoustic Birefringence Effects 340
  13.3.1 Voigt–Cotton–Mouton Geometry 341
  13.3.2 Faraday Geometry 343
13.4 Acoustical Activity 346
13.5 Non-reciprocal Surface Acoustic Wave Effects 347
13.6 Surface Acoustic Wave Effects
    in the Integral and Fractional Quantum Hall Effect 351

14 Ultrasonic Propagation
in Tunneling Systems 357
14.1 Crystalline Systems 357
14.2 Amorphous Systems 360
14.3 Recent Developments 363
14.4 Spin-Glass 364
14.5 Quasi-crystals 364

15 Conclusion and Outlook 367

Appendix 369

A Mass Systems and Units 369

B Wave Equation for Sound Waves 371

**C Elastic Constants and Symmetry Strains
for the Crystal Classes** 373

D g-Factor, Steven's Factors, CEF Operators 375
  D.1 $3d$ and $4f$ Ions: Landé g_J Factor, Steven's Factors 375
  D.2 Cubic CEF and Quadrupolar Operators 376

E Ultrasonic Attenuation in Metals 377

F Free Energy of Electrongas 379

G Order–Disorder Phase Transition 381

References 383

Index 419

Index of Materials 423

1 Introduction

Books on physical acoustics are numerous. A series of volumes on "Physical Acoustics, Principles and Methods" by W.P. Mason [1.1] and later editors have been available since 1964. In addition, there are monographs on the same topics: "Ultrasonic Methods in Solid State Physics" by Truell et al. [1.2], "Physical Ultrasonics" by Beyer and Letcher [1.3], "Microwave Ultrasonics in Solid State Physics" by Tucker and Rampton [1.4]. But since these latter books were written more than 30 years ago, it is time to present a new account of the field.

The precise aim of this book is to present a modern account of the field. The emphasis is also slightly changed. The aspects of traditional elasticity theory are only treated briefly since they can be found in the books and treatises mentioned above and in books devoted solely to elasticity. Examples of such books are Love [1.5], Kolsky [1.6] and the standard texts on theoretical physics such as Landau Lifshitz [1.7] etc. Physical acoustics embraces the measurements of ultrasonic velocity and attenuation. The elastic constants can be gained from the ultrasonic velocities. The elastic constants are thermodynamic derivatives, the second derivative of the free energy with respect to the strains. Therefore, they are connected with the atomic and molecular bonding in the crystal. They are important, together with the specific heat and thermal expansion, for the equation of the state of a material. On the other hand the attenuation, as a transport coefficient, is connected with dissipation. It is affected by defects and inhomogeneities of the solid and more fundamentally by conduction electrons, inner (f-)electrons, thermal phonons and other relaxation processes.

This book aims to show what can be learnt about "solid state physics" using ultrasonic waves. Therefore the main objects of investigation are fields such as electron–phonon coupling in metals and semiconductors, ultrasonic effects in superconductors, spin–phonon interaction in paramagnetic and magnetic systems, phonon coupling to collective excitations and sound propagation effects near phase transitions. Of course, modern aspects of all these effects are specially treated at the expense of older treatments. The symmetry aspects are emphasized whenever possible.

Since the late 1960s and early 1970s, new fields in solid state physics have appeared. Examples are "low dimensional physics", including "low dimensional spin physics", or "unconventional superconductivity", or new types of structural phase transitions like "charge ordering" and "cooperative Jahn–Teller transitions" or "mixed valency and heavy fermions" and finally the exciting field of two-dimensional electron systems, including the quantum Hall effect, and others. These new topics, in which ultrasonics has also played an important role, will be treated in detail. In semiconductors, there are the heterostructure materials GaAs/GaAlAs which exhibit integer and fractional quantum Hall effects. For the latter case, surface acoustic waves played an important role in characterising ground and excited states. Finally much work has been done in characterising non-crystalline solids with ultrasound. The important experiments done with ultrasonics for tunneling state systems in various materials, crystalline and non-crystalline, such as glasses will be discussed briefly.

All these new topics listed can be put together under the title "highly correlated electronic systems". Whereas in former years in Solid State Physics, single particle phenomena were mainly studied, in recent years the focus has shifted to correlated electronic systems. It is therefore important to know what can be learned about these systems with ultrasonic methods. Various chapters in this book focus especially on this topic (Chaps. 7, 9, 10, 12–14).

On viewing the table of contents, it can be seen that virtually all sections of solid state physics occur in this book, indicating that acoustic techniques play an important role in this part of physics. It should also be mentioned what has been left out of this book. From the title of the book, it is clear that of the whole area of condensed matter physics it is only the solid state which is being treated. Important fields, like the liquid state, have been left out. In the liquid state, there are various subjects where ultrasonics plays an important role: Liquid Helium as He^4 or He^3. In these quantum liquids, acoustic experiments played a decisive role in determining and interpreting the different phases (see e.g. Vollhardt and Wölfle [1.8]). Another important field in the liquid state is that of liquid crystals, where ultrasonics again made important contributions in the so-called nematic, cholesteric and smectic liquid crystals (see e.g. Stephen and Straley [1.9], de Gennes and Prost [1.10]). Other topics related to ultrasonics and microwave acoustics are the different methods of studying transport properties with the use of ballistic heat pulses. This technique has widespread applications, especially for well-characterised insulators. Contact with this field will be rare. Reviews on these topics are Wolfe [1.11]), Bron [1.12].

Non-dissipative sound attenuation in solids – i.e. sound scattering from imperfections like grain boundaries or diffraction losses from diffraction effects from the transducers or due to nonparallelicity of the crystal or due to mode conversion on the surface etc. – will only be mentioned occasionally. Also dislocation damping in all its different manifestations (point defect relaxation, Snoek relaxation, Zener relaxation, dislocation motion or eddy current effects) have been investigated in the early treatises and will not be covered in this text. See for the latter effects e.g. Nowick and Berry [1.13]).

The cgs-Gauss mass system is mostly used in this book since in the solid-state physics literature this mass system is used. Changes to the SI system or tables for units can be found in Appendix A.

An introduction to the field is given in the first three chapters. Experimental techniques are described in Chap. 2. The emphasis is especially on new developments that have not been covered in previous books on ultrasonics cited above. These new developments are e.g. resonant ultrasonic spectroscopy, ultrasonics in high pulsed magnetic fields, high resolution Brillouin scattering. Chapter 3 introduces elasticity emphasizing the symmetry aspects which are not covered in other books. Chapter 4 presents the necessary background of thermodynamics with the thermodynamic potentials and functions. The Landau theory of phase transitions and the range of its validity is also presented. The following Chaps. 5–14 cover the full range of the various applications in solid state physics.

This book is intended for all those who want to know what can be learned from ultrasonics about the solid state. I hope to give an updated account of this field. Graduate students and researchers in the field of physical acoustics or related fields, should also benefit from this book. Since the whole outlook of the book is based on Solid State Physics some books on this topic are listed for the reader: "Solid State Physics" by Ashcroft and Mermin [1.14], "Festkörperphysik" by Ibach and Lüth [1.15], "Introduction to Solid State Physics" by Kittel [1.16]. Recently a "Handbook of Elastic Properties of Solids, Liquids and Gases" (Editors: M. Levy, H.E. Bass, R.R. Stern) appeared placing emphasis on measuring methods and surveys on quite different materials (elements, novel materials, building materials). The outlook of this handbook is quite different to the one given in the present book.

2 Experimental Techniques

A discussion follows on the various experimental techniques used for ultrasonic investigations. In the last few decades, considerable progress has been made in the field of high resolution sound velocity and sound attenuation measurements. A wide variety of different methods are used in this field. A distinction can be made between techniques which use transducers and others which are contact free methods. To the latter ones belong the vibrating reed technique and the Brillouin scattering. Numerous books and review articles describe all these topics.

In the low frequency regime, where the sound wavelength is of the dimension of the specimen, the elastic moduli (Young's modulus E and shear modulus G) can be determined by a c.w. resonance method or by measuring flexural and torsional oscillations. At audio frequencies the vibrating reed technique is most frequently applied (Read et al. [2.1]). A brief account of this technique is given in Sect. 2.2.5. In the ultrasonic regime, the elastic constants c_{ij} can be determined by using pulse or c.w. techniques (Truell et al. [2.2], Bolef and Miller [2.3], Fuller et al. [2.4]). The most widely used method to measure sound velocity and attenuation is the pulse superposition technique or variations of it (Truell et al. [2.2], Fuller et al. [2.4]). A particularly interesting method is the shape resonance technique, also called resonant ultrasound spectroscopy (RUS), suitable for small crystals and low symmetry crystals (Migliori and Sarrao [2.5]). This technique will be discussed together with phase sensitive sound velocity and attenuation measurements in Sect. 2.2. In Sect. 2.3 a short account of phonon echoes will be given. In Sect. 2.4 ultrasonics in pulsed high magnetic field will be described. Apart from the low frequency and the ultrasonic regime, sound waves can also be generated in the microwave region (Tucker and Rampton [2.6]). A method which was used for various applications is the excitation of microwave sound at surfaces in microwave cavities. This will be discussed in Sect. 2.6. The excitation and detection of acoustic surface waves SAW will be discussed in Sect. 2.5. Finally, in Sect. 2.7, the Brillouin scattering will be mentioned. Thermal expansion and magneto-striction experiments are important to determine the length changes with temperature or magnetic field. Apart from this they are important thermodynamic functions in their own right. The experimental technique for these quantities and for measuring thermal conductivity will be discussed in Sect. 2.8. We begin with a discussion of ultrasonic transducers in Sect. 2.1.

2.1 Transducer

Several techniques are in use to generate and to detect ultrasonic waves. The most common one is to use piezoelectric transducers. But also magnetostrictive transducers or electromagnetic generation and detection of sound can be used in special cases. A brief review of the different techniques to generate and detect sound waves follows and some review articles where more details can be found are named.

With the strains and stresses defined in Chap. 3, the piezoelectric effect gives a stress–strain electric field relation

$$T_i = c_{ik}\varepsilon_k - e_{ij}E_j \ . \tag{2.1}$$

Here we use already the contracted Voigt notation, see Sect. 3.1 ($11 \to 1$, $22 \to 2$, $33 \to 3$, $23 \to 4$, $13 \to 5$, $12 \to 6$) for the stress tensor T_{ij} and the strain tensor ε_{ij}. c_{ij} are the elastic constants and e_{ik} the piezoelectric stress coefficients. The inverse relation reads

$$\varepsilon_i = s_{ik}T_k + d_{ij}E_j \tag{2.2}$$

with s_{ik} a component of the compliance tensor. The piezoelectric matrix in contracted notation is given by (e_{ij}) with $i = 1, 2, 3$ (3 directions) and $j = 1, \cdots, 6$ (six components of the stress tensor). For quartz (SiO_2) it has the form

$$[e_{ij}] = \begin{pmatrix} e_{11} & -e_{11} & 0 & e_{14} & 0 & 0 \\ 0 & 0 & 0 & 0 & -e_{14} & -2e_{14} \\ 0 & 0 & 0 & 0 & 0 & 0 \end{pmatrix}$$

with $e_{11} = 0.171 C/m^2$, $e_{14} = 0.0403 C/m^2$ with C dimension of Coulomb (Appendix A). For LiNbO$_3$ the corresponding components read:

$$[e_{ij}] = \begin{pmatrix} 0 & 0 & 0 & 0 & e_{15} & -e_{22} \\ -e_{22} & e_{22} & 0 & e_{15} & 0 & 0 \\ e_{31} & e_{31} & e_{33} & 0 & 0 & 0 \end{pmatrix}$$

with $e_{15} = 3.65 C/m^2$, $e_{22} = 2.39 C/m^2$, $e_{31} = 0.31 C/m^2$, $e_{33} = 1.72 C/m^2$ (Ogi et al. [2.7]).

For an X-cut quartz an electric field in the x-direction produces a strain in the same direction, therefore producing longitudinal waves. With the parameters for quartz, the longitudinal elastic constant $c_{11} = 8.6 \times 10^{11} erg/cm^3$, the piezoelectric constant $e_{11} = 0.171 \frac{C}{m^2}$ and with a typical applied electric field value of $E_x = 10V/mm$, the stress is given by $T_1 = e_{11}E_x = 1.71 CV/m^3$ and the strain by $\varepsilon_1 = \varepsilon_{xx} = T_1/c_{11} = 0.2 \times 10^{-6}$ – a rather small strain. This would be the maximum strain for a longitudinal sound wave in the x-direction for the typical applied ac electric field. Likewise, shear waves may be generated with a Y-cut crystal. The purest shear mode with a minimum of

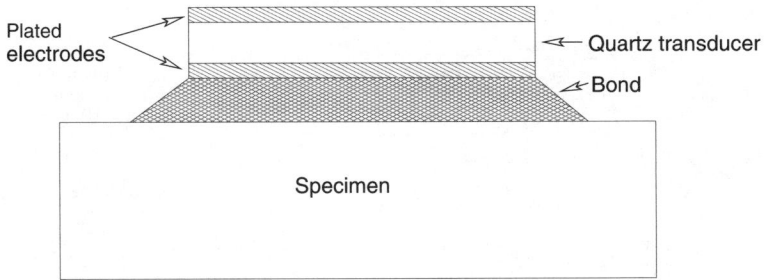

Fig. 2.1. Transducer-Bond-Specimen arrangement

coupling to other modes is the so-called AC-cut quartz crystal, a cut rotated by 31° about the x-axis.

Further details on these transducers can be found in Mason [2.8]. Another transducer material is $LiNbO_3$ with larger piezoelectric coupling constants as shown above. With these transducers, the resonance frequency or odd integer multiples thereof are used.

In addition, there are piezoelectric polymer foils with high efficiency. They can be used non-resonantly over a large frequency range. They are especially suited for longitudinal waves. All these transducers have to be bonded to the specimen with plane parallel polished faces (see Fig. 2.1). **Thiokol LP, GE** cement, **Nonaq** or **UHU** cement can be used as bond materials. If thin film technology is used, transducers, such as **ZnO** or **CdS**, can be evaporated or sputtered directly on to the specimen without a bond in between. Details on this fabrication technology can be found in Foster [2.9]. This thin film technique is also important for surface acoustic waves as discussed in Sect. 2.5. Problems arising with the different types of transducers for measuring ultrasonic velocity and attenuation are discussed in Sect. 2.2.

Apart from piezoelectric transducers, there have been studies also on magneto-strictive transducers and on electromagnetic generation of ultrasound. The latter ones have been carried out in metals, ferromagnets and various magnetic materials. Since these techniques have little technical application, no further details are given here but some relevant reviews are: Dobbs [2.10], Buchelnikov and Vasil'ev [2.11], Gorodetsky et al. [2.12].

2.2 Sound Velocity and Attenuation, Experimental Techniques

2.2.1 Simple Ultrasonic Set-Up

A simple ultrasonic system for producing stress waves is now discussed. In Fig. 2.1 we show the transducer–sample system. The (piezoelectric) transducer with electrodes on both sides is bonded to the specimen with parallel

8   2 Experimental Techniques

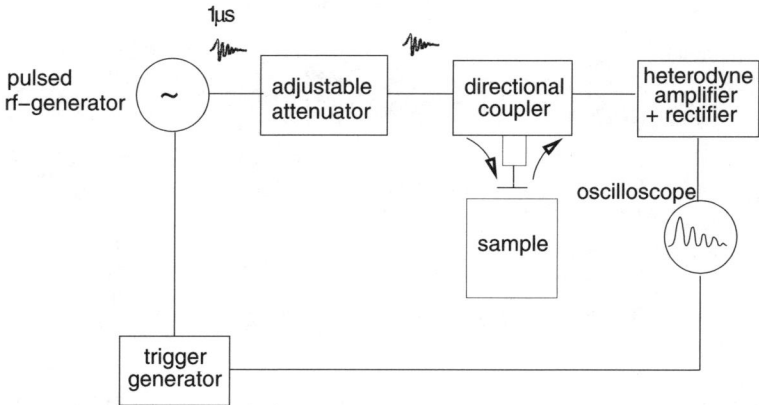

Fig. 2.2. Ultrasonic system to measure sound velocity and attenuation

end faces. For the choice of bond materials see Sect. 2.1. A pulsed electromagnetic signal of ∼ 1 µs duration, operating at the fundamental frequency of the transducer or at one of its odd harmonics, is applied across the transducer. With the piezoelectric effect, a stress wave is produced which propagates through the specimen.

In the single transducer arrangement of Fig. 2.2, the same transducer acts as a receiver. Upon reflection on the surface, a small amount of energy is converted back to an electric pulse. The major part of the pulse is reflected and propagates further, producing an ultrasonic echo pattern as shown e.g. in Fig. 2.5a. The converted electric pulse is amplified (with homodyne or heterodyne amplifiers) and shown on the oscilloscope. To investigate a wide frequency range, transducers can be used on both opposite faces of the sample (emitter and receiver) – this does not need a directional coupler. In the following, we discuss phase sensitive devices to measure ultrasonic velocity and attenuation with high accuracy.

A distinction can be made between pulse echo techniques and cw-techniques. In the latter case, the transmitting transducer is driven continuously and a resonant response is observed at frequencies which correspond to $L = n\frac{\lambda}{2}$ with L the sample length and λ the wavelength of the sound. The ultrasonic wave velocity is determined from the resonant frequencies with transducer corrections included. The attenuation follows from the quality factor Q of the resonance signals. In the following, we discuss a special phase-sensitive ultrasonic set-up, absolute sound velocity measurements, the so-called Resonant Ultrasonic Spectroscopy (RUS) and the vibrating reed technique.

2.2.2 Relative Sound Velocity and Attenuation Changes

Of the great variety of experimental techniques mentioned above, we describe an ultrasonic setup used successfully in our laboratory (Heil et al. [2.13], Lüthi et al. [2.14]). It is a slightly changed version of a setup described by Wallace and Garland [2.15]. Our apparatus allows us to perform simultaneous measurements of the ultrasonic velocity $\frac{\Delta v}{v_0}$ and attenuation α as a function of temperature or magnetic field. The frequency range of 5–500 MHz is covered and the duration of the ultrasonic echo pulse is 0.1–1 μs. The repetition rate depends on the available cooling power in the cryostat and lies between 100 Hz in the mK temperature range and a few kHz at higher temperatures. Figure 2.3 shows the electronic part of the ultrasonic setup for the simultaneous detection of ultrasonic velocity changes and attenuation.

With the frequency generator (1) a frequency between 10–500 MHz is chosen, depending on the choice of the transducers. The voltage divider (2) gives the signal for the specimen and for the reference channel. The diode switch (3) triggered by the pulse generator (4) gives a pulse modulated signal with a pulse duration of 0.1–1 μs and a typical repetition rate of 10 kHz. The high frequency pulses amplified with a power amplifier (5) are changed to ultrasonic pulses and vice versa with the transducers on both sides of the sample. The time for a transit through the sample is denoted by τ_0. The different echoes follow the transit signal in time intervals of $2\tau_0$ and with exponentially decaying intensity. Each transit gives a phase shift for the signal of $\Phi = kL_0$. The n-th echo therefore has a phase shift of

$$\Phi_n = kL_0(2n+1) = \frac{\omega}{v}L_0(2n+1) \qquad (2.3)$$

with respect to the reference signal. Here k is the wave number of the ultrasound, L_0 the length of the specimen, $\frac{\omega}{2\pi}$ the frequency and v the sound velocity. The received signal is amplified and split into two channels (8). In both mixers (11) a signal from the sample $B = B_0 \cos(\omega t + \Phi_n)$ is multiplied with the reference signal $A_1 = A_0 \cos(\omega t)$ and with the 90° reference signal from the hybrid (10) $A_2 = A_0 \cos(\omega t + \frac{\pi}{2})$. From these expressions

$$A_1 B = \frac{1}{2}A_0 B_0 [\cos\Phi_n + \cos(2\omega t + \Phi_n)]$$

and

$$A_2 B = \frac{1}{2}A_0 B_0 [\sin\Phi_n + \sin(2\omega t + \Phi_n)]$$

a low pass filter (12) gives the two signals

$$I_n = \frac{1}{2}A_0 B_0 \cos\Phi_n$$
$$Q_n = \frac{1}{2}A_0 B_0 \sin\Phi_n \ . \qquad (2.4)$$

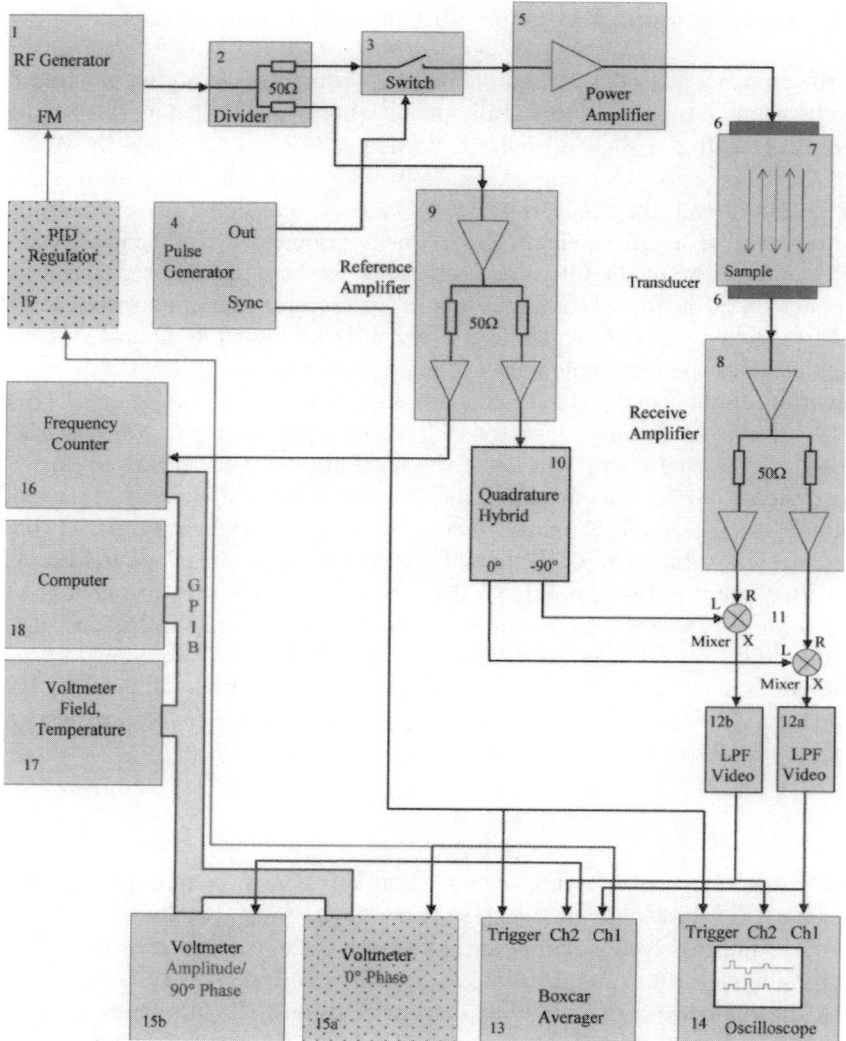

Fig. 2.3. Experimental set up for measuring sound velocity and attenuation. Here we show the apparatus with frequency feedback control. For the quadrature system the feedback loop is missing. For details see text (Lüthi et al. [2.14])

A boxcar (13) reads the two signals. A feedback loop with a PID regulator (19) keeps the phase of the signal I_n constant and changes accordingly the frequency. A computer is used to determine the frequency change and the amplitude of the 90° phase shifted signal Q_n. For the physical interpretation of these results (2.3) is taken into consideration and differentiated to give

$$\frac{\mathrm{d}\Phi_n}{\Phi_n} = \frac{\mathrm{d}\omega}{\omega} - \frac{\mathrm{d}v}{v} + \frac{\mathrm{d}L_0}{L_0} \,. \tag{2.5}$$

With the constant phase, the frequency change is directly proportional to the velocity change. Usually, the length changes due to temperature variation (thermal expansion) or due to magnetic field (magneto-striction) can be neglected in comparison to the velocity changes. The relative accuracy of this method is usually one part in 10^6 for $\frac{\Delta v}{v}$.

For the attenuation measurements, the 90° phase shifted signal Q is taken and the attenuation from the amplitude calculated. This procedure can be carried out for consecutive echoes. A simpler way is to fit echo trains, like the ones shown in Fig. 2.5a, with an exponential, according to the complex parameter k: $u = U \exp i(kx) = U \exp(-\alpha x + ik_r x)$ with $k = k_r + ik_i = k_r + i\alpha$. There are several pitfalls which can make attenuation measurements difficult. Amongst them are coupling losses in the transducers, diffraction effects, misalignment of the single crystal, phonon focusing etc. For a thorough discussion of these different effects see e.g. Truell et al. [2.2].

The attenuation is measured in Np/cm (Neper/cm) or in db/cm (decibel/cm). The relations between these units, with the sound wave amplitude $A(x)$ (see e.g. Fig. 2.2)

$$\text{db/cm} = 20 \frac{1}{x_2 - x_1} \log_{10} \frac{A(x_1)}{A(x_2)} \quad \text{and} \quad \text{Np/cm} = \frac{1}{x_2 - x_1} \ln \frac{A(x_1)}{A(x_2)}$$

and therefore db/cm = 8.6859 (Np/cm).

2.2.3 Ultrasonics at Very Low Temperatures

With the advent of top-loading dilution refrigerators, ultrasonics in the milli-Kelvin region became more easily feasible. The exchange of samples could be achieved within a few hours, which is indispensable because of transducer bonding and other problems. Many ultrasonic studies are now performed at temperatures below 1 K. We will discuss the results of such experiments especially in Chaps. 7, 8, 10, 13 and 14.

In Fig. 2.4 we show a schematic plot of a top-loading dilution refrigerator as developed by Oxford Instruments. The cooling power of this system is about 250 mW at 100 mK. Details on $He^3 - He^4$ dilution refrigerators can be found in books on "low temperature physics" as e.g. Pobell [2.16]. It is very important to use efficient ultrasonic transducers for low input power ($LiNbO_3$ or piezoelectric foils). The transducers have to be contacted with very fine 20 µm gold wires. The repetition frequency of the signal has to be as low as possible. With all these precautions, low temperatures below 50 mK can be achieved. For magneto-acoustic experiments, a specimen holder is often needed which can be rotated with respect to the magnetic field direction. The rotation mechanism has to be installed on top of the refrigerator at room temperature. Such a system is described by Wolf [2.17].

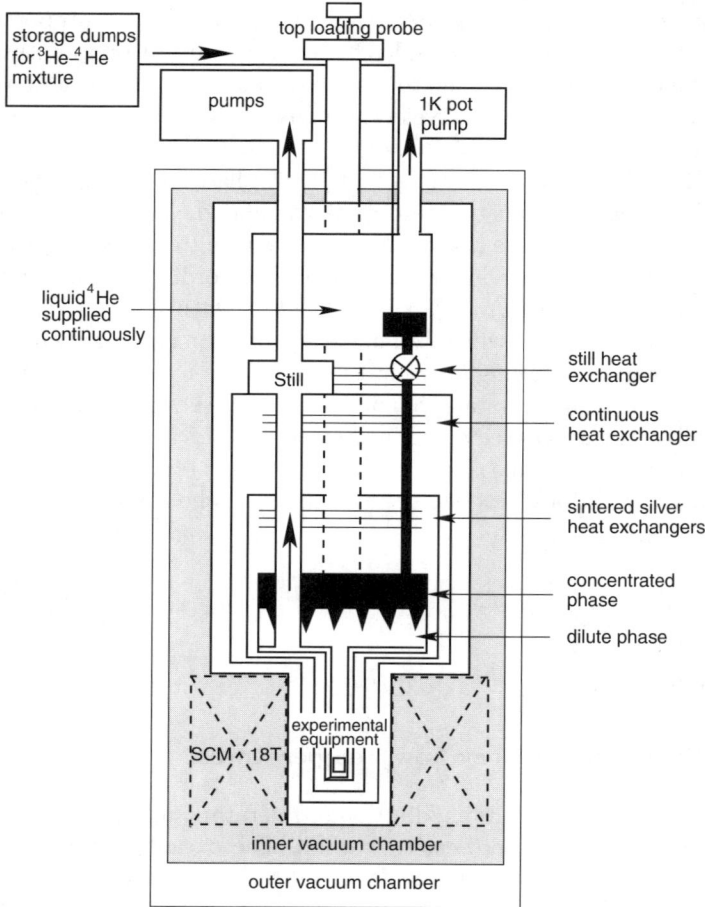

Fig. 2.4. Schematic plot of a top-loading dilution refrigerator as used for different experiments

2.2.4 Absolute Sound Velocity Measurements

An echo train, like the one shown in Fig. 2.5a, is usually taken for measuring the absolute sound velocity, the time interval between different echoes being measured. With the sample length, the velocity $v = L_0 \frac{2n+1}{t}$ can be determined for echo n. Care has to be taken that the starting points of the individual echoes are taken owing to echo shape deterioration. For echoes generated with piezoelectric films, this approach works well, as seen in Fig. 2.5b. For echo generation with transducers of finite thickness, the time delay within the transducer can affect the echo shape, especially for higher echoes. Getting the absolute phase velocity in this case is described in Kim et al. [2.18] and Niksch and Grill [2.19]. In Fig. 2.5a we show echo trains for a sample of URu_2Si_2 (Wolf [2.17]), one using quartz transducer (Fig. 2.5a) and the other using a

2.2 Sound Velocity and Attenuation, Experimental Techniques 13

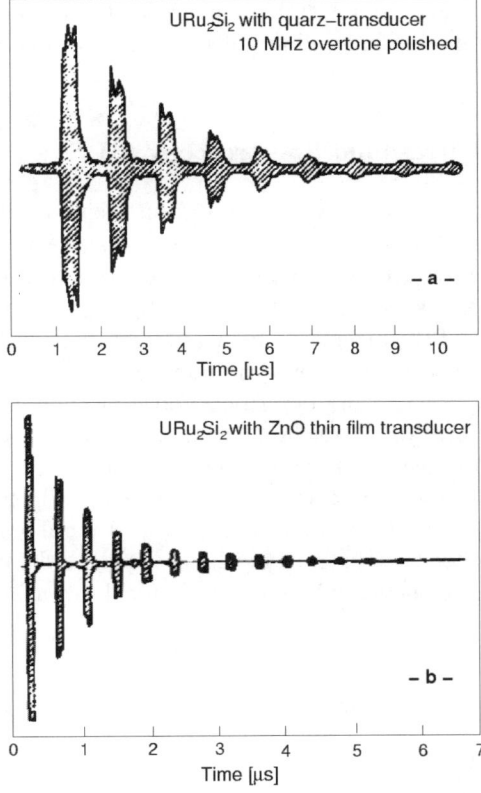

Fig. 2.5. Pulse echo train for ultrasonic waves in URu$_2$Si$_2$ (**a**) with Quartz transducers, (**b**) with ZnO–film transducers (Wolf [2.17])

ZnO–film (Fig. 2.5b). The quality difference of the successive echoes in favour of Fig. 2.5b is clearly seen. Another way to determine the absolute velocity is to measure the resonance frequencies of the system sample-transducer and using the relation $v = 2\Delta f L_0 (2n+1)$ where Δf is the frequency change for a phase shift of π for echo n. Phase shifts due to transducers and electronics have to be taken into account too. In all these methods it is necessary to know the length L_0 of the specimen. Furthermore, to convert from sound velocity to elastic constant, the mass density of the crystal must be known. The accuracy of these methods is about 1–3%. For a further sound velocity determination see also below the resonant ultrasound spectroscopy.

In the sound velocity measuring method described above, the phase velocity is measured. There are now new methods using phonon and ultrasonic imaging techniques, where the group velocity is measured, which for an elastic anisotropic medium is generally different in size and direction from the phase velocity. Phonon focusing and internal diffraction effects complicate the picture. This method is a promising tool for the future and can be already applied not only to insulators with high acoustic Q but also to quite different bulk specimen. An interesting and detailed account of this technique is given in Wolfe [2.20].

2.2.5 Resonant Ultrasound Spectroscopy, RUS

A completely different way to measure absolute sound velocities or elastic constants is the so-called shape resonance technique, also called resonant ultrasound spectroscopy (RUS) (Migliori and Sarrao [2.5]). This method consists of measuring the spectrum of resonance frequencies of a sample and a subsequent solution of the inverse problem of recovering all the components of the elastic constant tensor. In this technique, the plane wave approximation used in pulse echo experiments described above is abandoned. Instead, the normal modes of vibration of a solid specimen of known geometry is used to deduce the complete elastic tensor in a single measurement. The acoustic resonance frequencies can be determined as illustrated in Fig. 2.6. The specimen of known geometry, preferably in a form of a rectangular parallelepiped, is clamped between piezoelectric transducers as shown in the figure. One transducer is used to drive the sample with mechanical vibrations and the other

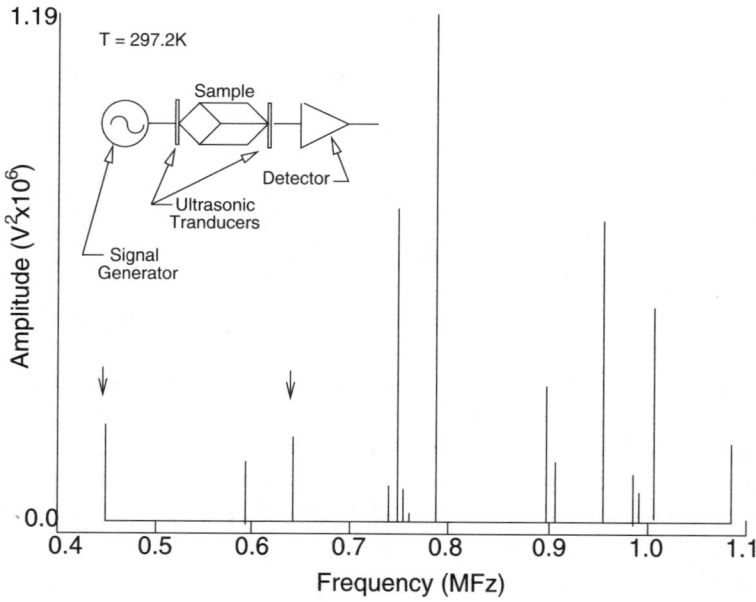

Fig. 2.6. Schematic set-up of the resonance spectroscopy with frequency spectrum (Migliori and Sarrao [2.5])

2.2 Sound Velocity and Attenuation, Experimental Techniques

is used to detect the mechanical response of the sample. A frequency sweep allows a rapid measurement of the resonance modes.

The acoustic resonance frequencies of a single crystal can also be calculated with given dimensions, crystal symmetry, mass density and elastic constants. Via a computational fitting procedure, unknown parameters, e.g. elastic constants, can be determined. With an iterative algorithm, the resonance frequencies calculated analytically can be matched with those measured experimentally. In crystals with a good acoustic Q, resonances ~ 100 in the frequency range of kHz to a few MHz can easily be determined with samples of $\sim$ mm dimensions. The fitting procedure gives elastic constant values with an accuracy better than 1%. One of the inaccurate parameters are the length measurements, which can also be included into the fitting procedure.

The great advantage of this method is the possibility of a complete determination of the elastic constant tensor in a single experiment. With pulse echo techniques, it is often necessary to take recourse of propagation directions which are not principal axes and for determining all elastic constants, (for lower symmetry than cubic) different propagation directions are usually required, i.e. often different crystals. This resonant ultrasonic spectroscopy is especially suited for relatively small crystals and for crystal symmetries lower than cubic, where there are more than three elastic constants (see Appendix C). The accuracy of elastic constant determination is high, diffraction and other interference effects are practically absent since there is no plane wave approximation. On the other hand for high accuracy relative elastic constant measurements with a resolution of ~ 1 part in 10^6, the phase sensitive techniques described earlier must be used. In addition, magnetic field dependent effects can only be measured for special field directions with this shape resonance technique. But temperature dependencies of elastic constants have been measured successfully with this technique.

With this new RUS method, the elastic constant tensor of a number of single crystal specimen have so far been investigated – mainly at room temperature: La_2CuO_4 (Migliori et al. [2.21]), $La_{1.86}Sr_{0.14}CuO_4$ (Migliori et al. [2.22]) $YBa_2Cu_3O_{7-\delta}$ (Lei et al. [2.23]), UPt_3, Sr_2RuO_4 (Paglione et al. [2.24]), $LiKSO_4$ and $YD_{0.1}$ (Leisure and Willis [2.25]), piezoelectric $La_3Ga_5SiO_{14}$ (Schreuer [2.26]). Use will be made of some of this data in Chaps. 9 and 10.

The mathematical methods used for this RUS spectroscopy consist of determining the elastic eigenfrequencies and eigenvectors for a simple geometry for a given material. These methods are explained in Migliori et al. [2.27] and Leisure and Willis [2.25]. The starting point is the Lagrangian L, $L = \int_V (KE - PE)dV$ with $KE = \frac{1}{2}\sum_i \rho(du_i/dt)^2$ the kinetic energy and $PE = \frac{1}{2}\sum c_{ijkl}\varepsilon_{ij}\varepsilon_{kl}$ the potential energy. Expanding the displacement vector in a complete set of functions $\{\Phi_\lambda\}$: $u_i = \sum_\lambda a_{\lambda i}\Phi_\lambda$ and substituting into L gives

$$L = \frac{1}{2}\sum \left[a_{\lambda i} a_{\lambda' j} \rho \omega^2 \int_V \delta_{ij} \Phi_\lambda(\mathbf{r}) \Phi_{\lambda'}(\mathbf{r}')\mathrm{d}V - \right.$$
$$\left. a_{\lambda_i} a_{\lambda' j} \int_V c_{iji'j'} \Phi_{\lambda,j}(\mathbf{r}) \Phi_{\lambda' j'}(\mathbf{r}')\mathrm{d}V \right]. \quad (2.6)$$

This expression is written in compact form as $L = \frac{1}{2}\rho\omega^2 \mathbf{a}^T \mathbf{E} \mathbf{a} - \frac{1}{2}\mathbf{a}^T \mathbf{\Gamma} \mathbf{a}$ where the expansion coefficient $\mathbf{a}$ is a vector and the integrals in (2.6) are the matrices $\mathbf{E}$, $\mathbf{\Gamma}$. With L an extremum with respect to the expansion coefficients gives the eigenvalue equation:

$$\mathbf{\Gamma a} = \rho\omega^2 \mathbf{E a}. \quad (2.7)$$

The eigenvalues are $\rho\omega^2$ and the eigenvectors $\mathbf{a}$ are the expansion coefficients. Polynomials – Legendre functions depending on the type of crystal and geometric form – are used for the set of functions Φ. There are standard programs to solve such equations. Symmetry arguments and suitable truncation procedure simplify the procedure. Further details can be found in the references given above.

2.2.6 Vibrating Reed Technique

In this technique the sample, a reed, is clamped on one side to a sample holder whereas its free end is electrostatically excited to vibrations. Its motion follows that of a driven oscillator. A second electrode on the other side detects electrostatically the vibrations of the sample. The measurements of the excitation frequency and that of the damping of the oscillator gives the change $\frac{\Delta v}{v}$ of the sound velocity $v = (\frac{E}{\rho})^{1/2}$ or Youngs modulus E and of the internal friction Q^{-1} of solids. The frequency range is typically from 0.1 to 10 kHz. Therefore thermoelastic relaxation effects can be studied together with ultrasound effects over a wide frequency range. Note that a piezo transducer is not needed for this technique. These techniques are described by e.g. Novick and Berry [2.28], Read et al. [2.1], Barmatz and Golding [2.29], Berry and Pritchet [2.30] and Esquinazi [2.31].

For a one ended clamped bar, the vibration frequencies are

$$f_n = \frac{n}{4L_0}(\frac{G}{\rho})^{1/2} \quad \text{and} \quad f_n = \frac{n}{4L_0}(\frac{E}{\rho})^{1/2}$$

with L_0 the length of the bar and for torsional and longitudinal modes respectively and somewhat more complicated expressions for flexural modes (Novick and Berry [2.28]). For a discussion of the elastic moduli E, G, K in an isotropic solid see Sect. 3.4. Experimental results for these low frequency modules are given in Sects. 6.1, 12.2.

2.3 Phonon Echoes

In the seventies and eighties analogous experiments to the nuclear spin echo experiment (Hahn [2.32]) were performed with acoustic waves. In MgO, doped with Ni^{2+}, echoes were observed with substituting the electromagnetic fields by phonon fields (Shiren and Kazyaka [2.33]). Another class of echo phenomena can be observed in two level systems as they are present in glasses. In such tunneling systems, which can be formally described in analogy to a spin $\frac{1}{2}$ system (pseudo-spin σ), such echo experiments were successfully performed (Graebner and Golding [2.34]). These experiments will be described briefly in Chap. 14. Another class of echo experiments were performed in piezoelectric and non-piezoelectric semiconductors like $LiNbO_3$, SbSI, CdS (Popov and Krainik [2.35], Thompson and Quate [2.36], Billmann et al. [2.37]), also in powder materials (see review articles by Kajimura [2.38] and by Melcher and Shiren [2.39]).

Another echo experiment was also performed with a material exhibiting a structural phase transition using an additional echo material (Fossheim and Holt [2.40]). It can be observed that a backward-wave echo is generated by the interaction of an acoustic wave (frequency ω) with a homogeneous electric field (frequency 2ω) giving a backward wave (frequency ω). This experiment, which is relevant to the physics discussed in Chap. 7, will be discussed in detail in Sect. 7.4.3. For a review of this topic see Fossheim and Holt [2.40].

All the various echo phenomena have the properties of time reversal and phase memory. Apart from the spin echoes mentioned above, photon echoes and, as discussed above, various kinds of phonon echoes have been observed. All echo phenomena are due to nonlinear coupling of the oscillations. As a simple example (Kajimura [2.38]) in a linear system a primary pulse $a \exp(-i\Omega t)$ and a secondary pulse $b \exp[-i\Omega(t-\tau)]$ add up in a superposition of the two pulses. Radiation from systems with many eigenfrequencies Ω gives a response $\sum_\Omega \{a \exp(-i\Omega t) + b \exp[-i\Omega(t-\tau)]\}$ which is essentially zero due to their phase cancellations. No coherent signal is thus produced. But nonlinear coupling gives a coherent echo signal at a particular time, as an example: $\sum_\Omega \{a^\star \exp(i\Omega t) b^2 \exp[-2i\Omega(t-\tau)]\} = \sum_\Omega a^\star b^2 \exp{-i\Omega(t-2\tau)}$ gives a coherent echo signal only for $t = 2\tau$.

Apart from the phonon echoes mentioned here, there exist memory echoes in piezoelectric or ferroelectric powders, in powders of ferromagnetic and superconducting metals produced with rf electric fields with extremely long memory times lasting up to days.

2.4 Ultrasonics in Pulsed Magnetic Fields

High Field Apparatus

Before giving details on the ultrasonic set-up in pulsed fields we describe first the apparatus for generating pulsed fields. For a general introduction to pulsed high magnetic fields see e.g. Herlach [2.41] and the series of books by Herlach and Miura [2.42]. In Fig. 2.7 we give a schematic of a pulsed field apparatus as developed by Oxford instruments and installed at the University of Frankfurt (Wolf et al. [2.43]).

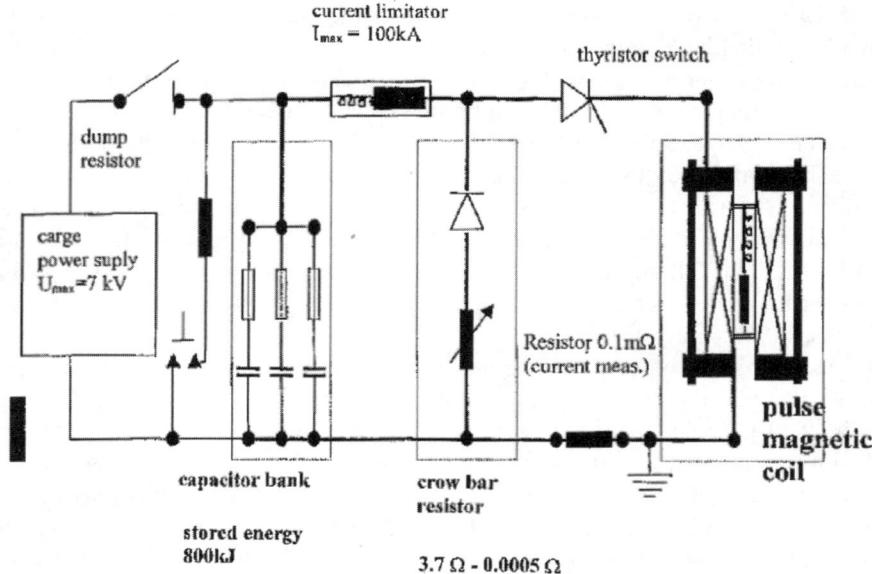

Fig. 2.7. High field apparatus (Oxford instruments)

It is a system for capacitor discharge through a specially designed coil. The capacitor bank has 32mF and the maximum stored energy is about 800 kJ. The discharge is triggered with thyristor switches. A crow bar resistor becomes active when the voltage changes sign, thus dissipating the energy and protecting the capacitors. The 50T coil (developed at the National High Field laboratory in Talahassee) has an inductance of $L = 1.81$ mH. The high field limitation of this liquid nitrogen cooled coil is not the energy of the capacitor bank but rather the warming up of the coil up to room temperature for a 50 T pulse. The rise time of this coil is about 8 ms and the complete discharge time is about 25 ms. The Copper-wire is reinforced with 1.5% Al_2O_3. Glasfiber together with epoxy resin and a 18 mm steel mantle casing give the necessary mechanical strength. The inner diameter of the coil is 24 mm. It should also be possible to use a 60T coil with this setup. With the inductance L and the capacitance $C = 16$ mF, the time for the maximum field $T_{incr} = \frac{1}{4}2\pi\sqrt{LC} = 8.5$ ms is close to the observed one. The decay time with a crowbar resistance ($R = 0.412\Omega$) is approximately $T_{dec} = \frac{2L}{R} = 8.8$ ms instead of 16 ms obtained experimentally because the back flow of current to the capacitor bank was neglected in this estimate.

Ultrasonic System

For experiments in pulsed magnetic fields we use essentially the same setup described in Sect. 2.2 for measuring ultrasonic velocity and attenuation. But

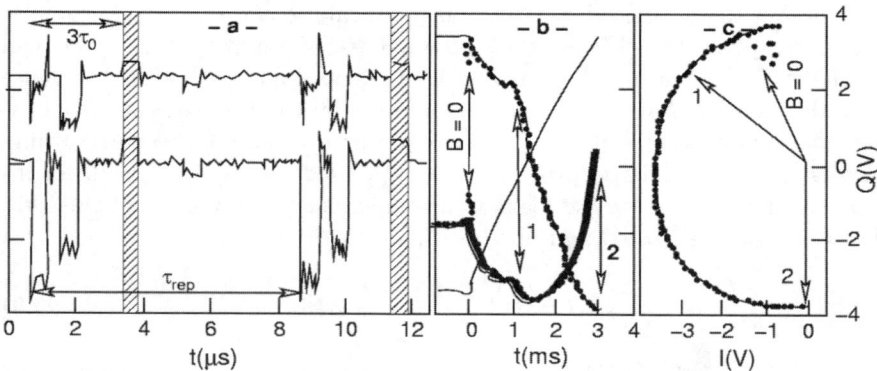

Fig. 2.8. (a) Echoes of an ultrasonic experiment, (b) t-dependence of I_n, Q_n with magnetic field, (c) Vector representation of I_n, Q_n. The arrows in (b) and (c) mark the same time. The shaded region is the position of the Boxcar gate

the feedback loop described there is too slow to follow the changes of the magnetic field. Therefore all the experiments have to be performed at a fixed frequency and the phase shifts (real and imaginary part) have to be determined numerically (Fig. 2.8). The change in the sound velocity is inversely proportional to the change in the phase (2.3) which must be determined from the data of the two channels:

$$\Phi_n = \frac{\omega}{v} L_0(2n+1) = \arctan\left(\frac{Q_n}{I_n}\right)$$
$$A_n = \sqrt{(I_n^2 + Q_n^2)}, \qquad (2.8)$$

where A_n is the amplitude of the echo n. In Fig. 2.8 we show the two signals I_n and Q_n for a typical echo pattern.

The repetition rate of the ultrasonic pulses has to be larger in the pulse field experiments. For a pulse length of the high field coil of 20 ms a repetition-frequency of at least 100 kHz is needed. With a standard sampling oscilloscope, 1 point per μs can be taken. Therefore with the repetition frequency given above. there are enough points to measure magnetic field and the two parameters I, Q. The resolution for sound velocity measurements for the equipment in Fig. 2.8 is usually of the order of 10^{-6}. In the pulse field mode this changes to $\sim 10^{-5}$. This homodyne set-up works very well and does not lack in resolution compared to a heterodyne set-up.

Disturbing Factors in Pulsed Magnetic Field Experiments

There are several disturbing factors in ultrasonic experiments with pulsed fields. Heating effects due to eddy currents are the most serious ones. Even for insulating samples care has to be taken with the metalized transducers. By using very thin metal electrodes, this effect can be reduced.

Another disturbing effect is the magneto-caloric effect as it occurs especially near phase transitions. Rough estimates of these disturbing effects can be made by comparing the sound velocity before and just after the pulse with the velocity change as a function of temperature. It is advisable to measure the sample temperature in order to get an estimate of these perturbing factors. For an adiabatic process $dS = (\frac{\partial S}{\partial T})_B dT + (\frac{\partial S}{\partial B})_T dB = 0$, an infinitesimal field increase dB leads to a temperature increase dT of the form $dT = -(\frac{\partial S}{\partial B})_T/(\frac{\partial S}{\partial T})_B dB$ and finally (Tishin et al. [2.44])

$$dT = -\left(\frac{\partial M}{\partial T}\right)_B \frac{T}{C_B} dB, \qquad (2.9)$$

where the magnetization $M(T, B)$ and the specific heat $C(T, B)$ are both field and temperature dependent. Equation (2.9) can be integrated from a starting field B_0 to the end field B_e. The magneto-caloric temperature increase is most pronounced for large $(\frac{dM}{dT})_B$ and small C. Experimental results of pulsed field ultrasonics are given in Chap. 9 and in Chap. 12. A study of the magneto-caloric effect in pulsed fields will be given in Sect. 12.1.1. Note that the magneto-caloric effect is disturbing measurements in pulsed fields under adiabatic conditions. On the other hand this effect has different important applications as e.g. producing low temperatures by magnetic refrigeration machines and for very low temperatures ($T < 1\,\mathrm{K}$) by the adiabatic demagnetization process, or for characterising phase transitions.

2.5 Surface Acoustic Wave Generation and Detection

Surface acoustic waves (SAW) are important in seismology, in non-destructive testing and especially in signal processing in electronic systems. The low acoustic velocity compared to the light velocity, together with the corresponding small acoustic wavelength make it possible to gain a reduction in size and weight compared to electromagnetic devices. There are many devices possible for signal processing applications such as delay lines, bandpass filters, ultrahigh frequency oscillator control elements, programmable devices for frequency and time domain filtering, frequency synthesizers and correlators etc. Apart from these technical devices, SAW also have many applications in solid state physics as shown in later chapters. The physical properties of SAW propagation, especially the so-called Rayleigh waves, will be presented in detail in Sect. 3.5.

Surface waves can be generated in quite different ways. Usually the same method can be used also for detection. There exist several reviews covering the different methods of generating and detecting SAW (White [2.45], Dransfeld and Salzmann [2.46], Oliner [2.47]). Here we briefly discuss the widely used single phase array and inter-digital SAW transducers, the latter one being the most fundamental component for SAW generation. A more detailed account can be found in the listed reviews.

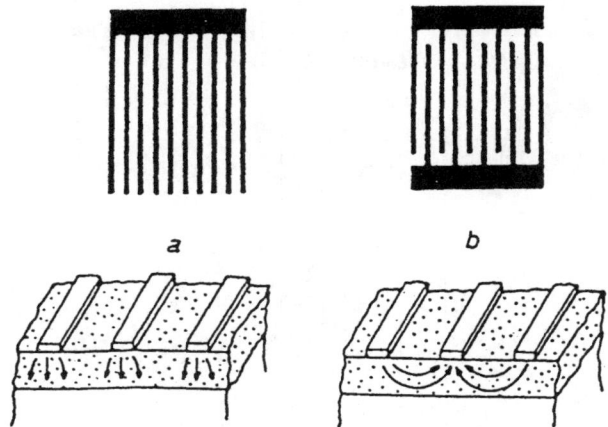

Fig. 2.9. Electrode structures for generation and detection of SAW: (**a**) single array, (**b**) inter-digital array

Figure 2.9 shows the electrode structure for SAW generation and detection. Consider a plane substrate: at the place of the structure, a thin metallic film as counter electrode and on top of it a piezoelectric film (CdS or ZnO) is evaporated or sputtered onto the substrate and finally the electrode structure is placed as shown in the figure. This structure is patterned lithographically. An electric field between the electrodes leads to strains in the piezoelectric layer and to the substrate. The field distribution in the layer is also shown in Fig. 2.9a,b for the two electrode-structures. Denoting with d the distance between two neighbouring stripes in the single phase array, the SAW velocity is $v_S = d\,f$, where f is the frequency of the hf field. The SAW has a wavelength λ with $d = n\lambda$, n integer. It radiates perpendicular to the structure in both directions. The inter-digital transducer has a higher efficiency than the single phase array. The possible frequency range of SAW generation and detection covers typically from 50 MHz to 5 GHz.

The wider field of surface phonons, i.e. not only low frequency and low k-vector, but the whole k range to the Brillouin zone can be investigated with different techniques: energy loss spectroscopy, scattering with He atoms. Reviews on these different techniques by Toennies and Ibach can be found in Kress and de Wette [2.48]. Applications of SAW to solid state physics problems will be given in Sects. 3.5, 6.2, 9.2, 10.6, 13.5 and 13.6.

2.6 Microwave Ultrasonics

If the sound wave frequency is increased up to the microwave frequency, say 10 GHz, the acoustic wavelength becomes very small. With a typical sound velocity of 5 km/s, the wavelength for 10 GHz is 5000 Å. This makes

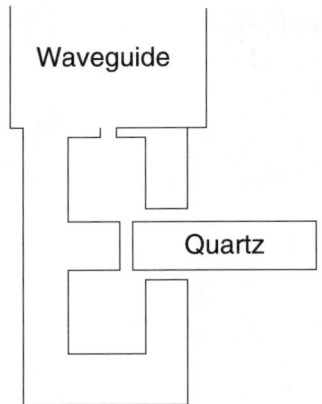

Fig. 2.10. Re-entrant co-axial resonant cavity used with a delay quartz crystal transducer

resonant ultrasonic emission and detection very difficult. Therefore a non-resonant method, using a surface generation is employed. The gradient of the piezoelectric stress is largest at the crystal surface and hence the ultrasonic longitudinal wave is (Jacobsen [2.49])

$$u_L = \frac{e_{11}E^0}{kc_{11}} \sin(\omega t - kx)$$

with e_{11} the piezoelectric stress constant and E^0 the amplitude of the microwave electric field.

The experiment is usually carried out with a re-entrant resonant cavity (Fig. 2.10). With the small wavelength quoted above, the crystal surfaces have to be flat and parallel within optical quality. In addition the microwave sound attenuation can be very high. Therefore only experiments with high acoustic quality crystals have been performed. With the cylindrical re-entrant microwave cavity a strong electric field at the polished end face of the sample is generated (Fig. 2.10) which leads, for instance, to a longitudinal stress (2.3) $T_{xx} = c_{11}\varepsilon_{xx} - e_{11}E_x$ and to a sound wave of the form given above (Jacobsen [2.49]). The first groups performing such experiments in the GHz range were Baranskii [2.50], Jacobsen [2.51] and Bömmel and Dransfeld [2.52], [2.53]. The latter group studied Quartz at 1 GHz. Pomerantz [2.54] used ferromagnetic films as transducers and studied at 9 GHz a number of high quality single crystals of CdS, GaAs, Ge, Si, CaF$_2$, Al$_2$O$_3$ and MgO. Lange [2.55] investigated rutile at 1 and 3 GHz and King and Rosenberg [2.56] performed experiments at 1 GHz in InSb and GaAs. The highest microwave acoustic frequency achieved with this cavity technique is 114 GHz in quartz (Iluker and Jacobsen [2.57]).

A detailed account of these types of experiments and what can be learned about phonon processes is given by Tucker and Rampston [2.6], Beyer and

Letcher [2.58] and Truell et al. [2.2]. Since, for the reasons listed above, the number of experiments and the number of investigated compounds is rather limited and since these experiments are widely discussed in the quoted monographs, only occasionally reference will be made to these and related experiments (Sects. 4.2.2, 5.5, 11.2, 13.3 and 13.4).

There are various different developments in microwave or more general high frequency ultrasonics. A widely known technique is the use of superconducting tunneling junctions as phonon generators and detectors as first shown by Eisenmenger and Dayem [2.59]. Depending on the superconducting tunnel junction, phonon experiments can be performed in a wide frequency range 70–800 GHz. For reviews on these techniques see Eisenmenger [2.60], Kinder [2.61]. Another widely used method is to produce heat pulses from metal films on insulators. The metal film may be ohmically heated with an electric current or optically heated with a laser beam. While many experiments, like phonon focusing and phonon imaging, were performed with such techniques, they lie outside the scope of this investigation. A good account of these techniques, mentioned already in Sect. 2.2 with many references, is Wolfe [2.20].

2.7 Brillouin Scattering

The wavelength of visible light is much larger than the lattice constants. Therefore in an ideal crystal, light can only be scattered by lattice vibrations or other elementary excitations. The scattering of light from acoustic phonons is called Brillouin scattering – from optic phonon branches it is the Raman scattering. Energy and momentum conservation laws give $\hbar\omega' = \hbar\omega \pm E_q$ and $k' = k \pm q$ with E_q the excitation energy of momentum q for a collective excitation, ω and ω' are the laser frequencies of the incoming and scattered light. For a given scattering geometry (k, k'), the frequency change $\omega' - \omega$ is measured interferometrically. The frequency changes measured conveniently lie in the GHz region. They are therefore complementary to the ultrasonic waves. There are now very sophisticated Brillouin-scattering apparatus available for transparent and opaque materials. Figure 2.11 shows an apparatus for 180° geometry (Sandercock [2.62]). The techniques are described in review articles (Pine [2.63], Sandercock [2.62]). For high resolution a Fabry–Perot interferometer is used. This device will transmit light of wavelength λ if the spacing L of the two mirrors is $L = m\frac{\lambda}{2}$ with m an integer. The ratio of maximum to minimum transmission, called the contrast of the interferometer, can be improved by placing two or more interferometers in series. A more elegant method is to pass the light two or more times through the same interferometer. For the design of such a multi-pass interferometer see Sandercock [2.62].

The Brillouin scattering technique being based on statistical data collection cannot have such a high resolution as the phase-sensitive ultrasonic

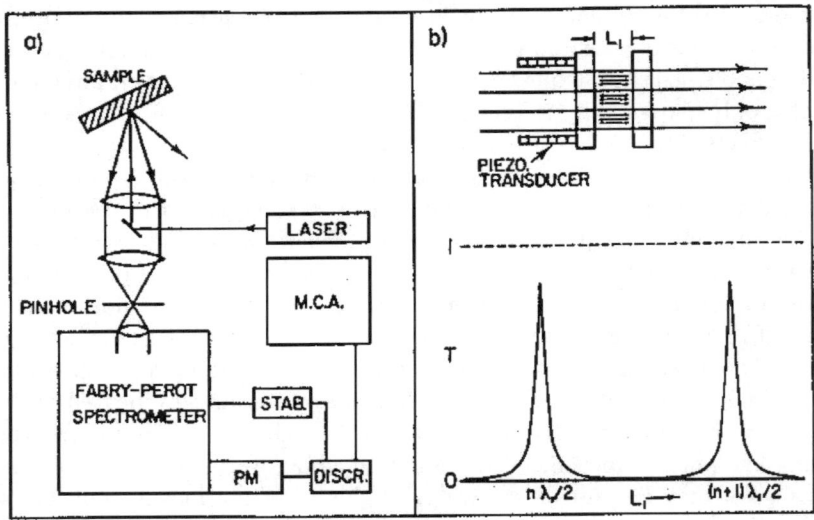

Fig. 2.11. (a) Experimental set-up for backscattering with a Fabry–Perot multipass spectrometer. PM photomultiplier, M.C.A. Multichannel analyser, (b) transmission of monochromatic light by scanning Fabry–Perot interferometer (Sandercock [2.62])

methods for velocity measurements discussed above. But its frequency range is in the Gigahertz region. Furthermore the method is contact free, a great advantage for many compounds. In the following we will use various Brillouin scattering results in addition to ultrasonic results. Examples are given in Sects. 7.2.2, 9.2, 10.4, 11.2.5, 12.6 and 13.5.

Using Brillouin scattering with a 180° backscattering geometry, Mock and Güntherodt [2.64] showed that bulk modulus and elastic constants can be determined for both metallic polycrystalline samples and single crystals. For a small penetration depth of the light, the spectrum consists of the SAW Rayleigh mode $\omega_R = v_R q_{//}$ and the longitudinal mode $\omega_L = v_L q_{//}$. These two modes determine the elastic moduli c_L, c_T and c_B for a polycrystalline metal. For single crystals various crystal directions can be used to get a complete set of elastic constants. For examples, see the references given above. Unlike the microwave phonon technique (Sect. 2.6) for the Brillouin scattering, the extreme high quality factors for the surface do not need to be considered. Therefore, the range of applications as far as materials are concerned, is much larger than for microwave ultrasonics, the relative accuracy is however lower as discussed above.

2.8 Thermal Expansion and Magneto-striction, Thermal Conductivity

2.8.1 Thermal Expansion and Magneto-striction

Length changes of specimen are important to know for making possible corrections for physical properties. This is especially true for compounds with giant magneto-elastic coupling constants, like heavy rare earth metals or the rare earth-iron Laves phase compounds RFe_2 (see Sect. 5.4). But thermal expansion and magneto-striction are also important thermodynamic derivatives, giving the same coupling constants as gained from elastic constant measurements. As a first derivative of the thermodynamic potential the coupling constants enter linearly (giving the sign) and not quadratically like for elastic constants.

For measuring relative length changes, a variety of different techniques can be used: strain gauges, interferometric methods, double grating with photocell arrangements and capacitive methods (White [2.65], Brändli and Griessen [2.66], Ott and Lüthi [2.67], Lang [2.68]). The most widely used method now is the capacitive method. It has the advantage of high accuracy even with small samples and it can be used without corrections in external magnetic fields. Therefore it is suitable for the study of both thermal expansion and magneto-striction, especially at low temperatures. Capacitance bridges of high sensitivity are commercially available (e.g. General Radio, Type 1615-A). Figure 2.12 shows a schematic drawing of a sample holder used for capacitive length measurements (Ott and Lüthi [2.67]). Such a capacitance dilatometer has a sensitivity of the order of 10^{-9} for relative length changes. It is therefore ideally suited for many applications such as length change corrections for elastic constant measurements, thermal expansion anomalies near phase transition temperatures, study of lattice effects of different phase transitions, etc.

Apart from corrections to sound velocities, the thermal expansion and magneto-striction results are discussed in Sects. 5.2, 9.2, 10.7 and Chap. 6. A thorough discussion of magneto-striction in ferromagnetic substances is given by de Lacheisserie [2.69].

2.8.2 Thermal Conductivity

Since thermal conductivity is discussed in several places (Sects. 4.5, 12.4) some remarks on the experimental techniques for measurements are in order. The thermal conductivity coefficient κ relates the heat flux $\frac{d\boldsymbol{Q}}{dt}$ and the temperature gradient by Newton's law $\frac{d\boldsymbol{Q}}{dt} = -\kappa \nabla T$. κ therefore has the dimension Watt/Km. To determine a tensor component of κ one usually takes a steady state method. The temperature difference ΔT along a bar of material is measured using thermocouples. ΔT varies typically about 1% of the absolute temperature, which is stabilised for each data point. Measuring $\frac{dQ}{dt}$

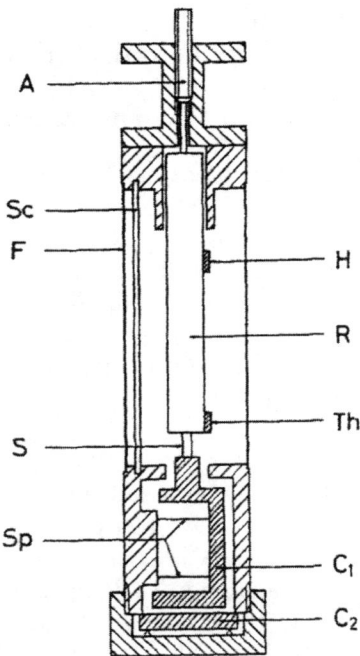

Fig. 2.12. Capacitance dilatometer: Schematic drawing of sample holder. A screw to adjust capacitor gap, C_1, C_2 capacitor plates, F copper frame, H heater, R copper rod, S sample, Sc screw fixing copper frame, Sp springs, Th thermometer (Ott and Lüthi [2.67])

of the heater and ΔT gives the thermal conductivity coefficient. The method has usually an accuracy of $< 5\%$. Details for this method can be found in Brüesch [2.70] and Berman [2.71].

3 Elasticity

The basic elements of the theory of elasticity, including the wave equation for sound waves, are first treated in this chapter. Detailed treatments of this continuum theory can be found in Landau–Lifshitz [3.1], Truell et al. [3.2], Beyer and Letcher [3.3], Wallace [3.4] and Ludwig [3.5]. We then discuss the symmetry aspects of the strains and elastic constants for different crystal classes Sect. 3.2 and the third order elastic constants Sect. 3.3. The elastic stability and the case of elastic isotropy will be treated in Sect. 3.4. Further on in Sect. 3.5, acoustic surface waves, especially the so-called Rayleigh waves, are covered. The concepts developed here will be used in later chapters. Finally, a short outline of the lattice dynamics and the Debye theory of lattice dynamics is presented in Sect. 3.6.

3.1 Strains, Stresses and Elastic Constants

Strains

In an elastic deformation, a volume element at the site $\bm{R}$ is shifted to $\bm{R}' = \bm{R} + \bm{u}(\bm{R}, t)$. In general, the displacement vector $\bm{u}(\bm{R}, t)$ can be both space and time dependent. If $\bm{u}$ is not dependent on $\bm{R}$ we have only a translation which does not lead to a real deformation. If we consider a second point, close to $\bm{R}$ and denoted by $\bm{R} + d\bm{R}$, this material point shifts also to $\bm{R}' + d\bm{R}'$ and we have $d\bm{R}' = d\bm{R} + d\bm{u}$. This is shown in Fig. 3.1.

For deformations, $\bm{u}$ is position dependent and we can express $d\bm{R}'$ by the partial derivatives

$$v_{ij} = \frac{\partial u_i}{\partial R_j}, \quad \text{i.e.} \quad dR'_i = dR_i + \sum_j v_{ij} dR_j \; .$$

$v_{ij} = \frac{\partial u_i}{\partial R_j}$ is a component of the deformation tensor. Here, a Lagrange or material description is chosen, with R_i denoting the material coordinate. (For a discussion of the Lagrange and Euler description see Thurston [3.6]). The deformation tensor (v_{ij}) need not be symmetric. If it is not position dependent, it is a homogeneous deformation. For some applications it is necessary to consider the finite strain tensor. This can be obtained by calculating the scalar product of the vectors

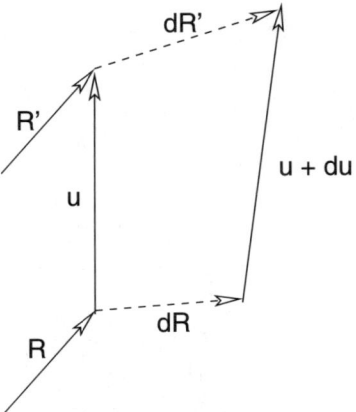

Fig. 3.1. Displacement vectors for the description of strains

$$d\boldsymbol{R}' \cdot d\boldsymbol{R}' - d\boldsymbol{R} \cdot d\boldsymbol{R} = \sum_{\alpha\beta} 2\eta_{\alpha\beta} dR^\alpha dR^\beta \ .$$

This gives the desired expression for η_{ij}

$$\eta_{ij} = \frac{1}{2}\left(v_{ij} + v_{ji} + \sum_\alpha v_{\alpha i} v_{\alpha j}\right) \ . \tag{3.1}$$

$\eta_{ij} \neq 0$ means that we are dealing with a pure strain deformation. If $\eta_{ij} = 0$, we are dealing with a pure rotation. Note that (η_{ij}) is a symmetric tensor. A general deformation in a solid can be built up by a pure strain deformation followed by a rotation, i.e.

$$(\boldsymbol{I} + \boldsymbol{v}) = \boldsymbol{R}(\boldsymbol{I} + 2\eta)^{1/2} \ . \tag{3.2}$$

Here $\boldsymbol{I}$ is the unit tensor and for the rotation tensor follows from (3.2)

$$\boldsymbol{R} = (\boldsymbol{I} + \boldsymbol{v})(\boldsymbol{I} + 2\eta)^{-1/2} \ . \tag{3.3}$$

Usually, only infinitesimal displacements are considered in elasticity theory. Therefore a component of the infinitesimal strain tensor reads, from (3.1)

$$\varepsilon_{ij} = \frac{1}{2}(v_{ij} + v_{ji}) \ , \tag{3.4}$$

and the rotation tensor $\boldsymbol{R}$ reduces by (3.3) to $R_{ij} = \delta_{ij} + \omega_{ij}$, where

$$\omega_{ij} = \frac{1}{2}(v_{ij} - v_{ji}) = -\omega_{ji} \tag{3.4a}$$

3.1 Strains, Stresses and Elastic Constants

is the antisymmetric part of the deformation tensor. The ω_{ij} are the components of a vector $\boldsymbol{\Omega} = \frac{1}{2}\mathrm{rot}\boldsymbol{u} = (\omega_{yz}, \omega_{zx}, \omega_{xy})$ which describes a bulk rotation of the body in first order for a homogeneous deformation. Usually, the small deformations are adequate for a description of sound waves and many of their properties. However there are effects where the finite strain tensor and the rotational tensor play an important part and have to be considered explicitly (see Sect. 13.2). The finite strain tensor of (3.1) may be written with (3.4), (3.4a) as

$$\eta_{ij} = \varepsilon_{ij} + \frac{1}{2}\sum_k (\varepsilon_{ki} + \omega_{ki})(\varepsilon_{kj} + \omega_{kj}) . \tag{3.1a}$$

Stress

Apart from the strain tensor we have to introduce the stress tensor. In the undeformed equilibrium state, the resultant forces on a volume element are zero. When a deformation occurs, forces will arise which tend to return the body to equilibrium. These internal forces, which occur when a body is deformed, are called internal stresses. Because of the short range of the molecular forces in a continuum approximation, the force on a volume element of a deformable body can be expressed as

$$\int F_i \mathrm{d}V = \int \sum_k \frac{\partial T_{ik}}{\partial R_k} \mathrm{d}V = \int \sum_k T_{ik} \mathrm{d}f_k , \tag{3.5}$$

where $\boldsymbol{F}\mathrm{d}V$ is the force on a volume element $\mathrm{d}V$, T_{ik} is an element of the stress tensor and $\mathrm{d}f_k$ is a component of the surface element vector $\mathrm{d}\boldsymbol{f}$. The components of the stress tensor of a rectangular parallelepiped are shown in Appendix B, Fig. B.1. On the volume element, $\mathrm{d}x\mathrm{d}y\mathrm{d}z$ act as stresses, which are defined as forces per unit area. On the surface $\mathrm{d}x\mathrm{d}y$, we have the normal stress T_{zz} and the two shear stresses T_{xz} and T_{yz}. From angular momentum or energy consideration follows that the stress tensor is symmetric, i.e. $T_{ij} = T_{ji}$ (Ludwig [3.5], Landau–Lifshitz [3.1]). In addition, molecules in the solid, can still experience external forces such as gravity, external pressure, magnetic volume force etc. These external forces are neglected in the following.

Elastic Waves

In order to treat elastic wave propagation in a solid, it is necessary to discuss the homogeneous deformation concept introduced above. In a sound wave, the strain components ε_{ij} and ω_{ij} are no longer constant and vary with time and position. If we consider only long wavelength sound waves, where the wavelength is much longer than the force range due to the internal stresses, than in a region of this force range the deformations are approximately homogeneous and the considerations from above can be used. This should hold true even for ionic crystals with Coulomb forces and for metals since charge neutrality is maintained over small regions. For shorter wavelengths, it is necessary to take into account the detailed structure of forces. This leads to lattice dynamics, a topic which is treated in Sect. 3.6. We will occasionally use the results of lattice dynamics (Sects. 5.4, 7.2.1 and 9.2).

The equations of motion for non-homogeneous deformations in an elastic continuum, as just discussed, can be obtained by considering the stresses acting on a volume element. Application of Newton's law gives (see Appendix B or references given above)

$$\rho \frac{\partial^2 u_i}{\partial t^2} = \sum_k \frac{\partial T_{ik}}{\partial R_k} \ . \tag{3.6}$$

Here T_{ik} is a component of the stress tensor and $\boldsymbol{u}$ the displacement vector introduced above.

With the linearized stress-strain relation, which is the phenomenological Hooke law

$$T_{ij} = \sum_{kl} c_{ijkl} v_{kl} \tag{3.7}$$

and for plane waves $u_i = U e_i \exp(i(\boldsymbol{k.r} - \omega t))$ (U amplitude, $\boldsymbol{e}$ polarization vector) we get the eigenfrequencies and normal modes (Appendix B)

$$\sum_{jlm} \left[\rho \omega_k^2 \delta_{il} - c_{ijlm} k_m k_j \right] e_l(k) = 0 \ . \tag{3.8}$$

Here ω_k is the eigenfrequency for the wave vector $\boldsymbol{k}$ and c_{ijkl} are the elastic stiffness constants. The components of the inverse tensor to the (c_{ijkl}) are denoted by s_{ijkl}, the components of the compliance tensor. Since the tensors are symmetric, the tensor components are independent of the indices ordering. Therefore they can be written in the contracted Voigt notation as c_{mn} (11 → 1, 22 → 2, 33 → 3, 23 → 4, 13 → 5, 12 → 6). In this way, Hooke's law reads (with $T_{ij} = T_n$, $\varepsilon_{ii} = \varepsilon_n$ for n = 1,2,3, $\varepsilon_{ij} = \varepsilon_n/2$ for n = 4,5,6)

$$T_n = \sum_m c_{nm} \varepsilon_m, \quad c_{nm} = c_{mn} \ . \tag{3.7a}$$

The inverse relation reads

$$\varepsilon_n = \sum_m s_{nm} T_m \ . \tag{3.7b}$$

Here the ε_n and s_{ij} are no longer components of the strain tensor (ε_{ij}) or of the compliance tensor (s_{ijkl}). Therefore it is better to work with 3.7. But we use the Voigt notation to label the elastic stiffness constants $c_{ij} = c_{ji}$.

Cubic symmetry, for example, has three elastic constants c_{11}, c_{12} and c_{44}. Consider, as an application, the c_{44}–mode. From Hooke's law (3.7),

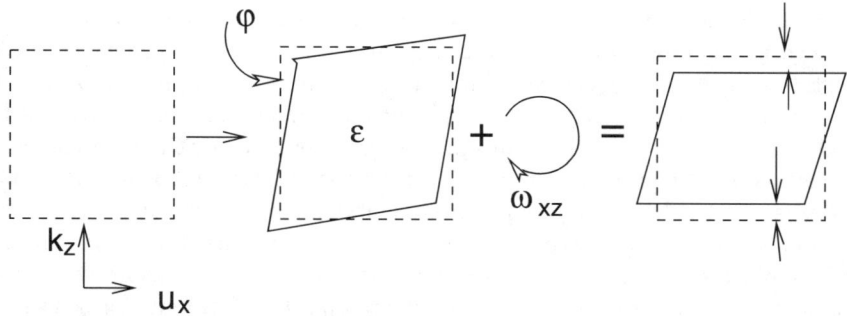

Fig. 3.2. Schematic diagram of the distortions associated with a transverse elastic wave c_{44} in a cubic crystal: this deformation is composed of a strain deformation followed by a rotation. In second order there is an additional deformation indicated in the last figure to the right (see text)

$T_5 = c_{55}v_5 = c_{44}v_5$ and from (3.4a), $v_5 = \varepsilon_5 = \frac{1}{2}\left(\frac{\partial u_x}{\partial R_z} + \frac{\partial u_z}{\partial R_x}\right) = \varepsilon_{xz}$. The plane wave with propagation direction in the z-axis then has the polarization direction along the x-axis: $u_x = Ue_x \exp[i(k_z z - \omega t)]$. Figure 3.2, shows a volume element with such a plane wave deformation $v_5 = v_{xz}$. For the angle φ in Fig. 3.2, $tg\varphi = 2\varepsilon_{xz}$, which is a measure of the shear strain. From (3.4) and (3.4a), follows $v_{xz} = \varepsilon_{xz} + \omega_{xz}$, which is illustrated in the figure schematically. But the detailed result of such a transformation is also shown. It is seen that second order contributions are present which can be expressed with the formalism given above. From (3.1) and (3.3) with $v_{xz} \neq 0$, a finite strain component $\eta_{xz} - \varepsilon_{xz} = E_{zz} = \frac{1}{2}v_{xz}^2$ is obtained. This is indicated in Fig. 3.2 (see Bonsall and Melcher [3.7]).

Equations (3.6) and (3.8) result in the following sound velocities in a cubic crystal (see (B-1) or references at the beginning of this chapter) with ρ the mass density as given in Table 3.1.

Table 3.1. Elastic constants and sound velocities for cubic crystals

mode	propagation direction	polarisation	elastic constant	sound vel.
long.	(100)	(100)	c_{11}	$v_L = \frac{\sqrt{c_{11}}}{\rho}$
transv.	(100)	$\perp$(100)	c_{44}	$v_T = \frac{\sqrt{c_{44}}}{\rho}$
long.	(110)	(110)	$c_L = c_{11} + c_{12} + 2c_{44}$	$v_L = \frac{\sqrt{c_L}}{\rho}$
transv. 1	(110)	(11'0)	$c' = \frac{c_{11}-c_{12}}{2}$	$v_{T1} = \frac{\sqrt{c'}}{\rho}$
transv. 2	(110)	(001)	c_{44}	$v_{T2} = \frac{\sqrt{c_{44}}}{\rho}$
long.	(111)	(111)	$c_L = \frac{c_{11}+2c_{12}+4c_{44}}{3}$	$v_L = \frac{\sqrt{c_L}}{\rho}$
transv.	(111)	$\perp$(111)	$c_T = \frac{c_{11}-c_{12}+c_{44}}{3}$	$v_T = \frac{\sqrt{c_T}}{\rho}$

A longitudinal and two transverse sound velocities are found in all symmetry directions. The transverse sound velocities can also be degenerate. Other crystal systems have been treated in the literature (Ludwig [3.5]). Occasionally, other crystal systems (tetragonal, orthorhombic and hexagonal) will be used. The symmetry elastic strains and elastic constants for the different crystal classes are given in Appendix C and discussed in the following Sect. 3.2. In Appendix C, the space group symbol is indicated for each system. Not all possibilities are listed there but only crystal classes with Inversion symmetry. As an example, there are symmetries for tetragonal crystals for which the elastic stiffness matrix also has components c_{16}. This happens for scheelite-structure materials (space group C_{4h}^6) LiYF$_4$ (Blanchfield and Saunders [3.8]). A coordinate transformation in the c-plane can eliminate this elastic component, reducing the number of elastic constants in this case again from 7 to 6. For a more complete list in appendix C, see Lines [3.9].

Elastic Energy

From (3.7), we can calculate the elastic energy density from the work done due to the deformation: $dE = \sum_{i,j} T_{ij} d\eta_{ij} = \sum_{ij} c_{ijkl} \eta_{ij} d\eta_{kl}$. This gives the elastic energy density

$$E_{el} = \frac{1}{2} \sum_{ijkl} c_{ijkl} \eta_{ij} \eta_{kl} \ . \tag{3.9}$$

where the elastic constants c_{ijkl} in cubic symmetry and Voigt notation read $c_{11} = c_{22} = c_{33}$, $c_{12} = c_{13} = c_{23}$, $c_{44} = c_{55} = c_{66}$.

Integrating dE gives – e.g. for cubic symmetry and small strains ε – from (3.9)

$$E = \frac{1}{2} c_{11}(\varepsilon_{xx}^2 + \varepsilon_{yy}^2 + \varepsilon_{zz}^2) + c_{12}(\varepsilon_{xx}\varepsilon_{yy} + \varepsilon_{xx}\varepsilon_{zz} + \varepsilon_{yy}\varepsilon_{zz}) + 2c_{44}(\varepsilon_{xy}^2 + \varepsilon_{xz}^2 + \varepsilon_{yz}^2) \ . \tag{3.9a}$$

The elastic energy density is quadratic in the strains.

We will compare this result in the next section with the analogous expression for the symmetry strains.

3.2 Symmetry Aspect

The components of the strain tensor ε can be grouped to form irreducible representations ε_Γ of the corresponding point group. For the necessary concepts of group theory, refer to some of the books on group theory (Tinkham [3.10], Burns [3.11], Inui et al. [3.12], Wagner [3.13]). Since ε is a symmetric second rank tensor, there are two vectors $\boldsymbol{A}$ and $\boldsymbol{B}$, $\boldsymbol{A} = \varepsilon \boldsymbol{B}$ and with the rotation $\boldsymbol{R}$, $\boldsymbol{A}' = \boldsymbol{R}\boldsymbol{A}$, $\boldsymbol{B}' = \boldsymbol{R}\boldsymbol{B}$ and $\boldsymbol{A}' = \varepsilon' \boldsymbol{B}'$ or $\varepsilon' = \boldsymbol{R}\varepsilon\boldsymbol{R}^{-1}$. With the orthogonality of $\boldsymbol{R}((\boldsymbol{R}^{-1})_{ij} = \boldsymbol{R}_{ji})$ we get $\varepsilon'_{\mu\nu} = D_{\mu\mu'}(\boldsymbol{R})D_{\nu\nu'}(\boldsymbol{R})\varepsilon_{\mu'\nu'}$. The last

3.2 Symmetry Aspect

Table 3.2. Character of the point group O

	E	8 C_3	3C_2	6C_2	6C_4	Basis function
$A_1\ \Gamma_1$	1	1	1	1	1	$x^2 + y^2 + z^2$
$A_2\ \Gamma_2$	1	1	1	-1	-1	
$E\ \Gamma_3$	2	-1	2	0	0	$x^2 - y^2, 2z^2 - x^2 - y^2$
$T_1\ \Gamma_4$	3	0	-1	-1	1	x, y, z
$T_2\ \Gamma_5$	3	0	-1	1	-1	xy, yz, zx
$\chi_V = T_1$	3	0	-1	-1	1	x, y, z
$[\chi_V(Q)]^2$	9	0	1	1	1	
$\chi_V(Q^2)$	3	0	3	3	-1	
$[\chi_V \times \chi_V]_S$	6	0	2	2	0	$\Gamma_1 + \Gamma_3 + \Gamma_5$

relation means that the elements of ε are Basis of product representations of the vector representation Γ_V i.e. $\Gamma_V \times \Gamma_V$. And the character of a symmetric tensor is given by (Tinkham [3.10])

$$[\chi_V(Q) \times \chi_V(Q)]_s = \frac{1}{2}[\chi_V(Q)]^2 + \frac{1}{2}\chi_V(Q^2) \tag{3.10}$$

where Q is a group element and χ the character of the representations. The components of the symmetric strain tensor transform according to the same representation as the basis functions of the symmetrized square of the representation of a polar vector. We consider the vector representation Γ_V and form the symmetric product of the corresponding character χ_V: $[\chi_V \times \chi_V]_S$ given by (3.10).

As a first example, these concepts are applied to the cubic point group O. In Table 3.2 we give the complete character table for the point group O with the additional symmetric products. It is seen that $[\chi_V \times \chi_V]_S$ decomposes into $A_1 + E + T_2$. Therefore we have the desired symmetry strains and symmetry elastic constants composed from ε_{ij} and from c_{11}, c_{12} and c_{44}

$$A_1(\Gamma_1): \varepsilon_v = \varepsilon_{xx} + \varepsilon_{yy} + \varepsilon_{zz} \quad \text{volume strain} \quad c_B = \frac{c_{11} + 2c_{12}}{3} \quad \text{bulk modulus}$$

$$E(\Gamma_3): \varepsilon_2 = \frac{\varepsilon_{xx} - \varepsilon_{yy}}{\sqrt{2}}$$

$$\varepsilon_3 = 2\varepsilon_{zz} - \varepsilon_{xx} - \varepsilon_{yy} \quad\quad \frac{c_{11} - c_{12}}{2}$$

$$T_2(\Gamma_5): \varepsilon_{xy}, \varepsilon_{yz}, \varepsilon_{zx} \quad\quad c_{44}\ . \tag{3.11}$$

The antisymmetric part of the strain tensor belongs to the representation $T_1(\Gamma_4)$: $\omega_{yz}, \omega_{zx}, \omega_{xy}$. A somewhat different derivation of these results can be found in Wagner [3.13]. The elastic energy density E_{el} for cubic symmetry can be constructed from these expressions. Equation (3.7) is used to calculate the work done for the strained crystal. This results in

34   3 Elasticity

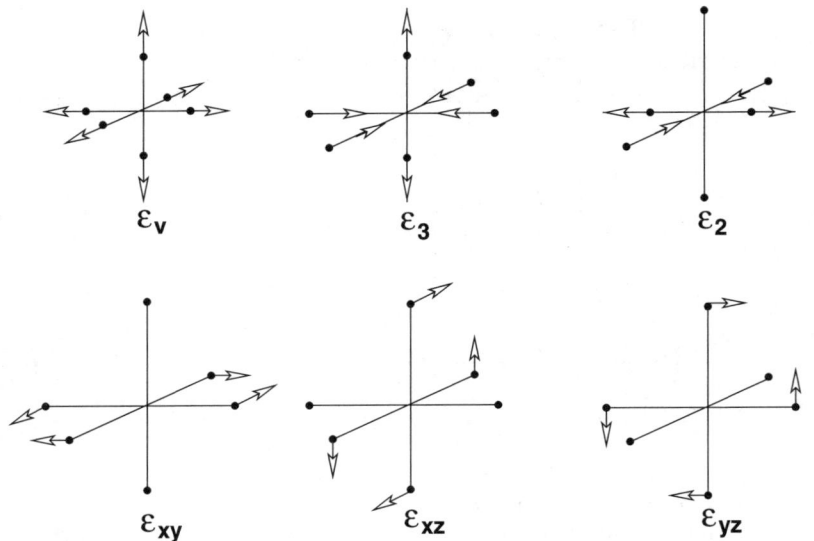

Fig. 3.3. Symmetry strains for cubic symmetry

$$E_{el} = \frac{1}{2}c_B\varepsilon_v^2 + \frac{c_{11} - c_{12}}{2}\left(\varepsilon_2^2 + \varepsilon_3^2\right) + 2c_{44}\left(\varepsilon_{xy}^2 + \varepsilon_{yz}^2 + \varepsilon_{zx}^2\right) . \quad (3.9b)$$

This result is identical with (3.9a) using the Cartesian strains $\varepsilon_{xx}, \varepsilon_{xy}, \cdots$

Figure 3.3 shows the various symmetry strains ε_Γ for cubic symmetry, taken from a paper by Van Vleck [3.14] which shows the local symmetry strains. As can be seen from this figure, the E and T_2 symmetry strains are volume conserving, while the A_1 volume-strain is not. The latter one will play an important role in systems with unstable ions (Chap. 9) and with superconductors (Chap. 10). The former volume conserving modes play an important role in the various structural transitions (Chap. 7).

As a second example, symmetry strains for tetragonal symmetry D_4 can be constructed in a similar way. The character table for the point group D_4 with the symmetry products is shown in Table 3.3.

The symmetry strains and symmetry elastic stiffness constants can again be determined from this table (Table 3.3):

$$A_1 \quad \varepsilon_\alpha^1 = \frac{1}{\sqrt{3}}\left(\varepsilon_{xx} + \varepsilon_{yy} + \varepsilon_{zz}\right) c_\alpha^1 = \frac{1}{3}\left(2c_{11} + 2c_{12} + 4c_{13} + c_{33}\right)$$

$$\varepsilon_\alpha^2 = \frac{\sqrt{2}}{3}\left(\varepsilon_{zz} - \varepsilon_{xx} + \varepsilon_{yy}\right) c_\alpha^2 = -\frac{\sqrt{2}}{3}\left(c_{11} + c_{12} - c_{13} - c_{33}\right)$$

3.2 Symmetry Aspect

Table 3.3. Character of the point group D_4

	E	$2C_4$	C_2	$2C_2'$	$2C_2''$	Basis function
$A_1\ \Gamma_1$	1	1	1	1	1	x^2+y^2, z^2
$A_2\ \Gamma_2$	1	1	1	-1	-1	z
$B_1\ \Gamma_3$	1	-1	1	1	-1	x^2-y^2
$B_2\ \Gamma_4$	1	-1	1	-1	1	xy
$E\ \Gamma_5$	2	0	-2	0	0	x, y, xz, yz
$\chi_V = A_2 + E$	3	1	-1	-1	-1	x, y, z
χ_V^2	9	1	1	1	1	
$\chi_V(Q^2)$	3	-1	3	3	3	
$[\chi_V + \chi_V]_s$	6	0	2	2	2	$2A_1 + E + B_1 + B_2$

$$B_1\quad \varepsilon_\gamma = \frac{1}{\sqrt{2}}(\varepsilon_{xx} - \varepsilon_{yy}) \quad c_\gamma = c_{11} - c_{12}$$

$$B_2\quad \varepsilon_\delta = \sqrt{2}\varepsilon_{xy} \quad\quad\quad\quad c_\delta = c_{66}$$

$$E\quad \varepsilon_\varepsilon^1 = \sqrt{2}\varepsilon_{zx} \quad\quad\quad\quad c_\varepsilon = c_{44}$$

$$\quad\quad \varepsilon_\varepsilon^2 = \sqrt{2}\varepsilon_{yz}$$

The elastic energy obtained from these symmetry strains is:

$$E_{el} = \frac{c_{11}}{2}\left(\varepsilon_{xx}^2 + \varepsilon_{yy}^2\right) +$$
$$c_{12}\varepsilon_{xx}\varepsilon_{yy} + \frac{c_{33}}{2}\varepsilon_{zz}^2 + c_{13}\varepsilon_{zz}\left(\varepsilon_{xx} + \varepsilon_{yy}\right) +$$
$$2c_{66}\varepsilon_{xy}^2 + 2c_{44}\left(\varepsilon_{zx}^2 + \varepsilon_{yz}^2\right)$$

As a third example, the symmetry strains for hexagonal symmetry D_6 are constructed. The character table for this symmetry is shown in Table 3.4. As in the case of tetragonal symmetry, this gives two one-dimensional symmetry strains for the bulk moduli $\varepsilon_1(\Gamma_6) = \varepsilon_{xx} + \varepsilon_{yy} + \varepsilon_{zz}$ and $\varepsilon_2(\Gamma_6) = 2\varepsilon_{zz} - \varepsilon_{xx} - \varepsilon_{yy}$ and two two-dimensional strains for the $(c_{11} - c_{12})$ - mode $\varepsilon_1(\Gamma_5) = \varepsilon_{xx} - \varepsilon_{yy}$, $\varepsilon_2(\Gamma_5) = \varepsilon_{xy}$ and for the c_{44}–mode $\varepsilon_1(\Gamma_6) = \varepsilon_{xz}$, $\varepsilon_2(\Gamma_6) = \varepsilon_{yz}$. There are two Γ_1 representations in this case which do not alter the symmetry, $\varepsilon_1(\Gamma_1)$ giving the volume change and $\varepsilon_2(\Gamma_1)$ changing only the c/a ratio. We will make use of this D_6 symmetry in Sects. 5.2.1, 7.2.3 and 9.3.

The last line of Table 3.4 shows that

$$[V]^2 = \Gamma_1(x^2+y^2+z^2) + \Gamma_1(2z^2 - x^2 - y^2) + \Gamma_6(xz, yz) + \Gamma_5(xy, x^2 - y^2)$$

which gives the symmetry strains listed above.

The symmetry strains and symmetry elastic constants for other crystal systems have been tabulated by Lines [3.9], Wallace [3.4] and others. A table is given in appendix C where we include the basis functions and the $[V]^2$

Table 3.4. Character of the point group D_6

	E	$2C_6$	$2C_3$	C_2	$3C_{2y}$	$3C_{2x}$	Basis function
$A_1\ \Gamma_1$	1	1	1	1	1	1	z^2, x^2+y^2
$A_2\ \Gamma_2$	1	1	1	1	-1	-1	z,
$B_1\ \Gamma_3$	1	-1	1	-1	1	-1	$y^3 - 3x^2y$
$B_2\ \Gamma_4$	1	-1	1	-1	-1	1	$x^3 - 3xy^2$
$E_1\ \Gamma_6$	2	1	-1	-2	0	0	(x,y) (xz,yz)
$E_2\ \Gamma_5$	2	-1	-1	2	0	0	xy, x^2-y^2
$\Gamma_2+\Gamma_6=\chi_V$	3	2	0	-1	-1	-1	x,y,z
$\chi_V^2(Q)$	9	4	0	1	1	1	
$\chi_V(Q^2)$	3	0	0	3	3	3	
$[\chi_V\times\chi_V]_s$	6	2	0	2	2	2	$2\Gamma_1+\Gamma_6+\Gamma_5$

representation. We will mostly use the three examples discussed above in the following.

3.3 Third Order Elastic Constants

The elastic constants discussed above are the so-called second order elastic constants. In the later Sect. 4.2, it will be seen that the temperature dependence of the second order elastic constants is due to an-harmonic effects. A harmonic crystal would have temperature independent elastic constants. Subsequently, the elastic energy can have terms of higher order than quadratic in the strain variables – see (3.9), (3.9a) and (3.9b). They allow the evaluation of the an-harmonic terms of the interatomic potential. We derive the higher order symmetry elastic constants below and give the further terms for the elastic energy of (3.9b). Generally, third order elastic moduli influence all anelastic phenomena such as thermal expansion, phonon–phonon interaction, thermal conductivity etc.

Starting with the Lagrangian symmetry strains (3.11), these span the irreducible representations $\Gamma_1, \Gamma_3, \Gamma_5$ for cubic symmetry and with the antisymmetric strains $\omega_{ij} \in \Gamma_4$. We consider the direct product of the irreducible representations of the point group O. For the ε_{ij} strains $\{\varepsilon_{ij}\varepsilon_{kl}\} \in 3\Gamma_1+3\Gamma_3+\Gamma_4+3\Gamma_5$. From the 36 strain products, there are 21 non commutative elements. Care has to be taken only to consider symmetrised products, see (3.10). Including the Γ_4 representation of the antisymmetric strains ω_{ij}, gives – for the direct product of the η_{ij} – the strains $\{\eta_{ij}\eta_{kl}\} \in 4\Gamma_1+\Gamma_2+5\Gamma_3+4\Gamma_4+6\Gamma_5$. Thus from $9\times 9 = 81$ products, there are 45 elements which are contained in these representations.

For the elastic energy of third order, there are contributions of these symmetrised strains labeled above with the third order elastic constants C_{ijk}.

In addition, there are also contributions from the higher order terms of the strains of (3.1) with the second order elastic constants c_{ij}. For cubic symmetry, the elastic energy of third order in the strains and rotations, in addition to the second order terms (3.9a), reads (see Rouchy et al. [3.15])

$$\begin{aligned}E^3_{el} = &\frac{1}{6}C_{111}(\varepsilon^3_{xx} + cycl.) + \frac{1}{2}C_{112}\{\varepsilon^2_{xx}(\varepsilon_{yy} + \varepsilon_{zz}) + cycl.\} \\ &+ C_{123}\varepsilon_{xx}\varepsilon_{yy}\varepsilon_{zz} + 2C_{144}\{\varepsilon_{xx}\varepsilon^2_{yz} + cycl\} \\ &+ 2C_{155}\{\varepsilon^2_{yz}(\varepsilon_{yy} + \varepsilon_{zz}) + cycl.\} \\ &+ 8C_{456}\varepsilon_{yz}\varepsilon_{zx}\varepsilon_{xy} + 2C_{177}\{\varepsilon_{xx}\omega^2_{yz} + cycl.\} \\ &+ 2C_{277}\{\omega^2_{yz}(\varepsilon_{yy} + \varepsilon_{zz}) + cycl.\} \\ &+ 8C_{489}\{\varepsilon_{yz}\omega_{zx}\omega_{xy} + cycl.\} \\ &+ 2C_{269}\{\varepsilon_{yz}\omega_{yz}(\varepsilon_{yy} - \varepsilon_{zz}) + cycl.\}\end{aligned} \quad (3.9c)$$

Here, the 10 third order elastic constants C_{ijk} are linear combinations of second order and true third order terms. The following symmetry relations were used:

$$C_{111} = C_{222} = C_{333}$$
$$C_{144} = C_{255} = C_{366}$$
$$C_{112} = C_{223} = C_{133} = C_{113} = C_{122} = C_{233}$$
$$C_{155} = C_{244} = C_{344} = C_{166} = C_{266} = C_{355}.$$

Third order elastic constantss can be determined by measuring the velocity change of various compressional and shear modes in different crystallographic directions as a function of hydrostatic pressure and uniaxial stress. Relevant papers explaining this method in detail are Thurston and Brugger [3.16], Brugger [3.17], Eastman [3.18]. We will only make occasional further use of it (Sects. 4.3.3 and 7.1).

3.4 Elastic Stability, Elastic Isotropy

Stability

From the considerations given above, stability criteria for the elastic solid can be stated. The strain energy has to be a positive definite quadratic function of the strain components (Born and Huang [3.19]). For cubic symmetry we get therefore from (3.9b) the conditions

$$c_{11} + 2c_{12} > 0, \quad c_{11} - c_{12} > 0, \quad c_{44} > 0. \quad (3.12)$$

We will use these stability conditions later on in several applications (Sect. 3.4 SAW, Chap. 7, Sect. 9.2 mixed valency).

Isotropy

In elastic isotropic media, there is one propagating longitudinal wave and one propagating shear wave. In elastically anisotropic media there are usually two propagating transverse waves. In cubic systems, these are, for example, the c_{44} and the $\frac{c_{11}-c_{12}}{2}$ modes. They coincide for $A = 2\frac{c_{44}}{c_{11}-c_{12}} = 1$, which is the condition for elastic isotropy. A is called the elastic anisotropy parameter. From (3.12) $A > 0$ but it can have values > 1 or < 1.

In an elastic isotropic system, the sound wave propagates in all directions with equal velocities. However longitudinal and transverse velocities can still be different. Therefore there are two elastic constants to characterise an isotropic elastic solid:

$$c_{12} = c_{13} = c_{21} = c_{23} = c_{31} = c_{32} = \lambda$$
$$c_{44} = c_{55} = c_{66} = \mu$$
$$c_{11} = c_{22} = c_{33} = \lambda + 2\mu.$$

Here λ and μ are the Lamé elastic constants. With an applied uniform stress in the x direction of a rod and with the lateral surfaces free from stress, Hooke's law (3.7) reads:

$$T_{xx} = (\lambda + 2\mu)\varepsilon_{xx} + \lambda(\varepsilon_{yy} + \varepsilon_{zz})$$
$$0 = (\lambda + 2\mu)\varepsilon_{yy} + \lambda(\varepsilon_{xx} + \varepsilon_{zz})$$
$$0 = (\lambda + 2\mu)\varepsilon_{zz} + \lambda(\varepsilon_{xx} + \varepsilon_{yy}).$$

Solving for ε_{ii} gives

$$\varepsilon_{xx} = \frac{\lambda + \mu}{\mu(3\lambda + 2\mu)}T_{xx} \quad \varepsilon_{yy} = \varepsilon_{zz} = -\frac{\lambda}{2\mu(3\lambda + 2\mu)}T_{xx}.$$

It follows from this that the bulk modulus is $c_B = \lambda + \frac{2}{3}\mu$ and the Poisson ratio ν as the ratio between lateral contraction and longitudinal extension of the material is

$$\nu = -\frac{\varepsilon_{yy}}{\varepsilon_{xx}} = \frac{\lambda}{2(\lambda + \mu)} = \frac{1}{2} \times \frac{3c_B - 2\mu}{3c_B + \mu}. \tag{3.12a}$$

The Poisson ratio can have values in the region $-1 < \nu < \frac{1}{2}$ with the end points -1 from $c_B = 0$ and $\frac{1}{2}$ from $\mu = 0$. The anomalous Poisson ratio $\nu < 0$ will be discussed later in Sect. 9.2.

There are other elastic moduli: Young's modulus E, the shear modulus G and the bulk modulus K (inverse of the compressibility). The Young's E modulus is defined by $E = \frac{T_{xx}}{\varepsilon_{xx}}$ and we get $E = \frac{\mu(3\lambda+2\mu)}{(\lambda+\mu)}$. Note that the longitudinal sound velocity for a long rod with no lateral stresses is given by $v_L^E = \frac{\sqrt{E}}{\rho}$. On the other hand, the longitudinal sound velocity in an isotropic medium is $v_L = \frac{\sqrt{(\lambda+2\mu)}}{\rho} > v_L^E$ and the transverse one $v_T = \sqrt{\frac{\mu}{\rho}}$. There is a relation between E, G, and K since only two moduli are independent, i.e.

$$\frac{3}{E} = \frac{1}{G} + \frac{1}{3K}. \qquad (3.12\text{b})$$

Equation (3.12b) follows from the expressions for E and K in terms of λ and μ given above and with $G = \mu$. For a general discussion of isotropic media see Kolsky [3.20] and also Sect. 2.2.6. Finally if $c_{12} = c_{44}$ the cubic-symmetry Cauchy relations are fulfilled. In general these Cauchy relations are fulfilled only for central forces of the crystal potential. For details see Wallace [3.4].

3.5 Surface Acoustic Waves, SAW

Apart from the technical importance, acoustic surface waves show interesting solid state physics effects. In metals, superconductors, magnetic materials, low-dimensional electronic systems and near phase transitions, various SAW phenomena can be investigated. The finite penetration depth and the characteristic particle motion enable special effects to occur using SAW. A brief outline of the fundamentals of SAW and treatment of the so-called Rayleigh waves for crystalline solids are given in this section. SAW applications in solid state physics will be presented in the following sections:

9.2 SAW propagation characteristics with negative Poisson ratio materials.
10.6 SAW attenuation for conventional superconductors.
10.4 An elastic constant determination for a high temperature superconductor involving bulk waves and SAW using Brillouin scattering.
13.5 A non-reciprocal SAW effect for a metal in the presence of magnetic field in Voigt geometry.
13.6 SAW experiments in the microwave frequency range for heterostructure materials GaAs/AlGaAs exhibiting the integral and the fractional quantum Hall effect.

Other SAW applications are mentioned in Sect. 5.2.2 and 6.2.

To be specific, we choose a crystal with cubic symmetry and a free surface being the x-y plane at z=0 (see Fig. 3.4). We use the method of Stoneley (Stoneley [3.21]) to derive the SAW equation. This implies a continuum treatment of the elastic body. Using (3.6,3.7) we get for the wave equation (3.6) in components (see (B-1)):

$$\begin{aligned}\rho\frac{\partial^2 u_x}{\partial t^2} &= \frac{\partial T_{xx}}{\partial x} + \frac{\partial T_{xy}}{\partial y} + \frac{\partial T_{xz}}{\partial z} \\ &= c_{11}\frac{\partial^2 u_x}{\partial x^2} + c_{12}\left[\frac{\partial^2 u_y}{\partial x \partial y} + \frac{\partial^2 u_z}{\partial x \partial z}\right] \\ &+ c_{44}\left[\frac{\partial^2 u_x}{\partial y^2} + \frac{\partial^2 u_y}{\partial x \partial y} + \frac{\partial^2 u_x}{\partial z^2} + \frac{\partial^2 u_z}{\partial x \partial z}\right]\end{aligned}$$

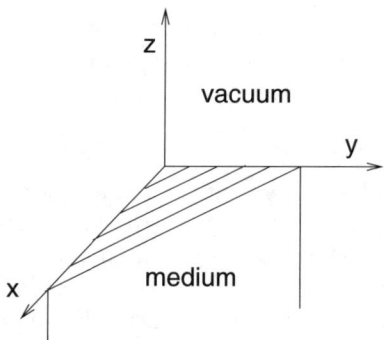

Fig. 3.4. Geometry for SAW. The surface lies in the x-y plane, the medium extends to z < 0

$$\rho\frac{\partial^2 u_y}{\partial t^2} = c_{11}\frac{\partial^2 u_y}{\partial y^2} + c_{12}\left[\frac{\partial^2 u_x}{\partial x \partial y} + \frac{\partial^2 u_z}{\partial y \partial z}\right]$$

$$+ c_{44}\left[\frac{\partial^2 u_y}{\partial z^2} + \frac{\partial^2 u_z}{\partial y \partial z} + \frac{\partial^2 u_x}{\partial x \partial y} + \frac{\partial^2 u_y}{\partial x^2}\right]$$

$$\rho\frac{\partial^2 u_z}{\partial t^2} = c_{11}\frac{\partial^2 u_z}{\partial z^2} + c_{12}\left[\frac{\partial^2 u_x}{\partial x \partial z} + \frac{\partial^2 u_y}{\partial z \partial y}\right]$$

$$+ c_{44}\left[\frac{\partial^2 u_x}{\partial x \partial z} + \frac{\partial^2 u_z}{\partial x^2} + \frac{\partial^2 u_y}{\partial y \partial z} + \frac{\partial^2 u_z}{\partial y^2}\right]. \qquad (3.13)$$

For the SAW we make the following assumption: $u_i = U_i \exp[\kappa z + i(k_x x + k_y y - \omega t)]$ i.e. a propagating wave in the xy–plane and an exponentially decreasing function in the $-z$ direction (see Fig. 3.4). The plane, given by the **k**-vector and the normal to the surface, the z–axis, is called the sagittal plane. The system of (3.13) then reads

$$U_x \left[\rho\omega^2 - c_{11}k_x^2 - c_{44}k_y^2 + c_{44}\kappa^2\right]$$
$$-U_y \left[c_{12}k_x k_y + c_{44}k_x k_y\right]$$
$$+U_z \left[ik_x \kappa c_{12} + ik_x \kappa c_{44}\right] \qquad = 0$$

$$-U_x \left[c_{12}k_x k_y + c_{44}k_x k_y\right]$$
$$+U_y \left[\rho\omega^2 - c_{11}k_y^2 - c_{44}k_x^2 + c_{44}\kappa^2\right]$$
$$+U_z \left[ik_y \kappa c_{12} + ik_y \kappa c_{44}\right] \qquad = 0$$

$$U_x \left[ik_x \kappa c_{12} + ik_x \kappa c_{44}\right]$$
$$+U_y \left[ik_y \kappa c_{12} + ik_y \kappa c_{44}\right]$$
$$+U_z \left[\rho\omega^2 - c_{44}k_x^2 - c_{44}k_y^2 + c_{11}\kappa^2\right] \quad = 0 \qquad (3.14)$$

3.5 Surface Acoustic Waves, SAW

As a first example, put $k_x = k, k_y = 0$ (propagation along (100) direction) and use $\frac{\omega}{k} = v$ (SAW velocity) and $\frac{\kappa}{k} = q$. The system (3.14) then becomes

$$U_x \left[\rho v^2 - c_{11} + c_{44}q^2\right] \qquad\qquad +U_z \left[iqc_{12} + iqc_{44}\right] \qquad = 0$$

$$U_y \left[\rho v^2 - c_{44} + c_{44}q^2\right] \qquad\qquad = 0$$

$$U_x \left[iqc_{12} + iqc_{44}\right] \qquad\qquad +U_z \left[\rho v^2 - c_{44} + c_{11}q^2\right] = 0.$$
(3.15)

The determinant of (3.15) gives the dispersion equation

$$\left[\rho v^2 - c_{44} + c_{44}q^2\right] \qquad\qquad (3.16)$$
$$\times \left[(\rho v^2 - c_{11} + c_{44}q^2)(\rho v^2 - c_{44} + c_{11}q^2) + q^2(c_{11} + c_{44})^2\right] = 0. \quad (3.17)$$

Now we introduce the boundary conditions. For a free surface at $z = 0$ the components of the stress tensor on the surface will be zero:

$$T_{zz} = T_{xz} = T_{yz} = 0 \quad \text{for} \quad z = 0.$$

With (3.7) this means for symmetric strains $T_{yz} = c_{44}\varepsilon_{yz} = c_{44}(\frac{\partial u_y}{\partial z} + \frac{\partial u_z}{\partial y}) = 0$. Since there is no y-dependence in the ansatz for the wave, thus $U_y = 0$ and (3.16) now reads

$$\left[(\rho v^2 - c_{11} + c_{44}q^2)(\rho v^2 - c_{44} + c_{11}q^2) + q^2(c_{11} + c_{44})^2\right] = 0. \quad (3.16a)$$

This is an equation of second order in q^2 with solutions q_1^2 and q_2^2. For general directions there would be three solutions (Gazis et al. [3.22]). Then the remaining wavelike deformations can be written as

$$u_x = \left[U_x^1 e^{\kappa q_1 z} + U_x^2 e^{\kappa q_2 z}\right] e^{i(kx - \omega t)}$$

$$u_z = \left[U_z^1 e^{\kappa q_1 z} + U_z^2 e^{\kappa q_2 z}\right] e^{i(kx - \omega t)}.$$

The second boundary condition reads

$$T_{xz} = c_{44}\varepsilon_{xz} = c_{44}(\frac{\partial u_x}{\partial z} + \frac{\partial u_z}{\partial x}) = 0$$

which gives

$$q_1 U_x^1 + q_2 U_x^2 + i(U_z^1 + U_z^2) = 0 \qquad (3.16b)$$

and the third boundary condition reads $T_{zz} = c_{11}\varepsilon_{zz} + c_{12}\varepsilon_{xx} = 0$ which gives

$$c_{11}(U_z^1 q_1 + U_z^2 q_2) + ic_{12}(U_x^1 + U_x^2) = 0. \qquad (3.16c)$$

With these three boundary conditions the problem is solved. q_1, q_2 can be eliminated from (3.16a,b,c) resulting in the final equation for the SAW velocity v with $R = \frac{\rho v^2}{c_{11}}$, i.e.

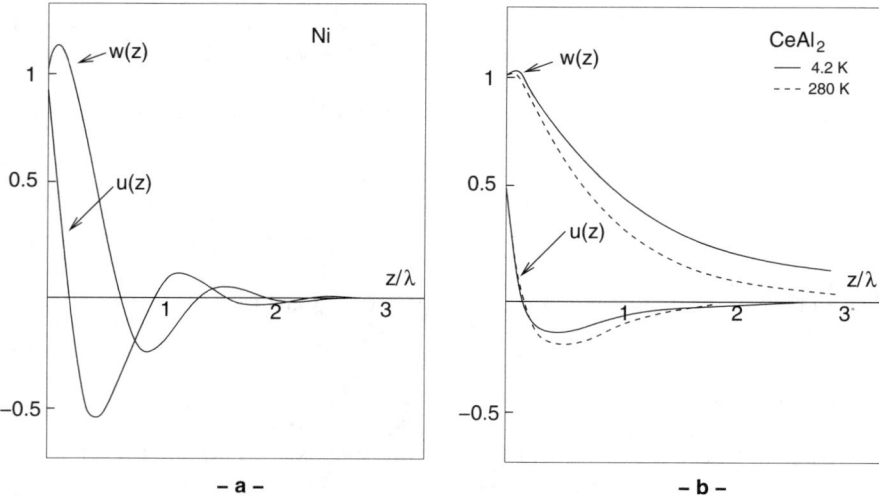

Fig. 3.5. Horizontal ($u_x = u(z)$) and vertical ($u_z = w(z)$) components of the SAW displacements as a function of z in units of the Wavelength λ, (**a**) Ni (generalised Rayleigh wave), (**b**) CeAl$_2$ (ordinary Rayleigh wave)

$$(1 - \frac{c_{11}}{c_{44}}R)\left[1 - \left(\frac{c_{12}}{c_{11}}\right)^2 - R\right]^2 = R^2(1-R) \; . \qquad (3.18)$$

It turns out that R < 1 always, the SAW velocity is always smaller than the bulk velocity, which in this case is $\sqrt{(\frac{c_{11}}{\rho})}$. From $q = \frac{\kappa}{k}$ follows for the penetration depth of the SAW $d = \kappa^{-1}$. The amplitudes $\frac{u_x}{u_z}$ are elliptically polarized as indicated in Fig. 3.5 for two cases. The particle displacements are in the sagittal plane.

An analogous calculation for propagation in (110) direction gives with $k_x = k_y = \frac{k}{\sqrt{2}}$ the following equation for $R = \frac{\rho v^2}{c_{11}}$ with $c_L = \frac{1}{2}(c_{11}+c_{12}+2c_{44})$:

$$\left(1 - \frac{c_{11}}{c_{44}}R\right)\left(\frac{c_{11}c_L - c_{12}^2}{c_{11}^2} - R\right)^2 = R^2\left(\frac{c_L}{c_{11}} - R\right) \; . \qquad (3.19)$$

Figure 3.5 presents two cases of SAW propagation for cubic crystals: CeAl$_2$ (Lingner et al. [3.23]) and Nickel (Lingner et al. [3.24]). In both cases the amplitudes $u_x = w(z)$ and $iu_z = u(z)$ are plotted as a function of the sound wavelength. Whereas the z-component decays exponentially with distance from the surface for CeAl$_2$, this is not the case for Ni. Here the exponential decay is modulated with a trigonometric function and both components disappear several times as a function of z. In this latter case, one speaks of a generalised Rayleigh wave, in the former case (CeAl$_2$) of an ordinary Rayleigh wave. The reason for this different behaviour is the character of the parameter q introduced above. Real values of q result in an ordinary Rayleigh wave, complex values in a generalised Rayleigh wave (Stoneley [3.21]).

3.5 Surface Acoustic Waves, SAW

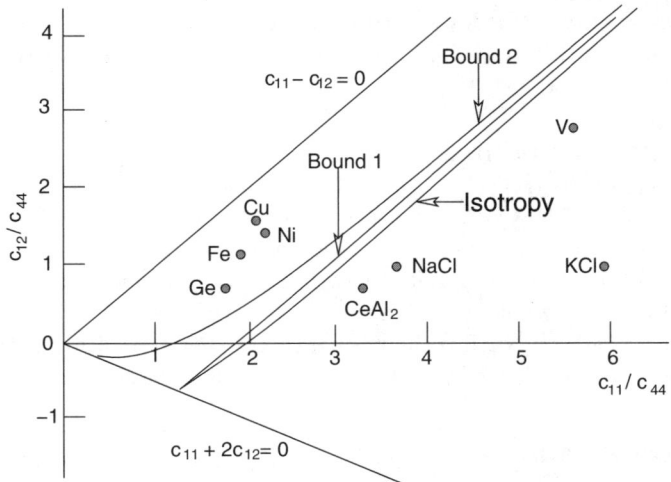

Fig. 3.6. c_{12}/c_{44} versus c_{11}/c_{44} with positions of various materials with respect to ordinary and generalised Rayleigh waves. Indicated are also different stability lines as discussed in the text (adapted from Gazis et al. [3.22])

These results can be presented graphically as shown in Fig. 3.6 for cubic crystals (Gazis et al. [3.22]). The axes denote $\frac{c_{11}}{c_{44}}$ for the x-axis and $\frac{c_{12}}{c_{44}}$ for the y-axis. This gives two bounds from elastic stability (3.12) $c_{11} - c_{12} = 0$ and $c_{11} + 2c_{12} = 0$. Only inside the region bounded by these conditions is the elastic body stable. The next line $A = 2\frac{c_{44}}{c_{11}-c_{12}} = 1$ is the line of elastic isotropy (see Sect. 3.4). As seen from Fig. 3.6, it lies close to the line called Bound 1. This line divides the generalised Rayleigh waves from the ordinary Rayleigh waves as defined above. Bound 1 is determined numerically (Gazis et al. [3.22]) for the (100)- and (110)-propagation directions and it is assumed to hold also for all propagation directions in between. Bound 2 in the region of generalised Rayleigh waves gives the condition for non-propagating SAW. All the materials corresponding to elastic constants lying above bound two are characterised by the existence of an excluded sector of directions of propagation in the vicinity of the (110)-direction. It should be noted, however, that a SAW can always propagate along the (110)-direction, although this direction is within the excluded sector. For details of these approximations and bounds see Gazis et al. [3.22]. Some examples like Al, Cu, Fe, Si, Ge, CeAl$_2$, Ni, NaCl, KCl, V are given in Fig. 3.6.

Apart from these ordinary and generalised Rayleigh waves, there exist other SAW modes. Leaky surface waves are possible for elastic anisotropic surfaces. This pseudo surface wave has an attenuation coefficient in the direction of propagation. Examples of other types of surface waves are the so-called Stoneley waves (two semi-infinite solids in contact), Lamb waves (in plates), Scholte waves (solid and liquid in contact), Bleustein-Gulyaev waves in piezoelectric media, electroacoustic waves etc. Details can be found in the reviews by White [3.25], Maradudin [3.26], Kress and de Wette [3.27].

3.6 Lattice Dynamics

In the derivation of the equation of elastic waves (Sect. 3.1, Appendix B), we have taken the wavelength of the sound wave $\lambda \gg a$ with a the lattice constant. In the lattice dynamics, we relax this condition and consider wave propagation in the whole wave length spectrum down to $\lambda \propto a$. Since the de Brogle wave length $\frac{h}{p}$, with p the momentum, is very small for lattice waves, the lattice wave problem can be treated essentially classically.

3.6.1 Phonon Dispersion

By defining the lattice displacement $\boldsymbol{u}_{s,l}$ (atom s in the unit cell l) we get the kinetic energy $KE = \sum_{s,l} \frac{1}{2} M_s (\frac{d\boldsymbol{u}_{s,l}}{dt})^2$ and for the potential energy $V(\boldsymbol{u}_{s,l})$ with an expansion in $\boldsymbol{u}_{s,l}$:

$$V = V_0 + \frac{1}{2} \sum_{s,s',l,l',j,j'} u_{s,l}^j u_{s',l'}^{j'} \left(\frac{\partial^2 V}{\partial u_{s,l}^j \partial u_{s',l'}^{j'}} + \cdots \right)$$

Here we have assumed that such a potential exists for the whole crystal and that it can be expressed in terms of the instantaneous positions of all the atoms (j = Cartesian component). Furthermore, because expansion is around an equilibrium position of the crystal, the first term linear in $\frac{\partial V}{\partial u_{s,l}^j}$ is zero. Newton's equation of motion gives with the force $-\frac{\partial V}{\partial \boldsymbol{u}_{s,l}}$

$$\frac{M_s d^2 u_{s,l}^j}{dt^2} = - \sum_{s',l',j'} \left[\frac{\partial^2 V}{\partial u_{s,l}^j} \partial u_{s',l'}^{j'} \right] u_{s',l'}^{j'} . \qquad (3.20)$$

This equation can be simplified by taking the translation symmetry into account through the Bloch form $\boldsymbol{u}_{s,l}(t) = e^{i\boldsymbol{q}\boldsymbol{r}} \boldsymbol{u}_{s,o}(t)$. This results finally in

$$\frac{M_s d^2 \boldsymbol{u}_{s,q}}{dt^2} = - \sum_{s^{\iota}} D_{ss^{\iota}}(q) \boldsymbol{u}_{s^{\iota},q} \qquad (3.21)$$

with $D(q)$ the Fourier transform of the V derivative. Equation (3.20) is a system of $3nN$ equations (n # of atoms in the unit cell, N # of cells in the lattice). But (3.21) has only $3n$ equations. With $\boldsymbol{u}_{s,q} = U_{s,q} e^{i\omega t}$ we finally get the eigenvalue equation

$$\sum_{s',j'} \left\{ D_{s,s'}^{jj'}(q) - \omega^2 M_s \delta_{ss'} \delta_{jj'} \right\} U_{s'q}^{j'} = 0 . \qquad (3.22)$$

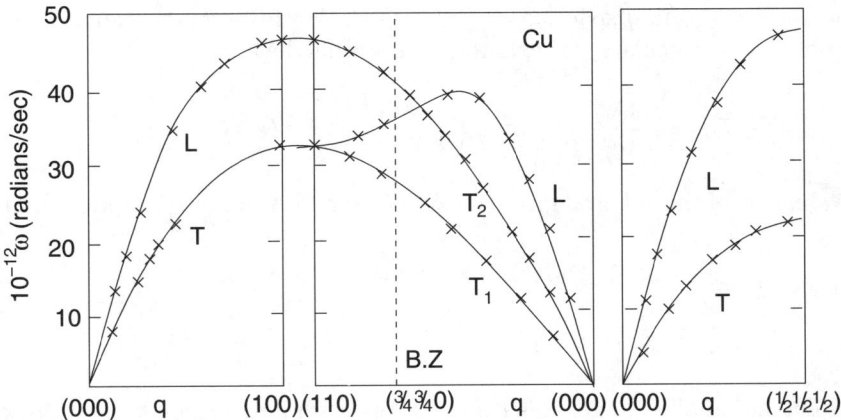

Fig. 3.7. Phonon dispersion spectra $\omega(\mathbf{q})$ for Cu for different $\mathbf{q}$–directions (Svensson et al. [3.28])

The determinant of (3.22) gives the eigenvalues $\omega(\mathbf{q})$. There are various models that give expressions for the force constants $D_{ss'}$, like the shell model, the force constant model etc. We will not discuss these topics any further but show the experimental results of $\omega(\mathbf{q})$ for copper Cu (Fig.3.7).

It is clearly seen that we have in general one longitudinal and two transverse acoustic branches. For $\omega \to 0$ and $q \to 0$ we come into the region of elastic waves with $\omega = v_s q$. If there are different atoms in the unit cell, as for instance in NaCl, optical branches (transverse TO and longitudinal LO ones) are also obtained. We will encounter one such case in Sect. 11.1.2. Other examples of lattice dynamics will be given in Sects. 5.4, 7.2.1, 9.2 and Chap. 11.

3.6.2 Debye Theory of Lattice Dynamics

Figure 3.7 shows that for small q near the Γ point (000), the phonon dispersion is linear and reads $\omega = v_s q$ with v_s the sound velocity in the corresponding direction. In the Debye theory of lattice dynamics, the actual dispersion curves $\omega(q)$ are replaced by the linear law $\omega = v_s q$ valid for long wavelengths and a cutoff frequency ω_D is introduced such that with $\hbar \omega_D = k_B \Theta$ the number of states are again 3nN. The Debye temperature Θ_D characterizes the thermodynamical properties of the lattice waves as expressed e.g. in the specific heat. This Debye temperature divides roughly the high temperature classical region $T > \Theta_D$ from the low temperature region $T < \Theta_D$ with quantum effects on the temperature dependencies of thermodynamic quantities. Only the acoustic modes are considered and the optic modes neglected. Instead of the Brillouin zone, a sphere with radius q_D and with the same volume as the first Brillouin zone is taken. With N points in the Brillouin zone and the density $\frac{V}{8\pi^3}$ in q space, $N = (\frac{V}{8\pi^3}) \frac{4\pi}{3} q_D^3$. With these approximations, it is possible to calculate, for example, the specific heat

$$C_V = \frac{3N}{4\pi q_D^3} \int 4\pi q^2 dq \left(\frac{\hbar \omega_q}{k_B T}\right)^2 e^{\hbar \omega / kT} (e^{\hbar \omega / kT} - 1)^{-2}$$

Evaluation of this integral gives the well-known laws in the appropriate limits:

$$\frac{T}{\Theta_D} \gg 1 \quad \text{Dulong – Petit law} \quad C_V = 3nNk_B$$

$$\frac{T}{\Theta_D} \ll 1 \quad \text{Debye law} \quad C_V = \frac{12}{5}\pi^4 k_B \left(\frac{T}{\Theta}\right)^3$$

Further application of this Debye theory with a determination of Θ_D will be given in Sect. 4.2.1. More detailed accounts of the lattice dynamics and Debye theory can be found in Born & Huang [3.19], Wallace [3.4], Ludwig [3.5], Brüesch [3.29] and in Solid State physics books given in the Introduction in Chap. 1.

4 Thermodynamics and Phase Transitions

Sound velocity and attenuation effects, as they occur at finite temperature, are discussed in this chapter. First, the thermodynamic potentials are introduced which will be used below. Then, the so-called background elastic constant, i.e. the temperature dependence of the elastic modes due to the interaction with thermal phonons, is discussed (Sect. 4.2). A brief discussion of sound attenuation due to phonons in insulators is also given. Afterwards, the order parameter concept for phase transition and the strain–order parameter coupling in the framework of the Landau theory is introduced (Sect. 4.3). The phase transition is characterised further by discussing scaling, thermodynamic fluctuations and the Landau–Khalatnikov theory. The Ginzburg criterion also characterises the phase transitions and gives the region of applicability of Landau theory (Sect. 4.4). Isothermal and adiabatic thermodynamic quantities as e.g. the elastic constants are also investigated (Sect. 4.5). The ultrasonic attenuation as a transport coefficient will be treated further, especially in the appropriate places (Sects. 4.3, 6.1, 7.2.5, 12.2, 12.3, 13.5). A general review on the theory of dynamical critical phenomena has been given by Hohenberg and Halperin [4.1].

4.1 Thermodynamic Potentials

For discussing temperature effects, it is necessary to use the appropriate thermodynamic potential, such as internal energy U, Helmholtz free energy F or free enthalpy or the Gibbs free energy G (Becker [4.2], Wallace [4.3], Callen [4.4]). In order to achieve this, we start with the work done on the system against the applied stress. Equation (3.7), for the anisotropic case (in contrast to the isotropic one $dW = -p\,dV$), gives the work done on the system

$$\delta W = V \sum_{ij} T_{ij} d\eta_{ij} \,. \tag{4.1}$$

The symbol δ instead of the differential d means that W and Q (heat) are not state variables and depend on the path.

For the internal energy U we get

$$dU = \delta Q + \delta W + \mu dN = TdS + V \sum_{ij} T_{ij} d\eta_{ij} + \mu\,dN \,. \tag{4.2}$$

Equation (4.2) is the first law of thermodynamics. Here $\delta Q = TdS$ is the reversible heat transfer and we have included µ, the chemical potential and N, the number of particles. Therefore U is a function of entropy S, strain η (Volume V) and particle number N. In a completely closed system without heat exchange, constant volume and no particle exchange, energy change is not possible, i.e. $dU = 0$. For the Helmholtz free energy $F = U - TS$, (4.2) gives

$$dF = -SdT + V\sum_{ij} T_{ij}d\eta_{ij} + \mu\, dN. \qquad (4.2a)$$

Here, the temperature, strain and particle number are free variables. This means that the system can now exchange heat when connected with a reservoir of heat. Finally for the Gibbs free energy $G = U - TS + pV$ the anisotropic case gives

$$dG = -S\,dT - V\sum_{ij} dT_{ij}\eta_{ij} + \mu\, dN. \qquad (4.2b)$$

It is easy to control the stresses (pressure) and temperature of this isobaric, isothermal system to levels most suitable for experiments.

Note the free variables for each thermodynamic potential: $U(S, \eta_{ij}, N)$, $F(T, \eta_{ij}, N)$ and $G(T, T_{ij}, N)$. Each thermodynamic potential belongs to a specific thermodynamic configuration as pointed out above. The Helmholtz free energy F are often used because it is easy to calculate isothermal thermodynamic quantities. More appropriate to an experimental situation is the Gibbs potential G where the temperature and stresses (pressures) can be handled more easily. The Legendre transforms connect the various thermodynamic potentials in a formal way (see Callen [4.4]).

The second law of thermodynamics states that in a closed system the entropy can only increase: $0 \leq dS$. From this follows for the other thermodynamic potentials that they only decrease: $dF \leq 0$, $dG \leq 0$. The equilibrium state is given by the minimum in F for the isochor-isothermal system and by the minimum in G for the isobaric-isothermal system. For example, for F(T,V) we consider the system with $dT = 0$ and $dV = 0$. As $dV = 0$ the system cannot exchange work with the surrounding, but it can exchange heat with the heat reservoir. With (4.2) $dU = \delta Q$ and for irreversible reactions $dU = \delta Q < TdS$ or $dU - TdS < 0$ from which follows for $dT = 0$ with $F = U - TS$ the desired relation $dF < 0$. The final equilibrium state is characterised by the minimum of F. In the following we will use these thermodynamic potentials in numerous applications.

4.2 Background Elastic Constant and Attenuation

4.2.1 Background Elastic Constant, Thermal Expansion and Specific Heat

We discuss the temperature dependence of the elastic constants and of the thermal expansion due to phonon–phonon interaction (Ludwig [4.5], Wallace [4.3]). In the harmonic approximation without phonon interaction, the elastic constants are temperature independent and the thermal expansion is absent. In the harmonic approximation, the Hamiltonian H_h has terms due to strains and due to harmonic phonons:

$$H_h = H_{el} + H_{ph} = \frac{1}{2}\sum_\Gamma c_\Gamma^0 \varepsilon_\Gamma^2 + \frac{1}{2}\sum_q \hbar\omega_q\left(n_q + \frac{1}{2}\right), \quad (4.3)$$

where c_Γ^0 is a symmetry elastic constant as discussed in Sect. 3.2 and $n_q = (\exp(\hbar\omega_q/kT) - 1)^{-1}$ is the thermal occupation number of the phonon q. The lowest order an-harmonic theory starts with a quasi-harmonic free energy (Leibfried and Ludwig [4.6], Wallace [4.3]) with $n_q = 0, 1, 2, \cdots$

$$\begin{aligned} F_{qh} &= F_0 - k_B T \ln Z \\ &= F_0 - k_B T \ln \sum_n \Pi_q e^{-(\hbar\omega_q/k_B T)(n_q + \frac{1}{2})} \\ &= F_0 - k_B T \ln \Pi_q \left(\frac{e^{(\hbar\omega_q/2k_B T)}}{1 - e^{-(\hbar\omega_q/k_B T)}}\right), \end{aligned}$$

this gives

$$F_0 + \sum_q \frac{\hbar\omega_q}{2} + k_B T \sum_q \ln\left(1 - e^{-(\hbar\omega_q/k_B T)}\right)$$

and which can now be written as

$$F_{qh} = F_0 + kT \sum_q \ln\left[2\sinh\left(\frac{\hbar\omega_q}{2k_B T}\right)\right], \quad (4.4)$$

where F_o is the elastic energy H_{el}. But now H_{el} and ω_q depend on $\mathbf{R}_i$, the lattice vectors. For equilibrium we have $\partial F_{qh}/\partial \varepsilon_{ij} = 0$, i.e.

$$c_{ij}^0 \varepsilon_{ij} + \frac{1}{2}\sum_q \frac{1}{\omega_q}\frac{\partial \omega_q}{\partial \varepsilon_{ij}} E(\omega, T) = 0 \quad (4.4a)$$

with $E = \hbar\omega(n + \frac{1}{2})$ the average energy of an oscillator.

Due to the third and fourth order an-harmonic terms in the crystal potential, there is a coupling between the homogeneous strains and the phonons. This an-harmonicity can be described by the strain dependence of the phonon frequencies via the phonon Grüneisen parameter

4 Thermodynamics and Phase Transitions

$$\gamma_{kl} = \frac{\partial \ln\langle\omega^2\rangle}{\partial \varepsilon_{kl}} \quad \Gamma_{klk'l'} = -\frac{\partial^2 \ln\langle\omega^2\rangle}{\partial \varepsilon_{kl}\partial \varepsilon_{k'l'}}. \quad (4.5)$$

For the volume thermal expansion coefficient α_v, by setting $\partial F_{qh}/\partial \varepsilon_v = 0$ in (4.4a)

$$\alpha_v = \frac{d\varepsilon_v}{dT} = \kappa \sum_q \gamma_q \frac{dE(\omega,T)}{dT} \quad (4.6)$$

with $\kappa = 1/c_B$ the compressibility. Equation (4.6) expresses the thermodynamic proportionality between the thermal expansion α and the specific heat $C = dU/dT = d\sum_q E(\omega,T)/dT$ with the average phonon Grüneisen parameter γ_{ph} and the compressibility κ the proportionality factor:

$$\alpha_V = \kappa \gamma_{ph} C.$$

At high temperature ($T > \Theta_D$ = Debye temperature), the thermal expansion is constant and at low temperature it follows a T^3 law like the phonon specific heat. These two limits correspond to the Dulong–Petit law and the Debye law respectively (see Sect. 3.6.2).

Analogously, for elastic constants, Equations (4.4) and (4.4a) with

$$c_{ij} = c_{ij}^0 + \frac{1}{2}\frac{\partial}{\partial \varepsilon_{ij}}\sum_q \frac{1}{\omega_q}\frac{\partial \omega_q}{\partial \varepsilon_{ij}} E(\omega,T) \text{ gives}$$

$$c_{ij} = c_{ij}^0 - \frac{1}{2}\Gamma U(T) + \frac{1}{4}\gamma_{ph}\left[U(T) - TC_V(T)\right]. \quad (4.7)$$

For $T > \Theta_D$, c_{ij} is proportional to $-T$ since $U \approx T$, they reach a constant value at $T = 0$ and for low temperatures c_{ij} is proportional to $-T^4$ ($C \approx T^3$, $U \approx T^4$). Fig. 4.1 shows a typical result for $c_{ij}(T)$ for cubic LaAl$_2$ (Schiltz and Smith [4.7]). With $\Theta_D = 374\,K$, a linear temperature dependence already for $T > 150\,K$ is observed, i.e. for $T > \Theta/2$ for all elastic modes and a bend over ($\sim T^4$) towards saturation for $T < 100\,K$.

The range of the T^4 dependence of the elastic stiffness constants is difficult to estimate. There are phenomenological expressions for the temperature dependence of the elastic constants, the best known is given by Varshny [4.8]. It reads

$$c_{ij} = c_{ij}^0 - \frac{s}{e^{t/T} - 1}, \quad (4.7b)$$

where s and t are constants which can be fitted to the experimental results. This empirical expression for the temperature dependence due to anharmonic phonon interaction describes the background elastic stiffness constants surprisingly well. It is therefore widely used. Some examples will be given in the following chapters.

Most of the crystals show a temperature dependence of its elastic constants as shown in Fig. 4.1. Below, some review articles and books listing such materials are given. Similar to the LaAl$_2$, there are other cubic Laves

4.2 Background Elastic Constant and Attenuation

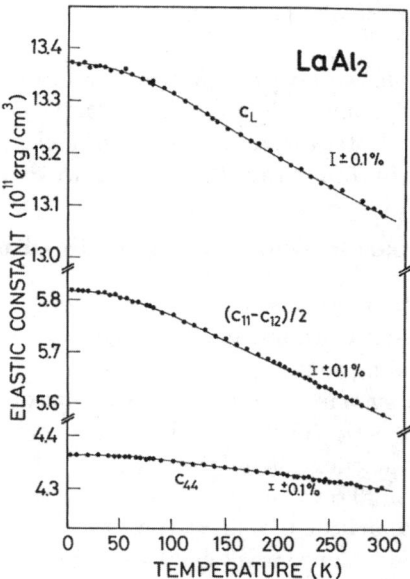

Fig. 4.1. Temperature dependence of elastic constants for LaAl$_2$ (Schiltz and Smith [4.7])

phase materials of the formula MX$_2$ with M and X metal atoms which exhibit similar temperature dependencies: CaAl$_2$, YAl$_2$, ZrCo$_2$, HfCo$_2$ (Schiltz and Smith [4.7], Shannette and Smith [4.9]). The Laves phase structure MX$_2$ is shown in Fig. 5.7. Mainly compounds which show anomalous effects, like anomalous temperature dependencies of elastic constants, are of course discussed in this book. Several Laves compounds also exist which show special properties: magnetic GdAl$_2$ (Sect. 5.1.3); superconducting CeRu$_2$ (Sects. 10.3, 10.5); crystal field effects and magnetic CeAl$_2$ (Chaps. 5.4, 13.1, 13.2).

Debye Temperature

From the sound velocities or elastic constants, a value for the Debye temperature can be obtained, labeled Θ_{el} in order to distinguish it from the thermal Debye temperature Θ_{th} gained from the specific heat. A brief outline of the Debye theory of lattice dynamics is given in Sect. 3.6.2. The simple formula for Θ_{el} can be given as follows: setting $k_B \Theta_{el} = \hbar \omega_D = \hbar v_m k_D$ gives (Anderson [4.10])

$$\Theta_{el} = \frac{h}{k_B} \left[\frac{3q}{4\pi} \frac{N\rho}{M} \right]^{1/3} v_m . \qquad (4.8)$$

Here N is the Avogadro number, ρ the density, M the molecular weight, q the number of atoms in a molecule and the average sound velocity $v_m = [\frac{1}{3}(2v_s^{-3} + v_L^{-3})]^{-1/3}$ with v_s the transverse and v_L the longitudinal sound velocity. Formula (4.8) is discussed and compared to Θ_{th} in Alers [4.11] and Anderson [4.12].

Finally we mention two reviews by Hearmon [4.13, 4.14], a book by Simmons and Wang [4.15] and a handbook of elastic properties (Levy et al. editors [4.16]) where elastic constants are listed of elements, compounds and composite materials. In some cases the temperature dependence is also given.

4.2.2 Sound Dissipation Due to Phonons in Insulators

In the previous section, we discussed the temperature dependence of the sound velocity or elastic constant due to phonon interaction. Here we briefly survey the corresponding sound attenuation due to phonon interaction. This obviously applies to systems where other attenuation mechanisms are not present, i.e. non-magnetic insulators like quartz or undoped semiconductors. For the sound attenuation in the MHz to GHz region, it is possible to distinguish three different temperature intervals. For low temperatures ($T \leq \Theta_D/10$) the attenuation α is due to general crystal imperfections and is temperature independent since practically no thermal phonons are present for sound scattering. In the next temperature region where $\omega\tau_{ph} > 1$ (τ_{ph} is the mean life time of phonons), the attenuation changes rapidly with temperature. Here the so-called Landau-Rumer mechanism describes the attenuation α. The mean free path of the thermal phonons is much larger than the sound wave length in this case. Finally at high temperatures, $\omega\tau_{ph} < 1$ and α is approaching a temperature independence. Here one can discuss the medium applying the phonon viscosity or working out the Akhiezer mechanism.

A detailed calculation of these various mechanisms is not given in this book but instead some references are cited where a full account of the various sound wave–phonon attenuation mechanisms can be found. In Sect. 2.6 on microwave ultrasonics, a list of experimental papers showing attenuation curves on a number of insulators is given. In addition, several books treating the same topic are also listed there. The review article by Maris [4.17] is also recommended.

4.3 Landau Theory, Strain–Order Parameter Coupling

We consider a phase transition between two phases exhibiting different symmetries. In the cases which we will treat in the following, we require that all symmetry elements of the low symmetry phase are contained in the high symmetry phase. This means that the low symmetry is characterized by a "broken symmetry" of the high symmetry phase.

In the following, we will treat various cases where the strain–order parameter coupling plays a dominant role. In the case of various structural phase transitions such as phonon induced transitions, cooperative Jahn–Teller transitions, charge order transitions etc, a Landau theory treatment involving the strain–order parameter coupling can be given. Therefore, the necessary unified background for the different cases to be treated in the various chapters (see Sects. 5.2, 5.3, 5.4 and 7.1, 7.2) as well as in Chaps. 10 and 12) is given here. A full presentation of Landau theory is given by Tolédano and Tolédano [4.18]. For more details of the Landau theory with strain coupling, refer to some review articles (Rehwald [4.19], Lüthi-Rehwald [4.20], Salje [4.21]).

4.3.1 Landau Theory for Second Order Phase Transitions

Landau theory, starts with a development of the free energy density in terms of the order parameter η (Kadanoff et al. [4.22]):

$$F = F_0 + \frac{1}{2}a(T)\eta^2 + \frac{1}{4}b\eta^4 + c(\nabla \eta)^2 \cdots \quad (4.9)$$

where it is assumed that $a = a'(T - T_0)$ in the vicinity of the phase transition temperature T_0. It is also assumed that η is small and that fluctuations (the gradient term) are neglected. Furthermore, with $\delta F/\delta \eta = 0$, $\eta(a + b\eta^2) = 0$ and, with the stability condition $\delta^2 F/\delta \eta^2 > 0$, this results in:

$$\begin{aligned} T > T_0 & \quad \eta = 0 \\ T < T_0 & \quad \eta = \sqrt{a'\frac{T_0 - T}{b}} \end{aligned} \quad (4.10)$$

Figure 4.2 shows the order parameter and the free energy for a second order phase transition. Here the order parameter diminishes continuously from $T = 0$ to zero at $T = T_0$ with $\eta \approx t^\beta$ close to T_0 with $t = (T_0 - T)/T_0$ and $\beta = 1/2$. η can be, for example, the magnetization, an electric polarization, a structural coordinate etc. For $T \sim T_0$ the free energy is very flat, suggesting large fluctuations in the order parameter close to T_0 (Fig. 4.2).

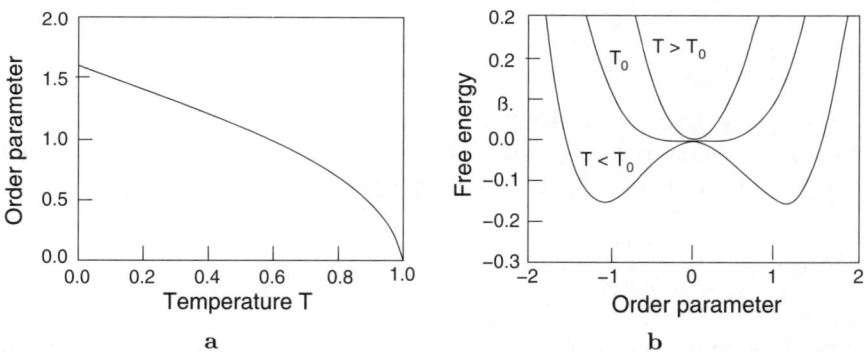

Fig. 4.2. (a) Order parameter η versus temperature for a second order phase transition, (b) free energy versus η for different temperatures: $T > T_0$, $T = T_0$, $T < T_0$

An additional Zeeman term $-h\eta$ can be introduced in (4.9). Here, h is a conjugate variable of the order parameter (like the magnetic field for the magnetization) in the free energy expression (4.9). The susceptibility can be calculated as a linear response of the order parameter to the applied field h:

$$\chi = \frac{d\eta}{dh} = \frac{1}{a'(T-T_0)} \quad \text{for } T > T_0$$
$$= \frac{1}{2a'(T_0-T)} \quad \text{for } T < T_0. \quad (4.11)$$

It is seen that χ diverges for $T \to T_0$. This means that the susceptibility as a measure of the fluctuations of η is large near the phase transition temperature in contrast to the assumption above. With the gradient term in (4.9), using linear response theory, the susceptibility is given by

$$\chi = \frac{\xi^2}{2c}$$

with ξ the correlation length which diverges as $\xi \sim (T-T_0)^{-\upsilon}$. We will discuss the range of validity of the Landau theory later in Sect. 4.4.

4.3.2 Scaling Relations

From (4.10) and (4.11), a distinct temperature dependence of the thermodynamic quantities near the critical point T_0 can be noticed. With $t = (T-T_0)/T_0$, the order parameter η, the susceptibility χ, the specific heat C, the magnetisation M and the correlation length ξ, the following power laws with corresponding exponents (for $T < T_0$ we have to take $|t|$) can be defined

$$\eta \sim t^\beta \quad \chi \sim t^{-\gamma} \quad C \sim t^{-\alpha} \quad M \sim h^{\frac{1}{\delta}} \quad \xi \sim t^{-\nu}. \quad (4.11\text{a})$$

For the Landau theory, from (4.10) and (4.11), $\beta = \frac{1}{2}, \gamma = 1, \delta = 3, \alpha = 0$ and $\nu = \frac{1}{2}$. Scaling relations exist between the different exponents (Kadanoff et al. [4.22]). With $d = $ "spatial dimensions"

$$\nu = \frac{2-\alpha}{d}, \quad \frac{d\gamma}{2-\eta'} = 2-\alpha, \quad 2-\alpha = \gamma + 2\beta = \beta(\delta+1), \quad (4.11\text{b})$$

where η' is an exponent introduced for the correlation function. The exponents from Landau theory given above do not fulfill the scaling equations (4.11b). Renormalization group theory can calculate such exponents. Experimentally, these scaling relations are remarkably well satisfied for quite different types of phase transitions (Kadanoff et al. [4.22]). These scaling relations will be used occasionally, especially in Chap. 6.

4.3.3 Landau Theory for a First Order Phase Transition

Very often the phase transitions are of first order. These can also be described with Landau theory. In the expansion of the free energy (4.9), an additional expansion term of sixth order $d\eta^6/6$, with d a constant, is added. This is usually done for ferro-electrics where the order parameter is $\eta = \boldsymbol{P}$, the electric polarization. Since $\boldsymbol{P}$ is a vector, for this case a term $d\eta^3/3$ would not be allowed. On the other hand, for nematic liquid crystals such a cubic term would be possible. Taking the sixth order term into account, with $\partial F/\partial \eta = 0$ and $\partial^2 F/\partial \eta^2 > 0$, results in the stable solutions $\eta = 0 (T > T_c)$ and

$$\eta_0^2 = -\frac{b}{2d}\left\{1 + \left[1 - 4a'db^{-2}(T - T_0)\right]^{\frac{1}{2}}\right\}$$

for $T < T_c$ with $T_c = T_0 + b^2/4a'd$. T_c is the temperature where $\partial^2 F/\partial \eta^2 = 0$, the stability limit. A first order transition requires that $b < 0$ and $d > 0$. At T_c, the order parameter has a finite value. Since the stability limits of the high temperature phase is T_0 and of the ordered phase T_c, which do not coincide, there is a first order transition. Figure 4.3 shows the free energy for different temperatures as a function of the order parameter. For temperatures below T_0 the free energy has two minima corresponding to $\pm\eta_0$. For temperatures above T_0 a further local minimum for $\eta = 0$ appears. Here the ordered phase is still the stable one. The free energy has three nodes for $T_c' = T_0 + 3b^2/16a'd$ namely for $\eta = 0$ and $\pm\eta_0(T_c')$. Finally for $T_c' < T < T_c$ the ordered phase is metastable and above T_c unstable. The order parameter $\eta(T)$ and the free energy $F(\eta)$ are presented in Fig. 4.3 together with T_0, T_c, T_c'. Note that the fluctuation range is strongly diminished compared with a second order phase transition (Fig. 4.2).

To summarise the analysis of $F(T, \eta)$ gives $\eta = 0$ stable for $T > T_c'$ and metastable for $T_0 < T < T_c'$; $\eta \neq 0$ stable for $T < T_c'$ and metastable for $T_c' < T < T_c$ as also seen from Fig. 4.3. This metastable region can serve as a model for superheat-ing or supercool-ing, an effect which can only occur for first order phase transitions.

If third order invariants of the order parameter η or of the strain ε are possible, the sixth order term in η is not required to produce a first order phase transition. In the case of cubic symmetry, third order terms in the strain components can always be constructed (Cowley [4.23]). These third order invariants can be determined by the symmetry of the irreducible representation of the elastic constant matrix. They are given in Chap. 3, (3.9c) by putting the rotational strains $\omega_{ij} = 0$. In terms of the symmetry strains (3.11), the expression reads as follows (Thurston and Brugger [4.24]):

56   4 Thermodynamics and Phase Transitions

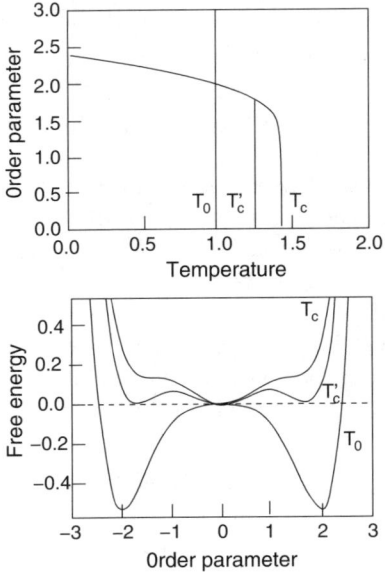

Fig. 4.3. (a) Order parameter η versus temperature for a first order transition. The various temperatures T_0, T_c, T_c' are indicated, (b) free energy F as a function of the order parameter η for different temperatures (T_0, T_c, T_c')

$$F_{el,3} = c_{aaa}\varepsilon_v^3 + c_{aee}\varepsilon_v\left(\varepsilon_2^2 + \varepsilon_3^2\right) + c_{att}\varepsilon_v\left(\varepsilon_{xy}^2 + \varepsilon_{yz}^2 + \varepsilon_{zx}^2\right)$$
$$+ c_{eee}\varepsilon_3\left(\varepsilon_3^2 - 3\varepsilon_2^2\right)$$
$$+ c_{ett}\left[\varepsilon_2\left(\varepsilon_{xz}^2 - \varepsilon_{yz}^2\right) + \frac{\varepsilon_3}{\sqrt{3}}\left(2\varepsilon_{xy}^2 - \varepsilon_{xz}^2 - \varepsilon_{yz}^2\right)\right]$$
$$+ c_{ttt}\varepsilon_{xy}\varepsilon_{yz}\varepsilon_{xz} \; .$$

Here the c_{ijl} are linear combinations of the C_{mnk} of (3.9c) and the ε_i are defined in Sect. 3.2, (3.11). With ε_i proportional to the order parameter the phase transition will be of first order. If the constants c_{ijl} are very small the transition can be approximately second order.

For the elastic anomalies we refer to the following Sect. 4.3.4.

4.3.4 Strain–Order Parameter Coupling

In a similar way to the expansion of the free energy in terms of the order parameter, the interaction energy density F_{int} between strain and order parameter can be written as an expansion in powers of the strain ε and the order parameter η:

$$F_{int} = g_\Gamma \varepsilon_\Gamma \eta_\Gamma + h_\Gamma \varepsilon_\Gamma (\eta^2)_\Gamma + k\varepsilon_\Gamma^2 (\eta^2)_\Gamma \cdots \quad (4.12)$$

where g, h and k are coupling constants.

4.3 Landau Theory, Strain–Order Parameter Coupling

Here we have included the symmetry label Γ to illustrate that the order parameter expression and the strain have to belong to the same irreducible representation. From this expression we can calculate the elastic constant. The order parameter moves more or less freely under the action from the ultrasonic strain field described by F_{int}.

Bilinear Coupling

If in the case of bilinear coupling $F_{int} = g\varepsilon\eta$ the order parameter can follow the strain, then $dF/d\eta = a\eta + b\eta^3 + g\varepsilon = 0$. From this follows $d\eta/d\varepsilon = -g/(a + 3b\eta^2) = -g(\partial^2 F/\partial\eta^2)^{-1}$ where the last term $(\partial^2 F/\partial\eta^2)^{-1}$ is an order parameter susceptibility. For the elastic constant follows for $T > T_0$

$$c = \frac{d^2 F}{d\varepsilon^2} = c_0 - \frac{g^2}{a'(T - T_0)} = c_0 \frac{T - T_0 - \frac{g^2}{a'c_0}}{T - T_0}.$$

Here c_0 is the background elastic constant. The expression above we can rewrite in the form

$$c = c_0 \frac{T - T_c}{T - T_0}, \qquad (4.13)$$

where $T_c = T_0 + g^2/a'c_0$ is the transition temperature in the presence of strain interaction (zero stress state) and T_0 is the transition temperature for zero strain (clamped state). (4.13) and the one before show that the elastic constant is a strain susceptibility in analogy to the magnetic susceptibility. This similarity will be further discussed in the next Sect. 5.2.2. In the case of this bilinear coupling the resulting equilibrium strain ε_{eq} for $T < T_c$ can be taken as the OP:

$$\varepsilon_{eq} = \frac{g}{c_0}\eta.$$

Many applications of (4.12) and (4.13) will be given later on (see chapters 7.1, 7.2). Fig. 4.4 shows c/c_0 from (4.13) as a function of T/T_c with T_0/T_c as a parameter.

It is seen from Fig. 4.4 that, for $T_0/T_c = 0$, the elastic constant follows a $1/T$ law. For $T_0/T_c = 1$, the elastic constant has no temperature dependence until T_c (clamped state) and for $0 < T_0/T_c < 1$, the temperature dependence lies between the two cases mentioned.

Quadratic Coupling

In the case of quadratic coupling $F_{int} = h\varepsilon\eta^2$ we get analogously $d\eta/d\varepsilon = 2h\eta/(a + 3b\eta^2) = 2h\eta(\partial^2 F/\partial\eta^2)^{-1}$ and for the elastic constant:

$$T > T_c \quad c_\Gamma = c_\Gamma^0 \quad T < T_c \quad c_\Gamma = c_\Gamma^0 - 2\frac{h^2}{b}. \qquad (4.14)$$

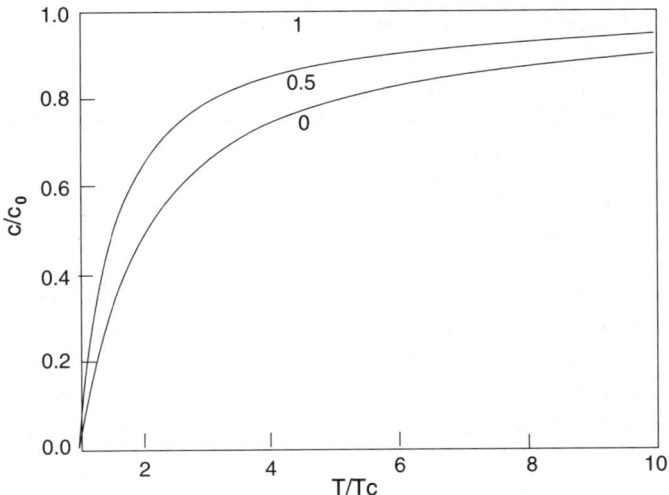

Fig. 4.4. Elastic constant as a function of reduced temperature T/T_c for $T > T_c$ according to (4.13). The parameter values are $T_0/T_c = 0, 0.5, 1$

The elastic constant exhibits a step function at T_c. This latter coupling applies to cases where the strain–order parameter coupling is not the primary interaction. We will treat such cases in Chaps. 7 and 10.

An extension of the theory to several order parameters or order parameter components has been given by Slonczewski and Thomas [4.25]. The corresponding formula for the two cases of bilinear and quadratic coupling given by (4.13) and (4.14) reads:

$$c_{ij} = c_{ij}^0 - \sum_{kl} \frac{\partial^2 F}{\partial \varepsilon_i \partial Q_k} R_{kl} \frac{\partial^2 F}{\partial \varepsilon_j \partial Q_l} \tag{4.14a}$$

with ε_i a given strain, Q_k the order parameter component and $R_{kl} = (\partial^2 F/\partial Q_k \partial Q_l)^{-1}$ is a generalised order parameter susceptibility. All these considerations are valid for second order phase transitions.

The first order phase transitions and the quadratic order parameter–strain coupling (second term in (4.12)) gives

$$\Delta c = c - c_o = -\frac{4h^2\eta^2}{a + 3b\eta^2 + 5d\eta^4}, \tag{4.14b}$$

which gives $\Delta c = 0$ for $T > T_c$ and $\Delta c = \frac{1}{2}h^2\eta^2/[a/2 + b\eta^2/4]$ for $T < T_c$ with $T_c = T_0 + b^2/4a'd$ defined above. Note that domain wall-stress effects can influence this step function Δc considerably (see Sect. 13.1).

Other Couplings

One can think of other strain–order parameter couplings such as $F_{int} = p\eta\varepsilon^2$ or $F_{int} = k\eta^2\varepsilon^2$ if symmetry is allowed. This would lead to $\Delta c = 0$ for $T > T_a$

and
$$\Delta c = c - c_0 = 2p\eta + 2k\eta^2 \tag{4.14c}$$

for $T < T_a$. But for superconductors, the first term $p\eta\varepsilon^2$ is not allowed because the order parameter is complex. This is also true for ferro electrics with the order parameter $\eta = P$, the polarization as a polar vector, the first term disappears. But the effect of (4.14c) could occur for structural transitions, where with the help of pressure or magnetic field, a single domain state can be made. Couplings of this type will be discussed in Sect. 7.2.2. Again these effects would only be observable for mono-domain phases for $T < T_a$ where domain-wall-stress effects are absent (see for details of the ΔE-effect in Sect. 13.1).

4.3.5 Fluctuation Effects

In the formulas above from Landau theory, fluctuation effects have been neglected. But fluctuations are important, as mentioned above, and seen from Fig. 4.2. In addition these fluctuations can give important contributions to elastic constants and ultrasonic attenuation. Theoretical treatments of transport coefficients incorporating these fluctuations are known as mode–mode coupling theories (Kawasaki [4.26]) and will be also discussed in conjunction with magnetic phase transitions (Sect. 6.1.2) and for structural transitions in Sect. 7.4.3. The physical picture behind it is that there is a whole spectrum of fluctuations of $\delta\eta = \eta(t) - <\eta>$, $\delta\eta(q)$ with different q-vectors centered around q_0, the wave vector of the order parameter. If it is assumed that a strain–order parameter coupling (4.12) of the form $F_{int} = h_{ijm}\eta_i\eta_j\varepsilon_m$ exists, two fluctuations $\delta\eta$ of nearly opposite wave vectors combine and produce, by an-harmonic interaction, a stress fluctuation of nearly zero wave vector. The total stress is the sum over the whole fluctuation spectrum

$$\delta\sigma_m = \sum_{q,i,j} h_{ijm}\delta\eta_i(q)\delta\eta_j(-q) \,. \tag{4.15}$$

From this equation, the elastic response using the fluctuation-dissipation theorem can be constructed (Landau and Lifshitz [4.27], Callen [4.4]), namely

$$c_m(T) = c_m^0 - \frac{1}{k_B T}\langle|\delta\sigma_m|^2\rangle$$
$$\alpha_m(T) = \frac{\omega^2}{2\rho v_s^3 k_B T}\int_0^\infty e^{i\omega t}\langle\delta\sigma_m(t)\delta\sigma_m(0)\rangle \mathrm{d}t \,. \tag{4.16}$$

Similar to Brownian motion, the sound attenuation coefficient can be expressed as the Fourier transform of time-dependent correlation functions of random stresses originating from (4.15). The fluctuation-dissipation theorem connects the attenuation, the imaginary part of the elastic constant to the Fourier transform of the time-dependent correlation function of the stress fluctuations $\delta\sigma_m$. An evaluation of these formulas for spin systems is discussed in Sect. 6.1.2. Evaluation of these formulas for structural transitions within molecular field theory can give terms like $c_\Gamma - c_\Gamma^0 \sim -\frac{T}{\sqrt{(T-T_c)}}$ (Henkel et al. [4.28]). Other treatments using dynamical scaling, mode–mode coupling and renormalization group analysis have been discussed for structural transitions by Schwabl [4.29] and Murata [4.30]. Further application of these results will be given in Sects. 6.1.2, 7.4 and 10.4.

It should be stressed that fluctuation contributions of quadratic order parameter-strain coupling can give similar temperature dependencies as the non-fluctuation formula (4.13) from bilinear order parameter-strain coupling. Therefore it is important to know exactly the actual order parameter-strain coupling.

4.3.6 Landau–Khalatnikov Theory

Another treatment of sound dispersion and attenuation is the so-called Landau–Khalatnikov theory (Landau–Lifshitz [4.27]) originally applied to sound propagation in liquid **HeII**. For the equations (4.13) to (4.14) the order parameter can follow the strain instantly. If this condition is relaxed, velocity dispersion and ultrasonic attenuation results. A physical property, such as the order parameter η which is affected by a strain wave $\varepsilon \sim \exp(-i\omega t)$, is considered. η can vary with time, tending towards the equilibrium value η_0: $d\eta/dt = -(\eta - \eta_0)/\tau$ with τ a typical relaxation time. η_0 can also be affected by the strain wave: $\eta_0 = \eta_{oo} + \eta_0'$ with η_{oo} the equilibrium value in the absence of a strain wave and η_0' the periodic time varying part. The true value is $\eta = \eta_{oo} + \eta'$. Then $d\eta/dt = d\eta_{oo}/dt + d\eta'/dt = d\eta'/dt = -(\eta_{oo} + \eta' - \eta_{oo} - \eta_0')/\tau = -(\eta' - \eta_0')/\tau$ and for a periodic distortion $-i\omega\eta' = -(\eta' - \eta_0')/\tau$ or

$$\eta' = \frac{\eta_0'}{1 - i\omega\tau}. \tag{4.17}$$

So if η_0' varies like $\exp(-i\omega t)$ then the actual η' has a phase factor i.e. it is out of phase with η_0' due to τ. Now we can calculate the elastic constant c or sound velocity v and attenuation α. First we get

$$\frac{d\eta}{d\varepsilon} = \frac{d\eta'}{d\varepsilon} = (1 - i\omega\tau)^{-1} \frac{d\eta_0'}{d\varepsilon} = (1 - i\omega\tau)^{-1} \frac{d\eta_0}{d\varepsilon}$$

and for the elastic constant

$$c = \rho v^2 = \frac{d\sigma}{d\varepsilon} = \left(\frac{d\sigma}{d\varepsilon}\right)_\eta + \left(\frac{d\sigma}{d\eta}\right)_\varepsilon (1 - i\omega\tau)^{-1} \frac{d\eta_0}{d\varepsilon}$$

$$= (1 - i\omega\tau)^{-1} \left\{ \left(\frac{\partial\sigma}{\partial\varepsilon}\right)_\eta - i\omega\tau \left(\frac{\partial\sigma}{\partial\varepsilon}\right)_\eta + \left(\frac{\partial\sigma}{\partial\eta}\right)_\varepsilon \frac{d\eta_0}{d\varepsilon} \right\}.$$

4.3 Landau Theory, Strain–Order Parameter Coupling

Now $\left(\frac{\partial \sigma}{\partial \varepsilon}\right)_\eta + \left(\frac{\partial \sigma}{\partial \eta}\right)_\varepsilon \frac{d\eta_0}{d\varepsilon}$ is the derivative of σ with respect to ε for processes which are slow so that the medium remains in equilibrium $= \left(\frac{\partial \sigma}{\partial \varepsilon}\right)_{eq}$. Therefore $\frac{d\sigma}{d\varepsilon} = (1 - i\omega\tau)^{-1} \{\left(\frac{\partial \sigma}{\partial \varepsilon}\right)_{eq} - i\omega\tau \left(\frac{\partial \sigma}{\partial \varepsilon}\right)_\eta\}$. With the definitions $c_0 = \left(\frac{\partial \sigma}{\partial \varepsilon}\right)_{eq}$ (zero frequency elastic constant) and $c_\infty = \left(\frac{\partial \sigma}{\partial \varepsilon}\right)_\eta$ (order parameter can no longer follow sound wave motion, high frequency elastic constant) we get for the real and imaginary part $(c = \rho v^2)$

$$v^2 = v_\infty^2 + \frac{v_0^2 - v_\infty^2}{1 + \omega^2 \tau^2}$$

$$\alpha = \frac{\omega^2 \tau}{1 + \omega^2 \tau^2} \frac{v_\infty^2 - v_0^2}{2 v_\infty^3}. \quad (4.18)$$

In this way, the frequency dependent velocity v and the attenuation α has been expressed for all frequencies and as a function of a general relaxation time. In Sects. 5.4, 5.5, 6.1 and 7.2, we will discuss such dispersive effects. For $\omega\tau \gg 1$, we get the background velocity v_∞ or the background elastic constant denoted by $c_\infty = c_0$, i.e. the background elastic constant is refined as c_0. Equation (4.18) can be rewritten using c_0, $c(\omega)$ and $c(\omega = 0) = c_0 - g^2\chi(T)$ with χ the order parameter strain susceptibility:

$$c(\omega) = c(\omega = 0) + \frac{g^2 \chi \omega^2 \tau^2}{1 + \omega^2 \tau^2}$$

$$\alpha(\omega) = \frac{g^2 \chi}{(2vc_0)} \frac{\omega^2 \tau}{1 + \omega^2 \tau^2}. \quad (4.18a)$$

This is the form used in various review articles (see e.g. Lüthi-Rehwald [4.20]). With $(c(\omega = 0) - c_0)/c_0 = \Delta c / c_0$ we get the simple relation for $\omega\tau \ll 1$

$$\alpha = -\frac{\omega^2 \tau}{2v_0} \frac{\Delta c}{c_0}. \quad (4.19)$$

By measuring the attenuation α and the relative elastic constant change $\Delta c/c_0$, the relaxation time τ can be determined by (4.19). It is the characteristic relaxation time for the response of the order parameter to an applied strain. It is related to the static susceptibility χ_η via the kinetic coefficient L (Landau–Lifshitz [4.27]):

$$\tau^{-1} = \frac{L}{\chi_\eta}. \quad (4.19a)$$

It is assumed that the order parameter susceptibility χ_η is divergent at the phase transition point and L is only weakly temperature dependent. L can be calculated again using mode–mode coupling methods (Kawasaki [4.26]). (4.19a) describes the phenomenon of critical slowing down by approaching the phase transition. It takes an increasingly long time for a system to reach thermal equilibrium as the critical point is approached.

Application of these formulas (4.19) and (4.19a) will be given in various places (Sects. 5.4, 6.1, 7.2, 9.3 and 12.2.1). The Landau–Khalatnikov theory applies to the bilinear, quadratic coupling of the strain–order parameter coupling of (4.12): $g\varepsilon\eta$ and the fluctuation expression above in Sect. 4.3.5 applies to biquadratic coupling of the order parameter to the strain (second term of (4.12): $h\varepsilon\eta^2$).

4.4 Ginzburg Criterion and Marginal Dimensionality

In the previous section we have introduced the Landau theory of phase transition which is equivalent to the molecular field theory. We have seen that the order parameter susceptibility χ diverges at the phase transition temperature T_0 in contrast to the assumption of negligible order parameter fluctuations. The range of validity of the Landau theory, i.e. the range of the relative temperature from T_0, $t_{cr} = |(T - T_0)/T_0|$ for which Landau theory is valid ($t > t_{cr}$) is called the Ginzburg criterion. Closely related with this is the concept of marginal dimensionality as will be explained below. this will be shown for the case of a magnetic phase transition. A pedagogical review of this topic is given by Als-Nielsen and Birgeneau [4.31].

For the mean square deviation of the order parameter or magnetization we have

$$\langle \Delta m^2 \rangle = \langle (m(r) - m_0)(m(r') - m_0) \rangle = g(r - r').$$

Here, $m_0(T)$ is the equilibrium magnetization at a given temperature. To discuss fluctuations, a fluctuation term $c\,|\nabla\eta|^2$ with $c > 0$ is introduced in the free energy (see (4.9)). Then for $r - r' = L$, we have (Kadanoff et al. [4.22]) $\langle \Delta m^2 \rangle = (kT_0/8\pi cL)e^{-L/\xi}$ and $m_0^2 = a'(T_0 - T)/b$ from (4.9). ξ is the so-called correlation length for the order parameter fluctuations. It diverges for $T \to T_0$ like $\xi = \xi_0 t^{-1/2}$. Therefore the magnetic susceptibility (4.11) is $\chi \sim \xi^2$. Now for the Landau theory to be valid, the fluctuations have to be small, i.e. $\langle \Delta m^2 \rangle \ll m_0^2$ or for t_c and for $L = \xi$ we get

$$t_c = \frac{k}{16\pi} \frac{e^{-L/\xi}}{L\Delta C\xi^2} = \frac{k}{16\pi}^2 \frac{e^{-2}}{\Delta C^2 \xi_0^6} \quad \text{or}$$

$$t_c \approx \left(\frac{a}{\xi_0}\right)^6. \tag{4.20}$$

For the different Landau parameters, this estimate uses experimentally observable quantities like the specific heat jump ΔC at T_o and $\xi(T)$. Equation (4.20) says that the critical temperature t_c depends on the sixth power of a/ξ_0 the lattice constant over the force range. For example, nearest neighbour exchange interaction leads to a large $t_c \sim 1$. In normal type I superconductors with a coherence length of $\sim 10^3$ Å, t_c can be as small as 10^{-14}.

The Ginzburg criterion can be formulated with the marginal dimensionality $d^\star$ instead of the critical temperature range t_c. The correlation function of the magnetization for arbitrary dimensions $g(r) \sim e^{-r/\xi}/r^{d-2} \sim \xi^{2-d}f(r/\xi)$, where d is the spatial dimension, is expressed by introducing $d^\star$. With the same reasoning as above, the marginal dimensionality is obtained

$$d^\star = \frac{\gamma + 2\beta}{\nu} - m,$$

where γ, β and ν are the critical exponents for the susceptibility, the magnetization and the correlation function respectively. m is a spatial anisotropy factor. For $d > d^\star$ molecular field theory is valid, for $d = d^\star$ there are logarithmic corrections and for $d < d^\star$ molecular field theory breaks down.

As an example let us take the ferromagnet with critical exponents $\gamma = 1$, $\beta = \frac{1}{2}$ and $\nu = \frac{1}{2}$ from Landau theory. Then $d^\star = 4$ and Landau theory should work for $d > 4$. If we take typical exponents for a three dimensional Heisenberg magnet $\gamma = 4/3$, $\beta = 1/3$, $\upsilon = 2/3$ then $d^\star = 3$. Elastic dominated phase transitions have been treated by Cowley [4.23] and Folk et al. [4.32]. Here, the anisotropy factor $m \neq 0$. For second order phase transitions: with c_{44} softening, $d^\star = 3$ and with $(c_{11} - c_{12})/2$ softening, $d^\star = 2$. These latter results for the elastic constants are of importance for the quadrupolar phase transitions treated in Sect. 7.2.

4.5 Adiabatic and Isothermal Quantities

In some cases we have to distinguish between adiabatic and isothermal thermodynamic quantities, especially for the elastic constants. With the help of the Maxwell relation (see quoted books on thermodynamics above), we can derive a simple relation between the adiabatic elastic constant c_S and the isothermal elastic constant c_T. The derivation here is given for the bulk moduli (longitudinal waves). With the stress σ and the strain ε we get

$$d\sigma = \left(\frac{\partial \sigma}{\partial \varepsilon}\right)_T d\varepsilon + \left(\frac{\partial \sigma}{\partial T}\right)_\varepsilon dT = \left(\frac{\partial \sigma}{\partial \varepsilon}\right)_S d\varepsilon + \left(\frac{\partial \sigma}{\partial S}\right)_\varepsilon dS$$

and for adiabatic conditions $dS = 0$ follows

$$\left(\frac{\partial \sigma}{\partial \varepsilon}\right)_S = \left(\frac{\partial \sigma}{\partial \varepsilon}\right)_T + \left(\frac{dT}{d\varepsilon}\right)_S \left(\frac{\partial \sigma}{\partial T}\right)_\varepsilon.$$

With the Maxwell relation

$$\left(\frac{dT}{d\varepsilon}\right)_S = -\left(\frac{\partial T}{\partial S}\right)_\varepsilon \left(\frac{\partial S}{\partial \varepsilon}\right)_T$$

we get

$$\left(\frac{\partial \sigma}{\partial \varepsilon}\right)_S = \left(\frac{\partial \sigma}{\partial \varepsilon}\right)_T - \left(\frac{\partial T}{\partial S}\right)_\varepsilon \left(\frac{\partial S}{\partial \varepsilon}\right)_T \left(\frac{\partial \sigma}{\partial T}\right)_\varepsilon.$$

This reads with the linear thermal expansion

$$\beta = \left(\frac{\partial \varepsilon}{\partial T}\right)_S, \quad \text{with} \quad \left(\frac{\partial \sigma}{\partial T}\right)_\varepsilon = -\left(\frac{\partial \varepsilon}{\partial T}\right)_S \left(\frac{\partial \sigma}{\partial \varepsilon}\right)_T = -\beta c_T$$

and with the specific heat at constant volume

$$C_V = T \left(\frac{\partial S}{\partial T}\right)_\varepsilon \quad \text{as}$$

$$c_S = c_T + T\beta^2 \frac{c_T^2}{C_V}. \tag{4.21}$$

The adiabatic elastic constant is larger than the isothermal one. For large thermal expansion (and consequently large specific heat), significant differences between adiabatic and isothermal elastic constants can occur even at low temperatures. For $T = 0$, $c_S = c_T$. A case with $c_S \neq c_T$ is encountered for various crystals, especially near phase transitions or for the heavy fermion compounds later (Chap. 7 and 9). The difference in c_S and c_T is given by the thermodynamic quantities β and C, both of which do not change the symmetry of the crystal. Therefore, the relation of (4.21) holds for fully symmetric $\Gamma_1(A_1)$ type elastic constants.

It is possible to estimate whether one is in the adiabatic or isothermal regime by considering the energy diffusion time τ_D. According to an argument by Maris [4.33] we are in the adiabatic region if the sound wave period $T_s \ll \tau_D$ or $\omega \gg 2\pi/\tau_D$. With $\tau_D^{-1} = q^2 \kappa/C = (4\pi^2/\lambda^2)\kappa/C$, κ the thermal conductivity and λ the sound wave wavelength and with the gas-kinetic expression for $\kappa = 1/3 C v^2 \tau_m$ (where C the specific heat, v the velocity due to phonons or electrons and τ_m a microscopic relaxation time), we get for adiabatic conditions

$$\omega \tau_m \ll 1 \quad \text{if phonons are responsible for the heat transport and}$$

$$\omega \tau_m \ll \frac{3}{2\pi} \left(\frac{v_s}{v_F}\right)^2 \quad \text{if electrons with Fermi velocity} v_F \tag{4.22}$$

are responsible for the heat exchange.

The same conditions are achieved by setting the diffusion length $L = \sqrt{(2D/\omega)} = \sqrt{(2\kappa/C\omega)} \ll \lambda$. We will use these conditions in Sect. 7.2.2 and later on by the discussion of the elasticity in heavy fermion compounds (Sect. 9.3). In general, isothermal conditions must again hold for quasi-static measurements ($\omega \to 0$)

On the other hand, for $\omega \tau_m > 1$, one is in the region of "zero sound". This is the region of lattice dynamics above the frequency regime of ultrasonics. In fact the sound velocities determined with inelastic neutron scattering can be different from the ultrasonically determined sound velocities. We will encounter examples in Chap. 7.

With the Maxwell relation above we get with the volume strain $\varepsilon_V = \Delta V/V$ an expression for the so-called Grüneisen parameter

$$\Omega = -\left(\frac{\partial \ln T}{\partial \varepsilon_V}\right)_S = \frac{1}{T}\left(\frac{\partial T}{\partial S}\right)_\varepsilon \left(\frac{\partial S}{\partial \varepsilon_V}\right)_T = \beta_T \frac{c_B}{C_V} \quad \text{i.e.}$$

$$\Omega = \beta_T \frac{c_B}{C_V}. \tag{4.23}$$

This relation is being used in lattice dynamics (phonon Grüneisen parameter) and in highly correlated electron systems (electronic Grüneisen parameter). The former case we have treated in Sect. 4.2, (4.5). For a volume strain ε_V we get for the phonon Grüneisen parameter γ_{ph}

$$\gamma_{ph} = \frac{\partial \ln \omega^2}{\partial \varepsilon_V} = 2\frac{\partial \ln \Theta}{\partial \ln V}$$

with Θ the Debye temperature (4.8). The electronic Grüneisen parameter Ω we will use in Chap. 9 to characterise the electron-phonon coupling for unstable moment compounds.

5 Acoustic Waves in the Presence of Magnetic Ions

The interaction of strain waves with magnetic ions is a topic which is still very much under investigation. There are different mechanisms of the so-called spin–phonon coupling and many widespread applications of it. In several chapters to follow, we will deal with different aspects of this interaction. In the present chapter, we wish to introduce first the basic interactions (Sect. 5.1), followed by some applications of crystal field systems for single ions involving several thermodynamic functions (Sect. 5.2). In Sect. 5.3, we investigate the orbital coupling between magnetic ions which will be used mainly in Chap. 7. Effects involving external and internal strains are dealt with in Sect. 5.4. Finally, we mention the spin-phonon effects with transition metal ions, investigated mainly by microwave ultrasonics (Sect. 5.5). A brief mention of the nuclear acoustic resonance is given in Sect. 5.6.

In this chapter, we deal mainly with single ion effects. There are many systems, like the ones diluted with magnetic ions, which show single ion effects. But also rare earth compounds with concentrated magnetic ions can show single ion effects because of the localised nature of the 4f wavefunctions. In 3d systems this is no longer the case. There, only diluted systems are investigated. But as already mentioned in Sect. 5.3 we introduce orbital interactions between ions which will be a topic in Chap. 7.

5.1 Strain–Magnetic Ion Interaction

In several chapters, we need the detailed coupling of phonons to magnetic ions (see Chaps. 6, 7, 9, 11, 12 and 13). Therefore we will discuss in this section the basic interaction between strains and magnetic ions. Historically, a distinction can be made between two kinds of phonon–magnetic ion coupling: the single ion coupling where the strain wave modulates the immediate surrounding (ligands, crystal electric field) of the ion (Van Vleck coupling, Van Vleck [5.1]) and the two ion coupling (Waller mechanism, Waller [5.2]). The two ion coupling arises due to the modulation of the distance between the ions, thus changing the exchange interaction. The latter mechanism was originally introduced for nuclear spins with dipolar coupling.

5.1.1 Magnetic Interactions

We first discuss the basic interaction leading to magnetic moments (Tinkham [5.3], Abragam and Bleaney [5.4], White [5.5]). Let us consider for the moment only localized magnetic ions with Russell–Saunders coupling whose ground state is given by the three Hund's rules: The ground state term can be characterized by $^{2S+1}L_J$.

The three Hund's rules can be stated as follows:

1. The total spin angular momentum S has the maximum value. This is due to the fact that the intra-atomic exchange is ferromagnetic.
2. The total orbital angular momentum L takes its maximum value in accordance with the exclusion principle.
3. The resultant L and S combine to form J the total angular momentum. It takes the values $|L - S|$ for less than half-filled shells and $L + S$ for more than half filled shells. This rule is based on the spin-orbit interaction between s_i and l_i of the same electron, as seen below.

For example, as shown in Appendix D, for 3d electrons, one can have only the following ground state terms: ^2D, ^3F, ^4F, ^5D, ^6S. Here S stands for $L = 0$, P for $L = 1$, D for $L = 2$, F for $L = 3$ etc.. For 4f electrons the ground state J can take the following values: $0, 5/2, 7/2, 4, 9/2, 6, 15/2, 8$. Because of the different strengths of spin-orbit and crystal fields in 3d- and 4f-compounds, L and J characterise the ground state level in these two cases (see below and Fig. 5.1). In U-compounds one often has the $j - j$ coupling (Abragam and Bleaney [5.4]). We shall discuss the ground states for the actinide compounds individually.

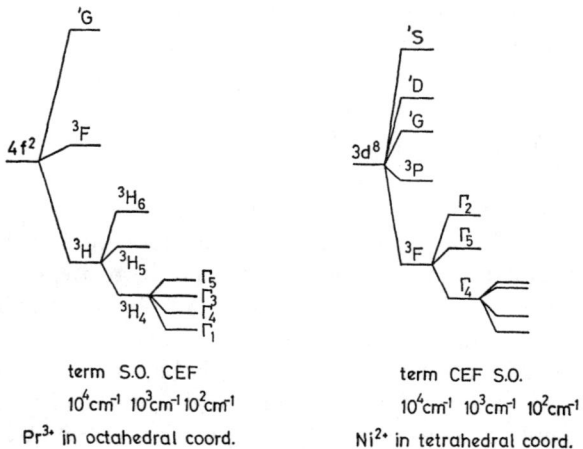

Fig. 5.1. Energy splitting for Pr^{3+} in octahedral symmetry and for Ni^{2+} in tetrahedral symmetry. Typical figures for overall splittings are given. Note the different strengths of spin orbit (s.o.) and crystal field (CEF) splitting for the two ions

The basic Hamiltonian (Spin Hamiltonian) which describes the properties of magnetic ions can be separated as

$$H = H_{Clb} + H_{so} + H_{CEF} + H_{ex} + H_{dip} + H_Z \, . \tag{5.1}$$

The first term is the intra-atomic Coulomb interaction, giving the term energy levels with the ground state determined by Hund's rules. For the spin-orbit interaction one can write $H_{so} = \lambda \boldsymbol{L} \cdot \boldsymbol{S}$ as long as the spin-orbit coupling is small compared to the term splitting. This interaction is strong for the heavy elements, i.e. for $4f$ and $5f$ ions. It gives rise to the J–multiplets in the case of $4f$ ions. For $3d$ ions, it is often less important partly due to quenching effects of the crystal field. In the following, the lowest J-level is considered for the $4f$ electrons whereas especially spin-orbit effects are mentioned whenever necessary for the $3d$ ions.

H_{CEF} is due to the crystalline Stark effect from the surrounding ions and from the $5d$ orbitals of the $4f$ magnetic ion (Williams and Hirst [5.6], Fulde [5.7]). Owing to the more extended $3d$ wave functions, this effect is more important for transition metal ions than for $4f$ compounds. As typical examples of term, spin-orbit and crystal field splitting, the case of $Ni^{2+}(3d^8)$ (tetrahedral coordination) and for $Pr^{3+}(4f^2)$ (octahedral coordination) for cubic symmetry is shown in Fig. 5.1. For a full exposition of CEF theory, refer to existing review articles (Hutchings [5.8], Wallace [5.9], Fulde [5.7]) and the books mentioned above. The electrostatic potential, which a $3d$ or $4f$ electron experiences due to the charge distribution of its surrounding, fulfills the Laplace equation $\Delta V(r) = 0$. Therefore $V(r)$ can be expanded in multipoles of the $3d$, $4f$ electrons. In the presence of unfilled $3d$ or $4f$ shells, the potential energy of the ion in the crystalline electric field is given by

$$H_{CEF} = e \sum_i V(r_i) \, , \tag{5.2}$$

where i runs over all $4f$ electrons. For the $4f$ ions according to Fig. 5.1, the calculation of the matrix elements of H_{CEF} can be restricted to the lowest J–multiplet. This restriction simplifies the calculations considerably as shown by Stevens [5.10]: The sum of the polynomials in x_i, y_i, z_i in (5.2) can be replaced by a sum of polynomials of J_x, J_y, J_z operators (Stevens' equivalent operators). There are rules for forming the symmetrized forms for products (Elliott and Stevens [5.11]). The CEF Hamiltonian can be written like

$$H_{CEF} = \sum_{l,m} B_l^m O_l^m \, . \tag{5.3}$$

For cubic point symmetry this can be written

$$H_{CEF}(\text{cubic}) = B_4 \left[O_4^0 + 5 O_4^4 \right] + B_6 \left[O_6^0 - 21 O_6^4 \right] \, , \tag{5.4}$$

where the O_l^m are polynomials of the angular momentum components J_i. The operators of (5.4) are listed together with others in Appendix D, they can also be found in Hutchings [5.8]. Especially for cubic symmetry, Lea et al. [5.12] should also be consulted. For B_i one can write

$$B_2 = A_2 \langle r^2 \rangle \alpha \quad B_4 = A_4 \langle r^4 \rangle \beta \quad B_6 = A_6 \langle r^6 \rangle \gamma. \tag{5.5}$$

Here $\langle r^n \rangle$ are radial integral over the wave functions and α, β, γ are the so-called Stevens factors due to the introduction of the equivalent operators. They are also listed in Appendix D. The coefficients A_n should be less variable across the rare earth series than B_n. Note that for calculating matrix elements with $4f$ ions, it it not necessary to go higher than the 6th degree polynomial ($2l = 6$) and for $3d$ ions not higher than the 4th degree ($2l = 4$). In the following, mostly cubic and hexagonal point symmetry systems will be used.

For the exchange interaction H_{ex} the most widely used form is the Heisenberg exchange Hamiltonian

$$H_{ex} = \sum_{ij}{}' J_{ij} \boldsymbol{S}_i \cdot \boldsymbol{S}_j \, , \tag{5.6}$$

where J_{ij} is the exchange integral, originating from the direct overlap of wavefunctions, or from super-exchange or from indirect exchange via conduction electrons (RKKY interaction, see White [5.5]). In many cases the isotropic form of (5.6) is not enough and one has additional terms (biquadratic exchange, anisotropic exchange, Dzyaloshinsky-Moryia mechanism etc.). These additional terms will be discussed where necessary (Chaps. 6, 9).

Finally, the dipolar interaction H_{dip} is usually small and is important only indirectly in the form of demagnetising fields, spinwave bands etc. We will mention, however, certain special cases where this interaction is important (see Sects. 11.2.1 and 13.2). The Zeeman term H_Z due to an applied magnetic field has been introduced in Sect. 4.3 and will be discussed below in more detail (Sect. 5.2).

5.1.2 Single Ion–Strain Interaction

The coupling of the elastic deformations and the magnetic ion are considered in this mechanism. It can be described as a strain coupling in the form of the strain derivative of the crystal field Hamiltonian or, in magnetically ordered systems, as the strain derivative of the anisotropy energy. A development of the multi-pole operators of a given symmetry can be given and combined with the symmetry strains of the same symmetry. Using symmetry strains of the irreducible representation Γ combined with quadrupole ($l = 2$) and higher order multi-pole operators ($l = 4, 6$ for $4f$ ions), the one-ion magneto-elastic Hamiltonian is given by

$$H_{me} = \sum_{\Gamma,n} g_{\Gamma,n} \varepsilon_\Gamma O_{\Gamma,n} \, . \tag{5.7}$$

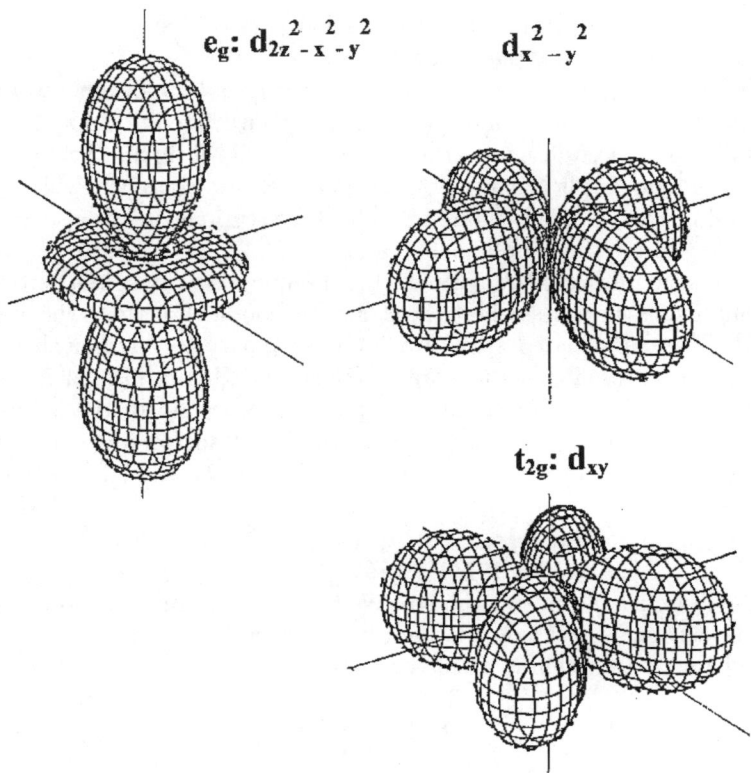

Fig. 5.2. Electron densities for d-electron orbitals e_g: $d_{2z^2-x^2-y^2}$, $d_{x^2-y^2}$, t_{2g}: d_{xy} (d_{xz}, d_{yz} not shown)

Here g_Γ is the magneto-elastic coupling constant and $O_{\Gamma,l}$ is the multipolar operator with l = 2,4,6. As an example, the symmetry strains ε_Γ for cubic symmetry derived in Sect. 3.2 are taken and combined with the quadrupolar operators $O_{\Gamma 2}$ for the same symmetry:

$$H_{me} = g_1 \varepsilon_v N + g_2 \sum_i \left(\sqrt{3}\varepsilon_2 O_2^2 + \varepsilon_3 O_2^0\right)_i + g_3 \sum_i \varepsilon_{xy} O_{xy,i} + \cdots . \quad (5.8)$$

The quadrupole operators O_{nm} can be expressed as linear combinations of Stevens operators as $O_2^0 = 2J_z^2 - J_x^2 - J_y^2$, $O_2^2 = J_x^2 - J_y^2$, $O_{xy} = \frac{1}{2}(J_x J_y + J_y J_x)$ etc. where the J_i are the angular momentum components (see also Appendix D).

Let us consider d-electron orbitals with $e_g(d_{z^2}, d_{x^2-y^2})$ and $t_{2g}(d_{xy}, d_{xz}, d_{yz})$ symmetry. Figure 5.2 shows the electron densities for these orbitals. They have the same symmetry as the strains for E_g and T_{2g} symmetry (see (3.11)) for the cubic point group and represent therefore a quadrupolar charge distribution. With the Stevens equivalent operators, the quadrupolar operators

O_{ij} listed above are obtained and the magneto-elastic Hamiltonian H_{me} (5.8) can be formed with J_i replaced by L_i.

The volume strain ε_V couples only to higher hexadecapole operators in cubic symmetry. Some of the multi-pole operators are listed in Appendix D and a complete list for cubic symmetry can be found in Morin and Schmitt [5.13]. There are many applications of this magneto-elastic coupling, which will be discussed in Sect. 5.2, 5.3, 5.4 and 7.2. In these sections, it will also be shown how to determine the coupling constants g_i.

The considerations above are valid for simple structures, like perovskites, NaCl and CsCl structures, where the strain can couple directly to the magnetic ion. In more complicated structures, it is necessary to distinguish between internal and external (elastic) strains (Born and Huang [5.14]). In general, the internal or ligand coordinates couple more strongly to the Jahn–Teller ion than a possible elastic strain coupling. This leads to soft elastic modes with renormalized coupling constants, as will be discussed later in Sect. 5.4.

5.1.3 Exchange Striction

For the second mechanism, the so-called two-ion magneto-elastic coupling, only the exchange striction mechanism which arises from a modulation of the exchange interaction is discussed (5.6):

$$H_{exs} = \sum_{ij} \left[J\left(\boldsymbol{\delta} + \boldsymbol{u}_i - \boldsymbol{u}_j\right) - J\left(\boldsymbol{\delta}\right) \right] \boldsymbol{S}_i \cdot \boldsymbol{S}_j . \tag{5.9}$$

Here $\boldsymbol{\delta} = \boldsymbol{R}_i - \boldsymbol{R}_j$ measures the distance between two magnetic ions and $\boldsymbol{u}_i$ is the displacement vector for the ion $\boldsymbol{R}_i$. With a sound wave given by $\boldsymbol{u} = U \boldsymbol{e}_q \exp(i(\boldsymbol{q} \cdot \boldsymbol{r} - \omega t))$ with $\boldsymbol{e}$ = polarization vector, U amplitude of the wave, ω frequency of the sound wave, the exchange striction term reads by expanding $\boldsymbol{u}_i$ (Stern [5.15])

$$H_{exs} = \sum_i \left[\frac{dJ}{dr} \frac{\boldsymbol{\delta}}{\delta} \times (\boldsymbol{u}_i - \boldsymbol{u}_j) \right] (\boldsymbol{S}_i \cdot \boldsymbol{S}_j) \text{ which gives}$$

$$H_{exs} = \sum_I \left(\frac{dJ}{dr} \frac{\boldsymbol{\delta}}{\delta} \cdot \boldsymbol{e}_q \right) (\boldsymbol{q} \cdot \boldsymbol{\delta}) \left(\boldsymbol{S}_i \cdot \boldsymbol{S}_{i+\delta} \right) e^{i(\boldsymbol{q} \cdot \boldsymbol{R}_i - \omega t)} . \tag{5.10}$$

Here we have expanded the exponential to first order because $q\delta \ll 1$ for sound waves of ca 100 MHz and δ of the order of a lattice constant. From (5.10) follows that for propagation along a symmetry axis, usually only longitudinal waves couple to the spin system because $\boldsymbol{q} \parallel \boldsymbol{\delta}$ or $\frac{\partial J}{\partial r} \boldsymbol{\delta} \parallel \boldsymbol{e}_q$. This exchange striction mechanism is responsible for a variety of effects, such as propagation of sound near magnetic phase transitions (Sect. 6.1) or for sound wave effects in low dimensional spin systems (Chap. 12). In ferromagnetic GdAl$_2$ (Schiltz and Smith [5.16]) and GdZn (Rouchy et al. [5.17]) large elastic effects in the temperature dependence of all elastic constants are observed for $T < T_C$. For Gd^{3+}-ion ($S = 7/2$, $L = 0$) these effects can be partly due to exchange striction and higher order effects, but also due to domain wall–stress effects. Apart from this exchange striction mechanism, other two-ion magneto-elastic couplings can formally exist.

We discuss briefly how to determine the coupling constant dJ/dr (5.10). Since a calculation of J is generally not possible, a theoretical determination of dJ/dr is like-wise not yet possible. In special cases, like the low dimensional spin systems (Chap. 12), one can determine this exchange striction coupling from the temperature and field dependence of the elastic constants (Sect. 12.2). Another way of estimating it is from the pressure dependence of the Curie temperature or from the thermal expansion anomalies at T_C, T_N. If one defines a magnetic Grüneisen parameter as

$$\Gamma_m = \frac{\partial \ln J}{\partial \ln V} = \frac{V}{J}\frac{\partial J}{\partial V} = \frac{\partial \ln T_C}{\partial \ln V} = \frac{V}{T_C}\frac{\partial T_C}{\partial V} = -c_B \frac{\partial \ln T_C}{\partial p}, \qquad (5.11)$$

with c_B the bulk modulus (inverse compressibility), it is apparent that an average exchange striction constant can be determined from the pressure dependence of the magnetic transition temperature. Values of Γ_m are usually of the order of 1–4 (Kamilov and Aliev [5.18]).

$\partial T_c/\partial p$ from (5.11) can be determined also from thermodynamic functions (thermal expansion $\alpha(T)$ and specific heat $C(T)$) using the Ehrenfest relation. For uniaxial pressure p_i follows (Lang et al. [5.19])

$$\left(\frac{\partial T_c}{\partial p_i}\right)_{p_i \to 0} = T_c V_{mol} \frac{\Delta \alpha_i}{\Delta C}. \qquad (5.11a)$$

Here V_{mol} is the molar volume and $\Delta \alpha$, ΔC the extra changes at T_c. A similar relation to (5.11a) in the form of a Grüneisen parameter has been used in Sect. 4.5 and will be used later in Sects. 9.2, 9.3.

5.2 Thermodynamic Functions for 4f–Rare Earth Ions in the Presence of Crystal Fields

Here we discuss thermodynamic properties such as specific heat, magnetic susceptibility, thermal expansion and elastic constants for crystals with magnetic ions in the presence of crystalline electric fields. We discuss here systems where other effects such as magnetic exchange interactions or quadrupolar interactions between the magnetic ions are negligible. These more complicated effects will be discussed in the following Sect. 5.3. Especially crystals with rare earth ions exist which show negligible interaction effects. Therefore, they can be discussed as single ion effects. Typical examples are PrSb and TmSb which do not exhibit any phase transition. The case of transition metal ions will be discussed in Sect. 5.5.

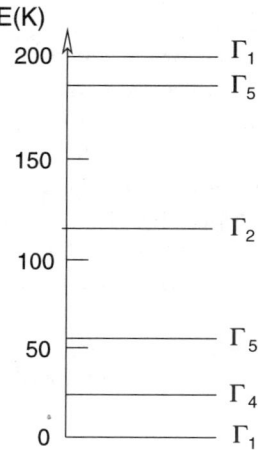

Fig. 5.3. Crystal field energy levels for TmSb, determined by inelastic neutron scattering (from Birgeneau et al. [5.20])

As a representative example, the thermodynamic effects for TmSb, which is a well-characterised substance, will be shown and discussed later on again in Sect. 5.5 and Sect. 13.2. This substance has the cubic NaCl structure. The Tm-ion has a stable valency $Tm^{3+}(4f^{12})$ and the ground state is (from Sect. 5.1) $^{2S+1}\boldsymbol{L}_J = {}^3H_6$. The crystal field levels of this ground state multiplet have been determined with inelastic neutron scattering (Birgeneau et al. [5.20]). The 13 fold degeneracy of the ground state multiplet is lifted with the lowest level being a singlet Γ_1, followed by a triplet Γ_4 at 25 K and a Γ_5 triplet at 56 K. All other levels lie higher than 100 K and do not contribute to the thermodynamic functions discussed below at low temperatures (see Fig. 5.3).

The CEF–levels from Fig. 5.3 can be fitted with CEF Hamiltonian of (5.4) using the tables of Lea, Leask and Wolf [5.12]: The CEF parameters (5.4) for TmSb are: $B_4 = 13$ mK, $B_6 = 0.028$ mK.

Before showing the experimental data, the necessary formula for the different thermodynamic derivatives are given. These are quite general and can then be applied to the case of TmSb. The free energy density of the system can be written as

$$F = F_{qh} + F_{CEF} = F_{qh} - N_s k_B T \sum_n e^{-\frac{E_n(\varepsilon_\Gamma)}{k_B T}}, \qquad (5.12)$$

where F_{qh} is the background free energy density, see equation (4.4), of the quasi-harmonic crystal and the crystal field free energy density is given by $F_{CEF} = -N_s k_B T \ln Z$ with Z the partition function. The energies E_n are strain-dependent. N_s is the number of magnetic ions per unit volume.

All the anomalous temperature dependencies of the different thermodynamic functions involving a coupling to the lattice (e.g. thermal expansion and elastic constants) arise because the asphericity of the 4f shell of the rare earth ions changes with temperature. The change in asphericity depends on the particular rare earth ion and its crystalline environment leading to the strain-dependent crystal electric field levels. The aspherical charge distribution for wavefunctions having a quadrupole moment is shown in Fig. 5.2 for Γ_3 and Γ_5 symmetry.

5.2.1 Specific Heat and Thermal Expansion

Using (5.12), the CEF part of the specific heat is given by

$$C_V = -T\frac{\partial^2 F}{\partial T^2} = \frac{N_s}{k_B T^2}\left\{\langle E^2\rangle - \langle E\rangle^2\right\}, \tag{5.13}$$

where the statistical average is defined as

$$\langle X\rangle = \frac{\sum_i X_i e^{-\frac{E_i}{k_B T}}}{\sum_i e^{-\frac{E_i}{k_B T}}}$$

Equation (5.13) gives rise to the well-known Schottky anomaly as shown in Fig. 5.4 with a maximum of C for temperatures around $\Delta/2$.

Likewise we can calculate the volume thermal expansion α_V due to CEF effects. With $F = F_{el} + F_{CEF} = c_B \varepsilon_V^2/2 + F_{CEF}$ follows for $\frac{\partial F}{\partial \varepsilon_V} = c_B \varepsilon_V + \frac{\partial F_{CEF}}{\partial \varepsilon_V} = 0$ the expression for the thermal expansion

$$\alpha_V = \frac{1}{V}\frac{\partial V}{\partial T} = \frac{d\varepsilon_V}{dT} = -\kappa V \frac{\partial^2 F}{\partial V \partial T} = \kappa \frac{N_s}{k_B T^2}\left[\langle E^2 \gamma\rangle - \langle E\rangle\langle E\gamma\rangle\right] \tag{5.14}$$

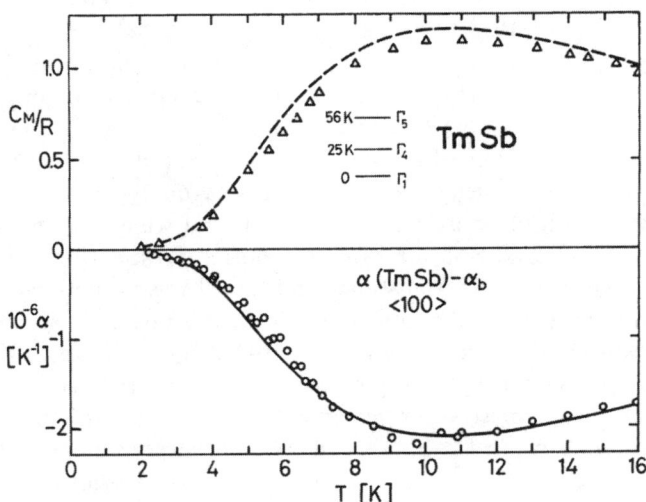

Fig. 5.4. Specific heat and linear thermal expansion in TmSb as a function of temperature together with a fit as described in the text (Ott and Lüthi [5.21])

Here, $\kappa = 1/c_B$ is the compressibility and $\gamma_i = -\frac{\partial \ln E_i}{\partial \ln V}$ is the Grüneisen parameter for the CEF level E_i. Analogously, (3.7b) with $T_m = \frac{\partial F}{\partial \varepsilon_m}$ could have been used (see Sect. 9.3.4). Equations (5.13) and (5.14) have a similar structure. α_V and C are proportional to each other as a function of temperature if all the γ_i which contribute to α_V are the same. This is the case if one has only two levels, as for example for $J = 5/2$ in cubic symmetry (e.g. SmSb, CeTe).

Figure 5.4 shows experimental results for the specific heat and thermal expansion for TmSb (Ott and Lüthi [5.21]). The magnetic part of the specific heat exhibits a well-defined Schottky anomaly where the dashed line is a fit of (5.13) with the experimentally determined energy levels given above (Fig. 5.3). The specific heat from the phonons and from conduction electrons is very small at low temperatures and can be estimated from measurements of LaSb.

In the same Fig. 5.3, we show the linear thermal expansion, which is $\alpha_V/3$ for cubic crystals. This result is a clear manifestation of CEF effects on thermal expansion. The full line is a calculation using $\gamma_{\Gamma 4} = \gamma_{\Gamma 5} = -1.2$ giving good agreement with the experiment. The same value for the γ implies the proportionality of α and C.

Including higher order contributions in (5.8) one gets for the volume strain ε_V and the bulk modulus c_B the following magneto-elastic Hamiltonian (Ott and Lüthi [5.21] and Appendix D):

$$H_{me}(c_B) = -g_1 N \varepsilon_v - g_{11} \sum_i \varepsilon_v (O_4^0 + 5 O_4^4)_i - g_{12} \sum_i \varepsilon_v (O_6^0 - 21 O_6^4)_i \, . \quad (5.15)$$

We can relate the CEF Grüneisen parameters γ given above with these magneto-elastic coupling constant. With the values for TmSb quoted above one gets $g_{11} = -16\,\text{mK}$.

Similar single ion effects for the specific heat and the thermal expansion have been observed for various other rare earth compounds, mainly the mono-pnictides (RX with X = N, P, As, Sb, Bi) and mono-chalcogenides (RX with X = S, Se, Te) all of which have a cubic structure (Ott and Lüthi [5.21]).

A particularly interesting case is PrNi$_5$, a hexagonal system which does not exhibit any low temperature phase transition and which can be interpreted with pure CEF effects. Specific heat, susceptibility and thermal expansion have been measured for this substance and the 4f–wave functions determined from these experiments (Andres et al. [5.22], Barthem et al. [5.23]). The thermal expansion data exhibit a minimum and maximum at 11 K for the c-axis and a-axis respectively. This data can readily be explained by noting that there are two Γ_1 symmetries for hexagonal symmetry (Lüthi and Ott [5.24]). This can be seen from Sect. 3.2 where the two symmetry strains read $\varepsilon_1(\Gamma_1) = \varepsilon_{xx} + \varepsilon_{yy} + \varepsilon_{zz}$ and $\varepsilon_2(\Gamma_1) = 2\varepsilon_{zz} - \varepsilon_{xx} - \varepsilon_{yy}$. In the hexagonal case, it is not necessary to go to the fourth order in the CEF operators as in the cubic case (5.15) but only to the second order. The magneto-elastic Hamiltonian reads therefore

5.2 Thermodynamic Functions for 4f–Rare Earth Ions

$$H_{me} = -B_V \varepsilon_1 O_2^0 + B_3 \varepsilon_2 O_2^0 \tag{5.15a}$$

and from this expression we can determine the thermal expansion coefficients α_a and α_c

$$\alpha_a = \frac{1}{a}\frac{\delta a}{\delta T} = \frac{d\varepsilon_{xx}}{dT} = (s_{11} + s_{12})\sigma_{xx} + s_{13}\sigma_{zz}$$
$$= \frac{1}{\Delta}[B_V(c_{33} - c_{13}) - B_3(c_{33} + 2c_{13})]f$$

$$\alpha_c = \frac{1}{c}\frac{\delta c}{\delta T} = \frac{d\varepsilon_{zz}}{dT} = 2s_{13}\sigma_{xx} + s_{33}\sigma_{zz}$$
$$= \frac{1}{\Delta}[B_V(c_{11} + c_{12} - 2c_{13}) + 2B_3(c_{11} + c_{12} + c_{13})]f ,$$
$$\tag{5.12c}$$

with s_{ij} the elastic compliances, $\sigma_{ij} = T_{ij}$ the components of the stress tensor (see Sect. 3.1) and

$$\Delta = c_{33}(c_{11} + c_{12}) - 2c_{13}^2 \quad \text{with} \quad f = \frac{1}{k_B T^2}\{\langle O_{2\Gamma\Gamma}^0 E_\Gamma \rangle - \langle O_{2\Gamma\Gamma}^0\rangle\langle E_\Gamma\rangle\} .$$

With these equations one can fit the experimental results very well, for details see Lüthi and Ott [5.24]. The different signs of α_a and α_c can be traced back to the different signs of the elastic compliances: $s_{11} + s_{12} = c_{33}/\Delta$, $s_{13} = -c_{13}/\Delta$, $s_{33} = (c_{11} + c_{12})/\Delta$. Similar behaviour in thermal expansion can be found in other hexagonal (UPt$_3$, Sect. 9.3.4) and tetragonal (SrCu$_2$(BO$_3$)$_2$, Chap. 12) materials. The elastic constants of PrNi$_5$ we discuss in Sect. 7.2 and 7.3.

Other systems showing crystal field effects on thermal expansion are rare earth aluminium and gallium garnets, R$_3$Al$_5$O$_{12}$ and R$_3$Ga$_5$O$_{12}$ (Kolmakova et al. [5.25]), rare earth phosphides RPO$_4$ (Kazei [5.26]) and RVO$_4$ (Morin and Schmitt [5.13]).

5.2.2 Magnetic Susceptibility and Elastic Constants

Next, the CEF magnetic susceptibility χ_m is discussed. It measures the response of a system of magnetic ions to an applied magnetic field B. The Zeeman Hamiltonian for 4f-ions can be written $H_Z = g_J \mu_B \mathbf{J} \cdot \mathbf{B}$. Here J is the total angular momentum (Sect. 5.1.1) and g_J is the Lande g-factor

$$g_J = \frac{3J(J+1) + S(S+1) - L(L+1)}{2J(J+1)} .$$

This g factor is tabulated for the rare earth ions in Appendix D. From the Zeeman energy and (5.4) one can determine the CEF energy levels $E(\Gamma_i, B)$. For the magnetic susceptibility χ_m we get

$$\chi_m = N_s \frac{\partial \langle J_z \rangle}{\partial B} = -\frac{\partial^2 F}{\partial B^2} = -N_s \left\{ \langle \frac{\partial^2 E}{\partial B^2}\rangle - \frac{1}{k_B T}\langle \left(\frac{\partial E}{\partial B}\right)^2\rangle + \frac{1}{k_B T}\langle \frac{\partial E}{\partial B}\rangle^2 \right\} .$$
$$\tag{5.16}$$

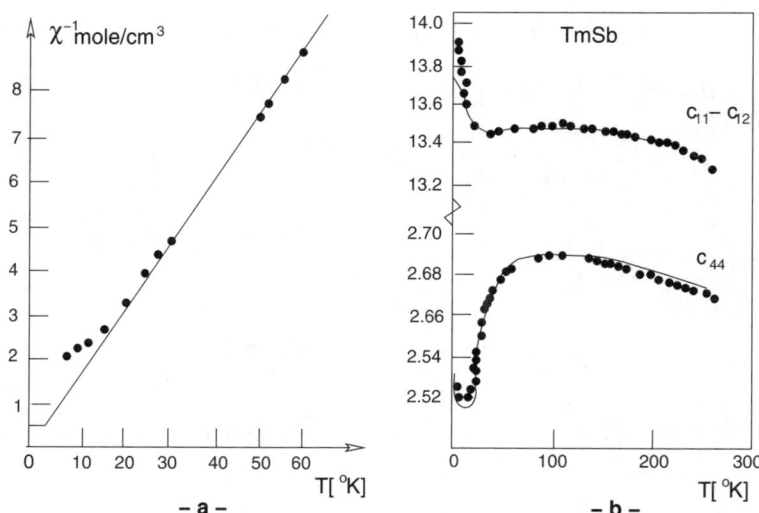

Fig. 5.5. (a) Magnetic susceptibility and (b) elastic constants $c_{11} - c_{12}$, c_{44} in 10^{11} erg/cm^3 for TmSb. For a discussion see text (after Mullen et al. [5.29])

In this expression the first term is the famous Van Vleck contribution, which probes the off-diagonal magnetic dipole matrix elements. The next two terms are the so-called Curie term showing a strong temperature dependence. They are due to the diagonal matrix elements and are particularly important for Kramers ions. Many examples of the CEF effect on the magnetic susceptibility are known (Van Vleck [5.27], Cooper [5.28], Wallace [5.9]). We discuss the magnetic susceptibility of TmSb below (Fig. 5.5).

In an analogous way, the strain susceptibility is the response of the structural order parameter $\langle O_\Gamma \rangle$ to an applied strain ε_Γ. The strain dependence of the CEF energy levels in second order perturbation theory is obtained from the magneto-elastic interaction in Sect. 5.1 equations (5.7) and (5.8) as follows

$$E_n(\varepsilon_\Gamma) = E_n - g_\Gamma \varepsilon_\Gamma \langle n | O_\Gamma | n \rangle + g_\Gamma^2 \varepsilon_\Gamma^2 \sum_{n \neq m} \frac{|\langle n | O_\Gamma | m \rangle|^2}{E_n - E_m}, \quad (5.16a)$$

where E_n are the unperturbed CEF energies. This perturbation theory is appropriate for small strains as used in ultrasonic measurements (see estimate in Sect. 2.1). This gives for the elastic constant $\Delta c_\Gamma = c_\Gamma(T) - c_\Gamma^0(T)$

$$\Delta c_\Gamma = \frac{\partial \langle O_\Gamma \rangle}{\partial \varepsilon_\Gamma} = \frac{\partial^2 F}{\partial \varepsilon_\Gamma^2} = -g_\Gamma^2 \chi_\Gamma^{str}$$
$$= N \left\{ \langle \frac{\partial^2 E}{\partial \varepsilon_\Gamma^2} \rangle - \frac{1}{k_B T} \langle \left(\frac{\partial E}{\partial \varepsilon_\Gamma} \right)^2 \rangle + \frac{1}{k_B T} \langle \frac{\partial E}{\partial \varepsilon_\Gamma} \rangle^2 \right\}. \quad (5.16b)$$

Equation (5.16b) can be interpreted in the same way as (5.16). c_Γ^0 is the background elastic constant discussed in Sect. 4.2. We can distinguish again between Van Vleck type terms ($\partial^2 E/\partial\varepsilon^2$) and the strongly temperature dependent Curie term. These CEF effects can be best observed in the paramagnetic phase for magnetic ions whose CEF states have strong quadrupolar matrix elements.

As a first simple example, an orbitally twofold degenerate CEF level which split under the strain ε_Γ like $E = \pm g\varepsilon_\Gamma$ is taken. According to (5.16b) this gives an elastic constant change

$$\Delta c_\Gamma = -\frac{ng^2}{k_B T}. \qquad (5.16c)$$

This is the orbital analogue of the well-known Curie susceptibility.

As another example we show in Fig. 5.5, experimental data for magnetic and strain susceptibilities for TmSb (Mullen et al. [5.29]). Because of the different symmetries of elastic waves the information from elastic waves on CEF is much greater than from magnetic susceptibility. From the high temperature magnetic susceptibility, we get an effective magneton number $p_{eff} = 7.46$ instead of a calculated $p_{eff} = g_J \sqrt{J(J+1)} = 7.56$ (with $g_J = 1.167, J = 6$). Because of the singlet ground state the susceptibility has a finite value for $T = 0$ with a value of $\chi_m(T=0) = 0.505 \text{cm}^3/\text{mol}$. The Van Vleck contribution of the lowest two states ($\Gamma_1 - \Gamma_4$ from Fig. 5.3) amounts to 0.57 cm^3/mol. The elastic constants can likewise be interpreted with the help of (5.16b). The singlet ground state contributes with the off-diagonal matrix elements to the Γ_4 and Γ_5 CEF states. These are the Van Vleck terms. The Γ_4 and Γ_5 states have diagonal Curie term. The agreement between experiment and theory for the ($c_{11} - c_{12}$) and c_{44} modes, as seen in Fig. 5.5, is again very good. For future comparison we note that the single ion magneto-elastic coupling constants g_Γ, obtained from a fit in Fig. 5.5, are of the order of 20–25 K. On the whole the single ion approach works very well for TmSb. This has been demonstrated by the different thermodynamic quantities given above and by the lack of dispersion in the CEF excitations as shown by inelastic neutron scattering (Birgeneau et al. [5.20]). This substance has been used for the study of other effects as will be shown later (Sect. 13.2).

Such CEF effects have been observed for many substances with rare earth ions as discussed below and in Sect. 7.2.2. In particular, single ion magneto-elastic effects, as discussed above for TmSb, have been observed for various intermetallic compounds such as PrSb, SmSb (Mullen et al. [5.29]). All these rare earth mono-pnictides have the NaCl structure (space group Fm3m,O_h^5). But in general two ion coupling constants have to be included. This will be discussed in the following Sect. 5.3 and many experimental results will be given in Sect. 7.2. For a similar effect for 3d-ions, see Sect. 5.5.

CEF effects on elasticity were also investigated by means of surface acoustic waves (SAW) in CEF -systems like CeAl$_2$ and SmSb (Lingner and

Lüthi [5.30]). Because of the penetration depth of the SAW of a fraction of a mm the analogous effects have been observed as for bulk sound waves.

5.2.3 Other Experimental Methods to Determine Magneto-elastic Coupling Constants

Apart from the elastic constant measurements there are other experimental methods to characterize the orbital degrees of freedom and to determine the magneto-elastic coupling constants. The elastic constants measure the strain susceptibility which is the response of the quadrupolar moments to the corresponding strain. Another method, the magneto-striction, or more precisely the para-striction, measures the quadrupolar field susceptibility which connects the quadrupolar moments to the magnetic field. A further method, the third order magnetic susceptibility corresponds to the second term (B^3) of the development of the magnetization as a function of the magnetic field. This term depends also on the quadrupolar interactions. The latter two methods have been widely applied for rare earth intermetallic compounds with interaction terms included and are described in detail in the review by Morin and Schmitt [5.13].

Yet another method for determining single ion magneto-elastic coupling constants is to measure magneto-striction in dilute magnetic compounds. A systematic study for cubic noble metals (Au, Ag) containing rare earth impurities below the 1% range have been performed by various authors. A brief review of these experiments for Dy, Tb, Ho, Er, Tm, Yb impurities can be found in Campbell and Creuzet [5.31].

5.2.4 Magneto-elastic Coupling Constants

The coupling constants for the Heisenberg exchange, J (5.6), for the CEF Hamiltonian, B_4^0, B_6^0 (5.4) and for the magneto-elastic interaction, g_i (5.7) and (5.8) can be calculated only in very rare cases and have to be determined from the experiment. For the B_i and g_i it is instructive to quote simple point charge calculations as done e.g. by Hutchings [5.8]. One obtains for cubic symmetry

$$B_4 = z_4 \frac{Ze^2}{a^5} \langle r^4 \rangle \beta \quad \text{and} \quad g_{2,3} = z_{2,3} \frac{Ze^2}{a^3} \langle r^2 \rangle \alpha$$

with α and β the Stevens' factors depending on the magnetic ion, which arise from the transformation ($x \to J_x$) to equivalent operators, z_i numerical factors depending on the ion coordination (e.g. octaeder, tetreader, cube etc.), Z the effective ligand charge, $\langle r^n \rangle$ is the r^n integral over the 4f or 3d wavefunctions and a the magnetic ion-ligand distance. The Stevens' factors and the g_J factors are listed in Appendix D. In Fig. 5.6a,b we show magneto-elastic coupling coefficients g_2/α and g_3/α for various intermetallic rare earth systems taken from the review by Morin and Schmitt [5.13]. The systems are

5.2 Thermodynamic Functions for 4f–Rare Earth Ions

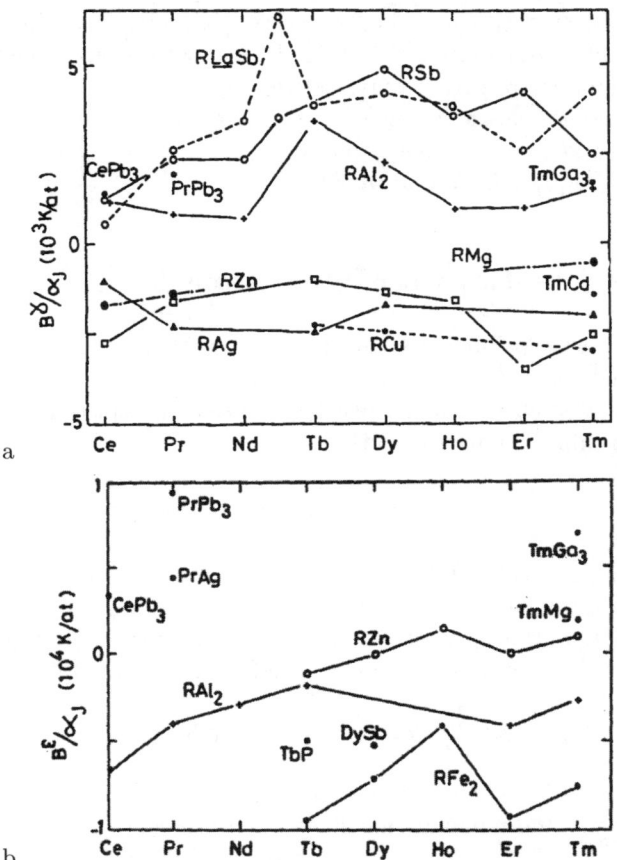

Fig. 5.6. (a) Magneto-elastic coefficients for tetragonal symmetry g_2/α for various series of rare earth cubic intermetallic compounds. g_2 corresponds to B^γ in the figure, (b) magneto-elastic coefficients for trigonal symmetry g_3/α for various series of rare earth cubic intermetallic compounds. g_3 corresponds to B^ε in the figure (Morin-Schmitt [5.13])

(R = rare earth): RSb, RAl$_2$, RZn, RAg, RFe$_2$, RPb$_3$ and others. They are cubic with NaCl, CsCl, Laves phase and Cu$_3$Au structures. The dependence of the individual rare earth ions is taken out by dividing with the Stevens' factor α. Note that these coupling constants g_i/α are rather constant across the lanthanide series. The coupling constants for dilute rare earth ions are very close to the ones gained from concentrated compounds. This can be seen by comparing e.g. RSb with RLaSb and from the values of Sect. 5.2.3.

Apart from the systems shown in Fig. 5.6a,b, there are other substances investigated in the same way. One class of materials is the rare earth hexaborides RB$_6$. They will be discussed in Sect. 7.2.2. Another class of materials are the rare earth metals. In the light rare earth metals, especially in Pr,

CEF effects in elasticity have been observed (Greiner et al. [5.32], Lüthi et al. [5.33], Palmer and Jensen [5.34]). In the heavy rare earth metals however, the high magnetic transition temperatures prevent the observation of similar CEF effects on elastic constants. The large magnetic anisotropy in these metals is a consequence of the strong CEF potential. This subject has been reviewed by Coqblin [5.35] and by Rhyne [5.36].

5.3 Susceptibilities with Interactions

The magnetic susceptibility changes from the single ion susceptibility, (5.16), if one includes the exchange interaction (5.6). If we denote the single ion magnetic susceptibility with χ_0 the total magnetic susceptibility (in molecular field or RPA approximation) reads

$$\chi_m = \frac{\chi_0}{1 - j\chi_0}. \tag{5.17}$$

Here, j is the $q = 0$ exchange constant which is > 0 for ferromagnets and < 0 for anti-ferromagnets. For a derivation of (5.17) see e.g. White [5.5].

In a similar way, an expression for the elastic constants in the presence of two-ion interactions can be obtained. The Hamiltonian is taken as

$$H = c_\Gamma \frac{\varepsilon_\Gamma^2}{2} + g_\Gamma \varepsilon_\Gamma \sum_i O_{\Gamma i} + K \sum_{i \neq j} O_{\Gamma i} O_{\Gamma j}. \tag{5.18}$$

Here, the elastic and magneto-elastic energy terms are taken from equations (3.11), (5.7) and (5.8). The last term gives the orbital interactions between the quadrupoles with a coupling constant K. Equation (5.18) in molecular field approximation reads

$$H = c_\Gamma \frac{\varepsilon_\Gamma^2}{2} + g_\Gamma \varepsilon_\Gamma \sum_i O_{\Gamma i} + K \langle O_\Gamma \rangle \sum_i O_{\Gamma i} = c_\Gamma \frac{\varepsilon_\Gamma^2}{2} + \zeta \sum_i O_{\Gamma i}, \tag{5.19}$$

with $\zeta = g\varepsilon + K\langle O_\Gamma \rangle$. For the free energy we get in molecular field approximation

$$F = c\frac{\varepsilon_\Gamma^2}{2} + KN_s \frac{\langle O_\Gamma \rangle^2}{2} - k_B T \ln \sum_i e^{-\frac{E_i}{k_B T}}. \tag{5.20}$$

Here the magnetic ion energies E_i depend on ζ as $E_i = E_i^0 + a_i \zeta + b_i \zeta^2$ like in (5.16a). Analogously to Sect. 4.3 we get the equilibrium condition $\partial F/\partial \langle O_\Gamma \rangle = K\langle O_\Gamma \rangle + \langle \partial E/\partial \langle O_\Gamma \rangle \rangle = 0$ from which follows $\langle O_\Gamma \rangle = -\langle \partial E/\partial \zeta \rangle$. With the single ion strain susceptibility $\chi_s = \mathrm{d}\langle O_\Gamma \rangle/\mathrm{d}\varepsilon$ (5.16b) we get $\mathrm{d}\zeta/\mathrm{d}\varepsilon = g/(1 - K\chi_s)$ and $c_\Gamma = \mathrm{d}^2 F/\mathrm{d}\varepsilon^2 = c^0 - gN_s\chi_s\mathrm{d}\zeta/\mathrm{d}\varepsilon$ or

$$c_\Gamma = c_\Gamma^0 - g^2 N_s \chi_{str} \quad \text{with} \quad \chi_{str} = \frac{\chi_s}{1 - K\chi_s}. \tag{5.21}$$

5.4 External and Internal Strains

This equation has the same structure as the magnetic susceptibility (5.17). The expression has wide spread applications in various fields: cooperative Jahn–Teller effect (Sect. 7.2); certain structural transitions; magnetic dimer–strain coupling (Sect. 12.2); etc. A similar derivation for the case of quadrupolar systems can be found by Allen [5.37], Kataoka and Kanamori [5.38], Levy [5.39], Mullen et al. [5.29], Gehring and Gehring [5.40], Melcher [5.41].

For $\chi_s \sim 1/T$ one obtains from (5.21) again (4.13) without CEF: $c_\Gamma = c_\Gamma^0 \frac{T-K-g^2/c_\Gamma^0}{T-K}$. For $K > 0$ one speaks of ferro-quadrupolar coupling. By lowering the temperature, the soft mode c_Γ can soften completely as shown in Fig. 4.4. For $K < 0$ one deals with anti-ferroquadrupolar coupling. Clearly in this case the temperature effect on c_Γ is small.

5.4 External and Internal Strains

As pointed out above, if $\chi_S \propto 1/T$ in the absence of CEF (5.21) gives again (4.13). For 3d compounds (4.13) is enough since the CEF splitting is very large, for 4f compounds one has to take (5.21). There is however a complication if the direct coupling of the elastic strain to the quadrupole is not the strongest coupling but rather a ligand–quadrupolar coupling, as mentioned briefly in Sect. 5.1.2. To illustrate this point we discuss a simple example. A full discussion is given by Thomas [5.42]. If we have an elastic strain–ligand coupling, the lattice Hamiltonian reads

$$H_L = c_\Gamma^0 \frac{\varepsilon_\Gamma^2}{2} + E_\Gamma Q_\Gamma \varepsilon_\Gamma + \frac{1}{2}\omega_\Gamma^2 Q_\Gamma^2, \tag{5.22}$$

where Q_Γ is the symmetry ligand coordinate and ω_Γ its frequency. The magneto-elastic coupling is taken again as

$$H_{me} = \sum_i g_\Gamma \varepsilon_\Gamma O_\Gamma + \sum_i A_\Gamma Q_\Gamma O_\Gamma(i), \tag{5.23}$$

i.e. we have a direct coupling of the elastic strain to the quadrupole operator and a coupling of the ligand ion to the quadrupole. The free energy in this case reads:

$$F = c_\Gamma^0 \frac{\varepsilon_\Gamma^2}{2} + E_\Gamma \varepsilon_\Gamma Q_\Gamma + \frac{1}{2}\omega_\Gamma^2 Q_\Gamma^2 - k_B T N_s \ln \sum_i e^{-\beta E_i}.$$

In this case we again get a formula like (5.16b) but with renormalized coupling constants

$$c_\Gamma = c_\Gamma^0 - \frac{E_\Gamma^2}{\omega^2} - \left(g_\Gamma - A_\Gamma \frac{E_\Gamma}{\omega^2}\right)^2 \chi_\Gamma. \tag{5.24}$$

In addition we get also a renormalisation of the elastic constant in the absence of a coupling to the quadrupoles ($g_\Gamma = 0$), only with a coupling

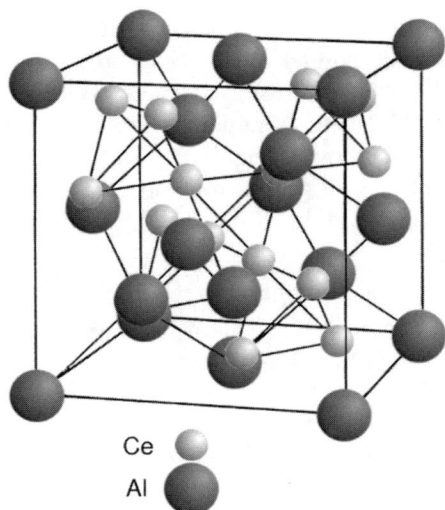

Fig. 5.7. Cubic unit cell for the Laves phase C15 structure: CeAl$_2$

of the elastic strain to the ligands. For the free energy from (5.22) one gets from $\partial F/\partial Q_\Gamma = 0$ the linear relation between external and internal strains: $Q_\Gamma = E\varepsilon_\Gamma/\omega_\Gamma^2$. The full symmetry analysis of the dependence of the internal strain from external pressure or strain has been developed by Freese and Döring [5.43]. If $AE < 0$ we get a strong enhancement of the effective coupling constant g_Γ. Applications of these concepts are given below (CeAl$_2$) and in Sect. 7.2.3 (UO$_2$). Inclusion of the quadrupole–quadrupole interaction into (5.24) as in (5.21) is straightforward, but we will not make use of it.

As an example where internal strains are important, consider the intermetallic Laves phase (C15) compound CeAl$_2$ (Lüthi and Lingner [5.44]). The unit cell of this compound (space group Fd3m) is shown in Fig. 5.7. The Ce ions have an almost integer 3+ valence state with $J = 5/2$. Due to the tetrahedral site symmetry this state should split by the CEF into a Γ_7 doublet and a Γ_8 quartet.

From the elastic constant data, especially the c_{44} softening (see Fig. 5.8), a splitting of the $\Gamma_7(0) - \Gamma_8(\Delta)$ of $\Delta \sim 100$ K was deduced. The c_{44} mode has Γ_5^g symmetry (see Sect. 3.2). In the Laves phase, an optical phonon exists with the same symmetry $\Gamma_5^g(\Gamma_{25}')$ (Cullen and Clark [5.45]). This arises due to the vibration of the two Ce ions in the unit cell. Therefore we can have a direct coupling of the external and the internal strains, $E_{\Gamma 5}Q_{\Gamma 5}\varepsilon_{\Gamma 5}$ (5.22) in addition to the coupling to the ionic quadrupolar moment $O_{\Gamma 5} = O_{xy}$. Therefore the fit for c_{44} can be made with (5.24). One gets for the effective coupling constant the large value $|g_{\Gamma 5} - A_{\Gamma 5}E_{\Gamma 5}/\omega^2| = 270$ K/ion. Looking at the temperature dependence of the various elastic modes in Fig. 5.8, we notice a strong softening of 20% only for the c_{44} mode and also for c_L the

5.4 External and Internal Strains

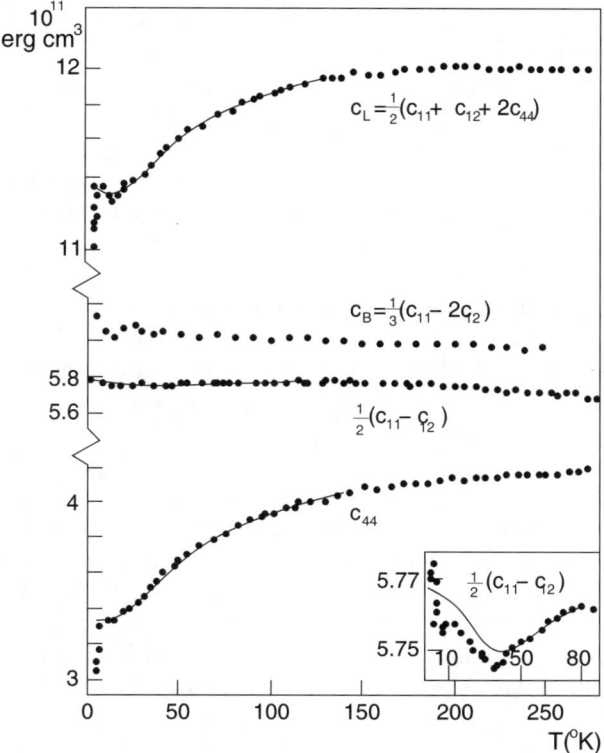

Fig. 5.8. Temperature dependence of elastic constants in CeAl$_2$. Full lines are fits as described in the text (Lüthi and Lingner [5.44])

longitudinal mode along (110) which contains c_{44} Tab. 3.1. The other modes like c_{11} and $(c_{11} - c_{12})/2$ do not show any softening. CeAl$_2$ undergoes an anti-ferromagnetic phase transition at $T_N = 3.8$ K. Below this temperature strong domain wall - stress effects are noticeable as seen in Fig. 5.8. Looking at the series of RAl$_2$ compounds (Lingner and Lüthi [5.46]), with $R =$ La, Ce, Pr, Nd, Gd, Tb, Tm and Yb, only CeAl$_2$ has such a large magneto-elastic coupling constant at least a factor 10 larger than the other compounds (see also Fig. 5.6b). The more extended 4f wave functions and the large Stevens factor α, together with the internal strain effect, for this substance serves as a qualitative explanation for the large magneto-elastic coupling constant. Also the typical single ion compound TmSb, discussed before (Sect. 5.2) has a coupling constant $g_{\Gamma 5}$ of 20 K, an order of magnitude smaller.

As to the other elastic constants, $(c_{11} - c_{12})/2$ exhibits only a slight softening with a minimum around 40 K, which can be nicely accounted for by the single ion strain susceptibility χ_s. The longitudinal mode c_L along the (110) direction contains c_{44} and therefor also shows strong softening as a function of temperature.

There are other effects due to the Γ_5 optical phonon in CeAl$_2$. An inelastic neutron scattering experiment of the $\Gamma_7 - \Gamma_8$ splitting unexpectedly gave – instead of a single excitation of ~ 100 K – a double peak structure (Loewenhaupt et al. [5.47]) of $\Delta_1 = 100 - 110$ K and $\Delta_2 = 180 - 200$ K. This was seemingly in conflict with the cubic symmetry of this compound. Inelastic neutron scattering for the phonon branches gave a large phonon density of states at an energy of 140 K with the same symmetry Γ_5 of the diamond type sublattice of the Ce ions (Reichardt and Nücker [5.48]). These facts were explained as bound states of CEF excitations and phonons (Thalmeier and Fulde [5.49], Thalmeier [5.50]). Again the strong magneto-elastic coupling (with the large off-diagonal quadrupolar matrix element with Γ_5 symmetry) is responsible for these effects. The presence of bound states has also been observed in Raman scattering (Güntherodt et al. [5.51]).

In addition to the strong quadrupolar effect, the c_{44} mode also exhibits around 10 K attenuation and velocity dispersion (Hampel and Blick [5.52]). The size of these effects depended on the crystal quality. They were interpreted phenomenologically with the Landau–Khalatnikov formulas (4.18 and (4.18a) and could be due to the coupling of the transverse sound wave mode to a collective magnetic excitation in the paramagnetic heavy Fermi liquid which is just being formed at and below 10 K (see Chap. 9).

It should be mentioned that other Laves phase compounds, namely the RFe$_2$ crystals, exhibit extremely strong magneto-strictive effects. For example, TbFe$_2$ which has its easy magnetization direction parallel to [111], develops giant zero temperature strains of 4×10^{-3} (Clark [5.53], [5.54]). The same arguments from above give an explanation of these effects (Cullen and Clark [5.45]).

The symmetry aspect of sound propagation for these type of crystals (TbFe$_2$, Tb$_{0.3}$Dy$_{0.7}$Fe$_2$) will be discussed further in Sects. 13.1 and 13.3.1. The magneto-elastic coupling constants of the heavy RFe$_2$ are shown in Fig. 5.6a,b. For TbFe$_2$ the magneto-elastic coupling constant g_3 is an order of magnitude larger than for TbAl$_2$ or TbZn indicating a similar enhancement as discussed for CeAl$_2$ above.

Apart from the examples given above (CeAl$_2$, TbFe$_2$) there are other materials which show internal rearrangement effects in the magnetic phase (Cooper [5.55]). Examples are UO$_2$ to be discussed in Sect. 7.2.3, UAs and CeBi. In the latter one elastic constant measurements show little effect for the c_{44} and $c_{11} - c_{12}$ modes above the Néel temperature $T_N = 24$ K but pronounced anomalies near the spin structure change $T^\star = 12.5$ K (Lüthi et al. [5.56]) where the magnetic structure changes from a type I antiferromagnet to a type IA.

5.5 Paramagnetic Spin–Phonon Interaction for 3d-Transition Metal Ions in Crystals

The interaction of sound waves with concentrated ions in magnetic crystals in the paramagnetic state has been dealt with in previous sections. There, mostly crystals with rare earth ions were considered. In most cases, the exchange interaction between the magnetic ions was not important. Experiments with diluted paramagnetic centers have also been performed, mostly with transition metal ions and with microwave ultrasonics. This is a nice complementary tool to electron paramagnetic resonance. Such experiments were carried out in the sixties and have given specific information on spin lattice relaxation and spin-phonon couplings. The experiments were exclusively performed on insulating perfect crystals with ideal acoustic properties. In addition saturation phenomena, phonon maser and double quantum detectors have been developed and studied. A number of reviews and books cover these topics: Tucker and Rampton [5.57], Bron [5.58].

First, an analogous effect as discussed in Sect. 5.2 for rare earth ions has also been found in a transition metal compound, $FeCl_2$ (Gorodetsky et al. [5.59]). In this case the cubic part of the crystal electric field splits the $Fe^{2+}(3d^6)$ ^5D level into an upper doublet and a lower triplet with a large separation. The triplet $^5T_{2g}$ level is, however, further split by an axial crystal field and spin-orbit coupling, giving rise to 3 singlets and 6 doublets between 0 and 500 K. The non-vanishing quadrupole matrix elements between these levels induce the CEF effect on the elastic constants. These effects can be quantitatively interpreted using (5.16b) for $T > T_N = 23.5$ K.

Ultrasonic paramagnetic resonance was performed for Ni^{2+} and Fe^{2+}-ions ($S = 1$) in MgO using longitudinal waves of 9.5 GHz at liquid Helium temperatures (Shiren [5.60]). Absorption and velocity changes as a function of magnetic field was detected near the resonance. A theoretical treatment of the ultrasonic dispersion for this case of ultrasonic paramagnetic resonance was developed by Jacobsen and Stevens [5.61]. A detailed symmetry treatment of selection rules for the paramagnetic acoustic resonance has been given by Dobrov [5.62].

Another related acoustic experiment was performed with Ni^{3+}-ions in α–Al_2O_3 (Sturge et al. [5.63]). The Ni ion is in octahedral environment and has an orbitally degenerate ground state ^2E, hence it acts as a Jahn–Teller ion (see Sect. 7.2). With 40 ppm Ni ion concentration, the strain–ion interaction can be treated as a single ion coupling. A Curie term elastic constant change is obtained for the transition metal ion, discussed following (5.16b) (with n the number of magnetic ions) $\Delta c = -ng^2/k_B T$. In the case of Ni^{3+}, only E_g symmetry strains give a Jahn–Teller coupling (5.8). Therefore the $(c_{11}-c_{12})/2$ mode shows a strong effect. Sound wave frequencies in the 2–200 MHz range give pronounced attenuation peaks at low temperatures (1–20 K). An example is given in Fig. 5.9. Plotted are the resonance frequencies and acoustic loss for pure Al_2O_3 and Ni - doped Al_2O_3 for low frequencies. Because the re-

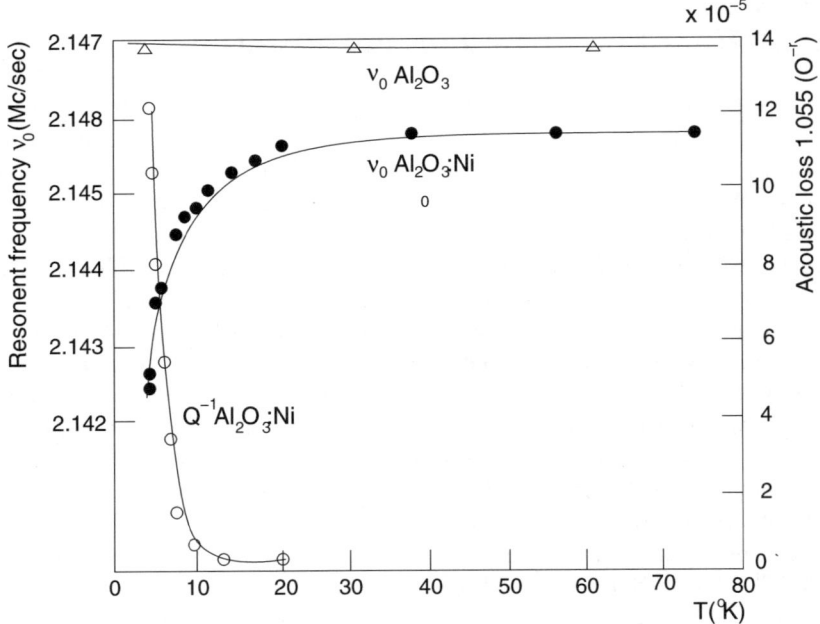

Fig. 5.9. Acoustic loss Q^{-1} and resonant frequency ν_0 as a function of temperature for Ni doped and undoped Al_2O_3 spheres (Gyorgy et al. [5.64])

laxation is not instantaneous, acoustic loss and dispersion is observed which can be described with (4.18) and (4.19). The responsible relaxation time is described by tunneling between the potential wells accompanied by simultaneous emission or absorption of a phonon.

In the following, some of these acoustic experiments performed in the ruby crystal (Cr^{3+} ions in corundum α–Al_2O_3) are reviewed. In Fig. 5.10 the energy level diagram for the $Cr^{3+}(3d^3)$ ion is shown schematically (Sugano and Tanabe [5.65]). The crystal electrical field is strongly cubic with a weak trigonal contribution. The ground state of Cr^{3+} is $^4F(S=3/2, L=3)$ which splits into a $^4A_2$, $^4T_1$ and $^4T_2$ states. The latter two form bands. The other important term is the ^2G which splits into the important ^2E state amongst others. This state splits via spin-orbit and trigonal CEF into the two Kramers doublets $R_1(E)$ and $R_2(2A)$ with an energy difference $\delta = 29 cm^{-1}$. For the ruby laser, the two bands are optically pumped and the laser emission occurs in the R lines.

For our purposes we have to consider the ground state $^4A_2$ and the R-lines more closely. The first mentioned state is also split by axial field and spin orbit coupling by $0.38 cm^{-1}$. It is further split by an external magnetic field into $\pm 1/2$ and $\pm 3/2$ levels. These levels were investigated by microwave longitudinal c_{33} modes of 9.1 GHz as a function of angular variation of magnetic fields

5.5 Paramagnetic Spin–Phonon Interaction

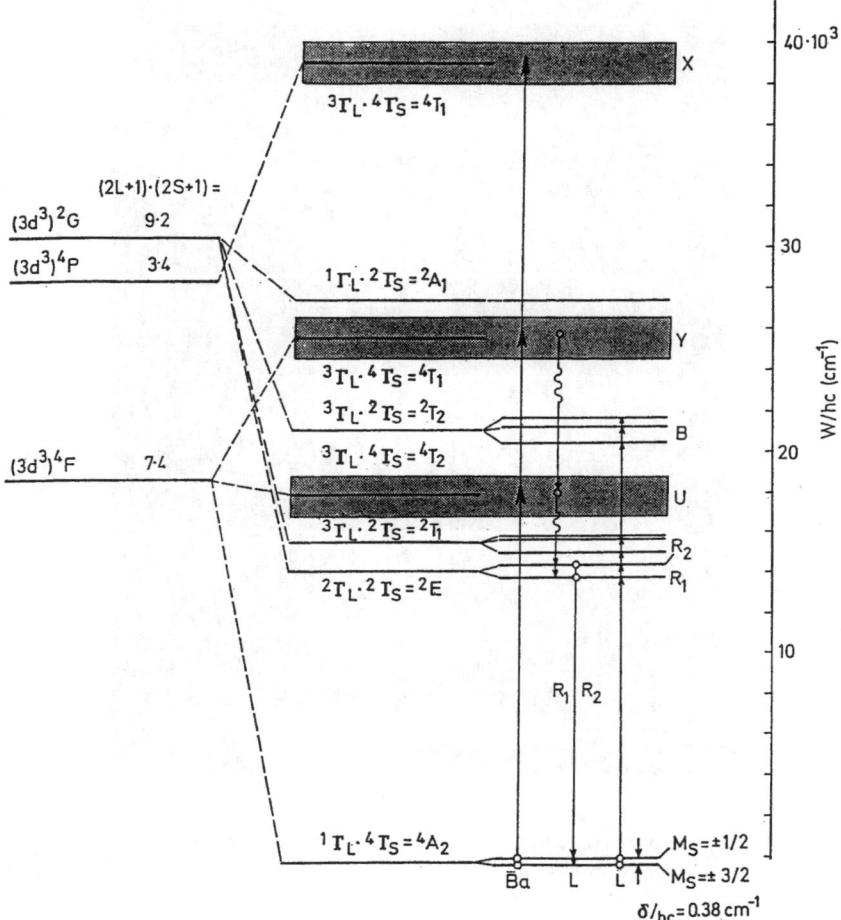

Fig. 5.10. Energy levels for Cr^{3+} in ruby (Sugano and Tanabe [5.65])

(Tucker [5.66]). Ultrasonic attenuation peaks due to strain–quadrupolar transitions are observed and quantitatively accounted for. In a second experiment (Tucker [5.67]) the field direction for the highest attenuation (60° between c-axis and field direction) was taken. In addition the same pump frequency of 24 GHz as for the ruby–Maser was taken to invert the spin level system. As shown in Fig. 5.11 acoustic amplification was observed in this case for 9.3 GHz microwave sound waves.

Finally, the question of sound velocity effects in non-equilibrium conditions is addressed, i.e. under laser pumping conditions. Such an experiment was attempted by Grill and Lüthi [5.68] again in ruby. Pumping the $^4T_2$ and $^4T_1$ bands (see Fig. 5.10) gives an occupation $n_{1,2}$ of the $R_{1,2}$ levels. With the strain dependence of these energies, where $E^0(R_2) - E^0(R_1) = \delta$:

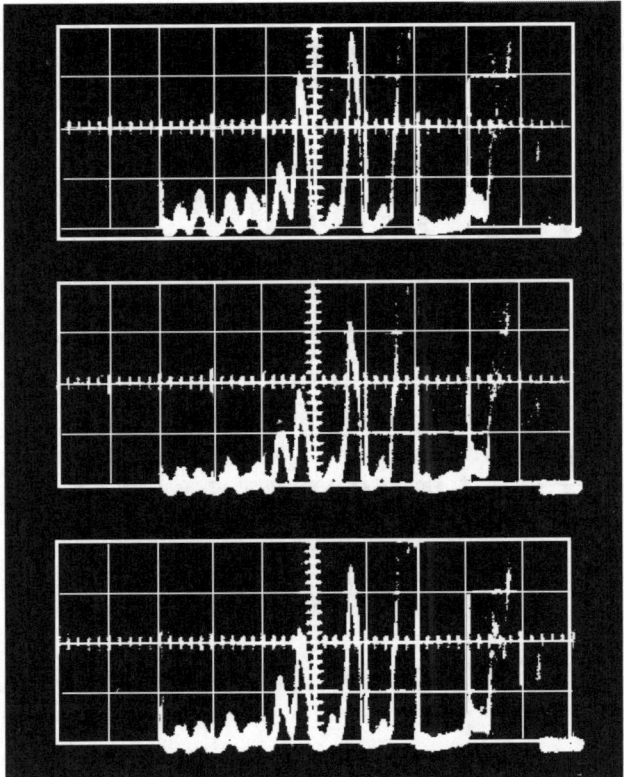

Fig. 5.11. Amplification of microwave sound by maser action in ruby (Tucker [5.67]). Top picture: magnetic field on resonance line and pump on. Center picture: magnetic field on resonance line, no pump. Bottom picture: magnetic field off resonance line. Ultrasound frequency is 9.3 Gc, time scale from right to left

$E(R_1) = E^0(R_1) - A\varepsilon_\Gamma$ and $E(R_2) = E^0(R_2) - B\varepsilon_\Gamma$ – as determined from pressure dependent R–line shifts (Feher and Sturge [5.69]), the adiabatic elastic constant c_L – with $n_1 + n_2 = n =$ constant and $\partial S/\partial \varepsilon_\Gamma = 0$, with S the entropy – is given by $c_L = c_l^0 - \partial n_1/\partial \varepsilon (A - B)$. Because of the linear strain dependence of the R-levels, no Van Vleck terms are present. Expressing n_1 by n gives

$$c_l = c_L^0 - n\frac{(A-B)^2}{k_B T} \frac{e^{-\frac{\delta}{kT}}}{\left(1 + e^{-\frac{\delta}{kT}}\right)^2}. \tag{5.25}$$

This formula is valid as long as there is thermal equilibrium between the excited levels n_1, n_2 during the sound wave period. Otherwise one has to include a dispersive term $(1 + \omega^2\tau^2)^{-1}$. Such a term is needed for the results in Fig. 5.12 at room temperature. The result of the experiment for 109 MHz longitudinal sound along the c-axis in ruby is shown in Fig. 5.12

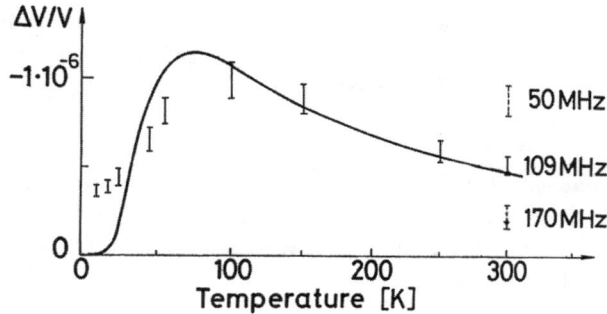

Fig. 5.12. Temperature dependence of relative change in sound velocity $\Delta v/v$ in ruby under laser pump conditions described in the text (Grill and Lüthi [5.68])

together with a fit without adjustable parameters from (5.25). The effect is of the order of 10^{-6} in the relative velocity change. The ruby crystal was pumped with a ruby laser pulse of 1ms duration and a repetition rate of 1s. A time resolved phase-sensitive detection scheme was used to resolve the small velocity changes.

It would be interesting to investigate with this method a material which would exhibit a structural transition under laser pumping condition. Several rare earth compounds investigated theoretically ($TmPO_4$, TmSb) gave at best a softening of 1% in the velocity at low temperatures. A case with a structural transition under optical pumping has not been realised yet. But a magnetic phase transition triggered by high power optical pumping has been observed in $EuCrO_3$ (Golovenchits et al. [5.70]). An interesting case would be also a material with a metastable state under irradiation. A typical example is $Na_2[Fe(CN)_5NO]\cdot 2H_2O$ (Woike et al. [5.71]). Another rapidly developing field is the quenching dynamics of anti-ferromagnets using femtosecond lasers, see e.g. Kimel et al. [5.72] and references therein.

5.6 Nuclear Acoustic Resonance

The interaction of sound waves with nuclear spins was studied in the period 1956–1966. The first type of experiment was a free nuclear induction decay experiment by acoustically saturation (Proctor and Tantilla [5.73]). A direct detection of acoustic energy absorption by the nuclear spin system was observed later on (see e.g. Bolef [5.74]). The coupling of ultrasound to the nuclei in non-magnetic materials is via the electric quadrupole interaction. For a metal, like Ta, such experiments could be performed in bulk samples (Gregory and Bömmel [5.75]).

Another type of acoustic NMR can be observed in magnetically ordered materials like $RbMnF_3$, a material to be discussed in more detail in Sect. 6.1.2. For magnetic ions, like Mn^{2+}, the interaction of nuclear spins with resonant ultrasonic oscillations is made possible via spin waves in materials with small magnetic anisotropy (de Gennes et al. [5.76]). The hyperfine coupling transfers the oscillations of the electron spins to the nuclear spins. The observation of nuclear acoustic resonance in the ordered region of $RbMnF_3$ can be easily observed because of this enhancement effect (see Sect. 6.1.2 and references). Acoustic nuclear spin echoes excited by acoustic pulses and electromagnetic echoes excited by a combination of electromagnetic and acoustic pulses in the ordered region of anti-ferromagnetic $KMnF_3$ and $RbMnF_3$ could also be observed (Bogdanova et al. [5.77]).

6 Ultrasonics at Magnetic Phase Transitions

Magnetic phase transitions are treated first in this chapter followed by structural phase transitions in the next chapter. A distinction has to be made between static properties and dynamical or transport properties at phase transitions. Experiments and theory are well developed for the static case. Critical exponents for thermodynamic quantities, scaling laws between these exponents and the so-called renormalization group theory have put the study of static phase transitions on a firm basis. It was found that quite different types of phase transitions, such as magnetic, liquid-gas, ferroelectric and structural ones behave rather similarly despite quite different interactions occurring between the particles. The critical phenomena arise from long-range correlations. The critical exponents depend on the dimension of the system and on the dimension of the order parameter, but not on the details of the forces. This is the universality law of critical phenomena (see Kadanoff [6.1]). The range of critical fluctuations can be estimated with the Ginzburg criterion, discussed in Sect. 4.4.

The case of dynamical critical phenomena is quite different. Here the dynamical properties depend on the details of the interaction. It is shown that in the case of magnetic phase transitions, longitudinal and shear waves behave quite differently with different mechanisms and different dynamical exponents for the attenuation. Also for structural transitions we will find in the next chapter a variety of different strain order parameter interactions.

In the case of magnetic transitions, we distinguish between sound wave effects at the magnetic transitions characterised by Curie or Néel transition temperatures and at a so-called spin reorientation transition respectively. In the first case, the transition cannot be described by Landau theory because, according to the Ginzburg criterion (Sect. 4.4), the range of critical fluctuation is large: $(T - T_c)/T_c \geq 1\%$. The marginal dimensionality d^* for a ferromagnet is four. In three dimensions, therefore, critical fluctuations dominate. On the other hand, spin reorientation phenomena have a large force range due to dipolar fields and can be described by Landau theory (Levinson et al. [6.2]).

We treat mainly magnets which can be described by a Heisenberg Hamiltonian. For such systems, the theory of sound wave phenomena is well developed and comparison with experiments is possible (Sect. 6.1). In the case of spin reorientation, we concentrate on rare earth orthoferrites, because they exhibit second order transitions at the corresponding spin reorientation phase transitions (Sect. 6.2). For sound propagation for spin density wave materials, we concentrate on the metal chromium Cr (Sect. 6.3).

6.1 Magnetic Phase Transition

The exchange interaction is the dominant interaction responsible for magnetic phase transition. See Sect. 5.1.1 for a short exposition of the various magnetic interactions. The isotropic Heisenberg exchange interaction (5.6) explains many different phenomena in magnetism, such as the occurrence of the phase transition or the elementary excitations called magnons or spin-waves. For a full discussion of magnetic concepts like exchange, magnetic excitation etc., we refer to books on magnetism, e.g. Martin [6.3], White [6.4]. The magneto-elastic interaction, i.e. the coupling of the lattice coordinates (strains or phonon coordinates) to the spin system plays only a secondary role. In the following, we first discuss the strain-spin coupling mechanism and afterwards the effect this coupling has on sound velocity and attenuation near magnetic phase transitions. We will see that two quite different mechanisms describe these effects.

6.1.1 Spin–Phonon Coupling Mechanism

We concentrate on S-state ions ($L = 0$) like Mn^{2+}, Fe^{3+}, Gd^{3+}, Eu^{2+} (see App. D). For these ions the single ion coupling constant g_i (5.7) is much smaller than the exchange striction coupling constant $\partial J/\partial \varepsilon$ (5.10). The case of orbital moment $L \neq 0$ ions are treated in Sect. 7.2. Fig. 6.1 shows attenuation results at the magnetic transition for FeF_2 (Ikushima and Feigelson [6.5]) and Gd (Lüthi and Pollina [6.6]). The propagation direction in both cases is along a symmetry direction; along the tetragonal axis for anti-ferromagnetic FeF_2 and along the hexagonal axis for ferromagnetic Gd. Therefore, according to (5.10), only longitudinal waves should couple to the spin system. This is clearly manifested in Fig. 6.1 with pronounced attenuation peaks for both materials FeF_2 and Gd. As shown for the case of Gd, dJ/dc is larger than dJ/da. dJ/dc is the coupling constant for propagation along the c-axis and dJ/da along the a-axis. There are a few notable exceptions to this result such as the metal Cr (Lüthi et al. [6.7]), CoO (Ikushima [6.8]) and heavy rare earth metals Tb, Dy, Ho (Pollina and Lüthi [6.9]) where the single ion magneto-elastic coupling is also important. In addition to the strong magneto-elastic coupling constants, these rare earth metals (Tb, Dy, Ho) have also a strong uniaxial (easy plane) anisotropy, which forces the magnetic moment and the spin fluctuations to the hexagonal plane. A strong polarization dependence for the critical ultrasonic attenuation is therefore expected. For a more detailed discussion of shear wave propagation of the rare earth metals with different propagation and polarisation directions, see Tachiki et al. [6.10].

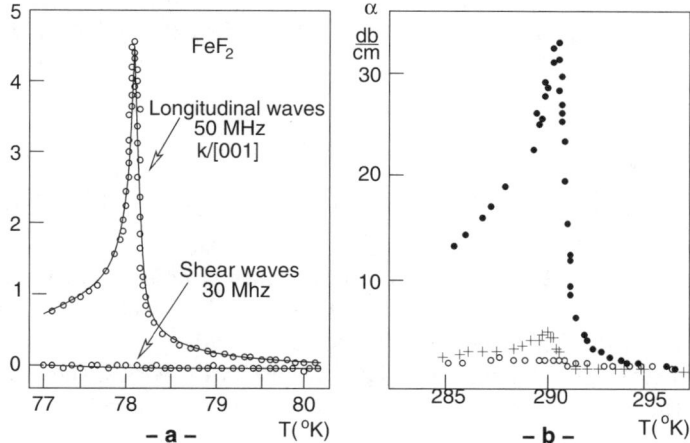

Fig. 6.1. Ultrasonic attenuation of longitudinal and shear waves along the c-axis near T_N and T_C for (**a**) FeF$_2$ (Ikushima and Feigelson [6.5]), (**b**) Gd (Lüthi and Pollina [6.6]). In **Gd** we show longitudinal sound waves along the c–axis (full circles) and along the a–axis (crosses) as well as shear waves along the c–axis (open circles), frequency 50 MHz

6.1.2 Critical Attenuation Coefficient

The critical effects that sound waves exhibit near magnetic phase transitions have been reviewed in many places (Garland [6.11], Lüthi et al. [6.7], Kawasaki [6.12], Lüthi [6.13], Kamilov and Aliev [6.14]). Critical effects on sound velocity and attenuation are expected because of the spin–phonon coupling, especially the exchange striction coupling (5.10). With this interaction, energy is transferred from the sound wave to the spin system, which has relaxation channels whose relaxation times can diverge at the critical temperature. Hence from expressions such as (4.16) and (4.18), it is seen that the sound attenuation α can diverge for $\omega\tau < 1$ and dispersive effects can occur for $\omega\tau \sim 1$. The relaxation time τ measures the time the system needs to come from a non-equilibrium state to the thermodynamic equilibrium. This time diverges as the critical point is approached, with the critical slowing down (see also discussion in Sect. 4.3.6 and (4.19a). This is the source of divergencies in transport coefficients, here the ultrasonic attenuation.

Another way to look at it is to make a comparison with Brownian motion (Reif [6.15]). A particle in a suspension is stopped by random forces, the attenuation being the space-time Fourier transform of random forces. The random force in our case is given as the strain derivative of the exchange energy (5.10), thus leading to time-dependent four-spin correlation functions. Therefore the attenuation coefficient can be determined by calculating the number of phonons with wave vector q absorbed minus those (induced) emitted (Stern [6.16]):

$$\alpha = \sum_{ij\delta\delta'} b(\boldsymbol{\delta},\boldsymbol{\delta}',\boldsymbol{q}) \int_0^\infty dt e^{i\omega t} e^{iq(Ri-Rj)} \langle \boldsymbol{S}_i(0)\boldsymbol{S}_{i+\delta}(0)\boldsymbol{S}_j(t)\boldsymbol{S}_{j+\delta'}(t)\rangle \quad (6.1)$$

with

$$b = \left(\frac{1}{\delta}\frac{\partial J}{\partial \delta}\right)^2 (\boldsymbol{\delta}\cdot\boldsymbol{e}_q)(\boldsymbol{\delta}'\cdot\boldsymbol{e}_q)(\boldsymbol{\delta}\cdot\boldsymbol{q})(\boldsymbol{\delta}'\cdot\boldsymbol{q})(1-e^{\hbar\omega/kT})(2eM\hbar\omega)^{-1}.$$

According to (6.1), the attenuation coefficient α is proportional to the space–time Fourier-Laplace transform of a four–spin correlation function. For the evaluation of this correlation function, the different theories can be divided into three groups: the so-called conventional theories, the mode–mode coupling theories and the coupling to energy fluctuations.

Conventional Theories

It is instructive to discuss first the conventional theories (Tani and Mori [6.17], Bennett and Pytte [6.18], Okamoto [6.19], Kawasaki [6.20]). The four–spin correlation function of (6.1) is first factorised. In the paramagnetic region, products of two–spin correlation functions remain:

$$\alpha = \sum_{\delta\delta'k} B(\boldsymbol{\delta},\boldsymbol{\delta}',q,k) Re \int_0^\infty e^{i\omega t} dt \langle S_k S_{-k}(t)\rangle \langle S_{k+q} S_{-q}(t)\rangle. \quad (6.2)$$

The two–spin correlation functions are Fourier transformed and B is the corresponding Fourier transform of b in (6.1). It is expected that this factorisation procedure will break down close to T_c (Bennett and Pytte [6.18]). As a next step, they use the hydrodynamic form for the two–spin correlation function: $\langle S_k S_{-k}(t)\rangle = \langle S_k S_{-k}\rangle e^{-t/\tau_k}$ which is valid only for $k\xi \ll 1$ with ξ correlation length of spin fluctuations and τ_k the characteristic decay time of the spin fluctuations. For an isotropic ferromagnet it has the form $\frac{1}{\tau_k} = Dk^2$ with D the spin diffusion constant (de Gennes [6.21]). With $\langle S_k S_{-k}\rangle \sim \xi^2$ (6.2) becomes $\alpha \sim \int \frac{d^3k\xi^4}{\tau_k}(\omega^2+\tau_k^{-2})^{-1}$. Finally with $\omega \ll \tau_k^{-1}$ the attenuation becomes

$$\alpha \sim \omega^2 \chi^{\frac{1}{2}} \tau_c, \quad (6.3)$$

where χ is the spin susceptibility, $\tau_c = \tau_{k<1/\xi}$ and where ω^2 originates from the linear q dependence of (6.1). (6.3) indicates that the critical slowing down of the spin fluctuations, $1/\tau_c$, enhances the singularity in the attenuation α on approaching T_c. Using the different expressions for τ_c from dynamical scaling (Halperin and Hohenberg [6.22]), the values of the exponent η in the expression $\alpha \sim \omega^2 t^{-\eta}$ are listed in Table 6.1. Here $t = (T-T_c)/T_c$. Tani and Mori evaluate the correlation function somewhat differently so we list their values separately.

6.1 Magnetic Phase Transition

Table 6.1. Theoretical predictions for the critical attenuation exponent η ($\alpha \sim \omega^2 t^{-\eta}$)

Spin model	Conventional theories	Tani and Mori	Mode–mode coupling ($\eta \sim z\nu$)	dynamical exponent z
isotropic ferromagn.	7/3	1	5/3	5/2
isotr. anti-ferromagn.	5/3	1	1	3/2
anisotr. ferromagn.	3/2		4/3	2
anisotr. anti-ferrom.	3/2		4/3	2

Mode–Mode Coupling Theories

Mode–mode coupling theories are discussed next (Kawasaki [6.20], Laramore and Kadanoff [6.23], Pawlak [6.24]). The conventional theories overestimate the effect of correlations. This can be shown, for example, by calculating the specific heat using the factorisation approximation, which gives $C \sim \chi^{1/2}$. In fact as pointed out by Bennett [6.25], if, correspondingly, $\chi^{1/2}$ is replaced by C in (6.3), then

$$\alpha_S \sim \frac{\omega^2 C \tau_c}{1+\omega^2 \tau_c^2}, \qquad (6.4)$$

which is essentially the result of the mode–mode coupling theories. In these theories, one hydrodynamic mode decays into several hydrodynamic modes, leading to a divergence in the transport coefficients. These theories retain the assumption $\omega \ll \tau_c^{-1}$ and the hydrodynamic form of the spin fluctuations but do not factorize the four–spin correlation function of (6.1). They make use of the static scaling laws in calculating the divergent part of the transport coefficients. The exponent η for the ultrasonic attenuation ($\omega \tau_c \ll 1$) for these theories is also listed in Table 6.1. According to (6.4) it is the exponent of $\tau_c \sim t^{-z\nu}$, neglecting the small exponent of the specific heat C. Here z is the dynamical critical exponent (listed in Table 6.1) and ν the critical exponent for the correlation length ($\xi \sim t^{-\nu}$, $\nu = 2/3$).

Energy Density Coupling

Finally, we discuss the energy density coupling mechanism. An important variant of the theories outlined above occurs if the exchange striction Hamiltonian H_{exs} is proportional to the exchange Hamiltonian: i.e. if in the corresponding expressions

$$H_{exs} = \sum_i A e^{i\mathbf{q}\mathbf{R}i} z_i' b_i (\mathbf{S}_0 \cdot \mathbf{S}_{Ri}) \quad \text{and} \quad H_{ex} = \sum_i z_i J_i \mathbf{S}_0 \cdot \mathbf{S}_{Ri}, \qquad (6.5)$$

each individual term of H_{exs} is proportional to the corresponding term in H_{ex}. Here A is the amplitude of the sound wave, z_i is the number of neighbours

(R_i) for a given site o. z'_i is generally, however, different from z_i, i.e. the sound wave couples in general only to part of the spin energy density. Two examples shall illustrate this (Lüthi et al. [6.7]), q = sound wave propagation vector):

1. EuO ($z_1 = 12$, $z_2 = 6$); $q = (100) : z'_1 = 8$, $z'_2 = 2$; $q = (110) : z'_1 = 2 + 8$ (different coupling strength), $z'_2 = 4$; $q = (111) : z'_1 = 6$, $z'_2 = 6$.
2. RbMnF$_3$ ($z_1 = 6$, $z_2 = 12$); $q = (100) : z'_1 = 2$, $z'_2 = 8$; $q = (110) : z'_1 = 4$, $z'_2 = 2 + 8$ (different coupling strength); $q = (1,1,1) : z'_1 = 6$, $z'_2 = 6$.

Therefore, only in the case of propagation along the (1,1,1) direction for RbMnF$_3$ and with the neglect of next nearest neighbour coupling, is the spin-phonon coupling truly a coupling to the energy density of the system. If this holds, then the attenuation is proportional to an energy correlation function instead of the four-spin correlation function discussed above:

$$\alpha_E \propto \omega^2 \int_0^\infty dt e^{i\omega t} \langle E_q E_{-q}(t) \rangle . \tag{6.6}$$

Here $\langle E_q E_{-q}(t) \rangle = \langle E_q E_{-q} \rangle e^{-t/\tau_{sl}}$ with $\tau_{sl}^{-1} = \gamma/C_m$ the spin-lattice relaxation time of the spin-energy density with γ a constant. The evaluation of (6.6) gives the same expression as for α_S (6.4) but with a weaker singularity like C_M for τ_{sl}. Another relaxation channel via energy diffusion $\tau_E^{-1} = \kappa q^2/C_M$, with κ the thermal conductivity, is less effective. This coupling to energy fluctuations was introduced by Lüthi and Pollina [6.26] and by Kawasaki [6.27]. A realistic calculation of the spin-lattice relaxation time was carried out by Huber [6.28] and Itoh [6.29].

Comparison with Experiment

In Table 6.2, the experimentally determined critical exponents η are listed for a wide variety of different compounds. We notice a large exponent $\eta \geq 1$ for all metallic compounds. We expect long range magnetic exchange for these compounds, especially the rare earth metals with RKKY exchange (see Sect. 5.1.1). Therefore the coupling to energy fluctuations is reduced and we deal with the results from the mode–mode coupling theories. Interestingly, the isotropic metallic ferromagnets Gd and Ni exhibit the largest exponents in agreement with theory ($\eta = 5/3$).

The insulators exhibit smaller exponents $\eta < 1$ in agreement with the energy density coupling concept. Although the energy density coupling is only approximately fulfilled, as shown by the examples for EuO and RbMnF$_3$, the exponent η is considerably smaller in insulators. There is one exception: Cr$_2$O$_3$ (Bachellerie and Frenois [6.30]). Here, interpretation of the spin wave spectra gave a rather long range magnetic exchange interaction (Samuelsen et al. [6.31]) and therefore the mode–mode coupling result should be taken for for comparison. But for the other insulating compounds, the exponents are

Table 6.2. Experimental values of critical exponents η for the attenuation

Material	Temperature region $t = (T - T_c)/T_c$	η		Ref.
Gd	$10^{-3} - 10^{-1}$	1.63 ± 0.1	metal, isotropic H-FM	Lü 70
Tb	$7 \times 10^{-3} - 7 \times 10^{-2}$	1.24 ± 0.1	metal. anisotropic H-AF	Lü 70
Dy	$3 \times 10^{-3} - 10^{-1}$	1.37 ± 0.1	metal, anisotropic H-AF	Lü 70
Ho	$3 \times 10^{-4} - 10^{-1}$	1.0 ± 0.1	metal, anisotropic H-AF	Lü 70
Ni	$3 \times 10^{-4} - 10^{-3}$	1.4 ± 0.2	metal, isotropic H-F	Gol 69
MnP	$5 \times 10^{-3} - 1.1 \times 10^{-2}$	0.9 ± 0.2	metal, anisotropic H-F	Gon 71
EuO	-	-	Insul. Isotropic H-F	Gor 71
RbMnF$_3$	$4 \times 10^{-4} - 4 \times 10^{-2}$	0.28 ± 0.05	Insul. Isotropic H-AF	Gor. 71
MnF$_2$	$10^{-4} - 3 \times 10^{-2}$	$0.13 - 0.16$	Insul. Anisotropic H-AF	Iku. 70
Rb$_2$CoF$_4$	$5 \times 10^{-3} - 2.5 \times 10^{-2}$	0.74 ± 0.05	Insul. 2dim.AF	Suz 81
Y$_3$Fe$_5$O$_{12}$	$1.8 \times 10^{-4} - 3 \times 10^{-2}$	0.50 ± 0.1	Insul. Isotropic H-Ferri	Al 89
Gd$_3$Fe$_5$O$_{12}$	$2.3 \times 10^{-4} - 5 \times 10^{-1}$	0.42 ± 0.1	Insul. Isotropic H-Ferri	Al 89
Cr$_2$O$_3$		1.3	Insul.	Ba 74

rather small. Taking the ratio of attenuation and velocity change (4.19), the spin-lattice relaxation time τ_{sl} is obtained. Calculations of this quantity for RbMnF$_3$, MnF$_2$ and EuO gave satisfactory agreement with the experimental values (Huber [6.28], Itoh [6.29]).

An interesting case is EuO. Here no critical attenuation was observed in the region of 50–575 MHz (Lüthi et al. [6.26], Gorodetsky et al. [6.32]) as shown in Fig. 6.2a, indicating $\omega\tau > 1$. This leads to $\tau_s^{-1} < 10^8 s^{-1}$. As shown in Fig. 6.2b, vibrating reed experiments with flexural vibrations gave a damping peak for 1.6 kHz (Golding et al. [6.33]). This leads to typical values of $\tau_s^{-1} \sim 10^6 s^{-1}$ as also estimated theoretically (Huber [6.28]).

In summary, the widely varying exponent η for the various materials can be well accounted for by the theories of energy density coupling, at least in a semi-quantitative fashion. For a more quantitative comparison, it is necessary to investigate the material properties experimentally in more detail and also to refine the theories by including dipolar contributions and other approximations.

Sound Attenuation in the Ordered Phase

In the ordered region, a number of new effects are observed. Apart from fluctuation contributions, which should be analogous to the ones in the paramagnetic region discussed above, strong domain wall–stress effects can be observed. This can be clearly seen in the case of EuO for $T < T_C$ and $\omega/2\pi > 10$ MHz in Fig. 6.2a. In some favourable cases (MnF$_2$, MnP), the Landau–Khalatnikov order parameter relaxation can be isolated (Bachellerie

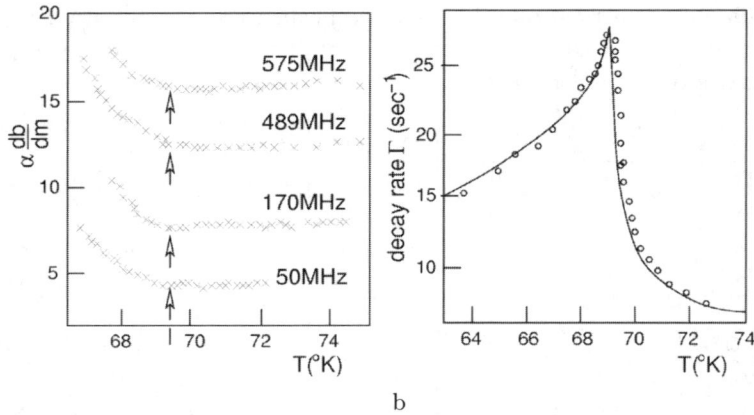

Fig. 6.2. Critical sound attenuation for EuO: (**a**) High frequencies (Gorodetsky et al. [6.32]), (**b**) low frequencies (Golding et al. [6.33])

et al. [6.34], Golding [6.35]). This can be done by measuring the frequency dependent maxima of the attenuation for $T < T_c, T_N$. The order parameter relaxation rates were found to be consistent with those from neutron scattering experiments. Other possible effects, such as nuclear acoustic resonance in the ordered region, were found and discussed for RbMnF$_3$ (Moran et al. [6.36], Jimbo and Elbaum [6.37], see also Sect. 5.6). They are in close agreement with direct nuclear acoustic resonance and nuclear magnetic resonance (see Jimbo and Elbaum [6.37]).

Sound attenuation as a function of magnetic field near magnetic phase transitions has been investigated mainly in rare earth metals (Moran and Lüthi [6.38], Tachiki et al. [6.39], Kamilov and Aliev [6.14]). Generally, in the paramagnetic region, a decrease in attenuation due to a suppression of spin fluctuations and an increase due to the polarisation of electronic subbands is observed. This latter effect occurs farther away from T_C, T_N. Experiments were carried out for Tb (Maekawa et al. [6.40]), for Dy (Treder and Levy [6.41]), for Ho (Tachiki et al. [6.39]) and for MnP (Komatsubara et al. [6.42]). For further details, see the review by Kamilov and Aliev [6.14] and Erdem [6.43].

Scaling Expressions

Finally, it is of interest to apply dynamical scaling concepts to the critical ultrasonic attenuation. In Fig. 6.3 the critical attenuation α/ω^{1+y} is plotted versus $ln(\omega\tau)$ for Gd (Aliev et al. [6.44]) and MnP (Golding [6.35]). Both substances have substantial critical attenuation for $T > T_c$ and order parameter relaxation for $T < T_c$. The fit was made with the scaling function $f(\omega\tau)$ so that the attenuation reads for $T > T_c(+)$ and $T < T_c(-)$ $\alpha^\pm = B^\pm \omega^{1+y\pm} f^\pm(\omega\tau^\pm)$. Note the rather different scaling for

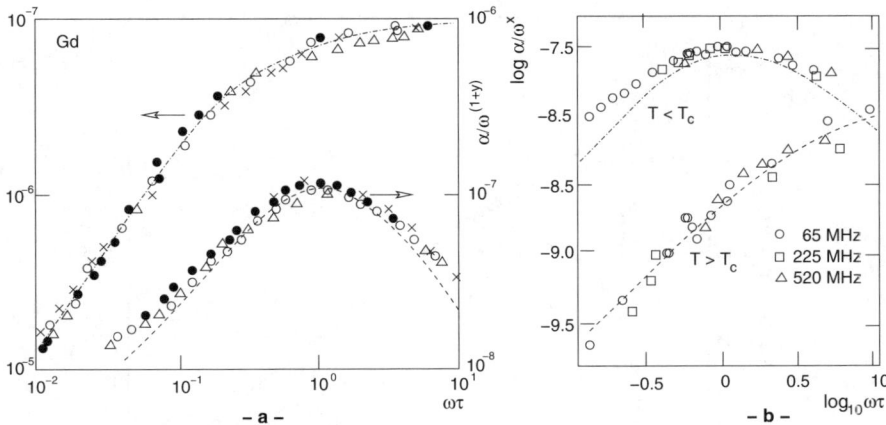

Fig. 6.3. Scaling plot for ultrasonic attenuation: α/ω^{1+y} vs $\omega\tau$, (**a**) Gd (Aliev et al. [6.44]) and (**b**) MnP (Golding [6.35])

$T > T_c$ versus $T < T_c$. The function f usually has the form of a Lorentzian in $\omega\tau$. The quasi two-dimensional Ising anti-ferromagnet Rb_2CoF_4 was analysed in a similar way (Suzuki [6.45]). For more details about the scaling parameters, see the review by Kamilov and Aliev [6.14]. For recent developments, see also Pawlak [6.24], [6.46].

6.1.3 Sound Velocity Effects near Magnetic Phase Transitions

So far we have only discussed the critical ultrasonic attenuation. But sound velocities also exhibit sharp dips at the magnetic phase transitions. An example of sound velocities for MnF_2 near T_N for different propagation directions (Kawasaki and Ikushima [6.47]) is shown in Fig. 6.4. The anomalies amount to $\Delta v/v_0 \leq \frac{1}{2}\%$. Experimentally, both quantities are usually measured together (see Sect. 2.2). As seen from the considerations given in Sect. 4.3 (4.18), for $\omega\tau \ll 1$ the sound velocity does not depend on the relaxation time τ and therefore it is not a transport coefficient in contrast to the attenuation. It is a thermodynamic quantity and it can be calculated by the static part of the spin fluctuations.

To get a feeling for the velocity anomaly, consider a magnetic free energy of the form $F_m = -Tf(T/T_c)$. As outlined in Sect. 6.1.1, the spin–phonon coupling is of the exchange striction type and the coupling constant is given by $dT_c/d\varepsilon$ and the longitudinal elastic constant is $c = c_0 + d^2F_m/d\varepsilon^2 = c_0 - (dT_c/d\varepsilon)^2(T/T_c)C_V$ with the magnetic specific heat $C_V = -Td^2F_m/dT^2$. Therefore we expect only a small divergence in the sound velocity, the same as the specific heat divergence and with no frequency dependence. This frequency and temperature dependence is also given by detailed calculations of the spin fluctuation contribution (Kashcheev [6.48], Bennett [6.25], Kawasaki

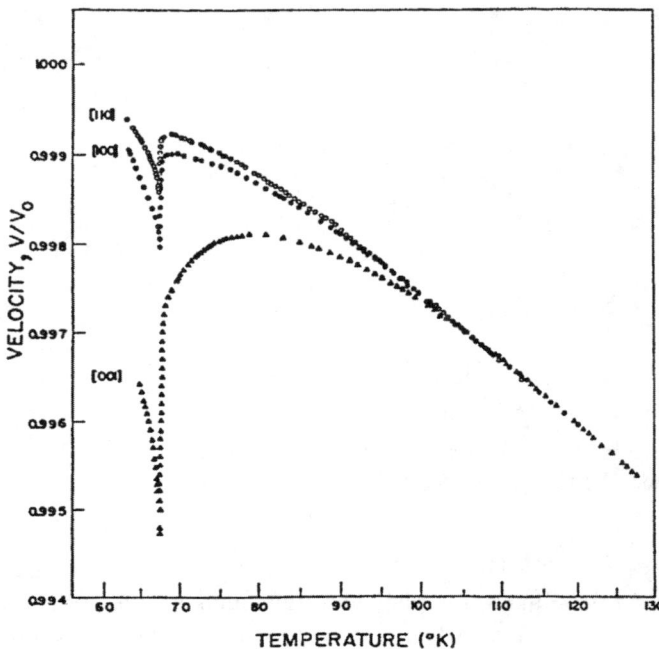

Fig. 6.4. Anomalous temperature dependence of the velocity of 10 MHz longitudinal sound in MnF$_2$ for three different directions (Kawasaki and Ikushima [6.47])

and Ikushima [6.47]). For the case shown in Fig. 6.4, the analysis gives a critical exponent ζ in

$$\frac{\Delta v}{v_0} \approx -\omega^0 (T - T_N)^{-\zeta} \tag{6.7}$$

of $\zeta \sim 0.12$ which is in agreement with the corresponding exponent α for the specific heat, namely $0 \leq \alpha \leq 0.18$ (Kadanoff et al. [6.49]).

Experiments on other substances confirm essentially the results discussed above, namely a logarithmic dependence of the velocity on $(T - T_{N,C})$ or with a small exponent as given above for MnF$_2$ (Kawasaki and Ikushima [6.47]) or RbMnF$_3$ (Golding [6.50], Lüthi et al. [6.7]) and Ho (Moran and Lüthi [6.51]). Other magnetic substances with small exponents are MnP, Fe$_3$O$_4$, Y$_3$Fe$_5$O$_{12}$, Gd$_3$Fe$_5$O$_{12}$ (Kamilov and Aliev [6.14]). Dispersive effects for $\omega\tau > 1$ have been observed in RbMnF$_3$ and MnF$_2$ in the frequency range of 30–390 MHz (Moran and Lüthi [6.36], Bachellerie et al. [6.34]) and in Ni (Golding and Barmatz [6.52]). Finally we refer again to (4.19) from which the spin relaxation time can be conveniently deduced.

The magnetic field dependence of the sound velocity and sound attenuation near magnetic phase transitions was investigated theoretically by Tachiki and Maekawa [6.53] and by Maekawa et al. [6.40]. They could explain the complicated field dependence of sound velocity changes $(v(B,T)-v(0,T))/v(0,T)$

for rare earth metals Gd, Tb, Dy and Ho in the paramagnetic region (Lüthi et al. [6.7]). See also the remarks on ultrasonic attenuation in a magnetic field (sound attenuation in the ordered phase) in the previous Sect. 6.1.2.

6.2 Spin Reorientation Phase Transition

An example of a spin reorientation transition is given because it constitutes one of the simplest examples of a phase transition in the solid state and because the strain–order parameter coupling effects have been studied in considerable detail for this case. Due to temperature-dependent anisotropy and magneto-elastic energies, the equilibrium spin configuration can also change as a function of temperature. Well-known examples of such spin reorientation phenomena are the spiral ferromagnetic spin structure changes in the heavy rare earth metals (Elliott [6.54]). Other examples are cubic magnetic crystals, where the easy axis of magnetization can change from, for instance, the (100) to (111) direction. Examples of this case are, e.g. magnetite Fe_3O_4 at 130K (see Sect. 7.1) and hematite Fe_2O_3 at 253K. In all these cases the spin reorientation occurs discontinuously at a given temperature, involving typically a first order orientational phase transition.

A particularly interesting spin reorientation phenomenon occurs in some rare earth orthoferrites (LnFeO$_3$, Ln = Er,Tm,Sm) (see Buchel'nikov et al. [6.55]) for a review. In these orthorhombic compounds, the competing Heisenberg and Dzyaloshinsky-Moryia exchange mechanisms produce a slight canting of the anti-ferromagnetically coupled sub-lattices. The resulting weak magnetic moment m can rotate from one symmetry direction to another under the influence of temperature dependent anisotropy energies. For example in ErFeO$_3$, m points along the orthorhombic a-axis for $T < T_1 = 87\,K$ and rotates continuously between $87\,K < T < 96\,K$ in the ac-plane. For $T > T_2 = 96\,K$, m points along the c-axis. This is illustrated in Fig. 6.5. This spin reorientation phenomenon can be described as a Landau-type second order phase transition (Horner and Varma [6.56], Levinson et al. [6.2], Hornreich and Shtrikman [6.57]), the marginal dimensionality being $d^* = 2.5$. The marginal dimensionality is defined so that for $d > d^*$ the system can be described by molecular field theory (see Sect. 4.4). The soft mode is the $\boldsymbol{k} = 0$ acoustic spin wave mode as shown in Fig. 6.5, which, because of its coupling to elastic modes, hybridises to a $\boldsymbol{k} = 0$ magneto-elastic wave, not shown in Fig. 6.5. The spin wave spectrum in Fig. 6.5 can be obtained from the free energy expression below (6.10). It was first calculated by Shane [6.58].

Figure 6.6 shows some elastic modes at the spin reorientation phase transitions for TmFeO$_3$, a compound with similar properties to ErFeO$_3$. It is clearly seen that we have a nice example of different strain–order parameter coupling (Gorodetsky and Lüthi [6.59]) as discussed in Sect. 4.3. For the longitudinal c_{33} mode we have a strain coupled to the square of the order parameter, for the c_{44} we have bilinear coupling and for the c_{55} we have no coupling at all.

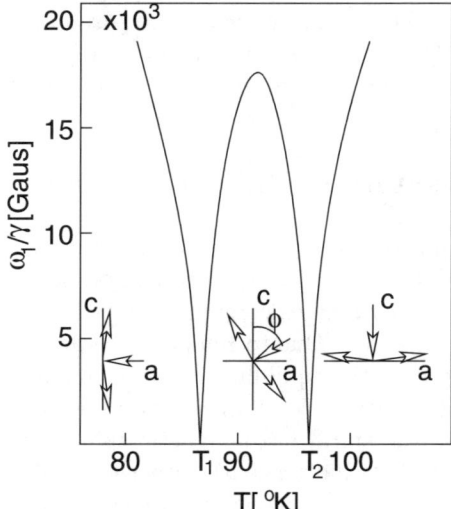

Fig. 6.5. Spin waves near reorientation phase transition. T_l and T_u correspond to T_1 and T_2 in the text

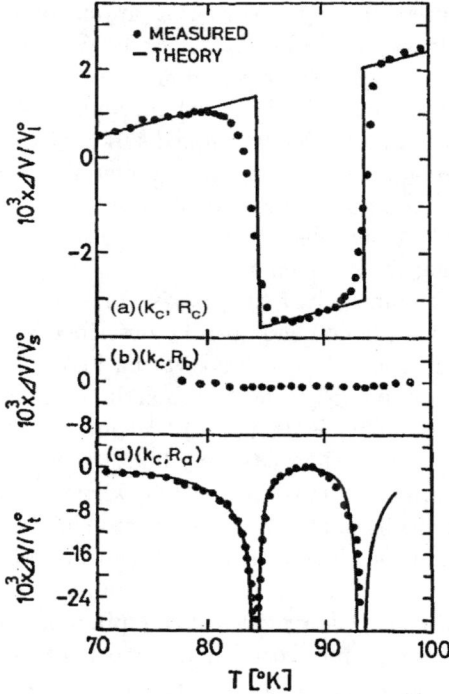

Fig. 6.6. Sound velocities of longitudinal and shear modes near spin reorientation of $TmFeO_3$ (adapted from Gorodetsky et al. [6.60])

6.2 Spin Reorientation Phase Transition

This behaviour can be described very simply with the following free energy density, with Θ being the angle between $\boldsymbol{m}$ and the c-axis in the $a-c$ plane:

$$\begin{aligned} F &= F_m + F_{me} + F_{el} \quad \text{with} \\ F_m &= \frac{1}{2}K_u \cos 2\Theta + K_b \cos 4\Theta - mB\cos(\Phi - \Theta) \\ F_{me} &= 2\varepsilon_{zz}(B_{33} - B_{31})\sin^2\Theta - \varepsilon_{xz}B_{55}\sin 2\Theta \,. \end{aligned} \tag{6.8}$$

Here F_m is the magnetic part, F_{me} the magneto-elastic part for sound propagation along the c-axis (with ε_{zz}, ε_{xz} and ε_{yz}) and F_{el} the elastic part of the free energy density. Φ denotes the angle between the magnetic field and the c-axis. From (6.8) and Fig. 6.5 obviously the order parameter can be taken as $\eta = \Theta$ for $T \sim T_1$ and $\eta = \frac{\pi}{2} - \Theta$ for $T \sim T_2$. From (6.8) we see that longitudinal waves (ε_{zz}) couple to the square of the order parameter whereas ε_{xz} couples linearly in the vicinity of $T_{1,2}$ and the other shear mode with ε_{yz} does not couple at all. The equations of motion with F_m give two magnon branches, an acoustical branch presented in Fig. 6.5 and an optical branch (Gorodetsky et al. [6.32]).

Likewise, the coupled magnon–phonon branches can be calculated using equations of motion for the two systems. Alternatively, the elastic constants or sound velocity changes can just be calculated from the free energy expressions

$$c_{ij} = \frac{\partial^2 F(\varepsilon_{ij}, \eta(\varepsilon_{ij}))}{\partial \varepsilon_{ij}^2} \quad \text{with} \quad \frac{\partial F(\varepsilon_{ij}, \eta)}{\partial \eta} = 0 \,. \tag{6.9}$$

As outlined in Sect. 4.3 the condition $\partial F/\partial\eta = 0$ gives the strain dependence of the soft mode coordinates. It is assumed that the order parameter can follow the strain dilation for the frequency range of the ultrasonic measurements therefore changing the elasticity of the crystal. The soft mode (acoustical branch) can be determined from the free energy (6.8).

$$\omega_m^2 = \frac{\partial^2 F}{\partial \Theta^2} = -2K_u \cos 2\Theta - 16 K_b \cos 4\Theta + mB \sin(\Phi - \Theta) \,. \tag{6.10}$$

This is represented in Fig. 6.5 and analogously for the elastic modes:

Longitudinal waves : $\quad \dfrac{\Delta v_L}{v_L} = -2\dfrac{(B_{33}-B_{31})^2 \sin^2 2\Theta}{\rho v_L^2 \omega_m^2}$

Shear waves for ε_{xz} mode : $\quad \dfrac{\Delta v_s}{v_s} = -\dfrac{B_{55}^2 \cos^2 2\Theta}{\rho v_s^2 \omega_m^2} \,.$ (6.11)

The full lines in Fig. 6.6 give a fit with (6.11) to the experimental results for TmFeO$_3$ for $B = 0$ (Gorodetsky et al. [6.60]). Magnetic field-dependent results are shown elsewhere (Gorodetsky et al. [6.32]). Note that the other c_{44} shear mode couples to the optical magnon branch and exhibits only very small effects (Gorodetsky et al. [6.59]). A thermodynamic description has been given for these magneto-acoustic effects which reflects the salient features of the experiments. With coupled equations of motion, essentially the same results are obtained (Gorodetsky and Lüthi [6.59]). A surface acoustic wave (SAW) study in ErFeO$_3$ gave essentially similar effects as discussed above (Gorodetsky and Shaft [6.61]). The differences are attributed to the effect of the surface boundary on the magnetic anisotropy.

Other rare earth orthoferrites that have been investigated acoustically are YbFeO$_3$ (Dan'shin et al. [6.62]) and HoFeO$_3$ (Dan'shin et al. [6.63]). A recent review on spin reorientation phenomena in orthoferrites, especially magneto-acoustics, has been given (Buchelnikov et al. [6.55]). Here the treatment of the magneto-elastic effects is presented in greater detail. A magneto-elastic gap in the spin wave spectrum, neglected in the treatment above, is shown to be small for the orthoferrites (see also Sect. 11.2.4). It is especially the influence of the rare earth sublattice that has been considered, however, and the low temperature phase transition at $\sim 4\,\mathrm{K}$ has been investigated in detail.

There are other cases of spin reorientation phase transitions. In magnetite Fe$_3$O$_4$, a spin reorientation from the (111) to the (100) direction in the cubic phase at 130 K is found (see also Sect. 7.1). A strong anomaly in the longitudinal c_{11} mode is observed which can be removed by application of a magnetic field (Moran and Lüthi [6.64], Schwenk et al. [6.65]). This is a first order transition. Another example is hexagonal Gadolinium which exhibits a spin reorientation at 220 K (Moran and Lüthi [6.38], Long et al. [6.66]). In this temperature range, the easy axis undergoes a continuous change from the c-axis to close to the basal plane (Graham [6.67]). Several spin reorientation transitions occur in Nd$_{2-x}$Ce$_x$CuO$_4$ for temperatures at 70 K and below (Henggeler et al. [6.68]). Strong magneto-elastic effects occur for the c_{66} mode and longitudinal modes (Zherlitsyn et al. [6.69] and [6.70]).

6.3 Sound Propagation in the Spin-Density Wave Anti-ferromagnet Chromium

Chromium (Cr) constitutes probably the best studied spin-density-wave antiferromagnet (SDW). This substance is discussed here because the magnetic ground state is more complicated than, for instance, simple ferromagnets or uniaxial and cubic anti-ferromagnets treated in Sect. 6.1 and because some interesting ultrasonic studies have been performed in Cr. Here the itinerant d-electrons form longitudinal and transverse SDW states, with a longitudinal one for $T < T_{SF} = 123\,K$ and a transverse one in the temperature region $T_{SF} < T < T_N = 311.4\,K$. A comprehensive review of the physical properties of Cr can be found in Fawcett [6.71].

The key to perform unique and reproducible ultrasonic studies in this crystal is to produce a single magnetic domain state. The magnetic structure consists of a SDW with a single $\boldsymbol{Q} = 2\pi/a(1 \pm \delta)$ along a cubic axis with the polarization vector $\boldsymbol{S}$ either parallel to $\boldsymbol{Q}$ (longitudinal SDW) or perpendic-

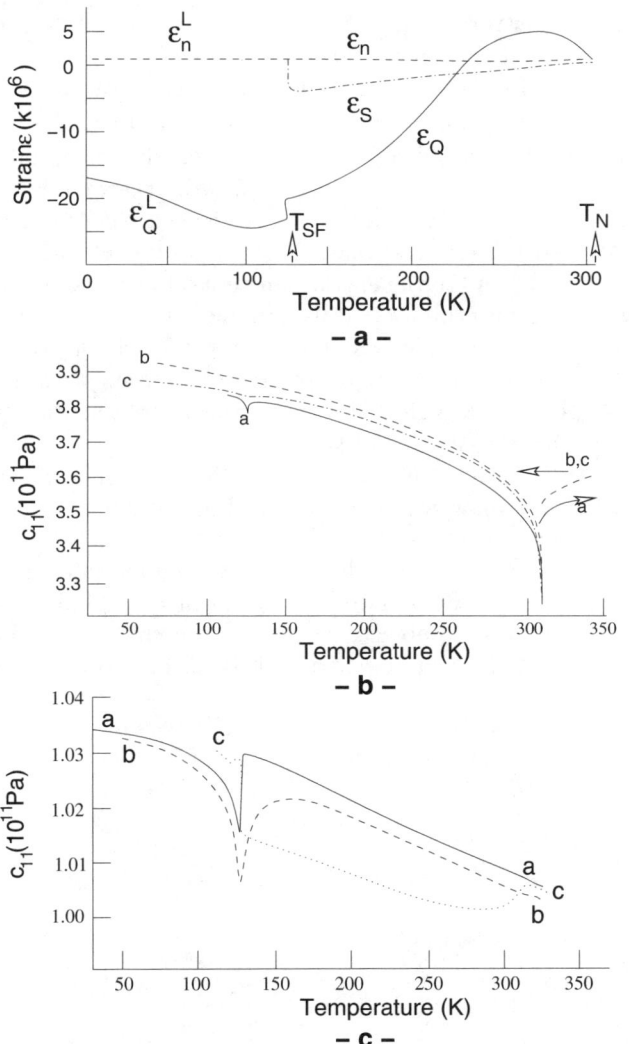

Fig. 6.7. Thermal expansion and elastic constants for Cr. (a) Thermal expansion of single-S single-Q Cr (after Steinitz et al. [6.73]), for details see text. (b) Temperature dependence of longitudinal elastic moduli a) c_{11}, b) c_{22}, c) c_{33} for the transverse SDW phase single-S single-Q Cr (after Muir et al. [6.74]), (c) temperature dependence of the shear elastic moduli a) c_{44}, b) c_{55}, c) c_{66} (after Muir et al. [6.74])

ular to Q (transverse SDW). δ and therefore $|Q|$ depend on temperature, from about 50 K ($\delta = 0.048$) to 320 K ($\delta = 0.037$). Single domain state can be obtained with field cooling and with tensile stresses. For example, cooling through T_N with a tensile stress along [100] results in a single Q_x domain, or application of a strong field $B \parallel [001]$ through T_{SF} of a single Q_z sample results in an irreversible change to a state containing Q_x and Q_y domains.

Typical results for thermal expansion and elastic constants are shown in Fig. 6.7. The strain ε_Q means the strain measured along $\boldsymbol{Q}$ with ε_Q^L for the longitudinal SDW phase. ε_n is taken along the field direction and ε_n^L along the field perpendicular to $\boldsymbol{Q}$. At low temperature, the thermal expansion coefficient $\alpha = \mathrm{d}\varepsilon_Q/\mathrm{d}T$ is negative leading to a large negative electronic Grüneisen parameter $\Omega \approx -10$. For a discussion of the electronic Grüneisen parameter, see Sects. 4.5 and 9.3. As to the elastic constants, Fig. 6.7 shows that the longitudinal waves exhibit typical critical anomalies at T_N, as discussed before in Sect. 6.1.3 for anti-ferromagnetic and ferromagnetic substances. On the other hand, transverse sound waves exhibit pronounced effects at the spin flip transition T_{SF}. A phenomenological theory for various physical properties of Cr close to T_N has been given by Walker [6.72].

There are no critical sound attenuation data for Cr in a single domain state. For a multi-domain state the attenuation near T_N is very large indicating strong domain wall - stress effects.

Apart from Cr, there are spin density wave ground states in organic compounds. For a review of these materials, see Grüner [6.75]. Since there are no sound wave experiments reported on these compounds, with one exception (Zherlitsyn et al. [6.76]), this topic will not be discussed any further.

7 Ultrasonics at Structural Transitions

There is hardly any phase transition where ultrasound is not affected in the form of velocity or attenuation anomalies. Magnetic, superconducting and of course structural transitions are detectable with ultrasound. Magnetic phase transitions have been described in the previous chapter. Here we give a survey of structural transitions. Charge order, orbital order (amongst them cooperative Jahn–Teller transitions) and phonon-induced structural transitions are treated. The case of superconductivity is considered in a separate chapter (Chap. 10). Examples of iso-structural phase transitions are given in Chap. 9.

A distinction can be made between displacive and order–disorder phase transitions (Papon et al. [7.1], Stanley [7.2]). In displacive transitions, atoms of the crystalline lattice move into new equilibrium positions. This is a collective effect and manifests itself usually through the existence of a soft mode. Order–disorder phenomena occur e.g. through ordering of binary alloys or through ordering of magnetic spin systems. Here the entropy-changes are noticeable and one can speak of an entropy driven phase transition.

A simple model illustrates the important differences between these two types of phase transitions (Blinc and Zeks [7.3]). If a single particle anharmonic potential with a local coordinate $Q: V = aQ^2 + bQ^4$ is considered, in the displacive type, $a = M\Omega_0^2$ with Ω_0 an effective frequency and $b > 0$. The interactions between different cells $\sum_{ij} v_{ij} Q_i Q_j$ can lead to a destabilization of the equilibrium position $Q_0 = 0$ with a soft mode $M\omega_q^2 \to 0$ at the phase transition. On the other hand, in the case of order–disorder transitions for $a < 0$ results in a double well potential with the equilibrium coordinate $Q_0 = -a/2b$ and the energy difference between maximum and minima $\Delta E = -a^2/4b$. For $\Delta E >> kT$ thermal fluctuations cannot overcome the potential barrier. The order–disorder transition has a transition entropy of $\sim k\ln 2$ per particle. If $\Delta E \ll kT$, thermal fluctuations can overcome the potential barrier which is again in the limit of the displacive phase transition region.

The Landau theory, with the strain–order parameter coupling as discussed in Sect. 4.3.4, is mainly used for the description of elastic constants. Depending on the type of coupling, the temperature dependence of the elastic constants can be described generally with (4.12) which results in (4.13) or (4.14). If the order parameter cannot follow the strain wave freely, dispersion and attenuation occur this being discussed in the framework of the Landau–Khalatnikov theory in Sect. 4.3.6. Applications of these effects have been given in Chap. 6 and will be given in the present Sects. 7.2.2 and 7.4.

Order parameter fluctuations often have to be considered which means that elastic constants and attenuation have to be evaluated with the help of (4.16). In the case of fluctuations, no generally accepted formula exists and specific models, as shown, e.g. in the present Sect. 7.4.3 and in Sect. 10.4, must be used. Usually one fits the elastic constant phenomenologically as $c_{ij} = c_{ij}^0 - \frac{A}{(|T-T_a|)^n}$. Examples are given in the present Sect. 7.4.3 and in Chap. 10.4.

Charge order transitions (Sect. 7.1) are discussed first below, followed by cooperative Jahn–Teller and quadrupolar phase transitions – commonly called orbital transitions (Sect. 7.2). These latter transitions include compounds with transition metal ions, with rare earth ions and with actinide ions. Further subjects in this chapter are ferro-elastic and martensitic transitions (Sect. 7.3) and other phonon driven phase transitions (Sect. 7.4).

7.1 Charge Order

Here, recent developments in the field of charge ordering for compounds containing $3d$ and $4f$ ions are discussed. Magnetic ions, with partly filled inner d- or f-shells, have orbital- and spin-magnetic moments. In addition, the charge must be taken into consideration. Charge ordering phenomena can be similar to the spin or orbital ordering. This means that – at high temperatures – the magnetic ions are in a mixed valence state with fluctuating charges. This fluctuation process can be arrested at a transition temperature where charge order sets in.

To specify this process, distinction is made between homogeneous mixed valence compounds and inhomogeneous ones. The former case deals essentially with a single ion property where the magnetic ion hybridises with the sea of conduction electrons and there is an exchange of an inner electron with the conduction band at the Fermi level. This topic will be discussed in Chap. 9. In the case of inhomogeneously mixed valency, the electrons can hop between the magnetic ions and, below a transition temperature, there is a distinct charge ordering. Well-known cases of such charge fluctuation phenomena are magnetite (Fe_3O_4) with Fe^{2+}–Fe^{3+} ions involved, $La_{1-x}Ca_xMnO_3$, Yb_4As_3, NaV_2O_5 and various Th_3P_4-structure materials.

It will be shown below that the strain–order parameter coupling in this case is pronounced and gives rise to strong softening effects of some symmetry elastic constants. The discussion of these phenomena relies heavily on a recent review article on "Charge fluctuation, Charge Ordering and Elastic Constants" (Goto and Lüthi [7.4]). As a description of the coupling of the symmetry strain and the charge order parameter we can take the Landau description as presented in Sect. 4.3. Because the Coulomb forces are

Table 7.1. Charge ordering of inhomogeneously mixed valence compounds

	Valence state	T_a	soft mode	charge fluctuation symmetry	mode symmetry change
Yb_4As_3	Yb^{2+} Yb^{3+}	292 K	c_{44}	T_{2g}	$T_d^6 - C_{3v}^6$
Fe_3O_4	Fe^{2+} Fe^{3+}	124 K	c_{44}	T_{2g}	O_h^7 - monocl.
NaV_2O_5	V^{4+} V^{5+}	34 K	c_{66}	B_{1g}	D_{2h}^{13}- monocl.
Eu_3S_4	Eu^{2+} Eu^{3+}	260 K	$c_{11} - c_{12}$	E_g	$T_d - D_4$

long ranged, the Landau theory should give an adequate description of these ordering phenomena (see the Ginzburg criterion in Sect. 4.4).

The change of the charge density across the charge ordering transition at T_c is described by the two terms $\rho = \rho_0 + \Delta\rho$. Here, the first term ρ_0 is an invariant part across the phase transition, i.e. ρ_0 has the total symmetric character A_{1g} in both space groups of G_0 (phase above T_c) and G_1 (ordered phase below). Usually G_1 is a subgroup of G_0 for a second order transition. The second term $\Delta\rho$ is the symmetry breaking part across the phase transition. It can be expanded in terms of the charge fluctuation mode $\rho_{\Gamma\gamma}$:

$$\Delta\rho = \sum Q_{\Gamma\gamma} \rho_{\Gamma\gamma} . \tag{7.1}$$

Here, $Q_{\Gamma\gamma}$ is the order parameter for the charge ordering in quenching of the charge fluctuation mode $\rho_{\Gamma\gamma}$. The order parameter changes from 0 above T_c to $\neq 0$ below T_c. The elastic constant with the same symmetry as the order parameter should exhibit strong softening in the high temperature phase on approaching T_c. This softening should be describable with formula 4.13. This is indeed the case for many substances. Table 7.1 gives all the substances investigated so far with relevant properties. Application of the projection operator technique gives the expected charge distribution below T_c (Goto and Lüthi [7.4]). The basis functions of the charge fluctuation modes can be projected with this group theoretical method.

Some typical examples for elastic constant effects near charge ordering transitions are given in below. As a first example, the case of Yb_4As_3 is treated (Goto et al. [7.5]).

Yb_4As_3

Above room temperature, Yb_4As_3 crystallizes in the cubic anti-Th_3P_4 structure. In the cubic phase, the Yb ions are in a mixed valence state with the ratio $Yb^{3+} : Yb^{2+} = 1:3$. At $T_c = 292$ K, Yb_4As_3 shows a charge ordering with a structural transition from the cubic to a trigonal structure (Ochiai et al. [7.6]). The angles of this rhombohedral structure are $\alpha = \beta = \gamma = 90.68°$. In the trigonal phase, Yb^{3+} ions order along one of the four diagonal chains, parallel to the (111)-direction (Kohgi et al. [7.7]). This substance has inter-

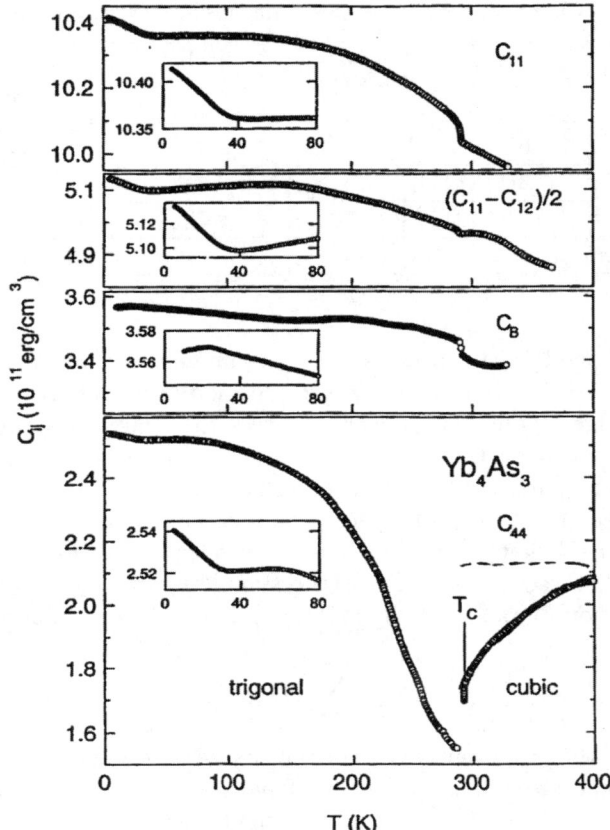

Fig. 7.1. Elastic constants of Yb$_4$As$_3$ as a function of temperature (Goto et al. [7.5])

esting low temperature properties, resembling heavy fermion compounds (for a review see Schmidt et al. [7.8]).

The emphasis here is on the charge order phase transition at room temperature. Elastic constant measurements have been performed by Goto et al. [7.5]. These results are shown in Fig. 7.1. It is seen that the c_{44} mode exhibits strong softening from 400 K to T_c. The other elastic modes exhibit distinct step functions at T_c but no anomalous temperature dependence for $T > T_c$. The temperature dependence of the c_{44} mode for $T > T_c$ follows closely (4.13) with $T_c = 247$ K (considerably smaller than $T_a = 292$ K, in Table 7.1) and $T_0 = 234$ K. From this and the rhombohedral angles follows for the spontaneous strains at T_c: $\varepsilon_{yz} = \varepsilon_{zx} = \varepsilon_{xy} = 6.46 \times 10^{-3}$. This cubic–trigonal transition is slightly of first order. The corresponding third order invariant of the order parameter reads (Sect. 4.3.3) $\delta Q_{yz} Q_{zx} Q_{xy}$. This gives a spontaneous strain directly at T_c.

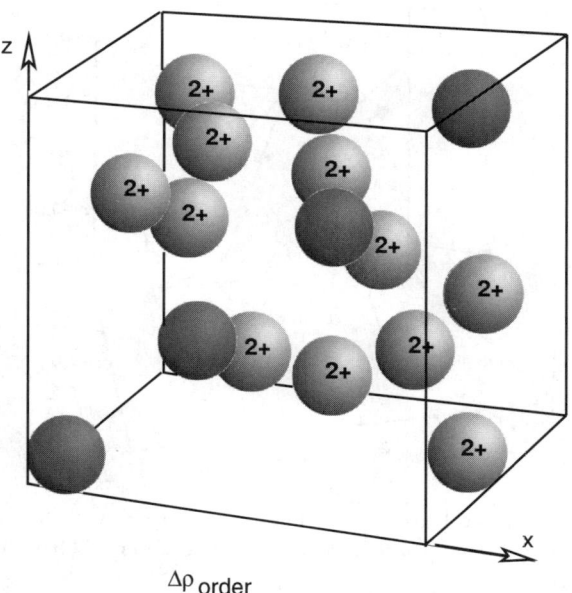

Fig. 7.2. Linear chain of Yb^{3+} ions below T_c. The dark spheres are Yb^{3+} and the labeled spheres Yb^{2+} ions. The Yb^{3+} ions align along the body diagonal (111)–direction (from Goto et al. [7.5])

A bilinear coupling of the charge fluctuation modes Q_{ij} is obtained with the strains ε_{ij} of the form (analogously to (4.12)) $F_{int} = g(Q_{yz}\varepsilon_{yz} + Q_{zx}\varepsilon_{zx} + Q_{xy}\varepsilon_{xy})$. The simultaneous condensation of the charge fluctuation mode $Q_{yz} = Q_{zx} = Q_{xy} \neq 0$ of the T_2 triplet gives rise to the linear chain of the Yb^{3+} ions along the (111)–direction as shown in Fig. 7.2.

Since the two ions Yb^{2+} and Yb^{3+} are non-magnetic (Yb^{2+}: $f^{14}, J = 0, L = 0, S = 0$) and magnetic ($Yb^{3+}$: $f^{13}, J = \frac{7}{2}, L = 3, S = \frac{1}{2}$) respectively, inelastic neutron scattering on the spin $\frac{1}{2}$ chain gave typical one-dimensional features of the magnetic spectrum (Kohgi et al. [7.7]). For more details, see Schmidt et al. [7.8].

Fe_3O_4

As a next example, magnetite is considered which undergoes a metal-insulator transition at $T_V = 124$ K (Verwey [7.9]). The nature of this transition was suggested to be connected to a charge ordering (Verwey [7.9]) although the details of this mechanism and of the low temperature structure are not yet settled, even today (Zuo et al. [7.10]). Magnetite undergoes a paramagnetic–ferrimagnetic transition at a much higher temperature of $T_c = 858$ K. The charge ordering at T_V is due to the $Fe^{2+}(S = 2)$ - $Fe^{3+}(S = 5/2)$ ordering on octahedral sites of the cubic inverse spinel structure (Chikazumi [7.11]). The magnetic formula can be written as $Fe_3O_4 = Fe^{3+}_{tet}[Fe^{2+}Fe^{3+}]_{oct}O_4$. In this

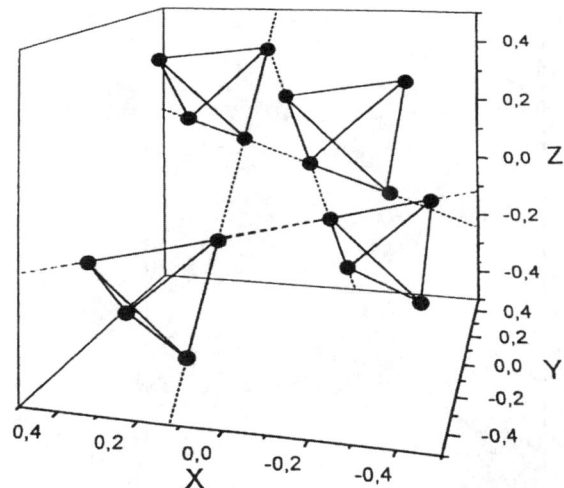

Fig. 7.3. Fe-ions on octahedral sites. They form four tetrahedra in the unit cell. Dotted lines connect the tetrahedra along the different (110) directions (from Schwenk et al. [7.12])

structure, the octahedral B-sites form a network of four separated, tetragons within a unit cell (Fig. 7.3). The four tetrahedrons form therefore a separate unit. The simplest view of the charge ordering on these tetrahedrons would be the one originally proposed by Verwey: Fe^{2+} (Fe^{3+}) cations alone occupy alternating (110) chains, as indicated by the dotted lines of Fig. 7.3.

For the elastic constants there is one mode, c_{44}, which becomes soft on approaching T_V from above. Figure 7.4 shows this mode for Fe_3O_4 and $Fe_{3-x}Zn_xO_4$ as a function of temperature (Schwenk et al. [7.12], Moran et al. [7.13]). Note that the spin reorientation transition at 130 K for Fe_3O_4 was suppressed with a magnetic field of 0.4T (see also Sect. 6.2). (4.13) gives a perfect fit to the experimental results for the three compounds with $x = 0, 0.02$ and 0.032, indicating that c_{44} is the soft mode. The fit parameters, T_c and T_0, result in $T_c = 66$ K and $T_0 = 56$ K, substantially lower than $T_V = 124$ K ($x = 0$), $T_V = 94$ K ($x = 0.02$), $T_V = 83$ K (x=0.32) due to the first order phase transition. The resulting charge distribution is like the one discussed above. If it is assumed that the low temperature phase is the orthorhombic subgroup C_{2v} of T_d, we would have a realisation of the order parameter $Q_{xy} \neq Q_{yz} = Q_{zx} = 0$ leading to the charge ordering discussed above. This is different from the case of Yb_4As_3 shown above. As pointed out above, however, the Verwey transition is more complicated. It leads from the cubic to a mono-clinic structure, the details of which are not yet known (Zuo et al. [7.10]). More recently, resonant X-ray absorption spectroscopy at the K-band did not find any charge ordering (Garcia et al. [7.14]) whereas high resolution X-ray and neutron powder diffraction data give a charge ordering,

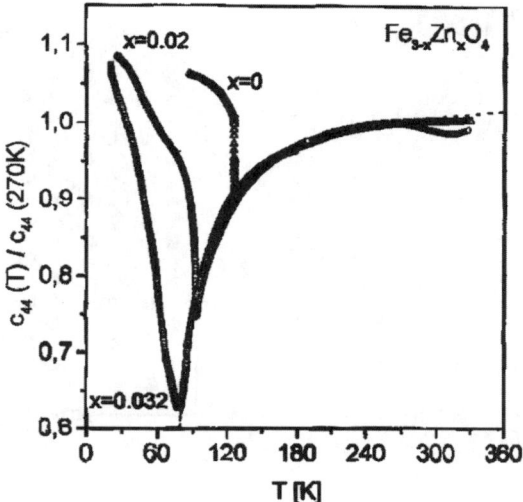

Fig. 7.4. Elastic c_{44} mode for $Fe_{3-x}Zn_xO_4$ for $x = 0, 0.02, 0.032$. The ratio of $c_{44}(T)$ and $c_{44}(T = 270\,K)$ is plotted versus the temperature T. The dotted line is a fit with (4.13) (from Schwenk et al. [7.12])

but different from the Verwey ordering (Wright et al. [7.15]). All these different contradictory investigations have hindered the solution of the problem of the strongly first order Verwey transition. More details on this charge order transition can be found in Goto and Lüthi [7.4].

NaV$_2$O$_5$

As a further example, the low dimensional spin system NaV$_2$O$_5$ is discussed. This compound exhibits a phase transition at 34 K. Above the transition temperature the magnetic susceptibility can be described with a spin $\frac{1}{2}$ chain model (Isobe and Ueda [7.16]). The space group at room temperature is the centro-symmetric space group Pmmn (D_{13}^{2h}) with only one Vanadium position in the unit cell (v. Schnering et al. [7.17], Smolinski et al. [7.18]). The average valence of the V ion is 4.5. Below the transition, NMR measurements clearly show the appearance of two different Vanadium positions, compared to one position above (Ohama et al. [7.19] and [7.20]). The positions are occupied by different charged V ions. This gives a charge ordering of V^{4+} and V^{5+} ions and leads also to a dimerization of the spin system with a non-magnetic ground state and an excited triplet state for V^{4+} as shown by ESR experiments (Vasilev et al. [7.21], Schmidt et al. [7.22], Hemberger et al. [7.23], Luther et al. [7.24]). The charge ordering is in the a-b plane and several low temperature X-ray measurements give, on the one hand, a charge ordering with $V^{4+}, V^{4.5+}, V^{5+}$ as shown in Fig. 7.5c (Lüdecke et al. [7.25] and de Boer et al. [7.26]) with a space group Fmm2, while on the other hand, there are reports indicating a complete charge order involving only V^{4+} and V^{5+}

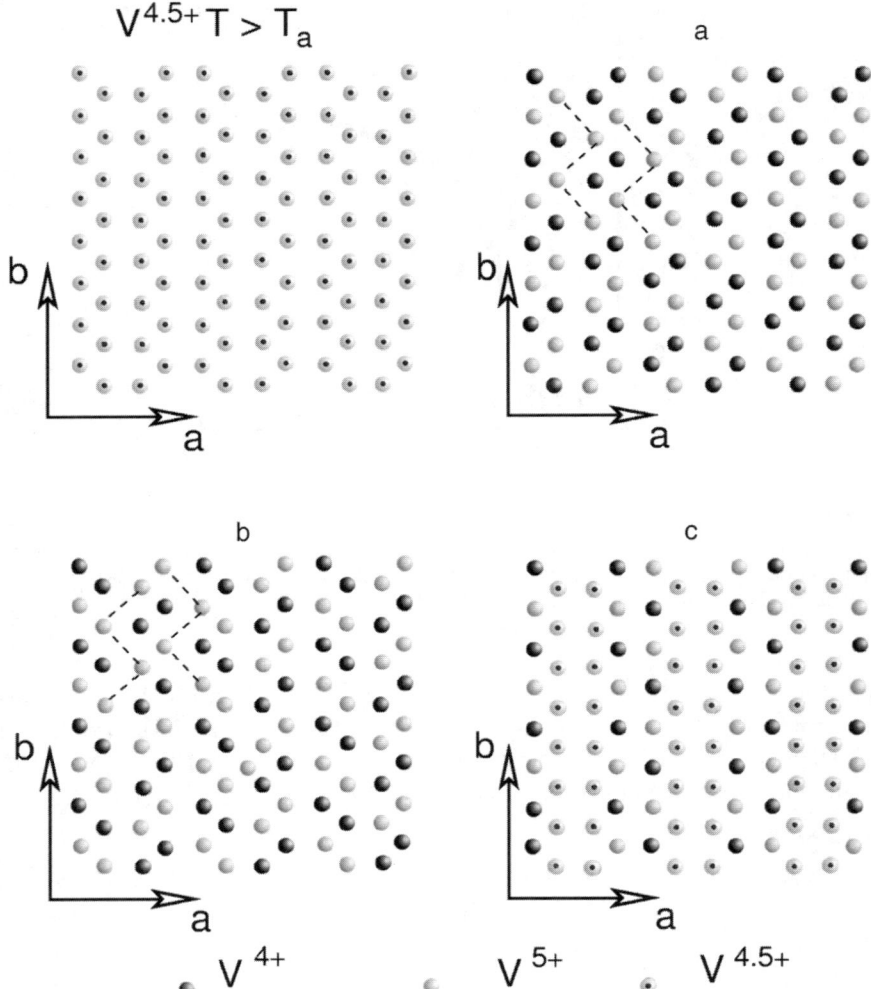

Fig. 7.5. Charge ordering pattern exhibiting various zig-zag structures (adapted from Riera and Poilblanc [7.30])

as shown in Fig. 7.5a,b (Nakao et al. [7.27], Sawa et al. [7.28]). In addition this latter group finds also a structural transition from the orthorhombic to a monoclinic phase below T_a. The thermodynamic properties of NaV_2O_5 have been discussed by Johnston et al. [7.29].

In this compound, a strong softening of the c_{66} mode from 80 K down to T_c of about 12 % can be observed (see Fig. 7.6) (Schwenk et al. [7.31]). This is a transverse mode propagating along the b-axis with polarization along the a-axis or vice versa. Clearly it has the characteristic of a soft mode.

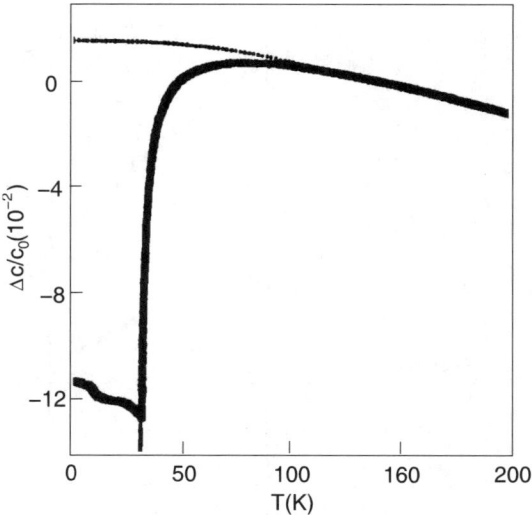

Fig. 7.6. Temperature dependence of c_{66} mode in NaV$_2$O$_5$ (Schwenk et al. [7.31]). $\frac{\Delta c}{c_0} = \frac{c(T)-c(200\,K)}{c(200\text{K})}$) is plotted versus the temperature. Dotted line is background fit, full line is calculated fit (see text)

The dotted line is the background c_0 calculated with the Varshny model (see Sect. 4.2). The full line gives a fit to (4.13) which again is quite perfect. We get for $T_c = 30.62$ K and for $T_0 = 30.34$ K. The strain–order parameter coupling expressed by $T_c - T_0 = 0.28$ K is very small. This is evident from the experimental results where the softening starts only for $T < 75$ K. Such a feature was discussed in Sect. 4.3.4 and Fig. 4.4. This soft mode has B_{1g} symmetry. While there is, at present, still some controversy about the detailed interpretation of the NMR and X-ray results in the structure of Fig. 7.5, our prediction also gives a zig-zag structure. It should be stressed however, that there is a cell doubling in the $a - b$ plane at the transition. This prevents a bilinear coupling of the ε_{xy} strain to the charge order parameter as in the case of Yb$_4$As$_3$ and Fe$_3$O$_4$. As pointed out above, there is, in addition to the charge ordering, also evidence for a structural transition from orthorhombic to monoclinic (Sawa et al. [7.28]). Such a structural transition with a slightly different transition temperature was observed before with thermal expansion experiments (Koeppen et al. [7.32]) and elastic constants (Schwenk et al. [7.31]). The low temperature structural determination (Sawa et al. [7.28]), with only two V-ions V^{4+}, V^{5+} and a monoclinic structure, could be the clue for the softening of the c_{66} mode. The monoclinic structure has $a \neq b \neq c$ and $\gamma \neq \pi/2$. The c_{66} softening with strain ε_{xy} can induce precisely such a structure (Lüthi et al. [7.33]).

Of interest are also other acoustic modes in this compound and a comparison to the related CuGeO$_3$, a so-called Spin-Peierls compound (see the

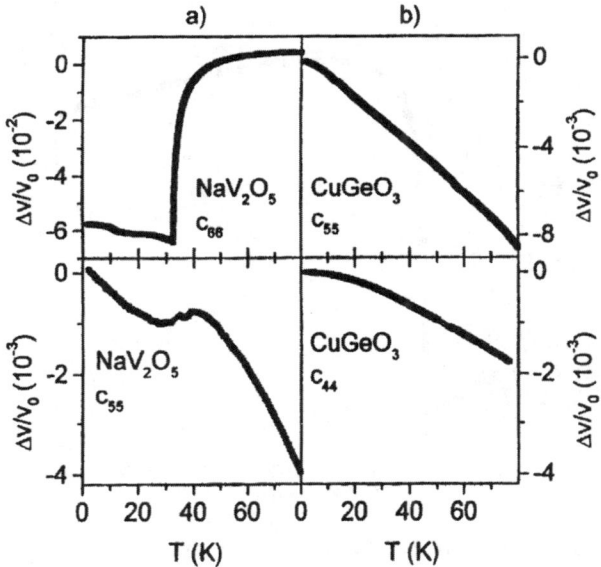

Fig. 7.7. Temperature dependence of the shear sound velocities for NaV$_2$O$_5$ and CuGeO$_3$ (see text) (Schwenk et al. [7.31])

short remarks in Sect. 12.5). Figure 7.7 shows the elastic modes c_{66} and c_{55} from NaV$_2$O$_5$ on the one hand and c_{55} and c_{44} for CuGeO$_3$ on the other hand (Schwenk et al. [7.31]). The chain axis for NaV$_2$O$_5$ is the crystallographic b-axis, while for CuGeO$_3$ the spins form chains along the c-axis. Therefore the relevant transverse modes for NaV$_2$O$_5$ are the $c_{66}(\varepsilon_{xy}$ strain) and $c_{55}(\varepsilon_{yz}$ strain) whereas for CuGeO$_3$ the interesting modes are the $c_{44}(\varepsilon_{xz}$ strain) and $c_{55}(\varepsilon_{yz}$ strain). From Fig. 7.7, we notice that only the c_{66} mode for NaV$_2$O$_5$ has soft mode character whereas the other modes have a normal temperature dependence (apart from c_{55} for NaV$_2$O$_5$ with a small anomaly of 0.1% for $T < T_c$). The anomalies of the two modes for CuGeO$_3$ at T_c are only of the order of 10^{-4}. More details on elastic constants for the spin-Peierls compound CuGeO$_3$ can be found in Sect. 12.5.

Finally, longitudinal waves exhibit a large fluctuation region of about 20 K above T_c (Schwenk et al. [7.31], Fertey et al. [7.34]). This is a rather similar effect to that seen in the specific heat (Schnelle et al. [7.35]). A fluctuation model, as discussed in Sect. 4.3 and 7.4.3, might explain these results. Small anomalies in the c_{33} mode in the vicinity of T_c could give indications of a structural and charge ordering transition.

La$_{1-x}$Sr$_x$MnO$_3$

The manganese oxides R_{1-x}Sr$_x$MnO$_3$ with R = La, Pr, Nd exhibit field-induced metal-insulator transitions and a colossal magneto-resistance CMR (Tokura et al. [7.36]). These materials possess the cubic perowskite structure.

The electron transfer between the Mn^{3+} and Mn^{4+} ions via the oxygen 2p orbitals leads to the so-called double exchange (de Gennes [7.37]). Apart from the possible charge ordered state, there is also orbital ordering due to the orbitally degenerate states of Mn^{3+}. The coupling of the lattice degrees of freedom with the orbital and spin degrees of freedom leads to various phases like the cooperative Jahn–Teller effect or charge ordering (Millis et al. [7.38]). For example: in $La_{1-x}Sr_xMnO_3$, with $x = 0.12$, a structural transition at $T_a = 290\,K$ with a soft $(c_{11} - c_{12})/2$ mode and a charge order transition at $T_{co} = 145\,K$ with soft $(c_{11} - c_{12})/2$ and c_{44} modes is found. This elastic mode softening can be well described again with (4.13). Further details can be found in Hazama et al. [7.39] and [7.40]), Goto and Lüthi [7.4].

Another case, the so-called "telephone" compound $Sr_{14}Cu_{24}O_{41}$, will be discussed in Sect. 12.4. This substance has many interesting properties amongst them also a charge ordering phenomenon. Other charge ordering phenomena can be found in the review article by Goto and Lüthi [7.4].

Charge Glass State
There are systems which do not exhibit a charge order transition, but rather they change gradually into a charge glass state. Prominent examples are the samarium chalcogenides Sm_3X_4 with $X = S$, Se, Te. These are cubic Th_3P_4 structure materials with the space group symmetry T_d^6. Instead of charge order, these substances find a random charge distribution, a charge glass. Ultrasonic dispersion effects give relaxation times which increase with decreasing temperature and lead the system into the charge glass state (Tamaki et al. [7.41]). Instead of transforming into a charge order state, as expected from the critical slowing down effect, the fluctuations of the 4f electrons in Sm_3X_4 freeze gradually into a charge glass state. This is characterized by a random distribution of Sm^{2+} and Sm^{3+} ions in space. A somewhat similar effect occurs in $Sr_{12}Ca_2Cu_{24}O_{41}$ at around 180 K (Schwenk [7.42]). Further details can be found in the review by Goto and Lüthi [7.4].

7.2 Cooperative Jahn–Teller Effect and Quadrupolar (Orbital) Transition

The Jahn–Teller theorem states that a symmetry degenerate electronic system of a molecule lowers its energy by lifting the degeneracy under the distortion of the molecular complex (Kaplan and Vekhter [7.43]). A magneto-elastic interaction like in (5.7) and (5.8) gives an energy lowering of the split term as $E = -gQ$, where Q is a local distortion (instead of the strain ε). In addition the elastic energy increases with distortion like $E_{el} = kQ^2/2$, with k a corresponding force constant. Therefore the total energy $E_{tot} = kQ^2/2 - gQ = \frac{k}{2}(Q - \frac{g}{k})^2 - \frac{g^2}{2k}$ shows a total energy reduction for $Q_0 = g/k$ of $\Delta E = -g^2/2k$ compared to the $Q = 0$ state. This is the essence of the static Jahn–Teller

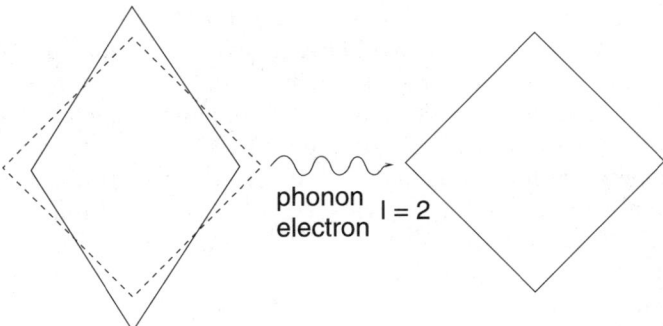

Fig. 7.8. Virtual exchange of a particle leading to a quadrupolar transition, symbolically shown for 2 octaeders

effect. Similar arguments can be given using the so-called internal strains (see Sect. 5.4)

Magnetic ions, in high enough concentration in a crystalline solid, with degenerate ground states can have interactions which lead to a structural transition. If the interaction is of magneto-elastic origin as in (5.8): $H_{me} = \sum_i g_\Gamma \varepsilon_\Gamma O_{\Gamma i}$ i.e. a strain coupling to the quadrupole moment of the magnetic ion, one speaks of a cooperative Jahn–Teller effect. On the other hand, if there is a bilinear quadrupolar-quadrupolar coupling (5.18): $H_Q = \sum_{ij} K_{ij} O_{\Gamma i} O_{\Gamma j}$ due to, e.g. phonon exchange (Orbach and Tachiki [7.44]) or conduction electron exchange (Fulde and Peschel [7.45]), one speaks of quadrupolar transitions (Levy et al. [7.46]). Figure 7.8 illustrates this exchange process. A spontaneous deformation of one molecular complex is mediated by a virtual exchange of a phonon or an electron with corresponding angular momentum. This exchange deforms the other molecule. In the following, we give a number of examples for such phase transitions for transition metal compounds, rare earth systems and actinide compounds. We emphasize of course the effects on elastic constants. There exist several complete reviews on this topic (Gehring and Gehring [7.47], Melcher [7.48], Thomas [7.49], Kugel and Khomskii [7.50], Morin and Schmitt [7.51], Kaplan and Vekhter [7.43]).

The case of a direct magneto-elastic coupling of the Jahn–Teller ion with the strain has just been discussed. But in cases where there are internal strains, these can couple more directly to the magnetic ion. Such a case is discussed in Sect. 5.4 where it was shown that symmetry elastic constants result in a renormalized coupling constant. The effect of internal strains for Jahn–Teller systems is important for the study of vibronic modes, as emphasised by Thomas [7.49]. The description with magneto-elastic and quadrupole-quadrupole interaction, as introduced above and in Sect. 5.3, is sufficient for crystals with simple structures (NaCl, CsCl, perovskite, spinel). Here the strain gives directly the deformation of the unit cell. For more complicated structures, the internal strains give the dominant coupling to the magnetic

7.2 Cooperative Jahn–Teller Effect and Quadrupolar (Orbital) Transition

ion and the coupling constants determined from equations like (5.21) or (4.13) would be renormalized ones.

7.2.1 Case of Transition Metal Compounds

The first evidence of cooperative Jahn–Teller effects were found in transition metal compounds with spinel structure (Dunitz and Orgel [7.52]). In addition, perovskite materials and MnF_3 and $KCuF_3$ structure compounds have been found which also exhibit such transitions. References to older work are given in Goodenough [7.53]. Theories for elastic constants have been developed by Pytte [7.54] and Kataoka and Kanamori [7.55]. Acoustic studies for transition metal compounds have been performed for $NiCr_2O_4$, $CsCuCl_3$ and Fe_2TiO_4. These we will survey here. In these three examples, negligible quadrupole–quadrupole interaction occurs, only the single ion magneto-elastic interaction (5.7) is important.

$NiCr_2O_4$
This substance forms a normal spinel lattice with the magnetic ion Ni^{2+} located at the tetrahedral site. The $Ni^{2+}(3d^8)$ has a $^3F_4$ ground state configuration with a CEF triplet Γ_4 ground state in this tetrahedral coordination (see Fig. 5.1) in contrast to a Γ_2 ground state for octahedral coordination. $NiCr_2O_4$ undergoes a first order cubic-tetragonal transition at $T_a = 299\,\text{K}$ and a Néel type anti-ferromagnetic transition at $T_N = 72\,\text{K}$ as shown in Fig. 7.9a,b (Kino et al. [7.56], [7.57], [7.60] and Kino and Miyahara [7.59]).

In Fig. 7.9a, the axial ratio $c/a - 1$ is plotted versus the temperature. At T_a there is a step in this order parameter. This is a consequence of the triplet state Γ_4 being a ground state. For a doublet state, the transition could be second order (Kataoka and Kanamori [7.55]). Also plotted in Fig. 7.9a is the sound velocity of the elastic mode $(c_{11} - c_{12})/2$. Due to the first order transition, the softening for $T > T_a$ is arrested at T_a. In the absence of CEF effects, the single ion strain susceptibility has a $1/T$ dependence (5.16c) and the total strain susceptibility $\chi_{str} \propto (T-\Theta)^{-1}$ from (5.21). Therefore we get for the symmetry elastic constant the identical expression to (4.13) (see also Sect. 5.3).

$$c_\Gamma = c_\Gamma^0 \frac{T - T_c}{T - \Theta}. \qquad (7.2)$$

Here, Θ is the transition temperature due to the two-ion (quadrupolar) interactions and $T_c = g^2 n/c^0 + \Theta$ is the structural transition temperature if the transition were second order. For $NiCr_2O_4$ we get $T_c = 274 \pm 5\,\text{K}$, $T_a = 299\,\text{K}$. For the Ni^{2+} triplet system, it is expected that the order parameter $c/a - 1$ will be diminished to 50% at T_a for the first order phase transition – in good agreement for $NiCr_2O_4$ (Fig. 7.9a). In addition, the calculated ratio $T_a/T_c = 1.082$ is very close to the experimental one of 1.09 (Kino et al. [7.56]). The analysis of the temperature dependence of the soft mode $c_{11} - c_{12}$ shows that the coupling responsible for the cooperative Jahn–Teller transition is of

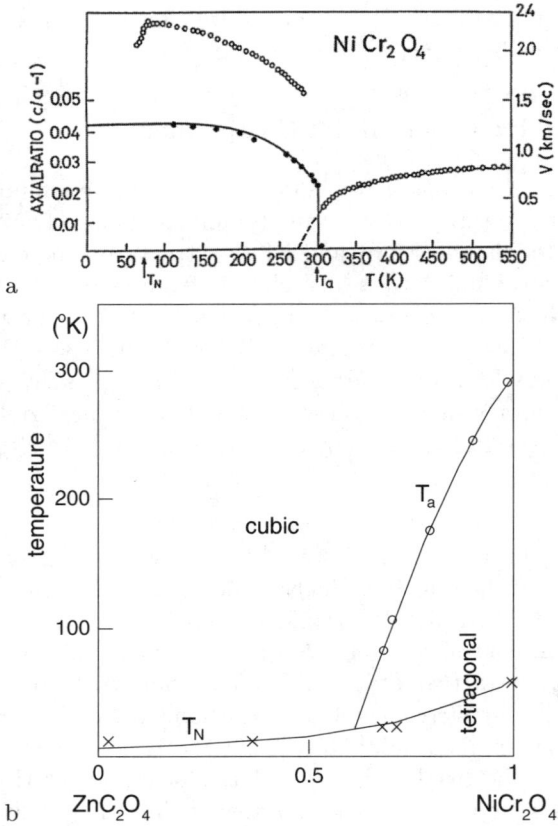

Fig. 7.9. (a) Lattice constant and sound velocity versus temperature of the Nickel-chromite system NiCr$_2$O$_4$ (Kino et al. [7.59, 7.60]), (b) phase diagram of the Nickel-Zinc-chromite system Ni$_x$Zn$_{1-x}$Cr$_2$O$_4$ showing the strong concentration dependence of the structural transition T_a and of the magnetic transition T_N (Kino et al. [7.56])

magneto-elastic origin and not due to two-ion interaction (quadrupolar coupling via acoustical or optical phonons) i.e. $\Theta \approx 0$. The ε_3 strain splits the Γ_4 ground state into a singlet and a doublet. The free energy without two-ion interaction reads (Kataoka and Kanamori [7.55])

$$F = c\varepsilon^2 - k_B T N \ln \left(2e^{-6g\frac{\varepsilon}{k_B T}} + e^{12g\frac{\varepsilon}{k_B T}} \right)$$

and the spontaneous strain is calculated from

$$\frac{\partial F}{\partial \varepsilon_s} = 0 \ .$$

This gives with

7.2 Cooperative Jahn–Teller Effect and Quadrupolar (Orbital) Transition

$$\xi = \frac{\varepsilon_s(T)}{\varepsilon_s(T=0)} \quad \text{and} \quad t = \frac{T}{T_c}$$

the equation

$$\xi = \frac{e^{3\frac{\xi}{t}} - 1}{e^{3\frac{\xi}{t}} + 2}.$$

From this follows the results given above: for $T = 0$, $\xi=1$ and for $T = T_a$, $\xi = 0.5$ and $T_a = 1.5\frac{T_c}{\ln 4} = 1.082 T_c$.

The study of the Nickel-Zinc-Chromite system $Ni_xZn_{1-x}Cr_2O_4$ shows that the structural transition temperature T_a diminishes very rapidly with increasing Zn content and T_a and T_N coincide for $x < 0.6$ (Kino et al. [7.56]). This is shown in Fig. 7.9b. The virtual phonon exchange mechanism, schematically shown in Fig. 7.8 (in this case with strain coupling) breaks down rapidly with the dilution of Ni-ions.

In the case of pure $ZnCr_2O_4$, the $c_{11} - c_{12}$ mode also exhibits some softening, indicating that, due to frustration, this compound is not a simple anti-ferromagnet (Kino et al. [7.57]). Recent studies (Martinho et al. [7.61], Lee et al. [7.62], Kino [7.58]) show that $ZnCr_2O_4$ undergoes a first order magnetic and a structural transition from cubic to tetragonal at T_N. Finally, one comment regarding the neglect of spin-orbit interaction for the Ni-Zn-Chromite system: the three-fold degeneracy of the ground state level Γ_4 implies a nonzero orbital moment so that spin-orbit interaction should be considered (Kugel and Khomskii [7.50]). But in the case of $NiCr_2O_4$, the Jahn–Teller effect can reduce the spin-orbit coupling appreciably through the vibronic reduction effect.

CsCuCl$_3$

Another interesting transition metal compound exhibiting a cooperative Jahn–Teller transition is $CsCuCl_3$. It belongs to the hexagonal perovskites AMX_3 with Jahn–Teller active ions $M = Cu^{2+}$, Cr^{2+} and alkali ions $A = Rb^+$, Cs^+ and $X = Cl$. $CsCuCl_3$ undergoes a cooperative Jahn–Teller transition at $T_a = 421$ K from the high temperature symmetry $P6_3/mmc$ to $P6_122$ (Kroese and Maaskant [7.63]). Figure 7.10 shows a schematic view of the high temperature structure (Graf et al. [7.64]). As indicated by the space groups, the compound remains hexagonal below T_a with the c-axis tripled, while the a-axis is almost unaltered. The $CuCl_3$ units in the low temperature phase are found tetragonally elongated with the long axis being helically arranged (Fig. 7.10). But the $CuCl_3$ octahedra are already dynamically distorted in the high temperature phase and the apparent high symmetry $P6_3/mmc$ is generated by averaging over this disorder, as seen by strong diffusive intensity around certain strong Bragg reflections (Graf et al. [7.64]). One would assume a soft mode for $q = (0, 0, 2\pi/3c)$ which was not found (Graf et al. [7.64]), probably due to strong anisotropy of the form $E_a = g\sum_I \cos 3\Phi_I$ (Khomskii [7.65]) or due to other 3rd order terms in the free energy (Hirotsu [7.66]).

For the elastic constants, the c_{44} mode shows strong softening on approaching T_a from above (Fig. 7.11) and a large step indicating a strong first

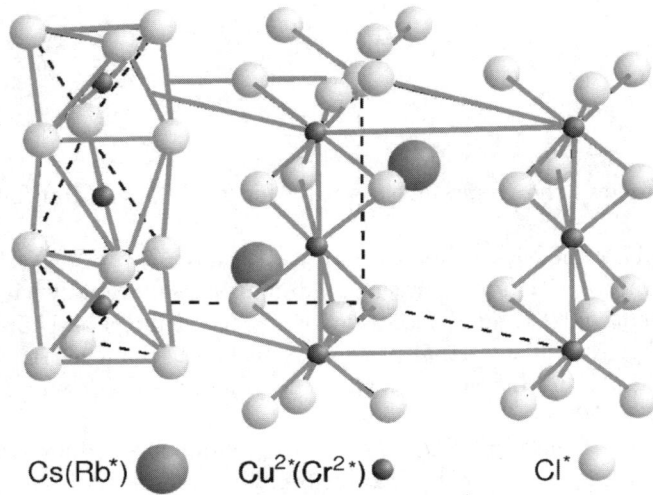

Fig. 7.10. High temperature structure of the hexagonal perovskite $CsCuCl_3$. The unit cell contains two formula units (Graf et al. [7.64])

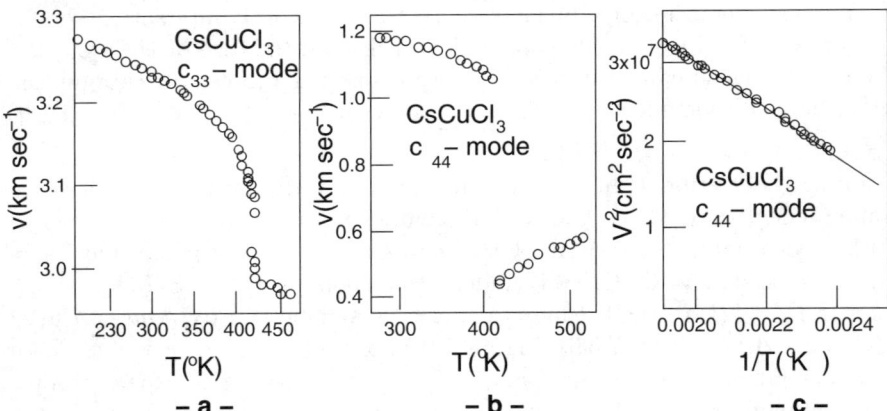

Fig. 7.11. Temperature dependence of the c_{33} and c_{44} sound velocities for $CsCuCl_3$. (a) $v(c_{33})$ vs. T, (b) $v(c_{44})$ vs. T, (c) $v^2(c_{44})$ vs. $1/T$ (Hirotsu [7.66], Lüthi unpublished)

order transition at T_a. This first order transition was also observed by optical rotation, X-ray and elastic neutron scattering experiments (Hirotsu [7.66]). The temperature dependence of the soft c_{44} mode for $T > T_a$ can be fitted again with formula (4.13) and (7.2) with $T_c = 329\,\text{K}$ and $\Theta \sim 0\,\text{K}$. This is indicated by the $1/T$-plot (Fig. 7.11). This means that, in this case, there are

7.2 Cooperative Jahn–Teller Effect and Quadrupolar (Orbital) Transition

again no quadrupole–quadrupole interaction, but a single ion magneto-elastic interaction which leads to $T_c = ng^2/c_{44}^0 = 329\text{K} < T_a = 421\text{ K}$ because of the first order nature of the transition. The longitudinal mode c_{33} has a normal temperature dependence for $T > T_a$ and a step-like behaviour for $T \sim T_a$ (Fig. 7.11).

Inelastic neutron scattering experiments for the phonon branches (Förster et al. [7.67]) revealed a strong Jahn–Teller type renormalization of the TA c_{44} phonon for $q_0 \sim 0.1(\text{Å})^{-1}$. The sound velocity for $q > q_0$ is 1057 m/s and changes to 579 m/s for $q < q_0$ at 434 K and reaches approximately the value of 440 m/s for $q \ll q_0$ in agreement with the value from Fig. 7.11. Interestingly, the TA phonon branch experiences also a hardening for $T < T_a$, indicating that the hardening in the elastic constants is not only a domain wall–stress effect, but rather intrinsic.

Fe_2TiO_4

This compound has the inverse spinel cubic structure containing only Fe^{2+} ions. It becomes weakly ferromagnetic at $T_c \sim 142$ K. Canted sublattices with magnetic moments of $4.2\mu_B$ for each sublattice lead to a spontaneous moment of $0.2\mu_B$ at 4.2 K (Ishikawa et al. [7.68]). Thermal expansion and elastic constant effects show dramatic effects from room temperature down to 142 K and below (Ishikawa et al. [7.69]). In particular, the $(c_{11} - c_{12})$-mode softens considerably in the paramagnetic phase. This leads to giant magneto-strictive length changes of the order of $\Delta L/L \sim 10^{-3}$, especially in the (100) direction with the magneto-striction constant $\lambda_{100} = -\frac{2}{3}\frac{g_1}{c_{11}-c_{12}}$. All these giant lattice effects are apparently due to the Fe^{2+}-ion on the tetrahedral site of the inverse spinel lattice. They are of a similar order of magnitude as the ones found in the Laves phase compounds RFe_2, with R = Tb or Dy (see Sect. 5.4).

7.2.2 Case of Rare Earth Compounds

Here, the richest variety of structural phase transitions due to the cooperative Jahn–Teller effect or due to quadrupolar coupling are found. Of special importance are the rare earth vanadates (RVO_4), the rare earth pnictides (RX with X = N, P, As, Sb, Bi) and some CsCl-structure material. Since the review articles, listed above, deal predominantly with these compounds, only a few typical examples are given and just some general physical problems discussed.

Rare Earth Insulators, Mostly Zirconates

The best documented cooperative Jahn–Teller transitions occur in the rare earth insulators with tetragonal zircon structure (Gehring and Gehring [7.47], Melcher [7.48]). A number of these materials have been investigated with

Table 7.2. Jahn–Teller transition for RMO_4 (R = rare earth, M = V, As, P)

table material	J-T trans.	Magn. T_N	Ground state - exc.states	soft elastic mode
TbVO$_4$	$T_a = 33\,\text{K}$	$T_N = 0.61\,\text{K}$	singlet-doublet (13 K)	$B_{2g} c_{66}$
DyVO$_4$	$T_a = 14\,\text{K}$	$T_N = 3.07\,\text{K}$	doublet-doublet (13 K)	$B_{1g} \frac{c_{11}-c_{12}}{2}$
TmVO$_4$	$T_a = 2.1\,\text{K}$		doublet-singlet (78 K)	$B_{2g} c_{66}$
TbAsO$_4$	$T_a = 27.7\,\text{K}$	$T_N = 1.5\,\text{K}$		
DyAsO$_4$	$T_a = 11.2$	$T_N = 2.44\,\text{K}$		
TmAsO$_4$	$T_a = 6.1\,\text{K}$			
TmPO$_4$	-	-	singlet-doublet (65 K)	$B_{2g} c_{66}$
TbPO$_4$	$T_a = 3.5\,\text{K}$	$T_N = 2.17\,\text{K}$		

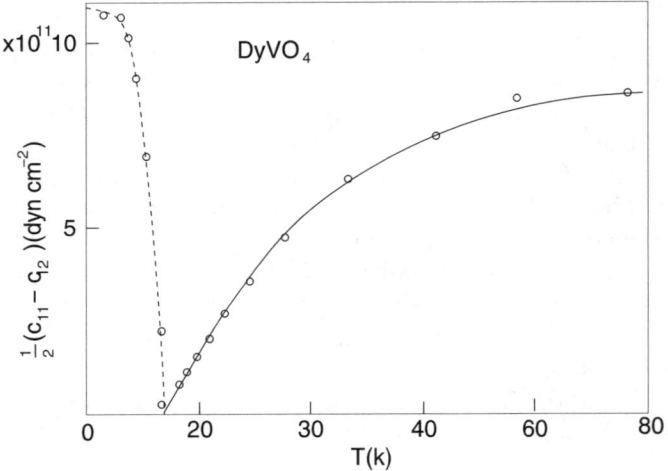

Fig. 7.12. Temperature dependence of the soft mode $(c_{11} - c_{12})/2$ for DyVO$_4$ (after Melcher and Scott [7.70])

ultrasound and Brillouin scattering: RVO$_4$ with R = Tb, Dy, Tm. Table 7.2 lists a number of these crystals with relevant physical properties.

DyVO$_4$

A typical example is shown in Fig. 7.12 for DyVO$_4$ (Melcher and Scott [7.70]). This substance has a structural transition at $T_a \approx 14\,\text{K}$. The energy levels of the Dy^{3+}-ion ($^6H_{15/2}$) split in the CEF of the tetragonal structure D_{4h}^{19} into Kramer's doublets. The lowest two separated by $2\Delta = 13\,\text{K}$, are well separated by $\sim 150\,\text{K}$ from the next higher ones, which can be neglected for the study of low temperature properties. Details of the RVO$_4$ substances can be found in Table 7.2.

For this tetragonal to orthorhombic transition, it is the $(c_{11} - c_{12})/2$ mode with B_{1g} symmetry which becomes soft. The crystal transforms consequently

7.2 Cooperative Jahn–Teller Effect and Quadrupolar (Orbital) Transition

from D_{4h} to D_{2d}. A total softening is observed which can be explained by (4.13) with only a small modification due to the existence of the two low-lying doublets. Therefore (5.21) is the appropriate expression with χ_s the single ion susceptibility determined from the two doublets. The fit with

$$\frac{c}{c_0} = \frac{\Delta - (\lambda + m)\tanh\left(\frac{\Delta}{kT}\right)}{\Delta - \lambda\tanh\left(\frac{\Delta}{kT}\right)} \qquad (7.3)$$

gives $\Delta = 6.5\,\text{K}$, $\lambda = -5.2\,\text{K}$, $m = 10.8\,\text{K}$. Of the various materials given in Table 7.2, only DyVO$_4$ has a sizeable quadrupole–quadrupole coupling, all others have a pure magneto-elastic coupling.

Other modes also give interesting effects in DyVO$_4$ (Gorodetsky et al. [7.71]). For the c_{44} mode (propagation in (001) direction), a splitting for $T < T_a$ can be observed. Applying a moderate magnetic field creates a single domain state. The mode splitting is due to a symmetry lowering with $c_{55} \neq c_{44}$. This follows from the interaction energy of the form $F_{int} = \frac{1}{2}Ku(\varepsilon_{xz}^2 - \varepsilon_{yz}^2)$ with K a higher order coupling constant and $u = \frac{1}{2}(\varepsilon_{xx}^0 - \varepsilon_{yy}^0)$ the static B_{1g} strain which is proportional to the order parameter. The elastic constants are $c_{44,55} = c_{44,55}^0 \pm Ku$. This expression describes the results quantitatively.

TmVO$_4$

This compound exhibits a softening of the $B_{2g}c_{66}$-mode (Melcher et al. [7.72]). In addition, it exhibits CEF effects similar to the ones discussed in Sects. 5.2 and 5.3. The specific heat at T_a can be fitted very well with a molecular field theory expression (Cooke et al. [7.73]). This is understood from the Ginzburg criterion (Sect. 4.4) indicating long range forces.

TbVO$_4$

A very interesting system from the acoustic point of view is TbVO$_4$. Here ultrasonic and Brillouin scattering experiments covered a wide frequency range from 14 MHz to 16 GHz (Sandercock et al. [7.74]). Therefore dispersive effects could be detected. Figure 7.13a shows ultrasonic and Brillouin scattering results (Sandercock et al. [7.74]).

In addition to the discussion in Sect. 4.3 and 4.4, not only do the isothermal and adiabatic range need to be taken into consideration but also five different elastic constants. These are:

1. The elastic constant c_I which applies to the "perfectly isolated" regime where the effective pseudo-spin population does not change under the action of the strain.
2. The elastic constant c_T denotes the isothermal elastic constant.
3. The "essentially isolated" regime with c_S is defined by allowing the spin system to reach internal thermodynamic equilibrium, whilst being unable to exchange energy with its surroundings.
4. In the adiabatic regime with $c_{S'}$ the spin system is allowed to come to local thermodynamic equilibrium with the lattice, although the temper-

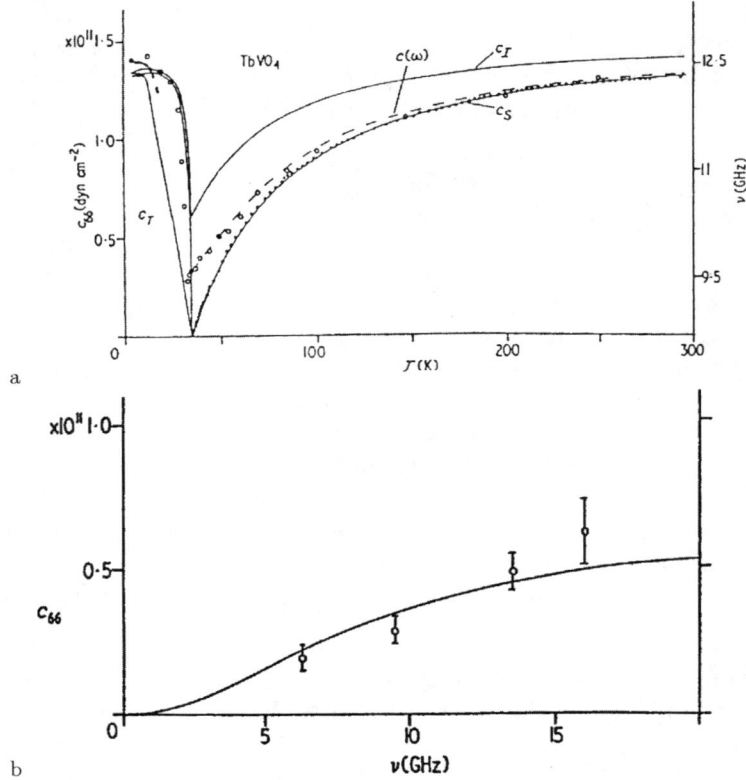

Fig. 7.13. (a) Brillouin scattering (open circles) and ultrasound (filled circles) for TbVO$_4$. For the different lines see the text, (b) Frequency dependence of the c_{66} mode at T_a. Full curve see the text (Sandercock et al. [7.74])

ature of the spin-phonon system varies from place to place as the elastic wave propagates.

5. Finally $c(\omega)$ is obtained from (4.18) with v_∞^2 replaced by c_I and v_0^2 by c_S.

In addition, the relaxation time τ is replaced by $\tau_S c_S/c_I$. The results in Fig. 7.13a clearly show the various regimes. The frequency dependence of the effect at T_a is given in Fig. 7.13b. As can be seen, it follows nicely the formula (4.18) with the modification stated above.

TmPO$_4$

This compound does not exhibit a structural phase transition. However the c_{66} mode shows considerable softening (Harley and Manning [7.75]). It can be accounted for again quantitatively using (5.21) with $K = -7.5$ K and $g^2/c_0 = 44.7$ K, indicating an anti-ferroquadrupolar interaction.

7.2 Cooperative Jahn–Teller Effect and Quadrupolar (Orbital) Transition   129

Other zircon structure materials have been studied magnetically ($HoVO_4$, $YbPO_4$, $PrVO_4$, $DyPO_4$, etc.) but not with ultrasonics. Therefore further analysis of these materials is not made.

$PrAlO_3$

Another isolating rare earth compound which exhibits cooperative Jahn–Teller transitions and which has been investigated with Brillouin scattering is $PrAlO_3$. At 205 K there is a first order rhombohedral-orthorhombic transition and at 151 K a second order rhombohedral-monoclinic phase transition, both investigated with inelastic neutron scattering (Birgeneau et al. [7.76]). In addition to these cooperative Jahn–Teller transitions, Brillouin scattering revealed a further pure elastic transition at 118.5 K (Fleury et al. [7.77]). Here it is assumed that this transition belongs to a class of phase transitions where the strain is the only order parameter (Anderson and Blount [7.78]).

Rare Earth Intermetallic Compounds

A large number of different metallic systems show typical Jahn–Teller and quadrupolar effects. Systems with CsCl-, NaCl-, $AuCu_3$-structure are amongst them. Table 7.3 lists those compounds which have a structural transition separated from a magnetic transition, i.e. pure cooperative Jahn–Teller or quadrupolar transition. We distinguish between ferro-quadrupolar and anti-ferroquadrupolar coupling, depending on the sign of the quadrupolar-quadrupolar interaction constant K (see Sect. 5.3). The list is constantly growing. Most of these substances have been thoroughly discussed in the review by Morin and Schmitt [7.51]. A few comments and additions will be made below.

Table 7.3. Intermetallic rare earth compounds exhibiting Jahn–Teller- or quadrupolar-transitions

Material		T_a	T_c T_N	Type of transition	Soft mode	Electr. ground state
TmCd	CsCl	3.16 K		Ferro-quadrupolar	$\frac{c_{11}-c_{12}}{2}$	Γ_5
TmZn	CsCl	8.55 K	8.12 K	Ferro-quadrupolar	$\frac{c_{11}-c_{12}}{2}$	Γ_5
$TmGa_3$	Cu_3Au	4.29	4.26 K			Γ_5
TmTe	NaCl	1.8	0.23–0.43	anti-ferroquadr.		Γ_8 or Γ_7
$PrPb_3$	Cu_3Au	0.37 K		anti-ferroquadr.		Γ_3
$PrCu_2$	orth.rh.	7.5 K		Ferro-quadrupolar	c_{66}	singlet
PrPtBi	cubic	1.35 K		Ferro-quadrupolar	$\frac{c_{11}-c_{12}}{2}$	Γ_3
CeB_6	cubic	3.35 K	2.31 K	anti-ferroquadr.		Γ_8
$Ce_3Pd_{20}Ge_6$	cubic	1.2 K	0.75 K	Ferro-quadrupolar	$\frac{c_{11}-c_{12}}{2}$	Γ_8
CeAg	CsCl	15.85 K	5.3 K	Ferro-quadrupolar	$\frac{c_{11}-c_{12}}{2}$	Γ_8

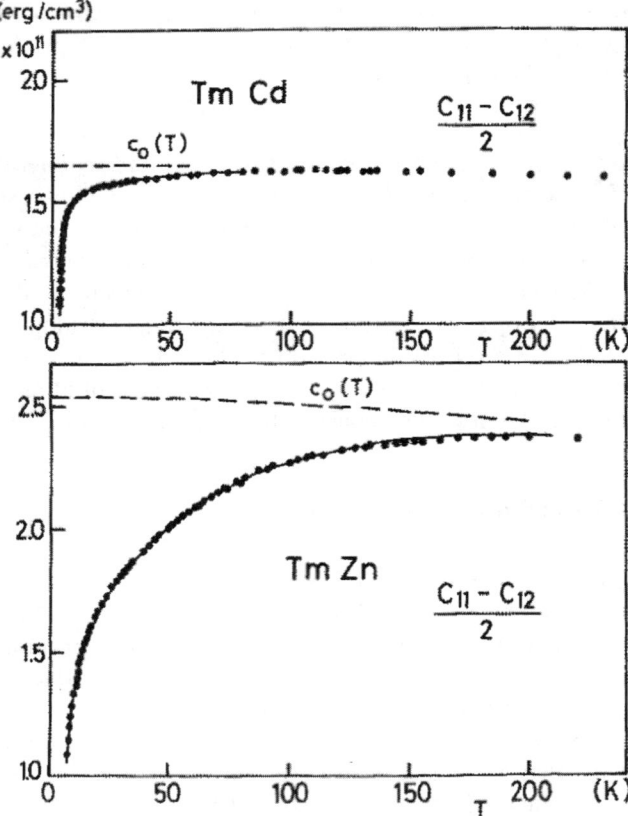

Fig. 7.14. Temperature dependence of the $(c_{11}-c_{12})/2$ mode for TmCd and TmZn. Full lines are fits with (5.21) (CEF effects included) (Lüthi et al. [7.79])

TmCd, TmZn

Figure 7.14 shows the temperature dependence of the soft $(c_{11}-c_{12})/2$ mode for the CsCl–structure materials TmCd and TmZn (Lüthi et al. [7.79]). According to Table 7.3, both have a ferro-quadrupolar phase transition. Using Fig. 4.4 as a guide, we see that TmCd has a larger ratio of inter-quadrupolar to magneto-elastic coupling constant than TmZn. In fact this ratio is 12.1 for TmCd and 6.7 for TmZn. Both compounds have a triplet Γ_5 ground state which splits up below the first order phase transition. The structural transition temperature T_a increases in a magnetic field along a cubic axis.

CeAg

This is another CsCl-type structure material which has a structural transition at 15.85 K and a ferromagnetic transition at 5.3 K. Elastic constant measurements revealed a pronounced softening of the $(c_{11}-c_{12})/2$ mode of about 24% (Takke et al. [7.80]). The main question here is the mechanism for the structural transition. $T_a = 15.85\,\mathrm{K}$ is rather large for 4f- orbital ordering. In addition, the magnetic properties cannot be quantitatively accounted for at low temperatures. The comparison with LaAg and LaAg$_{1-x}$In$_x$ (see Fig. 8.2, Knorr et al. [7.81]) suggests that an-harmonic interactions with a zone boundary M-point soft phonon is important. In fact the elastic mode for CeAg can be explained in this way. A measurement of the M-point phonon softening could decide whether the structural transiition is predominantly a martensitic–type transition (see Sect. 7.3) or rather a ferro-quadrupolar transition or a mixture of both. Such an inelastic neutron scattering experiment has not yet been performed.

CePd$_2$Al$_3$, Ce$_3$Pd$_{20}$Ge$_6$

Other more recent cases of ferro-quadrupolar ordering are CePd$_2$Al$_3$ (Nemoto et al. [7.82]) and Ce$_3$Pd$_{20}$Ge$_6$ (Suzuki et al. [7.83], Nemoto et al. [7.84]). In the latter case of this cubic material, a 50% softening of the $(c_{11} - c_{12})/2$ mode and apparently negligible quadrupolar pair interaction is observed; so this is a case of a pure Jahn–Teller phase transition. The former case (CePd$_2$Al$_3$) is interesting because this substance is iso-structural to the hexagonal heavy fermion superconductor UPd$_2$Al$_3$ (see Sect. 10.2). It is a so-called Kondo lattice compound with a fluctuation temperature $T^* \sim 20\,\mathrm{K}$ (see Sect. 9.3). While the c_{66} mode can be interpreted with the quadrupolar strain susceptibility in the whole temperature range, the longitudinal modes c_{11} and c_{33} show strong deviations for $T < T^*$ because of the Kondo singlet formation (Shimizu and Sakai [7.85]).

CeB$_6$

This cubic compound has an anti-ferroquadrupolar transition at $3.2\,\mathrm{K}$ followed by an anti-ferromagnetic transition at $2.3\,\mathrm{K}$. The Ce^{3+} ion has a Γ_8 ground state and a Γ_7 excited state at $\sim 540\,\mathrm{K}$ (Zirngiebl et al. [7.86]). Ultrasonic experiments revealed a softening of the symmetry elastic shear constants c_{44} and $(c_{11} - c_{12})/2$ of 1 - 3% with distinct anomalies at T_N and T_a (Lüthi et al. [7.87]). The temperature and field dependence of the elastic modes could be interpreted quantitatively. The fit gives evidence for the anti-ferroquadrupolar ordering at T_a. This substance is again widely discussed nowadays because of the renewed interest in orbital ordering and the interesting phase diagram. The anti-ferroquadrupolar ordering in CeB$_6$ was explained with the inclusion of octupolar moments and its competition with anti-ferromagnetic exchange (Sakai et al. [7.88], Sera and Kobayashi [7.89]). The $B - T$ phase diagram is anomalous with the paramagnetic–anti-ferroquadrupolar phase boundary increasing to higher temperature with increasing field. This can be explained apparently with the quadrupolar O_{xy}- and octupolar T_{xyz}–operators (Shiina et al. [7.90], Hanzawa [7.91]). It should be emphasized that CeB$_6$ exhibits also pronounced Kondo lattice properties (discussed in Chap. 9).

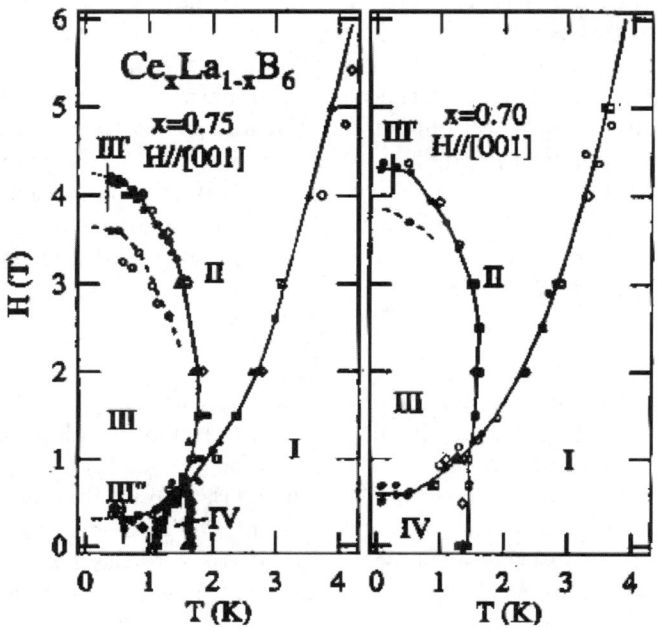

Fig. 7.15. B-T phase diagram for $Ce_xLa_{1-x}B_6$ for $x = 0.7, 0.75$ determined from ultrasonic measurements. $B//(001)$ (Akatsu et al. [7.93])

The alloy system $Ce_xLa_{1-x}B_6$ exhibits a complicated phase diagram, especially for $x = 0.65 - 0.75$ (Nakamura et al. [7.92], Akatsu et al. [7.93]), as shown in Fig. 7.15 for the concentrations $x = 0.7$ and 0.75. As for CeB_6, an anti-ferroquadrupolar phase II and an anti-ferromagnetic phase III are first found below the paramagnetic phase I. However, for $x = 0.75, 0.7$, a new phase IV is found characterised by a huge softening of 31% in c_{44} inducing phase IV. Although no magnetic order exists in phase IV, NMR and μSR experiments suggest strongly that time reversal symmetry is broken in this phase. This led Kubo and Kuramoto [7.94] to introduce octupolar interactions with a Γ_{5u} octupole symmetry to describe the $B-T$ phase diagram. It is possible to have anti-ferro-octupole order $\langle T_i^{5u} \rangle$, $i = x, y, z$ accompanying the ferro-quadrupolar moment $\langle O_{xy} \rangle = \langle O_{xz} \rangle = \langle O_{yz} \rangle$ necessary to account for the c_{44} softening.

RB_6

Apart from CeB_6, other rare earth hexaborides exhibit also pronounced quadrupolar effects (Nakamura et al. [7.95]). These are especially PrB_6, NdB_6, DyB_6, TbB_6 and HoB_6. Table 7.4 lists the RB_6 compounds with various physical properties. Large softening of the symmetry elastic constants $(c_{11} - c_{12})/2$ and especially c_{44} are observed. But the possible cooperative Jahn–Teller or quadrupolar transitions coincide with the magnetic Néel transitions T_N. Nev-

7.2 Cooperative Jahn–Teller Effect and Quadrupolar (Orbital) Transition

Table 7.4. Physical properties of rare earth hexaborides RB_6 (Nakamura et al. [7.95], S. Zherlitsyn et al. [7.96]). c_{ij} in $10^{11} \mathrm{erg/cm^3}$ at 200 K, Bulk modulus c_B, Poisson ratio ν, elastic anisotropy A, Debye temperature Θ_D and soft acoustic mode

RB_6	Density (g/cm^3)	c_{11}	c_{12}	c_{44}	c_B	ν	A	Θ_D	$T_N T_C$	soft mode
LaB_6	4.72	47.4	1.8	8.4	17	0.036	0.386	396 K		-
CeB_6	4.797	47.3	1.6	8.1	16.8	0.033	0.354	390		c_{44}
PrB_6		47.1		6.7					7.1, 4.2	c_{44}
NdB_6		46.9	7.3	5.9	20.5	0.134	0.298		8.7	c_{44}
SmB_6	5.06	43.4	6.7	6.4	10	-0.183	0.256	351		
EuB_6	4.895	41.5	4.2	6.0	16.7	0.092	0.322		15.5	
GdB_6		46.7	5.1	4.2	20.9	0.099	0.19		13	
TbB_6		50.6	5.5	3.9	20.5	0.098	0.18		17.8	c_{44}
DyB_6		54.2	7.8	3.2	23.3	0.095	0.138		32	c_{44}
HoB_6		47		3.1					6.4	c_{44}
YbB_6	5.53	33.5	8.1	4.1	16.6	0.195	0.322	274, 265		

ertheless, from the temperature dependence of the elastic constants, single-ion magneto-elastic coupling constants and anti-ferro-quadrupolar two-ion coupling constants could be deduced. Note that the soft acoustic mode for RB_6 is c_{44}, with the exceptions of LaB_6 (no $4f$ electron), GdB_6 ($L = 0$) and the mixed valent compound SmB_6. The softening amounts to 35% for PrB_6 and NdB_6 and to 70% for DyB_6 and HoB_6. The soft $c_{44}(T)$ can be interpreted quantitatively with the magneto-elastic strain–quadrupole coupling. The fact that, for the RB_6 compounds with orbital moments, predominantly the c_{44} mode softens is quite different from the case of rare earth pnictides, discussed below. There the soft mode and the strain symmetry of the ordered state and the easy axis of the magnetic state are governed by the CEF Stevens β-factor. This is the so-called Stevens-Pytte model. The reason why this model does not apply to the RB_6 series is not known. A more detailed investigation of CEF and phase transition analysis must first be done.

Other hexaborides investigated ultrasonically are LaB_6 (see Sect. 8.4.2), mixed valent SmB_6 (see Sect. 9.2), EuB_6 and YbB_6. From Nakamura et al. [7.95] and Zherlitsyn et al. [7.96] the following Table 7.4, showing useful data for RB_6, can be presented.

TmTe

Another case with renewed recent interest is TmTe. This belongs to the Tm–mono-chalcogenides TmX with $X = $ S, Se, Te with the NaCl-structure. TmSe will be discussed as a mixed valent material in Sect. 9.2. TmTe was found to have an anti-ferroquadrupolar structural transition at 1.8 K and an anti-ferromagnetic transition at ≈ 0.4 K (Matsumara et al. [7.97], [7.98]). The

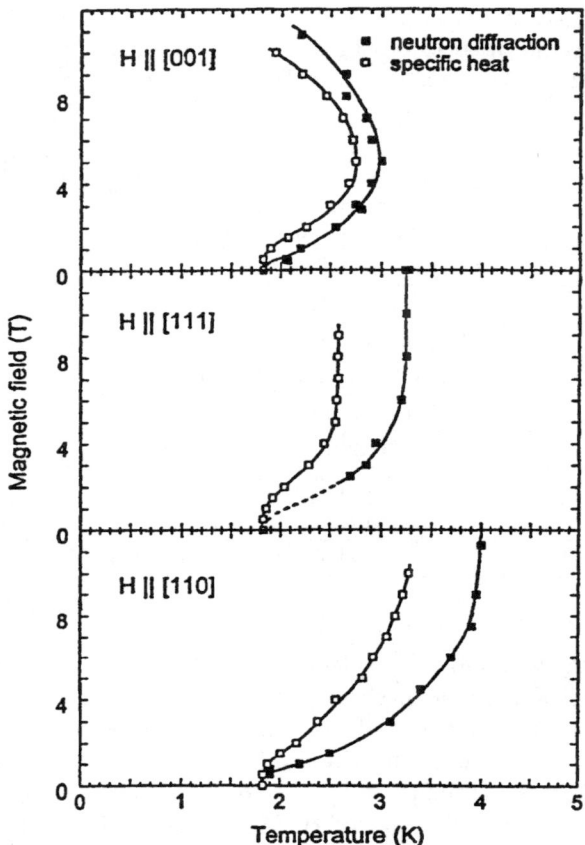

Fig. 7.16. Multipolar B-T phase diagram of TmTe for different orientations of the applied magnetic field B, showing the different estimates from neutron and specific-heat measurements (from Mignot [7.101])

overall CEF splitting $\{\Gamma_8 - \Gamma_7(10K) - \Gamma_6(16K)\}$ is less than $\Delta \approx 20\,\mathrm{K}$ as determined from specific heat, thermal expansion and elastic constants (Ott et al. [7.99]). Similar CEF splittings were measured by inelastic neutron scattering (Clementyev et al. [7.100]) and recent detailed elastic constant, specific heat and magnetic susceptibility experiments (Matsumara et al. [7.98]). These latter authors make detailed strain susceptibility fits and a comparison with CeB_6 as discussed above.

Figure 7.16 shows the B-T phase diagram as determined from specific heat (Matsumara et al. [7.97], [7.98]) and elastic neutron scattering (Mignot et al. [7.101]). This phase diagram can apparently only be interpreted with higher order multipolar (octupole) interaction taking into account (Shiina et al. [7.102]). However the CEF splitting was not considered in this calculation. For other theoretical development see Nikolaev and Michel [7.103].

PrCu₂

The CEF ground state of PrCu$_2$ with $J = 4$ splits in the orthorhombic structure (space group D_{2h}^{28}) into singlets. This compound was investigated with thermodynamic measurements (thermal expansion, elastic constants, specific heat, magnetic susceptibility) by Ott et al. [7.104] and later by Settai et al. [7.105]. It exhibits a ferro-quadrupolar order at 7.6 K with a soft elastic constant c_{66} propagating in the $a - b$ plane. The interesting point to be made here is that we deal with a singlet ground state quadrupolar order. This is the first intermetallic rare earth compound exhibiting this kind of order. This was found before for insulating compounds (Gehring and Gehring [7.47]). It must be remarked, however, that the lowest singlets are only a fraction of a degree apart. The coupling responsible for the phase transition is of magneto-elastic origin with the quadrupolar-quadrupolar coupling K being very small. We therefore deal here with a pure cooperative Jahn–Teller phase transition.

Other Metallic Singlet Ground State Systems

These were investigated ultrasonically by Bucher et al. [7.106]. The monopnictide TbP with a soft c_{44} mode is further discussed below with other monopnictide substances. Other cases of singlet ground state systems investigated are the cubic PrSn$_3$ and the cubic Th$_3$P$_4$–structure materials Pr$_3$S$_4$, Pr$_3$Se$_4$ and Pr$_3$Te$_4$.

PrPtBi

This compound, also listed in Table 7.3, has the cubic MgAgAs-type structure and exhibits a structural anomaly at $T_a = 1.35$ K. The elastic $c_{11} - c_{12}$ mode softens considerably due to a Γ_3 ground state. Magnetic susceptibility, magnetization, specific heat, elastic constants, electrical resistance and Hall effect have been measured (Suzuki et al. [7.107]). The latter quantity classifies PrPtBi as a low carrier system.

Rare Earth Monopnictides

These compounds have the NaCl structure. The cases of PrSb and TmSb which do not exhibit any phase transition have already been discussed in Sect. 5.2 in great detail. But RSb with R = Nd, Sm, Tb, Dy, Ho, Er and some other pnictides like TbP, DyP, HoP have anti-ferromagnetic phase transitions, accompanied with structural distortions at the same temperature T_N (Levy [7.108], Bucher et al. [7.109]). The tunneling model of Stevens and Pytte [7.110] can account nicely for these facts and can predict the kind of distortion and easy magnetic axis. They assume that the fourth order CEF potential with the Stevens factor β is important (5.4) and (5.5)) and that the effective potential has six minima in the (100) directions for $\beta < 0$ and eight minima in the (111) directions for $\beta > 0$. For a short discussion of the CEF potential and the Stevens factors see Sect. 5.1.1 and App. D. Therefore they predict tetragonal distortions for the Nd^{3+}, Dy^{3+} and Ho^{3+} compounds for which $\beta < 0$ and trigonal distortions for $\beta > 0$ for Tb^{3+} compounds (see Appendix D). Such distortions are indeed observed. In addition, elastic constant

Table 7.5. Rare Earth monopnictides: deformation and soft mode and Stevens factor β (see also App. D)

RX	Sign of β	Deformation	T_N(K) easy axis	soft mode	% change
NdSb	< 0	Tetragonal	< 100 >		
TbSb	> 0	Trigonal	15.5 < 111 >	c_{44}	10%
DySb	<0	Tetragonal	9.50 < 100 >	$c_{11}-c_{12}$	58%
HoSb	<0	Tetragonal	5.25 < 100 >	$c_{11}-c_{12}$	40%
ErSb	> 0		3.53 < 111 >		
TbP	>0	Trigonal	7.08 < 111 >	c_{44}	16%

investigations with corresponding soft modes ($c_{11} - c_{12}$ for tetragonal distortion and c_{44} for trigonal distortion) of these systems have been performed (Moran et al. [7.111], Mullen et al. [7.112], Bucher et al. [7.106], Nakanishi et al. [7.113]). In Table 7.5 we show the coincidence of soft mode symmetry and structural deformation symmetry. The agreement with the theoretical predictions is perfect.

A review on rare earth monopnictides is given by Hulliger [7.114].

The TbX monopnictides with X = P, As, Sb, Bi were analysed in detail by Kötzler and Raffius [7.115]. They could fit the sublattice magnetization, the specific heat, the soft elastic constant c_{44} and the trigonal distortion using one set of coupling constants for exchange and quadrupolar interactions for each TbX within molecular field theory. In addition, their fit showed that, within the TbX series, the magnetic transition is strongly first order for TbP, slightly first order for TbAs and second order for TbSb and TbBi as observed experimentally. This is of special interest because renormalization-group theories predicted first order phase transitions for Tb-monopnictides to type II anti-ferromagnetism, due to the large number of degrees of freedom of the order parameter ($n = 4$) (Mukamel and Krinsky [7.116], Brazovskii et al. [7.117]). But molecular field calculations including quadrupolar interactions seem to explain the phase diagram of the TbX monopnictides even better.

Other Metallic Rare Earth Compounds

Here we mention some more compounds which undergo quadrupolar phase transitions or show strong quadrupolar effects. In the SmX_3 compounds with X = Pd, In, Tl, Pb, strong softening of various elastic constants in the paramagnetic region towards the magnetic or structural phase transitions is observed (Endoh et al. [7.118]). In this cubic AuCu$_3$ structure, all SmX_3 crystals show anti-ferromagnetism with T_N ranging from 1.4 K (SmPd$_3$) to 14.8 K (SmIn$_3$) and ferromagnetism for SmTl$_3$ with a T_C = 8.6 K. In addition, some of these compounds exhibit structural transitions with $T_a > T_N, T_C$. These SmX_3 compounds have also Kondo lattice properties and are further discussed in Sect. 9.3.4 (see Table 9.1). Another system exhibiting quadrupolar effects is

RCu$_2$Ge$_2$ with R = Tb, Dy, Ho, Tm. Lattice constant measurements of these tetragonal compounds showed a strong increase of the lattice constant c below T_N (Sampathkumaran et al. [7.119]). GdCu$_2$Ge$_2$ showed no effect excluding the possibility of exchange striction effects. These materials showed likewise a strong deviation from the de Gennes scaling $T_N \sim (g_L - 1)^2 J(J+1)$ indicating the existence of such quadrupolar effects. Finally, a new class of materials undergoing quadrupolar phase transitions are the ternary compounds RB$_2$C$_2$ with R = Dy, Ho with a tetragonal structure of space group D_{4h}^5. HoB$_2$C$_2$ has been investigated with ultrasound (Yanagisawa et al. [7.120]). It has an anti-ferroquadrupolar transition together with an incommensurate AFM structure at 5.9 K, followed by an anti-ferromagnetic–anti-ferroquadrupolar order at 5 K. A description of the Ho^{3+}–CEF levels and a mapping of the complicated phase diagram could be obtained by ultrasonics.

Higher Order Multi-pole–strain Coupling

It has been shown that certain rare earth compounds like CeB$_6$ and TmTe exhibit higher order multi-pole (octupole) interactions between the magnetic ions. These ions Ce^{3+} and Tm^{3+} exhibit a quartet Γ_8 ground state for these compounds. With the direct product $\Gamma_8 \times \Gamma_8 = \Gamma_1 + \Gamma_2 + \Gamma_3 + 2\Gamma_4 + 2\Gamma_5$, magnetic dipole moments with Γ_4^- symmetry, electric quadrupole moments with Γ_3^+ and Γ_5^+ symmetry and magnetic octupole moments with Γ_2^-, Γ_4^- and Γ_5^- symmetry can occur (Kiss and Fazekas [7.121], Shiina et al. [7.90]). These latter ones are necessary to explain the $B - T$ phase diagram in CeB$_6$ and TmTe as pointed out above.

It should also be remembered that, for the interpretation of elastic constants, higher order couplings must be invoked. In addition to quadrupole moment–strain coupling one can think of hexadecapole moment–strain coupling. This is particularly the case for compounds with rare earth ions, possessing a large orbital momentum. Three cases are known up till now where such effects have been observed:

PrSb, PrPb$_3$, PrNi$_5$
These compounds all have Pr^{3+} ions with $L = 5$, $J = 4$, $S = 1$. In these cubic compounds the temperature dependence of some shear modes can only be interpreted including higher multipolar operators of the type O^4. The octupole operators like T_{xyz} mentioned for the case of CeB$_6$ and TmTe above do not couple to the strain tensor.

Following Callen and Callen [7.122], the magneto-elastic Hamiltonian equations (5.7) and (5.8) can be generalized, with the strains from (3.11), by including terms with $l = 4$ (hexadecapolar case, see Appendix D):

$$H_{me} = -\sum_I \{ \varepsilon_2 \left[g_{\Gamma 3} O_2^2 \sqrt{3} + g'_{\Gamma 3} O_2^4 \right]_i \qquad (7.4)$$
$$+ \varepsilon_3 \left[g_{\Gamma 3} O_2^0 + g'_{\Gamma 3} O_3^4 \right]_i + \varepsilon_{xy} \left[g_{\Gamma 5} O_{xy}^2 + g'_{\Gamma 5} O_{xy}^4 \right]_i + \cdots \} ,$$

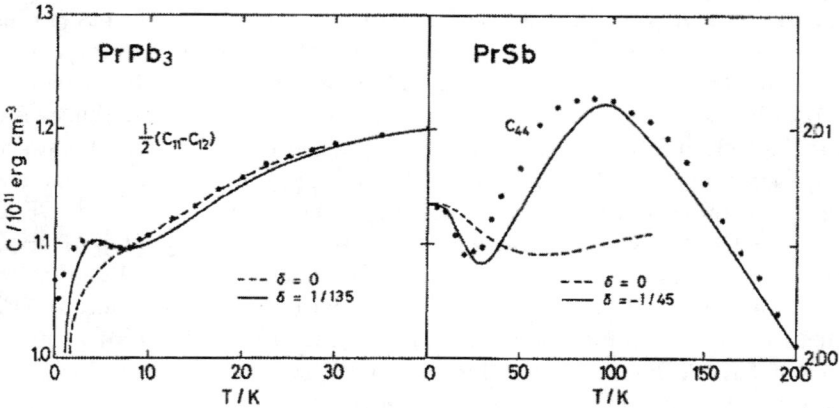

Fig. 7.17. Higher-order effects in PrSb and PrPb$_3$. Temperature dependence of the $(c_{11} - c_{12})/2$ and c_{44} modes. Dashed lines are fits with quadrupolar strain susceptibility, full lines with higher order susceptibility (Lüthi et al. [7.124])

where

$$O_2^4 = \frac{1}{4}\left\{\left[7J_z^2 - J(J+1) - 5\right](J_+^2 + J_-^2) + (J_+^2 + J_-^2)\left[7J_z^2 - J(J+1) - 5\right]\right\}$$

$$O_{xy}^4 = \frac{i}{4}\left\{\left[7J_z^2 - J(J+1) - 5\right](J_+^2 - J_-^2) + (J_+^2 - J_-^2)\left[7J_z^2 - J(J+1) - 5\right]\right\} .$$

The strain susceptibilities of these operators can be calculated with (5.16b). The ratio of the coupling constants $\delta = g'_\Gamma/g_\Gamma$ of (7.4) is now a new adjustable parameter. Figure 7.17 shows two elastic modes where these higher order couplings have been identified: the $(c_{11} - c_{12})/2$ mode for PrPb$_3$ with $\delta = g'_{\Gamma 3}/g_{\Gamma 3} = 1/135$ (Niksch et al. [7.123]) and the c_{44} mode for PrSb with $\delta = g'_{\Gamma 5}/g_{\Gamma 5} = -1/45$ (Lüthi et al. [7.124]). It is seen from Fig. 7.17 that these higher order couplings give an explanation for the occurrence of a minimum of the $(c_{11} - c_{12})/2$ mode in PrPb$_3$ and for the shift of the minimum in c_{44} for PrSb from 60 K to the observed one at 25 K. These effects occur in the paramagnetic region for both substances where other effects like interionic quadrupole interactions cannot explain these effects. These effects occur apparently for magnetic ions with a large orbital momentum like Pr^{3+}. Experiments on PrPb$_3$ establishing the $B-T$ phase diagram also found evidence for multipolar moment (octupolar) interaction (Tayama et al. [7.125]).

In PrNi$_5$, a hexagonal system exhibiting no low temperature phase transition where CEF effects have been observed (see Sect. 5.2.1), the elastic stiffness constant c_{66} exhibits a temperature dependence which can only be explained with the same higher order multi-pole operator O_{xy}^4 (Udagawa et al. [7.126]). In this case, $\delta = -0.03$. Note that PrSb and PrNi$_5$ are singlet ground state systems without any phase transition to the lowest temperatures measured and $PrPb_3$ has an orbital doublet Γ_3 ground state with a quadrupolar phase transition at 0.37 K (Table 7.3).

7.2.3 Case of Actinide Compounds

There exist very few uranium compounds exhibiting structural transitions due to magnetic ion quadrupolar interactions. Notable examples are UO_2, UPd_3, UCu_2Sn and $UNiSn$. Some physical properties of these compounds are listed in Table 7.6. Unlike the rare earth compounds, in the actinides crystal field effects are not so pronounced. In fact only for UPd_3 have CEF effects been observed with inelastic neutron scattering. The reasons for this are the more extended $5f$ wave functions compared to the well localised $4f$ wave functions. Therefore the hybridisation between the $5f$ and conduction electrons is stronger and leads to a partial itineracy of the $5f$ electrons. In addition the LS–coupling (Sect. 5.1.1) may no longer hold and be replaced by the jj–coupling. Also the valency may not be as well defined for the actinides. Heavy fermion actinide compounds suffer from the same uncertainties (Chap. 9). Here we discuss the four actinide compounds mentioned above which show typical Jahn–Teller and quadrupolar effects.

UO_2
This crystal was the first Jahn–Teller crystal investigated by ultrasound (Brandt et al. [7.127]). Here we have a $U^{4+}(5f^2)$ ($^3H_4$)–ion which is split by the cubic crystal field into a triplet ground state Γ_5 and higher lying states which do not influence the physics at low temperature. UO_2 has an antiferromagnetic phase transition at $T_N = 30.8$ K. This transition is strongly first order because of the coincidence of magnetic and structural transition. In Fig. 7.18 the elastic moduli for UO_2 are shown (Brandt et al. [7.127]). The soft mode c_{44} was interpreted with (4.13) by Allen [7.128] using the parameters $T_c = 26.4$ K and $T_0 = 20$ K indicating a quadrupolar and a Jahn–Teller coupling. Later on, it was found that the system does not transform into

Table 7.6. Actinide compounds, Physical Properties actinide compounds

Material	crystal structure	T_Q	T_N	T_C	soft mode	density(g/cm^3)
UO_2	Cubic		30.8	30.8	C_{44}	
UPd_3	Hexagonal	7.6, 6.8	4.4		C_{44}	
UCu_2Sn	Hexagonal		16.6		C_{66}	10.72
$UNiSn$	Cubic	42.9	42.9			
UN	Cubic		53			
U_3P_4	Cubic			138		10.05
$UCuP_2$	Tetragonal			75		9.01
$UCuAs_2$	Tetragonal			131		10.04

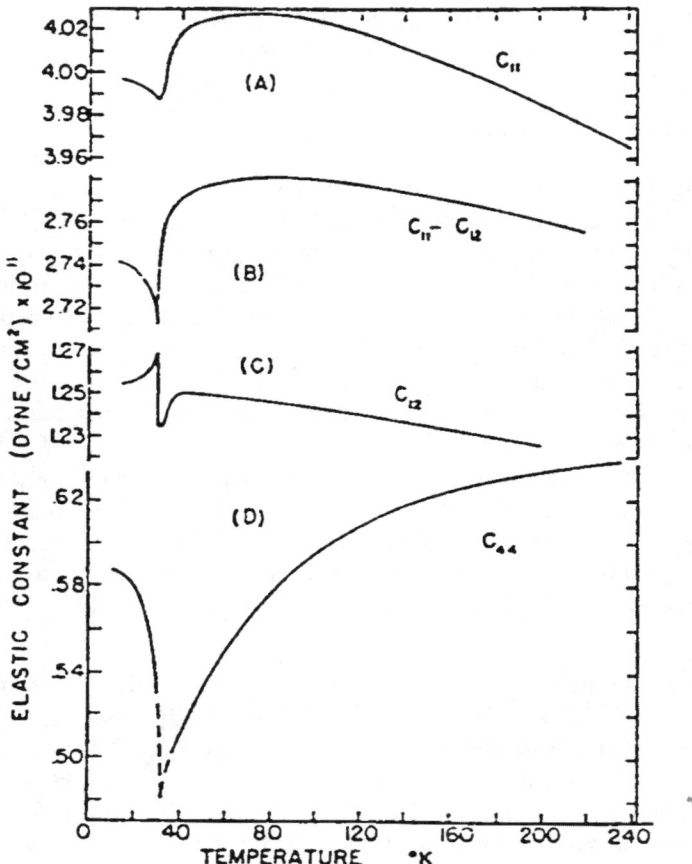

Fig. 7.18. Elastic constants versus temperature for UO_2 (Brandt and Walker [7.127])

a ferro-quadrupolar trigonal state but rather into a monoclinic phase with cell doubling and anti-ferroquadrupolar order (Faber et al. [7.129]). This new phase transition was analysed by Solt and Erdoes [7.130] using internal strains for the coupling and energy balance (see Sect. 5.4). The softening of the c_{44} mode by about 25% has not yet been explained in this new model.

UPd_3

This metallic compound is unique in several respects. It has the same space group $P6_3/mmc$ as the heavy fermion compound UPt_3, but it is a non-heavy fermion substance. It shows clear crystal field transitions for both sites in the double hexagonal close packed structure (the quasi-cubic and hexagonal sites) for the U^{4+}–ion with singlet ground states in inelastic neutron scattering (Buyers et al. [7.131]). This is very unusual for metallic actinide compounds. In the magnetically ordered state, the moment is very tiny ($\mu \approx 10^{-2}\mu_B$) –

7.2 Cooperative Jahn–Teller Effect and Quadrupolar (Orbital) Transition

much too small for induced singlet ground state systems with a $T_N = 4.4$ K (McEwen et al. [7.132]). It also shows several magnetic and quadrupolar phase transitions at low temperature first observed from specific heat (Andres et al., [7.133]) and thermal expansion (Ott et al. [7.134]) experiments. There are three reported phase transitions now at $T_2 = 4.4$ K, $T_1 = 6.8$ K and $T_0 = 7.6$ K (Lingg et al. [7.135]). T_2 is a Neel transition temperature and the others are quadrupolar structural transitions to be discussed below.

Ultrasonic studies for UPd$_3$ (Yoshizawa et al. [7.136], Amara et al. [7.137] and Lingg et al. [7.135]) show considerable softening of several symmetry elastic constants (see Fig. 7.19). We refer to Sect. 3.2 where we have listed the symmetry elastic constants for D_6 symmetry. From the three modes shown in Fig. 7.19, c_{33}, $c_{44}(\Gamma_6)$ and $c_{66}(\Gamma_5)$ are proper symmetry modes. It is seen from the figure that all modes exhibit considerable softening. A quantitative account of these temperature dependences can be given. Since only the cubic U–sites show CEF effects below room temperature, we have only to consider these ions for the strain susceptibilities of (5.16b), (5.21). Such a calculation for c_{44} (Amara et al. [7.137]) gives $g = 0.6$ K and $K = -6$ K, i.e. T_0, T_1 are anti-ferroquadrupolar transitions. This has been confirmed by elastic neutron scattering (McEwen et al. [7.132]). The temperature dependence of the three elastic modes in Fig. 7.19 can locate precisely the various transition temperatures T_0, T_1 and T_2. According to Lingg et al. [7.135] UPd$_3$ passes from a hexagonal structure to an orthorhombic one at T_0 and at T_1 from the orthorhombic to monoclinic. At $T_2 = T_N$ the magnetic structure is a trigonal triple–q state (McEwen et al. [7.132]). An unsolved problem is still the small magnetic moment for UPd$_3$. For singlet ground state systems with a $T_N = 4.4$ K and the lowest CEF state of 24 K, the reduction of the magnetic moment had to be of the order of 10% and not $< 1\%$ (Birgeneau [7.138]).

UCu$_2$Sn

This compound has again the hexagonal structure with space group P6$_3$/mmc as UPd$_3$, but with a single U–site. The U–ion lies in a layer sandwiched by the Cu layers. It exhibits a quadrupolar phase transition at $T_Q = 16.6$ K as observed from specific heat, susceptibility and ultrasound (Suzuki et al. [7.139]). Figure 7.20 gives the temperature dependence of the soft c_{66} mode together with a fit using (5.21). It clear from this figure that this is not a cooperative Jahn–Teller transition but rather a quadrupolar transition where $\mid K \mid \gg g^2/c_0$. It turns out that it is a ferro-quadrupolar transition. The U^{4+} ion has an orbital doublet CEF ground state. But the CEF –levels have not yet been looked for by inelastic neutron scattering.

UNiSn

Finally the interesting quadrupolar and magnetic phase diagram of UNiSn (Akazawa et al. [7.140]) is mentioned. Measurements of electrical resistivity, magnetization, thermal expansion and elastic constants gave a $B - T$ phase diagram shown in Fig. 7.21. Three phases can be discerned: a paramagnetic semiconducting phase (Phase I); Phase II is ferro-quadrupolar ordered and

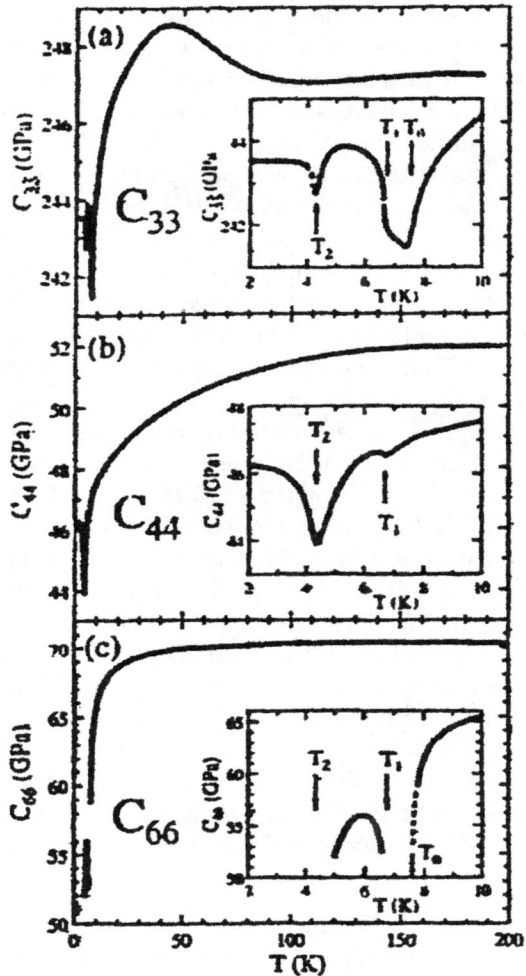

Fig. 7.19. Temperature dependence of the elastic constants c_{33}, c_{44}, c_{66} for UPd$_3$ (Lingg et al. [7.135])

metallic and it exists only for $B > 3T$; Phase III is a ferro-quadrupole ordered anti-ferromagnetic metal. This characterization was made for polycrystalline material. It is clear that single crystals are needed to show clearly the multipolar and magnetic order.

Table 7.6 lists important properties of the actinide compounds discussed above. They all show orbital transitions.

Uranium Pnictides

Table 7.6 also lists some physical properties of uranium pnictides. They all show ferromagnetic phase transitions (except UN) with T_C substan-

7.2 Cooperative Jahn–Teller Effect and Quadrupolar (Orbital) Transition

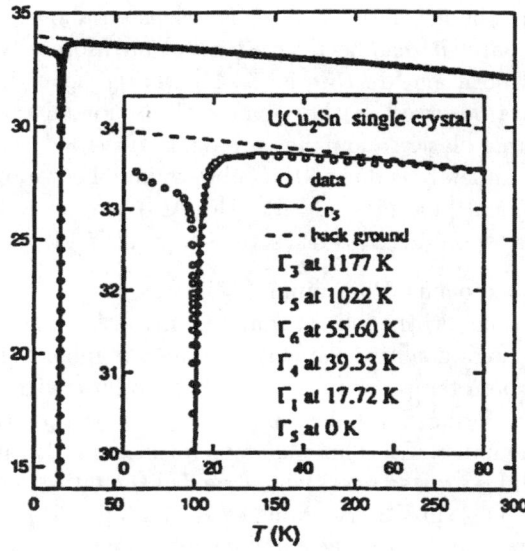

Fig. 7.20. Temperature dependence of the c_{66} mode for UCu$_2$Sn. The solid curve and the dashed line show the best fit for c_{66} and the background. The inset gives an expanded scale and the CEF level scheme (Suzuki et al. [7.139])

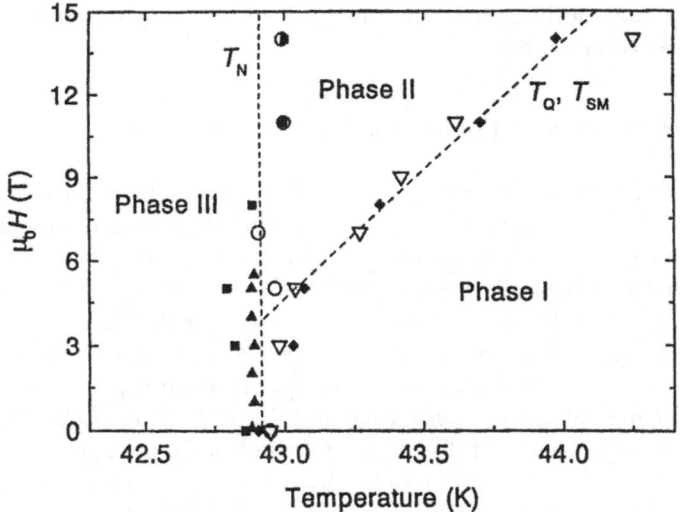

Fig. 7.21. B–T phase diagram for UNiSn (Akazawa et al. [7.140])

tially higher than the actinide compounds discussed above. Of the uranium mono-pnictides, only UN has been investigated ultrasonically (Yoshizawa et al. [7.136], van Doorn and du Plessis [7.141]). Large acoustic step function-like anomalies are observed in the vicinity of T_N for all the modes except c_{44}. In U_3P_4, large elastic anomalies are again observed, especially for c_{11} and c_{44}. The c_{44} anomaly is due to the cubic-rhombohedral distortion at T_C. The other tetragonal pnictides ($UCuP_2$, $UCuAs_2$) also show large ultrasonic effects at the corresponding T_C (Kaczorowski et al. [7.142]).

Other Actinide Compounds, Plutonium

For actinide compounds other than uranium, only lattice vibration results are known for Pu. This radioactive element exhibits a number of structural transitions between room temperature and the liquid phase with a melting point of 640°C. The fcc δ–phase can be stabilised at room temperature by alloying it with small amounts of Ga. This was done for inelastic X-ray scattering using highly brilliant X–ray sources, which yielded the lattice vibration spectra for (001), (011), (111) directions (Wong et al. [7.143]). These phonon spectra show some unexpected results: the small q elastic constants c_{11} and c_{44} are of similar magnitude ($c_{11} = 3.53 \times 10^{11} \text{erg/cm}^3$, $c_{44} = 3.053 \times 10^{11} \text{erg/cm}^3$) in good agreement with elastic constant data (Leadbetter and Moment [7.144]). The $(c_{11} - c_{12})/2$ mode ($= 0.49 \times 10^{11} \text{erg/cm}^3$) is about a factor of 6 smaller than the c_{44} mode. This leads to a large anisotropy parameter $A \sim 6.2$ (see Sect. 3.4), Pu is elastically very anisotropic. Furthermore a Kohn anomaly (see Sect. 12.5) for the T_1 branch in the (011) direction is observed and the transverse acoustic branch in the (111) direction shows softening at the L point of the Brillouin zone.

7.2.4 Effective Quadrupole–Quadrupole Interaction

The discussion of the experimental results of Sects. 7.2.1–7.2.3 has shown that, in most cases, nonmetallic substances exhibit a pure Jahn–Teller transition with $DyVO_4$ and UO_2 the rare exceptions. But intermetallic compounds usually show quadrupolar phase transitions. For a distinction between cooperative Jahn–Teller transitions and quadrupolar transitions, see introduction to Sect. 7.2. The reason for this different behaviour seems to be an effective interaction between conduction electrons and quadrupolar ions, the so-called aspherical Coulomb charge scattering (Fulde and Peschel [7.45], Teitelbaum and Levy [7.145], Hirst [7.146]). Especially ions with large orbital moments, e.g. Pr^{3+} ($J = 4, L = 5$), Tm^{3+} ($J = 6, L = 5$) give large effective quadrupolar coupling constants K (see Sect. 5.3, (5.18) and (5.21)). In Fig. 7.8 we show schematically the interaction between two quadrupoles, mediated by virtual phonon or conduction electron exchange.

Conduction electrons in metals with magnetic ions can make an exchange interaction between localised magnetic ions, the so-called RKKY interaction (White [7.147]). The aspherical Coulomb charge scattering constitutes an

7.2 Cooperative Jahn–Teller Effect and Quadrupolar (Orbital) Transition 145

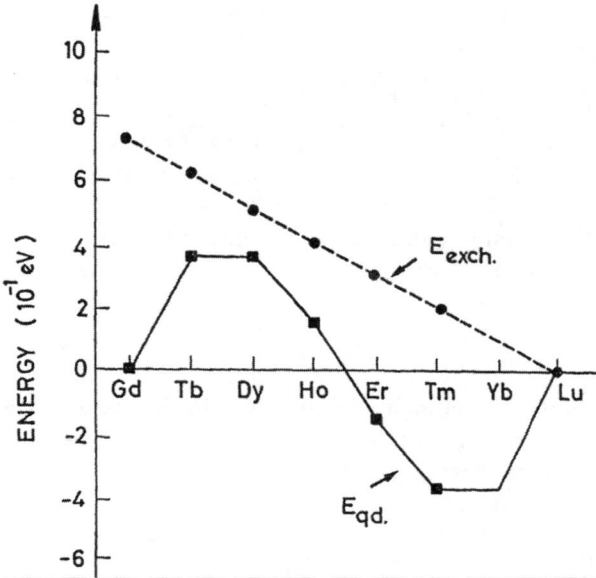

Fig. 7.22. Characteristic energies E_{exch} and E_{qd} as defined in the text for the heavy rare earth ions determined from magneto-resistance measurements (Fert et al. [7.148])

additional coupling. The scattering operators in spin and real space can be classified according to the orbital rank $0 \leq \Lambda \leq 6$ for f electrons with $l = 3$ and spin rank $\sum = 0, 1$. For the isotropic exchange, $(\Lambda, \sum) = (0, 1)$ and the aspherical Coulomb charge scattering gives $(\Lambda, \sum) = (2, 0)$ for rare earth intermetallics (Hirst [7.146]).

The importance of the aspherical Coulomb charge scattering for rare earth intermetallics was shown with magneto-resistance measurements for heavy rare earth impurities in Au (Fert et al. [7.148]). Measurements of the longitudinal and transverse magneto-resistance for rare earth impurities in Au and Ag gave an anisotropy which vanishes for Gd ($L=0$) and changes sign in the middle of the heavy rare earths between Ho and Er. This result shows that the resistivity anisotropy is due to quadrupole scattering. The characteristic energies plotted in Fig. 7.22 are $E_{qd} = 5L(S - 7/4)d$ and $E_{exch} = 5S\Gamma$ with d and Γ constant across the rare earth series. The sign change in E_{qd} occurs therefore between Ho^{3+}(S=2) and Er^{3+}(S=3/2). The characteristic energies of quadrupolar interaction E_{qd} (proportional to the coupling constant K) and of the isotropic exchange interaction E_{exch} (proportional to $(g_L - 1)J_{ex}$) are shown for the heavy rare earth ions. For the Tm^{3+} ion the quadrupolar interaction is larger than the exchange interaction.

This mechanism can also be partly responsible for the metallic U–compounds discussed in Sect. 7.2.3. On the other hand the virtual phonon exchange mechanism (Orbach and Tachiki [7.44]) can still be operative, not only in insulators but also in metals.

As discussed in Sects. 5.1.2, 5.2.2, 5.2.3 and 5.3, we also have the single ion magneto-elastic coupling constant g_Γ which can be determined with thermal expansion and elastic constant measurements. The g_Γ and the quadrupolar coupling constant K enter the molecular field equation for the soft elastic mode (4.13) and (5.21). Analogously to the discussion of K above, the magneto-elastic coupling constant g_Γ was discussed for the rare earth series in Chap. 5, Fig. 5.6a,b.

7.2.5 Summary

Many elastic constant results have been given in this section. Important coupling constants characterising the phase transitions have been obtained. Corresponding sound attenuation results are rare. There is one case, TmZn, where such a study was performed (Leung et al. [7.149]). It was found that the attenuation in the paramagnetic or high temperature phase follows a power law $\alpha \sim \omega^2 (T - T_a)^{-n}$ with T_a the structural phase transition temperature and $n = 1.5$ in good agreement with theory (Leung et al. [7.149], Leung and Huber [7.150], Becker et al. [7.151]).

Finally it should be remarked that the cooperative Jahn–Teller effect and the quadrupolar–quadrupolar phase transitions have been treated in molecular field theory (Kaplan and Vekhter [7.43]). This seems to be justified using the Ginzburg criterion (Sect. 4.4) because of the large force range of the elastic interaction and of the indirect aspherical Coulomb charge interaction.

In recent years, quadrupolar order has once more attracted increasing interest. But instead of ferro-quadrupolar or anti-ferro-quadrupolar order, one speaks today of orbital order. But the physics and especially the symmetry aspects are the same. The renewed interest is mostly confined to transition metal compounds, magnetic oxides, Mn-oxide compounds briefly discussed in Sect. 7.1 and low dimensional spin systems.

We have discussed several cases where higher multipolar order is important. Octupolar couplings seem to be important for CeB_6, TmTe and $PrPb_3$ as discussed in Sect. 7.2.2. Since the strain tensor components do not couple to these magnetic field-induced octupole moments (with Γ_u symmetry), elastic constant investigation is not so helpful for elucidating this problem. However strain–multipolar coupling effects have been observed in PrSb, $PrPb_3$ and $PrNi_5$. In this case a strain-hexadecapolar (l=4) coupling is effective as discussed also in Sect. 7.2.2 and 7.2.3. It is not accidental that these higher order couplings occur with Pr^{3+} ions since this ion has an orbital moment $L = 5$.

7.3 Ferro-elastic and Martensitic Phase Transitions

Martensitic phase transitionss are defined as diffusion-less solid state structural phase transitions with a deformation of the lattice such that a macroscopic strain results. The name martensite originates from the German microscopist A.Martens who first observed the microstructure in steel about 1890. If the macroscopic strain is the order parameter, a stress–strain relation in the ordered phase can arise which gives an elastic hysteresis. Such materials are called ferro-elastic. This term was introduced by Aizu [7.152]. According to his definition, a crystal is ferro-elastic if it has two or more orientation states in the absence of mechanical stress. It can be shifted from one to another of these states by a mechanical stress. The main interest in such phases is the study of the structural domains and domain walls (Barsch and Krumhansl [7.153]). With the spontaneous strain, which occurs at the transition, there arises a misfit between the phases, giving rise to a stress field. There is a wide variety of ferro-elastic crystals. In addition some of these materials exhibit shape-memory effects. These are important for different types of joints and various types of clamps in medicine. We give some representative examples which have been studied in some detail. Several materials exhibiting orbital phase transitions, discussed in Sect. 7.2, belong to the class of ferro-elastics.

In-Tl Alloys

Indium–Thallium alloys in the composition range 16–31 atomic % Tl undergo martensitic phase transformation from the face–centered tetragonal to the face–centered cubic structure. For example a 25% Tl sample has a transition temperature of 196 K, a 27% Tl sample at 127 K (Gunton and Saunders [7.154]) and for a 31% sample the transition temperature is ~ 0K (Brassington and Saunders [7.155]). In the former two concentrations, a soft $(c_{11} - c_{12})/2$ mode was observed. The first order transition of some of these compounds was determined by measuring third order elastic constants. Combinations of these determine development coefficients of third order invariants of the Landau free energy which are responsible for the first order transition as discussed in Sect. 3.3 (Liakos and Saunders [7.156]).

There are several other alloy compounds which undergo martensitic transformations: AgCd, Cu-Al-Zn, AuCuZn$_2$, AuAgCd$_2$, LaAg$_x$In$_{1-x}$ etc. The case of LaAg$_x$In$_{1-x}$ will be treated in Sect. 8.2. The mechanism for the transition can vary and is generally not specified. For a review of martensitic and ferro-elastic materials see Nakanishi [7.157], Nagasawa et al. [7.158] and Salje [7.159].

Shape-memory Effect in NiTi

This is a so-called shape-memory compound with an austenite (high temperature) phase of CsCl cubic structure ($Pm3m$). It transforms via intermediate so-called pre-martensitic phases to a low temperature martensitic phase of monoclinic symmetry ($P2_1/m$) (Bührer et al. [7.160]). The shape-memory

148   7 Ultrasonics at Structural Transitions

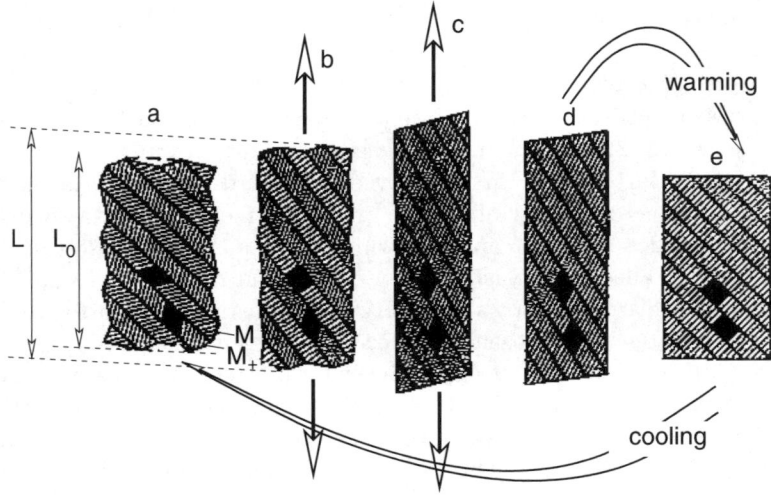

Fig. 7.23. Schematic presentation of the shape-memory effect (Müller [7.161])

effect accompanies the martensitic transition. It is an unusual mechanical deformation behaviour. It arises when an alloy plastically deformed at some low temperature reverts to its original shape upon heating to a temperature well above the martensitic phase transition. The physical mechanism is based on the occurrence of twin domains in the martensitic phase (Müller [7.161]). Figure 7.23a shows the low temperature martensitic twin domains M_+ and M_-. For small stresses (Fig. 7.23b), the domains are only elastically deformed. With larger stresses (Fig. 7.23c), those domains grow which are energetically favoured. This domain switching is the special process characteristic for the shape-memory effect. This is in contrast to ordinary metals where irreversible "slide- and dislocation" processes are the reaction to applied stresses. If the large stress is removed, the material stays strongly deformed (Fig. 7.23d). The additional temperature increase (Fig. 7.23e) leads to the mono-domain austenite phase. Cooling the sample again gives the martensitic domains as in Fig. 7.23a. The phase transition shows a pronounced temperature–hysteresis.

Elastic constants for the shape-memory compound NiTi as a function of temperature have been measured by various groups, the most comprehensive ones are due to Brill et al. [7.162]. They are presented in Fig. 7.24. They clearly exhibit the various phase transitions. These are indicated by arrows in the figure. Those below (above) are relevant for the cooling (heating) cycle only. In the high temperature phase, the $(c_{11} - c_{12})/2$ mode has the lowest value. This is typical for CsCl–structure materials which are unstable for $c_{11} - c_{12}$ strains (Zener [7.163]). It is also reflected in the high anisotropy value $A = 2c_{44}/(c_{11} - c_{12}) = 2$ in the high temperature β phase (see also Sect. 3.3). Since the soft mode for NiTi is a T_2A phonon with wave-vector

7.3 Ferro-elastic and Martensitic Phase Transitions

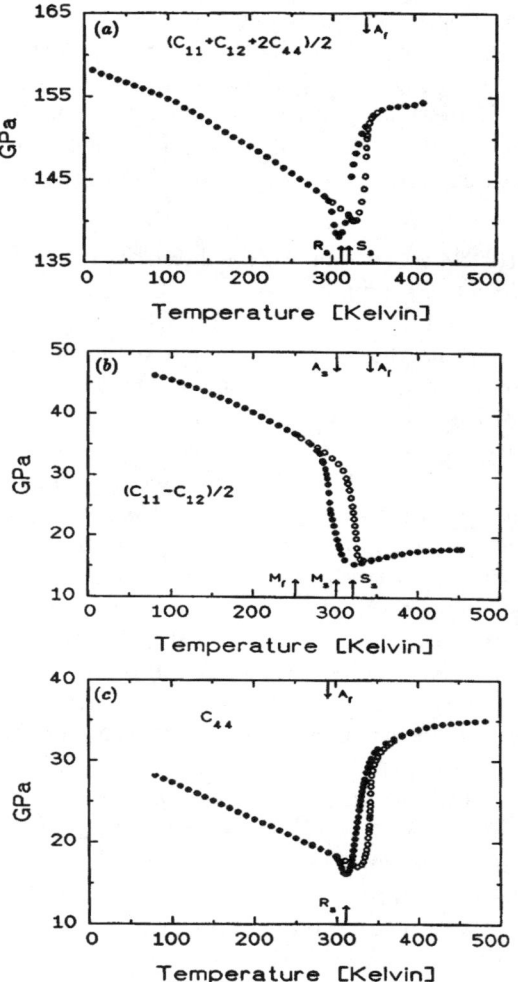

Fig. 7.24. Temperature dependence of the elastic constants of $Ni_{50.5}Ti_{49.5}$: (a) longitudinal mode $(c_{11}+c_{12}+2c_{44})/2$, (b) transverse mode $(c_{11}-c_{12})/2$, (c) transverse mode c_{44}. Full and open circles represent the cooling and heating cycle respectively. Arrows indicate phase transition temperatures (Brill et al. [7.162])

$q = 1/3[110]2\pi/a$, the ultrasonic strains are secondary order parameters and step function behaviour is expected at the phase transition as observed (see (4.14)). For a detailed description of the elastic modes and the phase transition, see Brill et al. [7.162] and references in it. Further analysis of the shape-memory effect can be found in Ren et al. [7.164]. A Landau theory of shape-memory alloys has been given by Falk and Konopka [7.165] and by Ishibashi and Iwata [7.166]. A microscopic description of the shape-memory effect has been given recently (Zhang and Guo [7.167]).

There are other shape-memory alloys exhibiting large reversible strains. One such class of compounds are the Heusler alloys, with a typical representative being the ferromagnetic Ni$_2$MnGa, which have been recently reviewed (Vasil'ev et al. [7.168]). Ni$_2$MnGa exhibits a martensitic phase transition (cubic-tetragonal) at $T_m \sim 220$ K. The transverse modes c_{44} and $(c_{11}-c_{12})/2$, as well as the lattice mode TA$_2$, exhibit considerable softening near T_m.

Invar Alloys

Fe-Ni alloys in the 30–50% region of Ni exhibit the Invar effect, i.e. an abnormally low coefficient of thermal expansion. For a modern review, see Wassermann [7.169]. These alloys have the cubic fcc structure and exhibit ferromagnetism. Apart from the Invar effect, these alloys have a very large forced volume magneto-striction, a large pressure dependence of the Curie temperature T_C and of the saturation magnetization, a temperature independent Young's modulus E for the 45% Ni alloy (Elinvar). All these properties stem apparently from the large volume dependence of the exchange energy. These alloys exhibit martensite transformations at low temperatures in the ferromagnetic region depending on the Ni concentration. The ferromagnetic alloy Fe$_{51.5}$Co$_{23.4}$Ni$_{25.1}$ constitutes a good invar system for room temperature, super-Invar is an alloy Fe$_{64}$Ni$_{31}$Co$_5$ which has an even smaller thermal expansion coefficient and the magnetic alloy Fe$_{50}$Ni$_{35}$Cr$_9$ is an often-used Elinvar alloy with a nearly constant E module around room temperature.

A thorough study of elastic constants and ultrasonic attenuation was carried out for the Fe–Ni alloy system (Hausch and Warlimont [7.170], Hausch [7.171]). They measured the c_{44}-, $(c_{11} - c_{12})/2$- and c_L modes along the (110) direction of the cubic single crystals in the concentration region from 31.5% to 51.4% Ni. A field of 0.6T was applied to saturate the crystals magnetically and to suppress domain wall–stress effects. Because of the strong exchange striction in these materials, leading to the physical properties listed above, they interpreted the temperature dependence of their results with $dJ/d\varepsilon$ and $d^2 J/d\varepsilon^2$, denoting the first and second strain derivative of the the exchange constant J (see Sect. 5.1.3). The Curie temperatures T_C and the martensitic start temperatures T_M varied with Ni concentration: for 31.5%Ni $T_C = 369$ K, $T_M = 239$ K, for 51.4%Ni $T_C = 773$ K, $T_M < 77$ K. The fitted exchange striction parameters fitted qualitatively the Bethe-Slater curve: exchange energy versus lattice constants. The best invar alloy corresponds to the inflection point of dJ/dr, for the Fe-Ni alloy for 35%Ni. Fe-Pt alloy was another system investigated with ultrasound (Hausch [7.172]). For disordered Fe$_{72}$Pt$_{28}$ with $T_c = 100$ K and a saturating field, the elastic constants exhibited again an anomalous temperature dependence $\Delta c \sim -(M/M_0)^2$.

7.4 Other Structural Phase Transitions

Phonon Induced and Order–Disorder

There are many other structural phase transitions which have been studied using ultrasonic waves. Amongst them, we find both examples where the strain plays the role of the order parameter and examples where the strain is not the order parameter, where the strain couples e.g. to the square of the order parameter (see the discussion in Sect. 4.3). In the first category, there are the piezoelectric ferro-electrics and other non-electronic substances like KCN, TeO$_2$. In the second category, we have the so-called phonon induced phase transitions and some ferro-electrics. Since many review articles report on these substances and since they have already been explored some time ago, we will make only brief remarks on these classes of substances.

7.4.1 Piezo-distortive Ferro-electrics

In these ferro-electrics, both a spontaneous strain and a spontaneous electric polarization develop in the ordered phase. A bilinear coupling between a symmetry strain component ε_i and the polarization P_i exist due to the piezo-electric effect. The best studied system is probably potassium di-hydrogen phosphate (KH$_2$PO$_4$ abbreviated KDP). The phase transition is triggered by the ordering of protons within the hydrogen bonds connecting adjacent PO$_4$ tetrahedra. Linearly coupled with this process are atomic displacements of K and PO$_4$ along the c-axis, producing the electric polarization P_3 and the shear strain ε_6 in the a–b plane. Many ultrasonic and Brillouin scattering experiments have been performed for this crystal. The soft mode is the c_{66} mode and the first order transition is from tetragonal D_{2d} to orthorhombic C_{2v}. Some physical constants of this compound, together with other substances, are listed in Table 7.7. Reviews covering piezo-distortive ferro-electrics are Garland [7.173], Rehwald [7.174], Lüthi and Rehwald [7.175].

7.4.2 Other Non-electronic Compounds

One famous example of a cubic material, which exhibits a structural transition with a soft c_{44} mode, is potassium cyanide KCN. Here the CN group can order with respect to its orientation. Therefore we deal here with an order–disorder phase transition. At 168 K, KCN transforms into a partially ordered orthorhombic structure (space group D_{2h}^{25}). In this phase the (CN)$^-$ anions have their long axis ordered along one (110) direction, but there is still disorder present with respect to head and tail. This final order is established below the phase transition at 83 K (space group D_{2h}^{13}). Ultrasonics (Haussühl [7.176], Rehwald et al. [7.177]) and Brillouin scattering (Krasser et al. [7.178]) found the c_{44} mode as the soft mode in the cubic phase.

Table 7.7. Materials for which strain is order parameter: ferroelectric and non-electronic transitions (adapted from Lüthi-Rehwald [7.175])

Ferroelectric	T_c(K)	Crystal classes	Spont. Strain	Symmetry	soft mode	order of transition
KH_2PO_4	122	$D_{2d} - C_{2v}$	e_6	B_2	c_{66}	1
RbH_2PO_4	145	$D_{2d} - C_{2v}$	e_6	B_2	c_{66}	1
CsH_2AsO_4	149	$D_{2d} - C_{2v}$	e_6	B_2	c_{66}	1
$KH_3(SeO_3)_2$	211	$D_{2h} - C_{2h}$	e_4	B_{3g}	c_{44}	
K-Na-tartrate	297	$D_2 - C_2$	e_4	B_3	c_{44}	2
	255	$C_2 - D_2$				2
Non-electronic						
KCN	168	$O_h - D_{2h}$	e_t	T_{2g}	c_{44}	1
	83	$D_{2h} - D_{2h}$				
NaCN	284	$O_h - D_{2h}$	e_t	T_{2g}	c_{44}	1
TeO_2	$p_c = 8.86$ kbar	$D_4 - D_2$	ε_3	B_{1g}	$c_{11} - c_{12}$	2

The temperature dependence of the soft c_{44} mode can be fitted again with (4.13). A microscopic calculation (Michel and Naudts [7.179]) as well as a Landau type treatment (Rehwald et al. [7.177]) give essentially the same result. Since we deal here with an order–disorder transition, the order parameter is related to the occupation probability of the CN anions along the different directions. The free energy consists here of the internal energy and a configurational entropy and the strain–order parameter coupling must be of the bilinear type, the molecular CN elastic dipole producing a stress field. A calculation of such a model is outlined in Appendix H. More details can be found in Table 7.7 and in a more recent review on the $K_{1-x}A_x$CN system (Höchli et al. [7.180]). In this review, the case of orientational quadrupolar glasses is also treated. If Br^- is substituted for CN^- in KCN at 43%, a new phase (for $T < 83$ K), the orientational glass is obtained.

Another interesting compound is para-tellurite TeO_2. Here the phase transition does not occur as a function of temperature but as a function of hydrostatic pressure. The $c_{11} - c_{12}$–mode shows a pressure variation characteristic for a piezo-distortive transition (Peercy et al. [7.181], McWhan et al. [7.182], see also the reviews cited above).

7.4.3 Strain Is Not Order Parameter

Here we discuss some displacive phase transitions, involving a soft phonon mode (see the discussion at the beginning of this chapter). This mode is usually a Brillouin zone mode with the temperature dependence $\omega_q^2 = A(T - T_c)$ for $T > T_c$. It can also be damped and a central peak can also appear for

$T \to T_c$. The vanishing of the soft mode frequency ω_q for $q = q_0$ at $T = T_c$ gives rise to a freezing of the corresponding normal mode displacement. The critical wave vector q_0 measures the phase difference of the atomic displacements from unit cell to the next. For $q_0 = 0$, the displacements are identical for all unit cells, for $q_0 \neq 0$, $\boldsymbol{q}_0$, as a vector, determines the new unit cell. Thus the order parameter of the transition can be taken as the eigenvector of the soft mode. In this case, the elastic symmetry strain couples quadratically to the order parameter as discussed in Sect. 4.3.

In this category, we have e.g. $SrTiO_3$, $KMnF_3$, $BaMnF_4$, the rare earth molybdate, quartz SiO_2, $NaNO_2$, ammonium halides and materials exhibiting charge density wave transitions. This list is by no means complete. It would be beyond the scope of this book to discuss all these cases. We refer to the review articles quoted above which give a complete survey. Instead, we just select one interesting aspect of such a phase transition: the investigation with the help of phonon echoes (Fossheim and Holt [7.183]). In $KMnF_3$, a structural transition from cubic to tetragonal occurs at $T_a = 187\,\mathrm{K}$. As in many structural transitions, the ultrasonic wave is often badly distorted from a plane wave due to sample quality and inhomogeneities. T_a can therefore vary slightly with position. The velocity, being strongly dependent on $T - T_a$, is therefore also a function of position. Strong interferences result in velocity and attenuation.

Attenuation Measurement with Phonon Echo Method

The phonon echo method, also called backward echo method, works with the electroacoustic effect (Fig. 7.25): A forward acoustic pulse (frequency ω, wave vector $\boldsymbol{k}$) $A_f \exp[-i(\omega t - \boldsymbol{k} \cdot \boldsymbol{r})]$ emitted by the transducer passes through the sample into the echo-active crystal bonded to it. This piezoelectric crystal, $Bi_{12}GeO_{20}$ (BGO), is located in a microwave cavity tuned to 2ω. Here a homogeneous electric field $E = E_0 \exp(i2\omega t)$ is applied to the forward wave. By a nonlinear mixing process, a backward wave $A_f E_0 \exp[i(\omega t + \boldsymbol{k} \cdot \boldsymbol{r})]$ is generated. This leads to a reversal of the wave vector, $\boldsymbol{k} \to -\boldsymbol{k}$, at all points of the wave front. Therefore, the echo wave front reconstructs continuously, replicating the original form during backward propagation, and finally reaches the transducer as a plane wave. The acoustic amplitude can then be measured. This is an example of a time-reversal experiment in solid state physics. It is also mentioned in the symmetry Chap. 13 and in Sect. 2.3.

In Fig. 7.26, a result for such an echo experiment is shown for $KMnF_3$. The interference pattern of the ultrasonic attenuation for a conventional experiment is compared with the attenuation determined with a backward echo experiment. In this latter experiment, no disturbing interference effects are observed and the critical exponent for the attenuation can be reliably determined (Fossheim and Holt [7.183] and [7.184]).

Using experimental attenuation results from such echo experiments, a dynamical scaling function could be obtained for a set of longitudinal wave

154   7 Ultrasonics at Structural Transitions

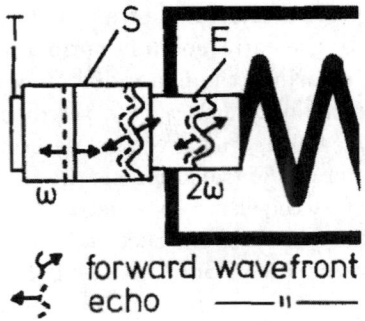

Fig. 7.25. Setup for echo experiment (Fossheim and Holt [7.184])

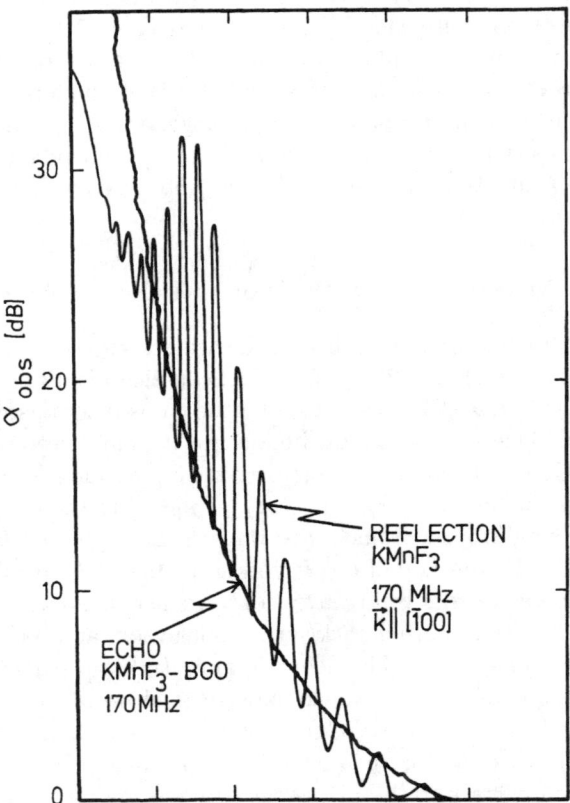

Fig. 7.26. Conventional and echo experiment for the attenuation in KMnF$_3$ (Fossheim and Holt [7.184])

frequencies, similar to the case of the magnetic phase transitions (Sect. 6.1.2 and Fig. 6.3). This echo method has not been used so far for other substances. A general account of this phonon echo technique can be found in Fossheim and Holt [7.184].

In the prototype structural phase transition for $SrTiO_3$, numerous ultrasonic and Brillouin scattering investigations have characterised the structural cubic-tetragonal phase transition at 105 K. By applying biaxial pressure elastic measurements could be performed in a practically single domain state (Fossheim and Berre [7.185]). Critical effects could be observed in this way. This phase transition has also been treated theoretically using a dynamical renormalization group theory (Schwabl and Iro [7.186]). The fluctuations of the TiO_6 octahedra rotation give rise to the soft mode, diverging ultrasonic attenuation and EPR linewidth and a central peak phenomenon.

Fluctuation Contribution to Elastic Constants

In cases where the strain is not the order parameter, or proportional to the order parameter, Landau theory gives a step function behaviour for the elastic constant at the structural transition temperature T_a (4.14). This behaviour is observed very rarely. Examples are the spin reorientation phase transition (Sect. 6.2), some superconductors (Sect. 10.3,7) and some ferroelectric phase transitions. In the majority of cases, fluctuation effects predominate as described by (4.16).

For structural phase transitions occurring in materials such as CeAg (Sect. 7.2.2), $LaAg_{0.78}In_{0.22}$ (Sect. 8.2), the high temperature superconductors $La_{2-x}Sr_xCuO_4$ (Sect. 10.4), K_2SnCl_6 and others, fluctuation models based on (4.16) can describe in principle the temperature dependence of the elastic constants and the attenuation. Writing the soft elastic constant as follows

$$c_\Gamma = c_\Gamma^0 - A_\Gamma \, |T - \Theta|^{-n}$$

gives: for the $c_{11} - c_{12}$ mode for CeAg, $n = 1$ and $\Theta = -50\,K$ (Takke et al. [7.80]); for $LaAg_{0.78}In_{0.22}$, $n = 1$ and $\Theta = 58.4\,K$ (see Sect. 8.2.1); for the rare earth elpasolites Rb_2NaHoF_6 and Rb_2NaTmF_6, $n = 0.5$ and $\Theta = T_a$ (Selgert et al. [7.187]) and for K_2SnCl_6, $n = 0.5$ (Henkel et al. [7.188]). These last two results were fitted to theoretical fluctuation expressions by Henkel et al. [7.188]. The anomalous fluctuation contribution to the elastic constants in the high temperature superconductor $La_{2-x}Sr_xCuO_4$ will be discussed in Sect. 10.4. Another case of fluctuation contribution to elastic constants will be encountered for TMMC in Sect. 12.2.1.

In some cases, critical effects predominate, i.e. a logarithmic temperature dependence of the elastic constant and a power law dependence of the attenuation. To this class belong certain materials exhibiting order-disorder transitions, such as the ammonium halides and some ferro-electrics. In the latter case, for, e.g. di-calcium propionate ($Ca_2Sr(C_2H_5CO_2)_3$), in the longitudinal elastic constant, critical sound attenuation and velocity effects on top of a step-like behaviour were found (Kameyama et al. [7.189]). In addition, velocity dispersion is observed in this case even in the low frequency range of 3–300 MHz, similar to the effects observed at magnetic phase transitions (Sect. 6.1.3). For more details, especially also detailed tables, see the reviews by Garland [7.173] and Lüthi and Rehwald [7.175].

7.4.4 Conclusion

We have kept the discussion of structural transitions due to phonons relatively brief. The reason for this is that most of this topic has already been discussed in the books listed in the introduction, Chap. 1, more than 30 years ago. We have concentrated on new developments like the phonon echo method to determine the attenuation near a structural phase transition, or on the fluctuation dominated temperature dependence of elastic constants. In addition, other structural phase transitions appear in other chapters. Examples are the iso-structural transition in unstable moment compounds treated in Chap. 9, structural transitions occurring in high temperature superconductors, such as $La_{2-x}Sr_xCuO_4$, in Sect. 10.4 or transitions in low dimensional spin systems in Chap. 12 etc.

8 Metals and Semiconductors

Ultrasonics has, in the past, played a key role in the investigation of the Fermi surface in metals. A number of important effects like geometric oscillations, magneto-acoustic de Haas van Alphen effect, Doppler-shifted cyclotron resonance and various other magneto-acoustic effects have been discovered and used in pure single crystal material of metals and semiconductors. All these effects rely on the interaction of sound waves with the conduction electrons. Usually, it is important to have electron mean-free paths, l_e, which are comparable to or larger than the sound wave length, $ql_e > 1$, with q the wave vector of the sound wave. On the other hand, there are cases where this condition is not important but where there is still a strong electron–phonon coupling like the one observed in heavy fermion substances. In this case and for intermetallic compounds, one is usually in the limit $ql_e < 1$. In the following sections, we discuss the deformation potential coupling, basic effects for pure metals and intermetallic compounds. We emphasize the effects occurring in intermetallic compounds more at the expense of $ql_e > 1$ effects for pure metals, which have been widely discussed (see Gudkov and Gavenda [8.1] and references therein). Some typical effects for semiconductors are discussed at the end of this chapter.

8.1 Deformation Potential Coupling

For a sound wave of wave vector q and frequency ω and an electron gas with mean free path l_e and relaxation time $\tau = l_e/v_F$, the magnitude of ql_e and $\omega\tau$ play a crucial role. If $ql_e \gg 1$, the mean free path of the electrons is much larger than the wavelength of the sound wave and the electron can absorb energy from the sound wave. But usually $\omega\tau = ql_e v_s/v_F \ll 1$ because the sound velocity v_s ($\sim 10^5$cm/s) is much smaller than the electron Fermi velocity v_F ($\sim 10^8$cm/s). For pure metals like Al, Cu, Ga etc. $ql_e \gg 1$ for low temperatures, but for heavy fermion compounds like UPt$_3$ and CeCu$_2$Si$_2$ $ql_e < 1$. But in the latter case, there is a strong electron–phonon coupling as discussed later (Chap. 9).

Deformations of the lattice caused by sound waves or phonons modify the charge distribution and lattice potential which leads to a coupling between conduction electrons and phonons. It causes a temperature dependence of elastic constants, it provides an ultrasonic attenuation mechanism and it is responsible for the various magneto-acoustic effects listed above. It is very difficult to calculate the matrix elements of this interaction for phonons of arbitrary wavelength (Ziman [8.2]). In the long wavelength limit, however, the influence on electronic states can be treated within a "deformation potential" approach. In this case, the deformation can be described by a slowly varying local strain. Therefore the electronic energy bands $E_\nu^0(\boldsymbol{k})$ (where ν is the band index) are changed adiabatically to

$$E_\nu(\boldsymbol{k}, \boldsymbol{R}_i) = E_\nu^0(\boldsymbol{k}) + \sum_{k,l} d_{kl}^n \varepsilon_{kl}(i) \ . \tag{8.1}$$

These position-dependent energy bands are a useful concept only if the sound-wave length is much larger than the lattice constant. In this case, the deformation potential $d_{kl}(k)$ is the same as that for homogeneous strains. It can be split into two contributions (Pippard [8.3], Fawcett et al. [8.4]). Consider a constant energy surface $E(k)$ in k-space: the induced strain, (ε_{kl}), causes an adiabatic change $\Delta k_n^{(1)}$ (where n is the normal component) of wave vectors lying on the Fermi surface. The resulting surface is, in general, not yet a constant energy surface and has still to relax by an amount $\Delta k_n^{(2)}$ at every point. This can happen if $\omega\tau \ll 1$ holds for the sound wave. Then the total change $\Delta k_n = \Delta k_n^{(1)} + \Delta k_n^{(2)}$ of every normal wave vector component can be written as $\Delta k_n = \sum_{kl} \lambda_{kl} \varepsilon_{kl}$ with $\lambda_{kl}(k) = -k_k n_l + r_{kl}(k)$. Here n_l are components of the surface normal vector and the two contributions for λ_{kl} correspond to $\Delta k_n^{(1)}$ and $\Delta k_n^{(2)}$ respectively. The parameters λ_{kl} can be determined in principle by strain-dependent de Haas van Alphen experiments (Fawcett et al. [8.4]). With $\Delta E = \hbar v_n(k) \Delta k_n$, where v_n is the normal electron velocity component, the deformation potential is given by

$$d_{kl}(\boldsymbol{k}) = \frac{\partial E(\boldsymbol{k})}{\partial \varepsilon_{kl}} = \hbar v_n(\boldsymbol{k}) \left[-k_k n_l + r_{kl}(\boldsymbol{k}) \right] \ . \tag{8.2}$$

For symmetry strains ε_Γ linear combinations d_Γ of the components of (8.2) can be formed. For isotropic bands with $E(k) = E(|k|)$, the deformation potential $d_{\Gamma 1}$ of a volume strain $\varepsilon_{\Gamma 1} = \varepsilon_v = \varepsilon_{xx} + \varepsilon_{yy} + \varepsilon_{zz}$ has no relaxational part r_{kl} because ε_v only changes the radius of the constant energy sphere and not their shape. Therefore (8.2) (Ziman [8.2]) leads to

$$d_{\Gamma 1} = -\frac{n}{N(E_F^0)} = -\frac{2}{3} E_F^0 \tag{8.3}$$

with n the electron density and E_F^0 the Fermi energy of the unstrained crystal. Equation (8.3) holds only for parabolic bands.

8.2 Elastic Constants and Ultrasonic Attenuation, Case of $ql_e < 1$

Here we discuss the redistribution mechanism between strain split bands which gives a useful formula for the symmetry elastic constant (8.8). Afterwards, the so-called Alpher–Rubin effect for sound velocity and attenuation changes in a magnetic field is introduced.

8.2.1 Electronic Redistribution Mechanism

To study the influence of a deformation potential coupling on elastic constants, it is necessary to compute the strain dependence of the conduction electron free energy. Here band energies are always measured from the strain dependent Fermi energy $E_F(\varepsilon_\Gamma) = E_F^0 + \langle d_\Gamma \rangle \varepsilon_\Gamma$ where $\langle d_\Gamma \rangle$ is an average of $d_\Gamma(\bm{k})$ over the Fermi surface. With $E(\bm{k}) = E_{\bm{k}}^0 + d_\Gamma \varepsilon_\Gamma$ we get $E(\bm{k}) - E_F = E^0(\bm{k}) - E_F^0 + (d_\Gamma(\bm{k}) - \langle d_\Gamma \rangle)\varepsilon_\Gamma$. For an isotropic band $D_\Gamma(\bm{k}) = d_\Gamma(\bm{k}) - \langle d_\Gamma \rangle$ vanishes at the Fermi surface. Therefore the deformation potential coupling to a single isotropic band does not influence elastic constants.

However the deformation potential coupling leads to observable effects in the following cases:

1. Degenerate isotropic bands (spherical Fermi surfaces) with different coupling constants d_Γ^n (n = band index). This is called the "band Jahn–Teller effect" (Labbe and Friedel [8.5]).
2. A single anisotropic band with anisotropic deformation potential coupling $d_\Gamma(\bm{k}) \neq \langle d_\Gamma \rangle$. This case is illustrated in Fig. 8.1a,b.

The deformation potential is subject to restrictions required by crystal symmetry (Gray and Gray [8.6]). This gives

$$\sum_{\{\bm{k}\}} d_\Gamma(\bm{k}) = 0$$
$$\text{for } \Gamma \neq \Gamma_1 \quad (8.4)$$
$$\langle d_\Gamma \rangle = 0 \ .$$

Here $\{\bm{k}\}$ denotes the star of the $\bm{k}$-vector and Γ_1 is the fully symmetric representation. As an example for cubic symmetry, $\varepsilon_{\Gamma 1} = \varepsilon_V$ is the only volume changing strain (see Sect. 3.2). Thus (8.4) holds for any volume conserving strain (tetragonal or trigonal) in cubic systems. An illustration for the star sum rule of (8.4) is shown in Fig. 8.1a,b.

For the further discussion of ultrasonic attenuation and elastic constants, various limits, as pointed out at the beginning of Sect. 8.1, must be considered. Here we discuss first the case of $ql_e < 1$, encountered in many compounds, especially rare earth and actinide compounds. For example, in intermetallic rare earth compounds, the residual resistivity is of the order of

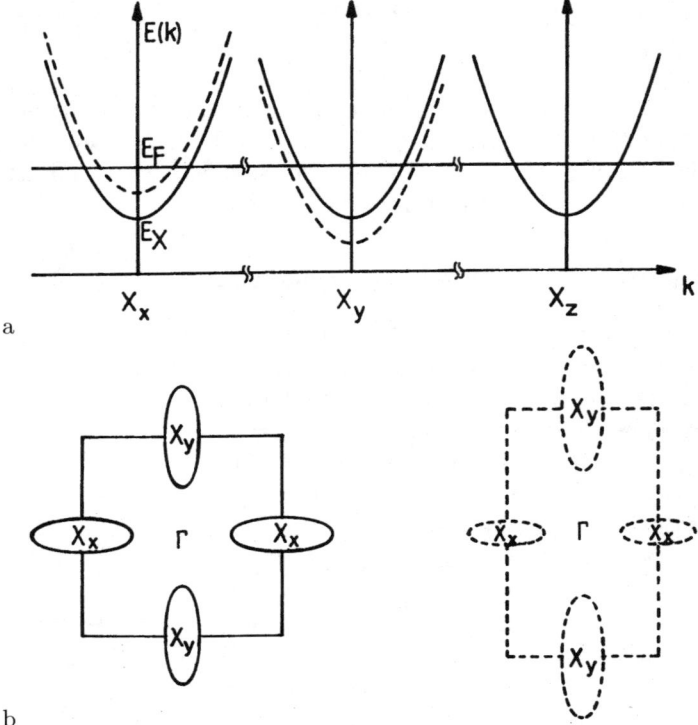

Fig. 8.1. (a) Illustration of the effect of an orthorhombic strain on electron pockets at the three X–points in e.g. LaAg. Full lines, unstrained state; dashed lines, strained state. E(k) for the three X-points, (b) X-point Fermi surface pockets (Niksch et al. [8.7]).

$1\mu\Omega$cm, which leads to relaxation times smaller than 10^{-12}s. For a sound wave propagating in such a material, with frequency $\omega/2\pi < 100\,\mathrm{MHz}$ this means that $\omega\tau \ll 1$ and $ql_e \ll 1$. In these limits the ultrasonic attenuation for both longitudinal and transverse waves is proportional to $q^2 l_e$. Ultrasonic attenuation in metals has been calculated by many groups and listed in reviews and books, see e.g. Akhiezer [8.8], Mason [8.9], Steinberg [8.10], Holstein [8.11], Morse [8.12], Pippard [8.3], Chambers [8.13] Rayne and Jones [8.14], Abrikosov [8.15]. A brief outline of a calculation for $ql_e < 1$ using the viscosity of the electron gas is given in Appendix E. In addition we have given attenuation formulas for arbitrary ql_e (E.3). The results for the attenuation of longitudinal and transverse ultrasonic waves for a free electron gas in the limit $\omega\tau \ll 1$ and $ql_e < 1$ are from (E.2), (E.3):

$$\alpha_L = \frac{4}{15} nm \frac{v_F}{\rho v_L} q^2 l_e \,, \qquad \alpha_t = \frac{1}{5} nm \frac{v_F}{\rho v_t} q^2 l_e \,. \tag{8.5}$$

8.2 Elastic Constants and Ultrasonic Attenuation, Case of $ql_e < 1$

The deformation potential coupling constant enters here via (8.3) for the longitudinal attenuation. The expressions for α_L and α_t are very similar. With typical values of $\tau = 10^{-12}$s, $v_F = 2 \times 10^8$cm/s, $n = 10^{22}/$cm^3, $v_L = 10^5$cm/s, $\omega = 6 \times 10^8$s^{-1}, we get $\alpha_L \approx 10^{-6}$cm^{-1} a negligible attenuation, especially compared to the background attenuation due to impurities, inhomogeneities, etc. Sound wave attenuation due to conduction electrons can therefore only be observed at low temperatures for relatively pure substances with $ql_e > 1$. See, for example, the normal state data for Sn shown in Fig. 10.3 in Chap. 10. By introducing the electrical conductivity $\sigma = ne^2\tau/m$, the attenuation reads $\alpha_L = \frac{4}{15}\omega^2 m E_F \sigma/(\rho v_L^3 e^2)$ an expression which was used to fit the attenuation data for Pb in the normal state from 2–12 K where ql_e changes from > 1 to < 1 (Morse [8.12]).

The elastic constants, however, can show observable effects in the limit $ql_e < 1$. Within this limit, the temperature of the lattice and the electron gas is the same. The electrons experience a quasi-static strain induced by the sound wave. The resulting energy shift $d_\Gamma(\mathbf{k})\varepsilon_\Gamma$ is the net shift with electron relaxation processes already included, as discussed above (Sect. 8.1). One can calculate the isothermal elastic constants using the free energy density of band electrons (Ziman [8.2] and Appendix F)

$$F_{el} = nE_F - k_BT \sum_k \ln\left(1 + e^{\frac{E_F - E(\mathbf{k})}{k_BT}}\right), \quad (8.6)$$

where n is the number of electrons per unit volume. With the conserved particle number

$$\frac{\partial n}{\partial \varepsilon_\Gamma} = \frac{\partial}{\partial \varepsilon_\Gamma} \sum_k f_\mathbf{k} = 0 \text{ results in } \langle d_\Gamma \rangle = \frac{\partial E_F}{\partial \varepsilon_\Gamma} = \frac{\sum_k f_\mathbf{k}(1 - f_\mathbf{k}) d_\Gamma(\mathbf{k})}{\sum_k f_\mathbf{k}(1 - f_\mathbf{k})}$$

and with $\frac{\partial F_{el}}{\partial \varepsilon_\Gamma} = \sum_k f_\mathbf{k} \frac{\partial E(\mathbf{k})}{\partial \varepsilon_\Gamma}$ the isothermal elastic constant is

$$c_\Gamma = c_\Gamma^0 + \sum_k f_\mathbf{k} \frac{\partial^2 E(\mathbf{k})}{\partial \varepsilon_\Gamma^2}$$
$$- \frac{1}{k_BT} \sum_k d_\Gamma^2(\mathbf{k}) f_\mathbf{k}(1 - f_\mathbf{k}) \quad (8.7)$$
$$+ \frac{1}{k_BT} \frac{[\sum_k d_\Gamma(\mathbf{k}) f_\mathbf{k}(1 - f_\mathbf{k})]^2}{\sum_k f_\mathbf{k}(1 - f_\mathbf{k})}.$$

Here $f_\mathbf{k} = \left[1 + exp\left(\frac{E(\mathbf{k}) - E_F}{k_BT}\right)\right]^{-1}$ is the Fermi distribution function and c^0 denotes the background elastic constant. Equation (8.7) has the same structure as (5.16b) of Sect. 5.2. The first term with the second derivative $\partial^2 E(\mathbf{k})/\partial \varepsilon_\Gamma^2$ corresponds to the Van Vleck type term and the other terms with the strong $1/T$ temperature dependence to the Curie term. From particle conservation, we got the strain dependence of the Fermi level $\partial E_F/\partial \varepsilon_\Gamma$ and the elastic constant (8.7).

To evaluate (8.7), additional assumptions have to be introduced. We will briefly mention two models: the so-called rigid band model and the Grüneisen parameter model. The latter will be discussed in the next Chap. 9 for unstable magnetic ions. In the rigid band model it is assumed that $d_\Gamma(\mathbf{k}) = \langle d_\Gamma \rangle$ is independent of $\mathbf{k}$, i.e. a given band shifts as a whole under the influence of ε_Γ. Therefore the Van Vleck term does not contribute. In such a rigid single band model, the deformation potential coupling does not lead to any change in the elastic constant, as already mentioned above, because $D_\Gamma(\mathbf{k}) = 0$ which leads to a cancellation of the last two terms in (8.7). The reason is that for a single band, due to particle conservation, the energy band and the Fermi energy shift correspondingly.

A rigid two-band model (A,B) with constant density of states N_A, N_B gives

$$c_\Gamma = c_\Gamma^0 - (\langle d_A \rangle - \langle d_B \rangle)^2 N_A N_B f_A f_B / [N_A f_A + N_B f_B]$$

and for the special case

$$N_{A,B} = N_0, \langle d_A \rangle = -\langle d_B \rangle = d \text{ and } n = 2N_0(E_F - E_0)$$

one obtains the so-called band Jahn–Teller formula

$$c_\Gamma = c_\Gamma^0 - 2d^2 N_0 \left[1 - e^{-(E_F - E_0)/k_B T}\right]. \tag{8.8}$$

E_0 is the bottom of the conduction band and N_0 the density of states at E_F. The name band "Jahn–Teller effect" suggests that one deals with degenerate bands of say e_g or t_{2g} symmetry. Such a degeneracy, however, is not necessary. What is needed are only identical Fermi surface pieces for the star $\{\mathbf{k}\}$ of the $\mathbf{k}$ vector, which deform differently under the action of a symmetry strain. This electronic redistribution mechanism between strain split and shifted itinerant energy bands was first used for heavily doped semiconductors (see Sect. 8.5.1). A typical example is shown in Fig. 8.1a,b for a two-dimensional cross-section of the Fermi surface and the change of the Brillouin zone under the symmetry strain $\varepsilon_{\Gamma 3} = \varepsilon_{xx} - \varepsilon_{yy}$. This orthorhombic strain has an opposite effect on the Fermi surface pieces A and B (or X_x and X_y in Fig. 8.1a,b). In the limit $\omega\tau \ll 1$ the different Fermi surface pieces are in equilibrium and (8.8) can be interpreted without using band Jahn–Teller concepts. It is a consequence of the sum rule (8.4).

An example of elastic constant data for LaAg and LaAg$_x$In$_{1-x}$ is shown in Fig. 8.2. This alloy system shows a martensitic type of phase transition for an In concentration $> 1\%$ (Ihrig et al. [8.16]). Consequently for the interpretation of the temperature dependence for the $(c_{11} - c_{12})/2$ mode, an an-harmonic term must be added to the deformation potential term of (8.8) (Knorr et al. [8.17]). We have discussed the martensitic phase transition previously (Sect. 7.3). Here we concentrate on the deformation potential term

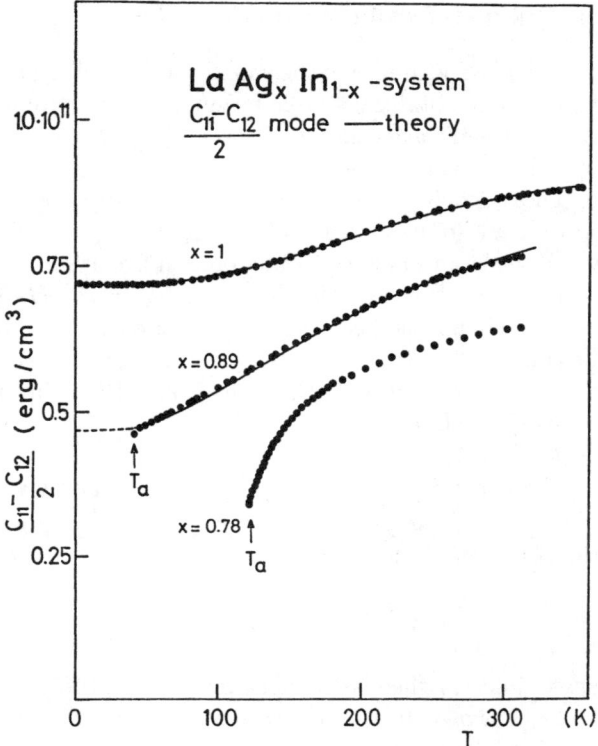

Fig. 8.2. Temperature dependence of the $(c_{11} - c_{12})/2$ mode for LaAg and LaAg$_{1-x}$In$_x$ (Knorr et al. [8.17]). Fit parameters for

$x = 1$: $E_F - E_0 = 285\,K$, $\quad 2d^2 N_0 = 0.375 \times 10^{11}\,\text{erg/cm}^3$, $A = 0$

$x = 0.89$: $E_F - E_0 = 222\,K$, $\quad 2d^2 N_0 = 0.49 \times 10^{11}\,\text{erg/cm}^3$,

$\hspace{5.5em} A = 1.78 \times 10^{11}\,\text{erg/cm}^3$

$x = 0.78$: $A = 24.64 \times 10^{11}\,\text{erg/cm}^3$, $T_c = 58.4\,K$

(8.8). A good description of the experimental data for LaAg for this formula is given by the following parameters: $E_F - E_0 = 285\,K$, $2d^2 N_0 = 0.375 \cdot 10^{11}\,\text{erg/cm}^3$. From this fit, it follows that this deformation potential coupling is too weak to induce a phase transition. For the compound with $x = 0.89$, a deformation potential term and an an-harmonic term of the form $A/(T - T_c)$ was used. Finally, for $x = 0.78$, only the an-harmonic term is needed. This an-harmonic term describes phenomenologically the coupling to the soft zone boundary phonons. Such an-harmonic or fluctuation effects were discussed in Sects. 4.3.5 and 7.4.3. The notion of deformation potential coupling for LaAg will be substantiated in the discussion of magneto-acoustic quantum oscillations (Sect. 8.4.2).

8.2.2 Alpher–Rubin Effect in Magnetic Fields

For the magnetic field dependence of the sound wave velocity and attenuation in the case of $ql < 1$ one obtains a simple formula (Alpher-Rubin [8.18]). In a conducting solid, a sound wave generates a periodic deformation of the lattice and forces the charge carriers to follow the motion if they are tightly bound to the lattice. These moving charges are subject to the Lorentz forces and therefore produce a periodic current. With a finite resistivity, this leads to dissipation which gives an enhanced ultrasonic attenuation and dispersion. One considers the Lorentz force on the electrons: $\boldsymbol{F}_B = \boldsymbol{j} \times (\boldsymbol{B}+\boldsymbol{b})$ with $\boldsymbol{j}$ the current, B the static magnetic field and b the dynamical field from the ionic motion. In the limit $\omega\tau \ll 1$, $ql_e < 1$ the electron can be considered to be coupled strongly to the ions through the collisions and $\boldsymbol{F}_B$ can be considered to act on the ions. Therefore with

$$\boldsymbol{j} = \sigma\left(\boldsymbol{E} + \frac{1}{c}[\boldsymbol{v}\times\boldsymbol{B}]\right) \quad \text{and} \quad \boldsymbol{v} = i\omega\boldsymbol{u}$$

one obtains the equation of motion

$$\rho\frac{\partial^2 \boldsymbol{u}}{\partial t^2} = c_{l,t}q^2\boldsymbol{u} + \boldsymbol{F}_B$$

with ρ the density and $c_{l,t}$ the elastic constant for longitudinal or transverse waves respectively. For $\boldsymbol{q} = (0,0,q)$ and $\boldsymbol{u}, \boldsymbol{b} \propto \exp i(\omega t - qz)$ and eliminating $\boldsymbol{E}$ and $\boldsymbol{b}$ with Maxwell's equations: $\mathrm{rot}\boldsymbol{E} = -1/c\,\partial\boldsymbol{b}/\partial t$ and $\mathrm{rot}\boldsymbol{b} = (4\pi/c)\boldsymbol{j}$ and $\mathrm{div}\boldsymbol{b} = 0$, the following dispersion equation is obtained:

$$(\omega^2 - v_{l,t}^2 q^2) = \left(\frac{q^2 B_{x,y}^2}{4\pi\rho}\right)\left(1 + \frac{c^2 q^2}{4\pi\sigma i\omega}\right).$$

With the dimensionless parameter $\beta_{l,t} = c^2\omega/4\pi\sigma v_{l,t}^2$, the final result for relative velocity change and the attenuation is given by:

Longitudinal waves   Transverse waves

$$\frac{\Delta v_l}{v_l} = B^2\frac{\sin^2\Theta}{8\pi\rho v_l^2}\left(1+\beta_l^2\right)^{-1} \qquad \frac{\Delta v_t}{v_t} = B^2\frac{\cos^2\Theta}{8\pi\rho v_t^2}\left(1+\beta_s^2\right)^{-1} \qquad (8.9)$$

$$\alpha_l = B^2\sigma\beta_l^2\frac{\sin^2\Theta}{2\rho v_l c^2}\left(1+\beta_l^2\right)^{-1} \qquad \alpha_t = B^2\sigma\beta_t^2\frac{\cos^2\Theta}{2\rho v_t c^2}\left(1+\beta_t^2\right)^{-1}$$

θ is the angle between the magnetic field $\boldsymbol{B}$ and propagation vector $\boldsymbol{q}$ of the sound wave. The dimensionless parameter β can be related to the skin depth $\delta = c/\sqrt{2\pi\sigma\omega}$ as $\beta = \frac{1}{2}q^2\delta^2$. This Alpher–Rubin effect occurs always if the displacement vector $\boldsymbol{u}$ experiences a Lorentz force, i.e. for q_z the u_l from B_x and u_t from B_z. For $\sigma = (1\mu\Omega\mathrm{cm})^{-1}$, and with $\beta = 0.5$, $v = 10^5\mathrm{cms}^{-1}$ and $\omega/2\pi = 10\,\mathrm{MHz}$, results in a $\Delta v/v$ of 0.4% at 10T. Such effects can always be present and have to be considered for many effects. An example for

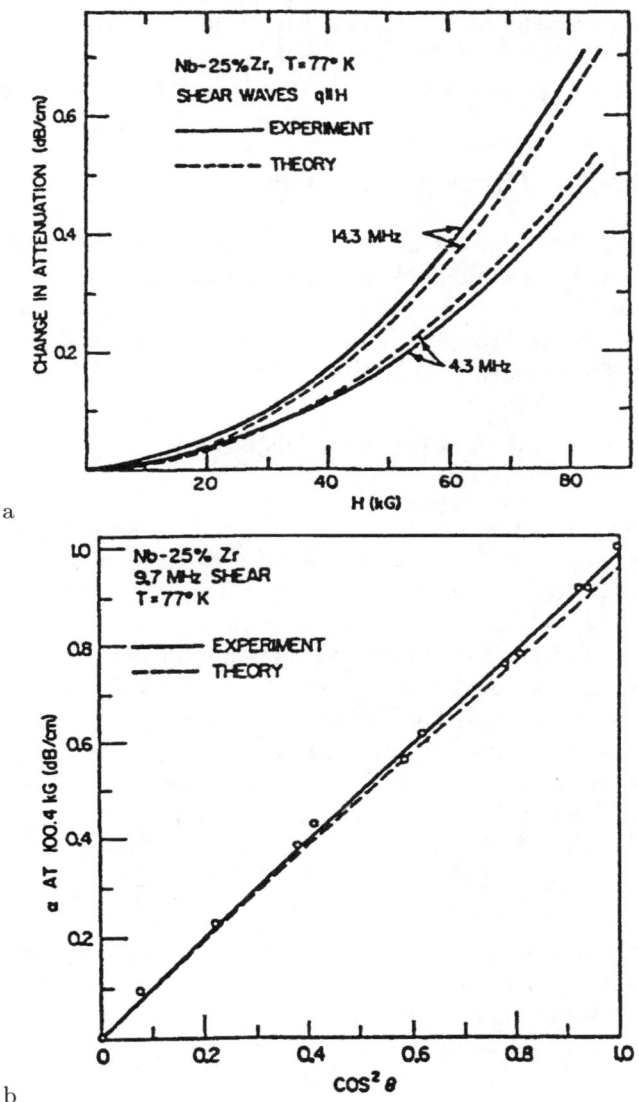

Fig. 8.3. (a) Magnetic field dependence of the attenuation of shear waves in Nb-25%Zr at 25 at 77 K, $B \parallel q$, dashed curves are calculated with (8.9) for the two frequencies 43, 143 MHz, (b) angular variation of attenuation for 9.7 MHz at B = 100.4 kG and 77 K. Dashed line calculated with (8.9) (Shapira [8.19])

this Alpher–Rubin effect is shown in Fig. 8.3a,b (Shapira [8.19]) for the alloy Nb-25%Zr for $T = 77$ K, far above the superconducting transition. Equation (8.9) fits the experiment very well for both the velocity and the attenuation. Another example is LaAg, discussed above (Knorr et al. [8.17]). Further examples occur for many metallic compounds in a magnetic field discussed in later chapters. A detailed theoretical account for this effect using the Boltzmann equation is given by Rodriguez [8.20].

8.3 Ultrasonic Propagation, Case of $ql_e > 1$

The case of ultrasonic attenuation in pure metals has been treated in great detail in many reviews and books as listed in Sect. 8.2.1. In view of this, only qualitative arguments for the different effects are given here and we refer for any derivation to the literature. Most experiments and calculations deal only with the ultrasonic attenuation (see below).

8.3.1 Ultrasonic Attenuation and Dispersion

The basic absorption mechanism for $ql_e > 1$ is due to the fact that the sound wave loses energy to the conduction electrons if the electrons ride in the field of the sound wave. This happens of course if $v_s = v_F \cos \Theta$ or $\boldsymbol{q} \cdot \boldsymbol{v}_F \approx \omega$ with v_s the sound velocity, v_F the Fermi velocity of the conduction electron, ω the frequency of the sound wave and Θ the angle between $\boldsymbol{q}$ and $\boldsymbol{v}_F$. Because usually $v_F \gg v_s$, the relevant electrons move almost perpendicular to $\boldsymbol{q}$. For the attenuation, the calculation consists of solving the Boltzmann transport equation or equivalent methods. Since we will not make strong use of this result, we just quote the necessary formula. In appendix E we have given the Pippard formula for longitudinal and transverse ultrasonic attenuation (Pippard [8.3]) for arbitrary ql_e. They reduce for $ql_e \gg 1$ to:

$$\alpha_L = \frac{\pi^2}{3}\left(\frac{Nmv_F}{\rho v_L}\right)q, \qquad \alpha_T = \frac{2}{\pi}\left(\frac{Nmv_F}{\rho v_T}\right)q. \qquad (8.10)$$

Unlike the case of $ql_e < 1$, here the attenuation no longer depends on the mean free path l_e. This is obvious since the formula is valid for $ql_e \gg 1$. It furthermore shows a linear dependence on q or the frequency. Comparison of (8.5) and (8.10) shows that $\alpha(ql_e < 1) \sim \alpha(ql_e > 1) \times ql_e$. The transition from $ql_e < 1$ to $ql_e > 1$ can be observed in the temperature dependence of the ultrasonic attenuation (8.5) and (8.10) as discussed in Sect. 8.2 for Sn and Pb and an example is given in Chap. 10 for superconducting substances (Fig. 10.3). It shows that the ultrasonic attenuation follows closely the electrical conductivity $\sigma = ne^2\tau_e/m = ne^2l_e/mv_F$ as already discussed above in Sect. 8.2.1. This is also borne out by the calculations listed above.

For the calculation of sound velocity and dispersion in the limit of long mean free paths of conduction electrons, the contribution to the elastic forces becomes nonlocal. The same ineffectiveness approach, i.e. only electrons with $\boldsymbol{q} \cdot \boldsymbol{v}_F \sim \omega$ interact strongly with the elastic wave, is then valid also for the change in sound velocity. Therefore the effects depend strongly on the particular shape of the Fermi surface. Sound wave dispersion and attenuation depend strongly on the direction of $\boldsymbol{q}$. They increase in critical directions when the electrons on a local piece of Fermi surface take part in the synchronous interaction with the wave. This is similar to the effect in strong magnetic fields, the so-called high field tilt effect (see e.g. Spector [8.21]). For a review of these effects for $ql > 1$ for sound attenuation and dispersion see Kontorovich [8.22].

8.3.2 Geometric Resonances

We turn now to sound wave effects in a magnetic field. Apart from the conditions $\omega\tau < 1$, $ql_e > 1$ we have now also to consider $qR_c > 1$ with R_c the cyclotron radius of the electron orbit in a magnetic field. We first neglect the quantization of the electron orbit, i.e. $\hbar\omega_c < k_B T$ with ω_c the cyclotron frequency, but we still have $\omega_c \tau > 1$. One striking sound wave effect, the so-called geometric resonance, occurs for $\boldsymbol{B} \perp \boldsymbol{q}$. So if we take $\boldsymbol{B} = (0,0,B)$ and $\boldsymbol{q} = (q,0,0)$ the sound absorption depends whether the orbit diameter is a whole number of wavelengths $(2R_c = n\lambda)$ or $2R_c = (n+1/2)\lambda$ with $n =$ integer (Pippard [8.23], Cohen et al. [8.24]). This can be seen in Fig. 8.4 which shows a transverse acoustic wave and two cyclotron orbits in real space.

Because of $v_t \ll v_F$ the electrons on these orbits see a practically stationary internal electric field. If the orbit diameter AB fulfills $AB = \lambda(n+1/2)$, there is a big influence of the sound wave field on the electrons because they move along the internal field and take energy from the sound wave field. On the other hand, for $AB = \lambda n$, the electrons at A move parallel and at B antiparallel to the field and they compensate each other. The two conditions give maximum and minimum acoustic attenuation and a corresponding large change in the elastic constants. As a function of magnetic field, maximum attenuation is obtained for

$$AB = \lambda\left(n+\frac{1}{2}\right) = 2R_n = 2\hbar\frac{k_F}{eB}, \tag{8.11}$$

since only extremal orbits contribute. Thus (8.11) gives a B^{-1} periodic dependence with the oscillation period $\Delta(1/B) = e\lambda/2\hbar k_F$ from which one can determine the Fermi momentum k_F. The conditions for observing geometric oscillations are $\omega_c\tau > 1$, $qR_c > 1$ and $ql_e > 1$. A typical set of oscillations is shown in Fig. 8.5 for Mg (Ketterson and Stark [8.25]). For a quantitative theory of geometric oscillations see Abrikosov [8.15] and Cohen et al [8.24].

The geometric resonances occur at relatively small fields. At higher fields, quantization effects (de Haas van Alphen effect) start to dominate. These will be discussed in the next Sect. 8.4. Another case of geometric resonances with non-reciprocal surface acoustic waves will be discussed in Sect. 13.5.

168   8 Metals and Semiconductors

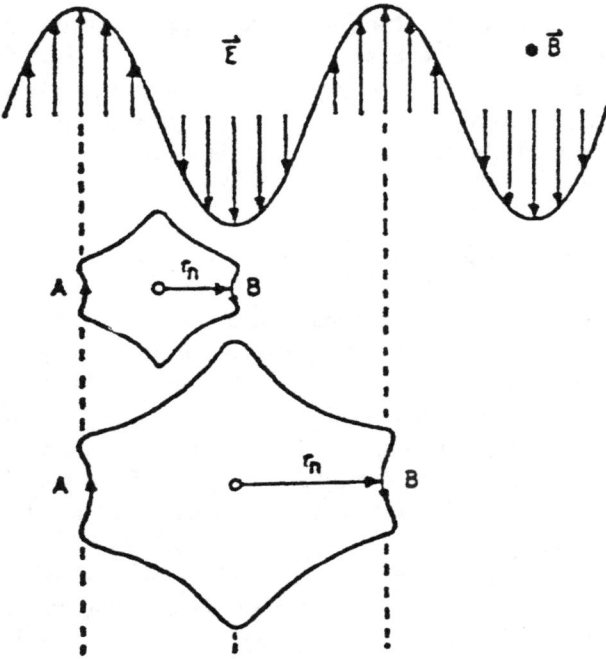

Fig. 8.4. Cyclotron orbits and sound wave electric field in real space for the two cases $2R_c = \lambda/2$ and $2R_c = \lambda$

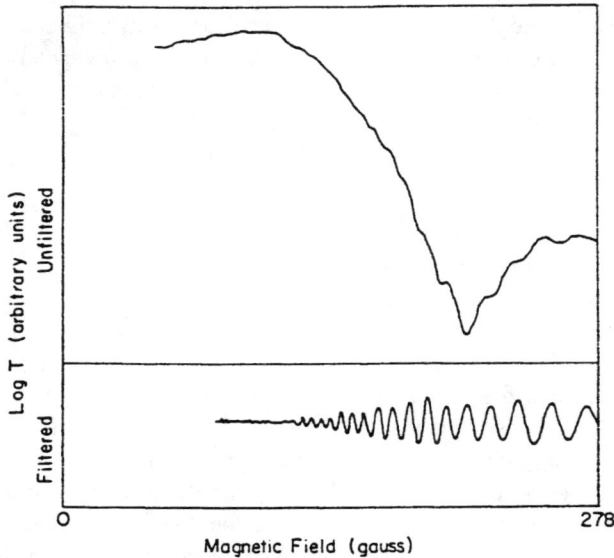

Fig. 8.5. Magneto-acoustic signal for Mg showing geometric oscillations. Upper trace are unprocessed data, lower trace are electronically filtered data (Ketterson and Stark [8.25])

8.4 Magneto-acoustic Quantum Oscillations

8.4.1 Theory

Apart from the magnetic field effects observed for $ql_e < 1$ (Alpher–Rubin effect, Sect. 8.2.2) and for $ql_e > 1$ (geometric resonances, Sect. 8.3.2), effects due to the quantization of the electron orbits can be observed. This effect, first observed in the magnetic susceptibility, is called de Haas van Alphen effect. A thorough presentation is given by Shoenberg [8.26]. The effect on ultrasonic attenuation and sound velocity we call magneto-acoustic quantum oscillations (MAQO). These effects have been reviewed by Roberts [8.27], Testardi and Condon [8.28], Shoenberg [8.26], Abrikosov [8.15], Onuki et al. [8.29], Gudkov and Gavenda [8.1].

The clearest way to understand this Landau quantization and to give conditions for the observation is the Onsager quantization for a crystal (Onsager [8.30]). The starting point is the Bohr-Sommerfeld quantization rule

$$\oint \boldsymbol{p} d\boldsymbol{r} = (n+\gamma)h \,. \tag{8.12}$$

For $B = 0$, $\hbar \boldsymbol{k}$ plays the role of the crystal momentum. Therefore in the crystal and with $B \neq 0$ we replace $\boldsymbol{p} - e\boldsymbol{A}/c$ by $\hbar \boldsymbol{k} - e\boldsymbol{A}/c$, with $\boldsymbol{A}$ the vector potential. With $\hbar d\boldsymbol{k}/dt = e/c[d\boldsymbol{r}/dt \times \boldsymbol{B}]$ we get

$$\oint \left(\hbar \boldsymbol{k} - \frac{e}{c}\boldsymbol{A}\right) d\boldsymbol{r} = \oint \frac{e}{c}[\boldsymbol{r} \times \boldsymbol{B}] d\boldsymbol{r} - \frac{e}{c}\oint \boldsymbol{A} d\boldsymbol{r}$$

$$= \frac{e}{c}\boldsymbol{B}\oint [\boldsymbol{r} \times d\boldsymbol{r}] - \frac{e}{c}\oint \boldsymbol{A} d\boldsymbol{r}$$

$$= eB2\frac{F_r}{c} - \frac{e}{c}\oint \boldsymbol{A} d\boldsymbol{r} \,.$$

Here F_r is the area of the orbit in real space. With $\oint \text{rot}\boldsymbol{A} d\boldsymbol{f} = \oint \boldsymbol{A} d\boldsymbol{r} = \oint \boldsymbol{B} d\boldsymbol{f} = BF_r$ follows

$$\oint \boldsymbol{p} d\boldsymbol{r} = \oint \left(\hbar \boldsymbol{k} - \frac{e}{c}\boldsymbol{A}\right) d\boldsymbol{r} = \left(\frac{eB}{c}\right) F_r = (n+\gamma)h \,. \tag{8.13}$$

Equation 8.13 says that the magnetic flux $\Phi = BF_r$ is quantized so that

$$\Phi = (n+\gamma)h\frac{c}{e} \,. \tag{8.14}$$

The elementary flux quantum is $hc/e = 4.13 \times 10^{-7} \text{G/cm}^2$. The orbit in real space with the area $F_r = A_r$ corresponds to the area in k-space $F_k = A_k$, rotated by 90° and $A_k = A_r(eB/\hbar c)^2$ or

$$A_k = (n+\gamma)\frac{eB}{\hbar c} \,. \tag{8.15}$$

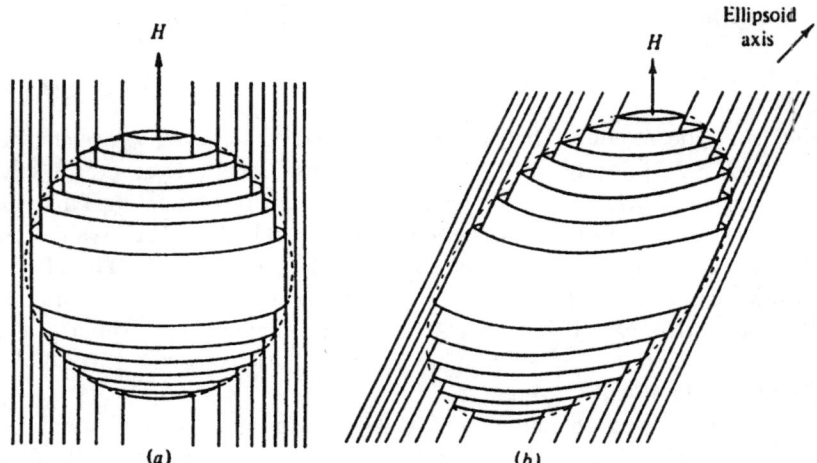

Fig. 8.6. Schematic sketches of Landau tubes for spherical and ellipsoidal surfaces of constant energy (Shoenberg [8.26])

This quantization of the orbits in k-space is shown in Fig. 8.6.
The allowed energies lie on so-called Landau tubes with energies of

$$E = (n+\gamma)\hbar\omega_c + \hbar^2 \frac{k_z^2}{2m}, \qquad (8.16)$$

where n is an integer indicating the corresponding Landau level and $\omega_c = eB/mc$ the cyclotron frequency. To observe quantum effects, the conditions

$$\omega_c \tau > 1 \quad \text{and} \quad \hbar\omega_c > k_B(T_D + T) \qquad (8.17)$$

have to be fulfilled. The first condition is the same as for the observation of geometric resonances, making the electron orbiting the constant energy path many times without being scattered by collisions. The second one separates the energy levels sufficiently not to be disturbed by temperature and collision broadening. The Dingle temperature $k_B T_D = \hbar/\tau$ is, together with $k_B T$, just a measure for such a broadening $k_B(T_D + T)$. Furthermore, the use of the Bohr-Sommerfeld quantization rule valid for $n \gg 1$ excludes the treatment of the so-called quantum limit ($n \sim 1$).

Given the energy spectrum, the free energy for quantized electrons can be constructed and from this all thermodynamic quantities such as magnetic susceptibility or strain susceptibility (elastic constants) can be calculated. The free energy formula due to Lifshitz and Kosevich [8.31] is derived in many books, e.g. Ziman [8.2], Abrikosov [8.15], Shoenberg [8.26].

Here we quote the oscillatory part of the free energy density:

$$F_{osc} = G\cos\left(\frac{F}{B} + \gamma\right) \qquad (8.18)$$

8.4 Magneto-acoustic Quantum Oscillations

with
$$F = c\hbar \frac{A_{ext}}{e}$$

and
$$G = 2k_B T \left(\frac{eB}{c\hbar}\right)^{3/2} (A''_{ext})^{-1/2} \frac{e^{2\pi^2 k_B T_D/\hbar\omega_c}}{\sinh\left(2\pi^2 k_B \frac{T}{\hbar\omega_c}\right)}.$$

A_{ext} is the extremal Fermi surface cross-section normal to the magnetic field vector, i.e. $A_{ext} = (A_k)_{ext}$ from above. We have already used the fact that only extremal Fermi surface cross sections contribute to the energy. Here $A'' = |\partial^2 A/\partial k_z^2|_{ext}$ near an extremal cross sectional area A_{ext}, $\omega_c = eB/m^*c$ is the cyclotron frequency and the Dingle temperature $T_D = \hbar/k_B\tau$ characterises the scattering time τ. There are also neglected higher harmonics concentrated on a single orbit with area A_{ext}. From (8.18) follows that Landau levels pass successively the Fermi level at a constant period in $1/B$.

Differentiation obtains any thermodynamic quantity, e.g. magnetization, susceptibility, elastic constant etc. The isothermal elastic constant change is

$$c_{osc} = g(B,T) \cos\left(\frac{F}{B} + \gamma\right) \tag{8.19}$$

with
$$g(B,T) = \left(\frac{\partial A_{ext}}{\partial \varepsilon_\Gamma}\right)^2 \left(\frac{c\hbar}{e}\right)^2 \frac{G}{B^2}.$$

For this, it is assumed that the strain dependence of the extremal Fermi surface cross-sectional area $\partial A/\partial \varepsilon_\Gamma$ is the coupling constant. For volume conserving strains one has (Fawcett et al. [8.4]) $(\partial A/\partial \varepsilon_\Gamma)_{Ef} = \int d_\Gamma(\mathbf{k})/\hbar v_n(\mathbf{k}) dl_k$ where dl_k is the line element of an extremal orbit. For parabolic bands this formula leads to

$$\frac{\partial A}{\partial \varepsilon_\Gamma} = 2\pi \frac{m^\star}{\hbar^2} \frac{\partial E_F}{\partial \varepsilon_\Gamma} = \left(2\pi \frac{m^\star}{\hbar^2}\right) <d_\Gamma>. \tag{8.20}$$

The above equations (8.19) and (8.20) show that useful information can be obtained from MAQO (extremal Fermi cross-section, effective masses, electronic relaxation times) and the strain derivative of the cross-sectional area for different symmetry strains. Combined with the de Haas-van Alphen susceptibility measurements results in the dimensionless quantity

$$\left(\frac{\partial \ln A_{ext}}{\partial \varepsilon_\Gamma}\right)^2 = \left(\frac{\partial \ln F}{\partial \varepsilon_\Gamma}\right)^2 = \frac{1}{F^2}\left(\frac{\partial F}{\partial \varepsilon_\Gamma}\right)^2 = -\frac{c_\Gamma^{osc}}{B^2 \chi^{osc}}. \tag{8.21}$$

From (8.21) it follows that the oscillatory elastic constant tends to favour smaller frequencies F, or smaller Fermi surface pieces, than the de Haas-van Alphen frequency ($c^{osc} \sim A_{ext}^{-2}$). A detailed account of oscillatory elastic constants and magneto-striction in the effective mass approximation has been worked out (Kataoka and Goto [8.32]). They can relate the coupling constants $\partial A_{ext}/\partial \varepsilon_\Gamma$ to the different components of the effective mass tensor for the different symmetries Γ.

8.4.2 Applications

The MAQO of some compounds is discussed below. The elemental metals have been discussed in previous books and reviews such as Shoenberg [8.26], Testardi and Condon [8.28], Roberts [8.27]. Thus the concentration is here on compounds of recent interest, like intermetallic rare earth compounds, heavy fermion compounds and Sr_2RuO_4, in short strongly correlated electron systems. For these materials one usually has the condition $ql < 1$, but the conditions of (8.17) have still to be obeyed.

Heavy fermions are discussed in detail in Chaps. 9 and 10. The large Fermi surface pieces have heavy masses of the order of $m^\star/m \sim 100$ as determined experimentally by de Haas van Alphen measurements (Taillefer and Lonzarich [8.33]). Renormalized band structure calculations involving resonant phase shifts from scattering of electrons on f-centers can explain the experimental results adequately (Zwicknagl [8.34]). Since MAQO couples only to small Fermi surface pieces, as discussed above, information cannot be gained about heavy masses from such experiments. Therefore these new developments on renormalized band structures are not discussed here.

LaAg
A first case, **LaAg**, illustrates the coupling aspects. Figure 8.7 shows the temperature dependence of the symmetry elastic constants together with MAQO at low temperature for the same modes. It is seen that of the various modes, it is the $(c_{11} - c_{12})/2$ mode which exhibits the biggest effect both in the temperature dependence and in the MAQO. In the temperature dependence, the other modes show no anomalies. The c_L mode ($k \parallel 110$) exhibits small MAQO due to $c_L = c_B + (c_{11} - c_{12})/6 + c_{44}$ containing $(c_{11} - c_{12})/2$. In Fig. 8.2 of Sect. 8.2, we have already analysed the temperature dependence of this $(c_{11} - c_{12})/2$ mode for **LaAg** using the redistribution formula (8.8). The analysis of the MAQO in this case gave good agreement with a scalar relativistic band structure calculation with the spin-orbit interaction included (Niksch et al. [8.7]). Furthermore it was possible to determine the deformation potential coupling constants by measuring both magnetic susceptibility and elastic constants on the same sample (see (8.21)). They vary between $d_{\Gamma 3} \sim 5000$ K and 38000 K. The fact that only the Γ_3 strain has an appreciable electron-phonon coupling can be understood from the simple coupling argument in Sect. 8.2, Fig. 8.1a,b. The observed Fermi surface pieces at the X and M points of the Brillouin zone are sensitive to Γ_3 strains but not to Γ_5 strains (Niksch et al. [8.7]).

8.4 Magneto-acoustic Quantum Oscillations

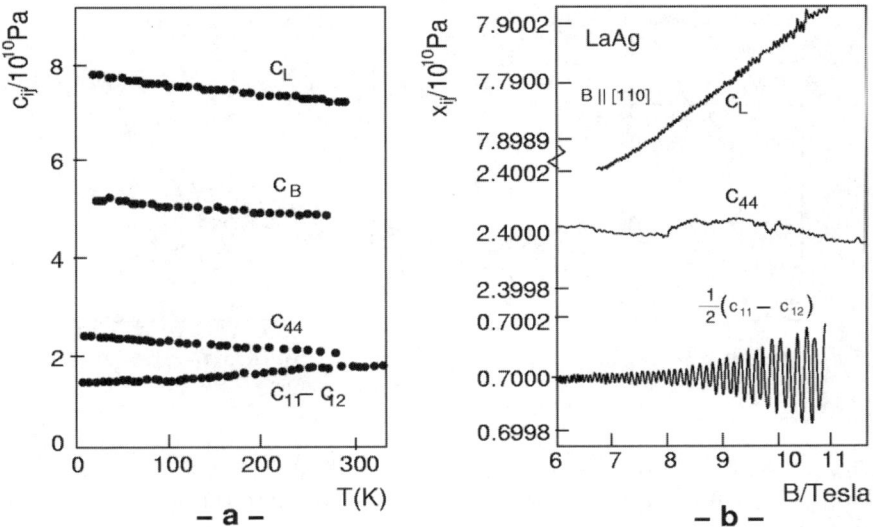

Fig. 8.7. Temperature and magnetic field dependence of various elastic modes in LaAg. (a) Temperature dependence, (b) field dependence for $B \parallel [110]$ at 1.3 K (Niksch et al. [8.7])

LaB$_6$

A particularly interesting system for ultrasonic determination of MAQO is the compound LaB$_6$. It has a cubic structure and it is the background substance of the RB$_6$ compounds which are again widely discussed because of many interesting properties for the various rare earth ions (see Sect. 7.2.2). Very small, nearly spherical Fermi surface pieces are located at the X-point for LaB$_6$. These can be determined with MAQO as shown in Fig. 8.8. Typical MAQO frequencies are as low as $F \sim 5$–6T. The agreement for the angular dependence of the MAQO with band structure calculations is very good (Suzuki et al. [8.35]). A detailed analysis of the electron–strain interaction for the different orbits has been given more recently (Matsui et al. [8.36]).

CeCu$_2$Si$_2$

MAQO also have been observed in the heavy fermion superconductor CeCu$_2$Si$_2$ (Wolf et al. [8.38]). The frequencies were found in the A and B phase (see Sect. 10.7.1 for a discussion of the various phases in CeCu$_2$Si$_2$) and were of magnitude $F \sim 128$–157T. They agreed nicely with de Haas van Alphen measurements (Hunt et al. [8.39]). The effective masses are of the order of 5, corresponding to the small frequencies F. The large Fermi surface pieces, giving rise to large frequencies F, have not been observed yet in the de Haas van Alphen effect.

Other correlated electron Ce systems exhibiting MAQO are CeSb (Settai et al. [8.40]), CeCu$_2$ (Settai et al. [8.41]), CeB$_6$ (Goto et al. [8.42]), CeCu$_6$

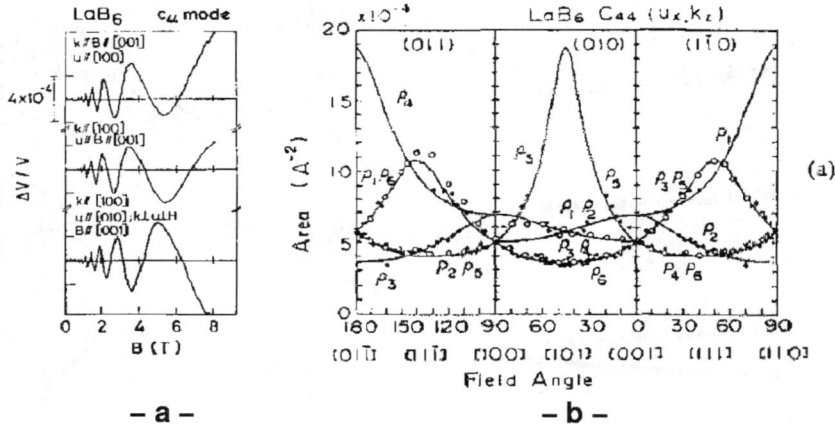

Fig. 8.8. LaB$_6$: (a) MAQO for c_{44} mode (Suzuki et al. [8.37]), (b) angular dependence of MAQO frequencies together with band-structure calculation (Suzuki et al. [8.35])

(Goto et al. [8.43]). In the latter case, a meta-magnetic like elastic constant anomaly was observed at 35 mK. Further discussion of CeCu$_6$ is given in Sect. 9.3. A review of these materials is presented by Onuki et al. [8.29].

TmSb

This cubic monopnictide material has been presented in this book as a model substance for magneto-elastic interaction (Sect. 5.2) with small exchange and small quadrupolar interaction and also as a model substance for rotationally magneto-elastic interactions (Sect. 13.2). Thus recent MAQO experiments performed in this compound (Nimori et al. [8.44]) are discussed briefly. For this low carrier substance, this acoustic technique is ideally suited to make Fermi surface investigations. The dHvA frequencies were smaller than 2000T and the Fermi surface consists of three branches α, β, γ with the α being the most important one as ellipsoids at the X points of the Brillouin zone. MAQO experiments gave small electron masses $m^*/m_0 \sim 0.1$–0.2 and enhanced Dingle temperatures 2–5 K. Exchange split bands were observed, with a typical exchange interaction of f electron-conduction electron of ~ 0.05 eV, and indications for rotational terms in the c_{44} mode were given.

U-compounds

For the uranium compounds MAQO were observed in UBe$_{13}$ (Wolf et al. [8.45]). However these are probably due to Al inclusions due to the Al flux growth (Corcoran et al. [8.46]). More recently MAQO were observed above and below the meta-magnetic transition ($B_c = 20$T) of UPt$_3$ (Feller et al. [8.47]). The de Haas-van Alphen frequency is $F = 450$T with an effective mass of $m^*/m = 15$, significantly lighter than the ones from larger orbits in this compound (Taillefer and Lonzarich [8.33]). Figure 8.9 shows Δc_{osc} for

8.4 Magneto-acoustic Quantum Oscillations

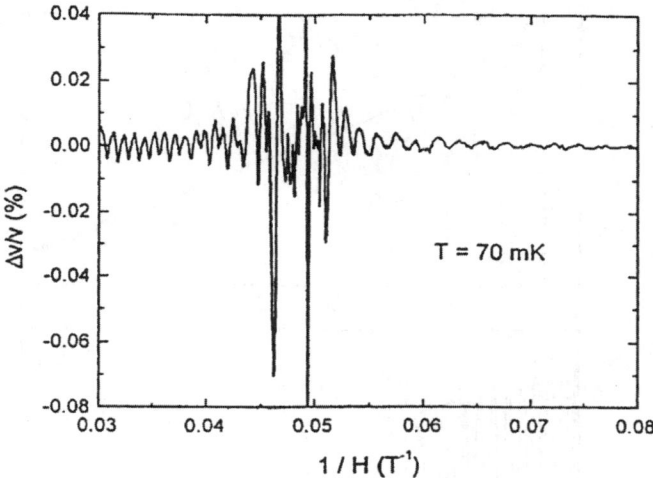

Fig. 8.9. MAQO in UPt$_3$ at 70 mK near the meta-magnetic transition for longitudinal sound (q//b axis//B) with background subtracted (Feller et al. [8.47])

the c_{11} mode ($\boldsymbol{q} \parallel b-$axis $\parallel \boldsymbol{B}$ at 70 mK below and above the meta-magnetic transition $B_c = 20T(1/B_c = 0.05)$.

Sr$_2$RuO$_4$

As a final example we discuss Sr$_2$RuO$_4$. This substance has attracted much attention in recent years because of its unconventional superconductivity. This topic will be discussed in Sect. 10.7.2. The crystal structure is of the same tetragonal layered perovskite as the high T_c superconductors. In contrast to the CuO$_2$ plane in high T_c cuprates, the alternating RuO$_2$ planes are responsible for the strongly correlated two-dimensional electronic states. Three cylindrical Fermi surfaces, labeled α, β and γ have been observed by various magnetic quantum oscillations in resistivity (Shubnikov–de Haas effect), magnetization (de Haas van Alphen effect) and ultrasonic velocity (MAQO) (Yoshida et al. [8.48], Mackenzie et al. [8.49], Matsui et al. [8.50], [8.51]). These Fermi surface pieces consist of bands whose origin is the hybridisation of anti-bonding Ru-4d and O-2p orbitals.

Ultrasonically the c_{11}, c_{33}, c_{44}, c_{66} and $(c_{11} - c_{12})/2$ modes were investigated (Matsui et al. [8.51] and [8.50]). Of the three Fermi surface pieces mentioned above, the c_{33} mode couples strongly to the α surface via a A_{1g} volume coupling, the $(c_{11} - c_{12})/2$ mode couples to the α and β pieces. The γ Fermi surface piece could not be observed by acoustic means because of the large A_{extr} (see (8.21)) and the large effective mass of $17m_e$ as determined from de Haas-van Alphen experiments. In Fig. 8.10 we show MAQO for the $c_{11} - c_{12}$ mode at 1.4 K together with the frequency spectrum.

The analysis of the MAQO in Sr$_2$RuO$_4$ gives the following parameters: Dingle temperature $T_D \sim 0.4 - 0.7$ K for the 3 Fermi surface pieces. Cou-

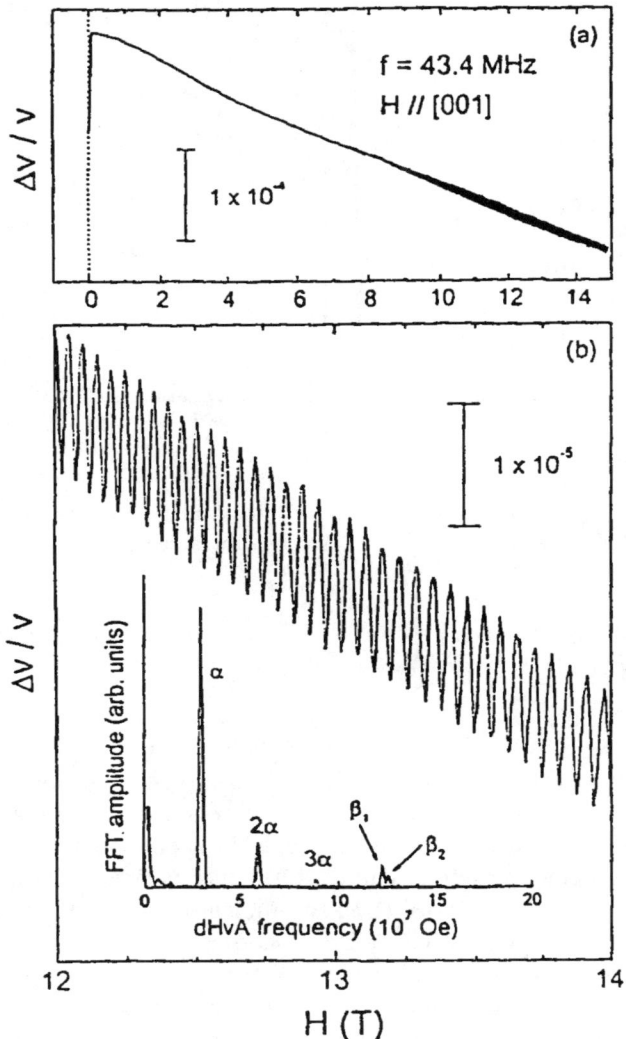

Fig. 8.10. MAQO for Sr_2RuO_4 for the $c_{11} - c_{12}$ mode at 1.4 K with 43.4 MHz sound waves (Matsui et al. [8.50]) (**a**) acoustic dHvA oscillations, (**b**) enlarged portion for 12–14 T, inset Fourier transform spectrum

pling constants $\partial \ln A_{extr}/\partial \varepsilon_\Gamma \sim 20$ for α and $\Gamma_{1g}(c_{33})$, ~ 0.8 for α and $B_{1g}(c_{11} - c_{12})$, $\sim 0.3 - 0.2$ for $\beta_{1,2}$ and $B_{1g}(c_{11} - c_{12})$. Further discussion of this compound, especially its superconducting properties will be given in Sect. 10.7.2. A complete determination of the elastic tensor at room temperature using resonant ultrasonic spectroscopy (RUS) has been given recently (Paglione et al. [8.52])

8.5 Ultrasonics in Semiconductors and Semimetals

In semiconductors, many ultrasonic effects were studied in the fifties and sixties. These include inter-valley scattering as a mechanism for attenuation and velocity changes of ultrasonic waves, magneto-acoustic quantum oscillations, the acousto-electric effect and amplification in photoconductive semiconductors. Since a number of reviews on these topics exist (e.g. Einspruch [8.53], Keyes [8.54]), only a short account of the most important effects in these fields is given here. Since some of the considerations are also valid for semi-metals a certain amount of overlap to previously discussed topics is possible.

8.5.1 Inter-valley Scattering Effect

Free carriers in heavily doped multi-valley semiconductors can produce significant contributions to certain symmetry elastic constants. The basic mechanism is the same as discussed previously for semi-metals (Sect. 8.2 and Fig. 8.1a,b). A small strain wave that shifts the equivalent valleys of the conduction band (in the n-type case), or shifts the heavy and light hole bands (in the p-type case) with respect to each other, causes a redistribution of carriers between the relevant band extrema. This results in a reduction of the electronic contribution to the free energy. Therefore the effective elastic constant for the applied strain is smaller for the doped than for the intrinsic material. With a static strain, the redistribution of the carriers is large and the electronic contribution to the elastic constants is considerable.

While the intra-valley scattering time is very short, as in the case of the metals, the inter-valley scattering time is longer and can be determined from (4.19) by measuring elastic constant changes and attenuation. In Ge the c_{44} mode lifts the degeneracy and equivalence of the valleys since the valleys lie on [111] axis. Therefore the elastic modes involving c_{44} are strongly attenuated in the heavily doped case. For As doping of $10^{18}/cm^3$ in Ge, an inter-valley scattering relaxation time of $\tau = 4 \times 10^{-13} s$ was found (Mason and Bateman [8.55]). Microwave hypersonic waves of 8.9 GHz gave nice agreement with these descriptions in pure and heavily doped Ge (Pomerantz et al. [8.56]).

8.5.2 Acousto-electric Effect

This effect happens by the transfer of momentum from the ultrasonic wave to the electrons. It is due to bunching of electrons in the potential minima of the periodic electric field created and accompanying the sound wave. The electrons create an acousto-electric current by following the wave motion. An acousto-electric field E_{ae} is measured which is related to the attenuation α of the sound wave (Weinreich [8.57]) $\alpha = (eNv_s/2I)E_{ae}$ with v_s the sound velocity and I the intensity of the acoustic wave. Here the doping level is smaller than in the relaxation case discussed above.

If one applies a dc electric field E_{dc} parallel to the sound wave direction, the electrons gain a drift velocity $v_d = \mu E_{dc}$ with μ the mobility. If the drift velocity exceeds the sound velocity v_s, the electrons move ahead of the sound wave and will be further accelerated and de-accelerated depending on the sign of the sound wave field. This leads on average to an amplification as described below.

8.5.3 Sound Wave Amplification in Piezoelectric Semiconductors

Light-sensitive ultrasonic attenuation has been observed in photoconductive CdS. These effects result from the relaxation interaction of mobile charge carriers with the strong longitudinal electric fields of piezoelectric origin accompanying the acoustic wave in the material (Hutson and White [8.58]).

The possibility of acoustic amplification has been realised in photoconductive CdS by applying a dc electric field (Hutson et al. [8.59]). If the dc electric field is so large that the drift velocity of the charge carriers exceeds the sound velocity, acoustic amplification has been observed. Figure 8.11 shows the acoustic apparatus for this experiment. A shear wave with propagation direction perpendicular to the hexagonal c-axis and with polarization along the c-axis has a frequency of 15 or 45 MHz. An Indium film acts as an electrode for the drift field pulse and as a bond for the delay line buffers. The light wavelength is 5770/5790 Å. In the dark, the crystal is insulating.

In Fig. 8.12, the sound wave attenuation for the two frequencies used is shown as a function of drift electric field. It is seen that the acoustic attenuation changes to acoustic gain at a field strength of about 700 V/cm. In order to observe this effect, the 5μsec drift-field pulse had to overlap the 3.5μsec transit time of the 1μsec ultrasonic pulse in the sample. It was further

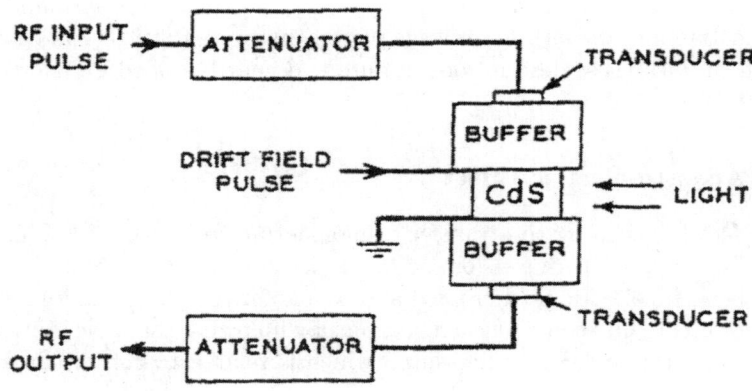

Fig. 8.11. Acoustic amplification in piezoelectric semiconductors. Experimental arrangement (Hutson et al. [8.59])

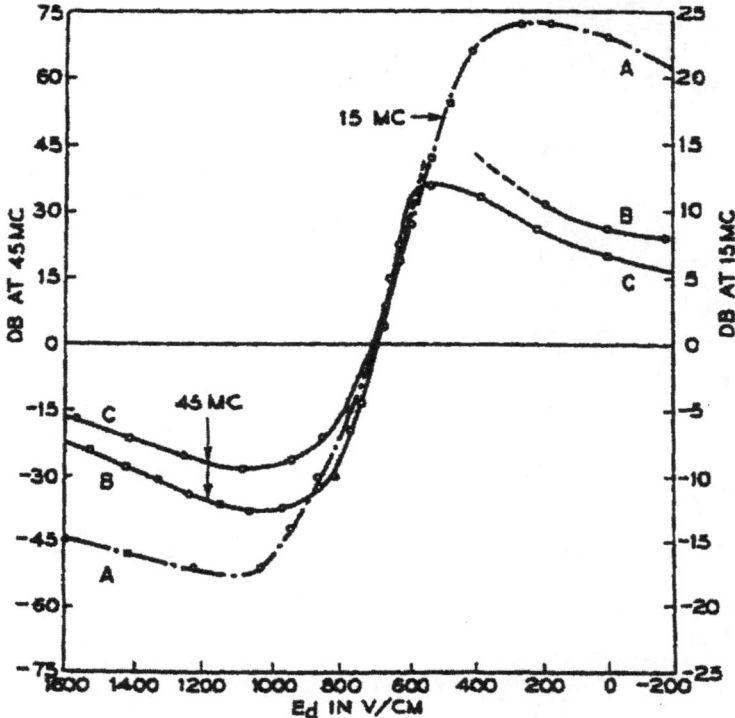

Fig. 8.12. Observed attenuation as a function of drift field (Hutson et al. [8.59])

observed that, with no input signal in the presence of the drift field, acoustic oscillations were produced by amplification of noise in the crystal.

The new field of semiconductor–heterostructure materials, including substances exhibiting the quantum Hall effect, will be discussed in Sect. 13.6.

8.5.4 Ultrasonic Amplification in Bismuth

Ultrasonic Amplification, using an electric drift field has been achieved also in the semi-metal Bi (Toxen and Tansal [8.60], Walther [8.61]). Originally a kink in the I-V characteristic has been observed for perpendicular magnetic fields of $B > 1$T at drift fields of the order of a few Volts/cm (Esaki [8.62]).

With $B = B_z$, $E = E_y$, the drift velocity $v_x = cE_y/B_z$ is independent of the electron mass and velocity. It turns out that $v_x \sim v_s$ = sound velocity at the kink field. Therefore for $v_x > v_s$ one can observe amplification of sound. The maxima of the attenuation in the giant quantum oscillations for shear waves changes to minima of attenuation by application of an appropriate drift field (Walther [8.61]).

9 Unstable Moment Compounds

The instability of magnetic moments in intermetallic rare earth and uranium compounds represents a very interesting many-body phenomenon in condensed matter physics. It is a very common phenomenon in cerium compounds but exists also in Yb, Eu, Sm and U compounds. Many of these compounds have a non-magnetic or nearly non-magnetic ground state, despite the existence of free $4f$ or $5f$ moments at elevated temperatures. This effect has given the name to this class of compounds, which will be discussed in this chapter. In a simplified view, the Curie–like high temperature susceptibility saturates below a fluctuation temperature $T^\star$ in an enhanced Pauli–type susceptibility characteristic for a non-magnetic Fermi liquid ground state. $T^\star$ gives the relevant energy scale for these systems. The microscopic origin of this behaviour lies in the interplay between the hybridisation of localized $4f$ (or $5f$) and itinerant $5d6s$ electrons and the intra-site Coulomb repulsion of the f electrons (Anderson [9.1]). Unstable moment systems can be roughly divided into two classes: Kondo lattices and mixed valent compounds with various compounds lying in between. It has to be emphasized that for the rare earth compounds, this division can be justified somehow, but for the uranium compounds, the Kondo lattice aspect (or better the heavy fermion ground state) is not yet understood. For a recent development for U–compounds see the dual model of Zwicknagl et al. [9.2]. This division into Kondo lattice and mixed valence effects can be best visualised with Fig. 9.1 from Thalmeier and Lüthi [9.3].

For Kondo lattices, the $4f$ level lies well below the Fermi level, charge fluctuations are suppressed and the $4f$ occupation n_f is almost integer. The fluctuation temperature $T^\star$ is then given by the single ion Kondo temperature T_K, which is of the order of 1–20K in Ce compounds. A spin singlet is formed below T_K at each site of the magnetic ions between the f electrons and the conduction electrons. The energy gain due to this singlet formation is given by $-k_B T^\star$. Below $T^\star$, coherent heavy electron bands may develop in the Kondo lattice case with large effective masses up to $m^\star \sim 10^2 m_0$, (m_0 = free electron mass) and a band width of T_K. There are also compounds where heavy band formation is intercepted by magnetic order of already reduced moments. The heavy bands can also experience a spin splitting with very small resulting magnetic moments ($\mu < 0.1\mu_B$) below T_K. Generally speaking, with heavy fermions, there is a competition between long range magnetic order (RKKY

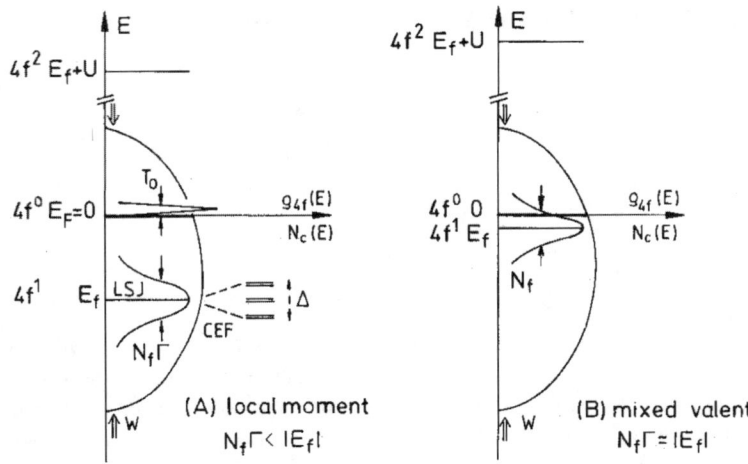

Fig. 9.1. Schematic energy diagram for electronic states in Cerium intermetallic compounds (Thalmeier and Lüthi [9.3])

mechanism discussed in Sect. 5.1.1) and a local screening of the f-moments by the conduction electrons (Kondo effect). Heavy fermion systems have been intensively investigated in the last few decades and a number of review articles exist: Brandt and Moshchalkov [9.4], Stewart [9.5], Lee et al. [9.6], Fulde et al. [9.7], Ott [9.8], Grewe and Steglich [9.9], Hess et al. [9.10], Hewson [9.11].

In mixed valence compounds, also represented in Fig. 9.1, the 4f level is closer to the Fermi level and real charge fluctuations lead to non-integer occupation n_f. Homogeneous valence fluctuations are dealt with here in contrast to the case of inhomogeneous ones which have been discussed in Sect. 7.1. Roughly speaking, in the case under consideration, we are dealing essentially with a single ion property where the magnetic ion hybridises with the sea of conduction electrons and one has an exchange of an inner $(4f)$ electron with the conduction band at the Fermi level. T^* ranges here from a few hundred degrees in the weakly mixed valence compounds to a few thousand degrees in the strong ones. Reviews on mixed valency are e.g. Varma [9.12], Jayaraman [9.13], Lawrence et al [9.14], Wachter [9.15].

The crossover from mixed valent to heavy fermion behaviour can be seen by considering the low temperature magnetic susceptibility χ and the specific heat $C = \gamma T$ expressed with the energy scale T^* : $\chi(T=0) \sim \mu_B^2 \langle n_f \rangle / T^*$ and $\gamma \sim \pi^2 k_B^2 \langle n_f \rangle / T^*$. For heavy fermion systems the f-occupancy $\langle n_f \rangle$ is close to 1, but in mixed valent compounds < 1. As seen from Fig. 9.1, increasing hybridization increases T^* thus lowering χ and γ, whereas small T^* leads to high density of states or high effective masses for heavy fermions. The ratio of χ and γ leads to the so-called Wilson ratio R discussed later (9.14).

For unstable moment compounds, the pressure is an important experimental parameter. Volume reduction changes the hybridisation strength between the f-electron and the conduction electrons. This affects the heavy mass of the quasi-particles and therefore the Kondo temperature T_K or generally the fluctuation temperature $T^\star$. As experimentally shown, a pressure of 10 GPa can change T_K from 10K to $\approx$ 100K, i.e. it shifts the system from the heavy fermion state to the mixed valence regime (Schilling [9.16], Thomas et al. [9.17]).

We begin with an experimental characterisation of mixed valent and Kondo lattice compounds (Sect. 9.1), followed by a detailed exposition of structural and acoustic effects of the two systems. The electron-lattice coupling in mixed valence systems is manifested by a strong softening of the bulk modulus (inverse compressibility) and anomalous Poisson ratios. The various mixed valent compounds can be characterised by the so-called charge fluctuation rate. These topics are the content of Sect. 9.2. For heavy fermions the volume dependence of the $s-f$ hybridization strength, manifested by the volume dependence of the characteristic fluctuation temperature $T^\star$, is the important electron lattice coupling (Grüneisen parameter coupling). This will be discussed in Sect. 9.3.1–9.3.4. Furthermore, meta-magnetic transitions in high magnetic fields (Sect. 9.3.5) and non-Fermi liquid effects near magnetic quantum critical points (Sect. 9.3.6) are other characteristic features of heavy fermion compounds.

9.1 Experimental Characterisation

9.1.1 Mixed Valence Compounds

One of the most striking properties of mixed valence compounds and a first rough way to estimate its average valency is its dependence on the lattice constant, commonly called Végard's law. This law assumes a linear dependence of the valence from the lattice constant and a knowledge of the lattice constants a(n) and a(n-1) for the rare earth configurations $4f^n$ and $4f^{n-1}$. In Fig. 9.2 we plot the lattice constant of the series of rare earth compounds called rare earth mono-chalcogenides (RTe, RSe, RS) (Bucher et al. [9.18]). They form in the NaCl structure and the R^{2+} ions have a larger radius than the R^{3+} ions. Note that the Sm, Eu, Yb chalcogenides have 2+ ions and are insulators while the others with 3+ ions are metals. If the compound is in the mixed valence state, the lattice constant is an average of the 2+ and 3+ lattice constants suitably weighted. This is realized for SmS and TmSe as seen in Fig. 9.2. Such studies have been performed for many mixed valence compounds (see reviews listed above and Penney [9.19]); examples are the Ce-compounds (α–Ce, CeBe$_{13}$, CePd$_3$, CeNi), the Sm-compounds (SmS, SmB$_6$), TmSe and the Yb–compounds (YbAl$_2$, YbInCu$_4$). Related to the anomalous lattice constants, the thermal expansion also shows pronounced

184   9 Unstable Moment Compounds

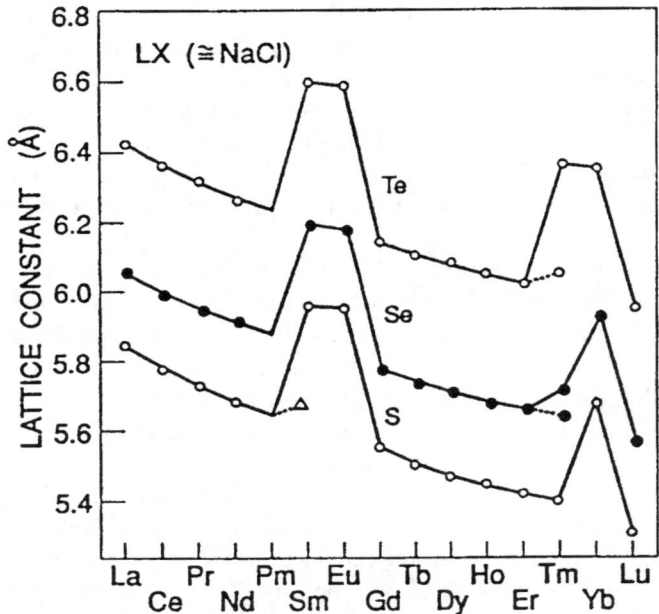

Fig. 9.2. Lattice constants for the rare earth mono-chalcogenides (Bucher et al. [9.18]). Note the intermediate lattice constants for SmS and TmSe

anomalies in the same temperature range as the magnetic susceptibility. This will be discussed in detail in Sect. 9.2, Fig. 9.9. Of great importance in this field are spectroscopic investigations. Here a distinction is made between high energy spectroscopy (photoemission with X-ray and ultraviolet sources and bremsstrahlung isochromate spectroscopy) and low energy spectroscopy (Mössbauer effect). In the former case, the frequency of the spectroscopic tool is larger than the charge fluctuation rate ω_c, thus, in principle, the valencies of the ions can be distinguished. In the latter case, the frequency is smaller than ω_c. Isomer shifts in Mössbauer measurements help to determine the valence and indicate that the valence distribution is uniform through the sample and homogeneous (Nowik [9.20]). The quasi-elastic line widths in neutron scattering give a measure of the fluctuation temperature (Loewenhaupt and Fischer [9.21], Murani [9.22]). Raman scattering can also help to identify the charge fluctuation rate (see Sect. 9.2).

9.1.2 Kondo Alloys and Kondo Lattices

The Kondo effect was discovered originally for 3d–substances. The resistance minimum, observed already in the nineteen-thirties, was explained by Kondo [9.23] with a higher order exchange scattering of conduction electrons by a magnetic impurity. Interaction effects between the Kondo ions are

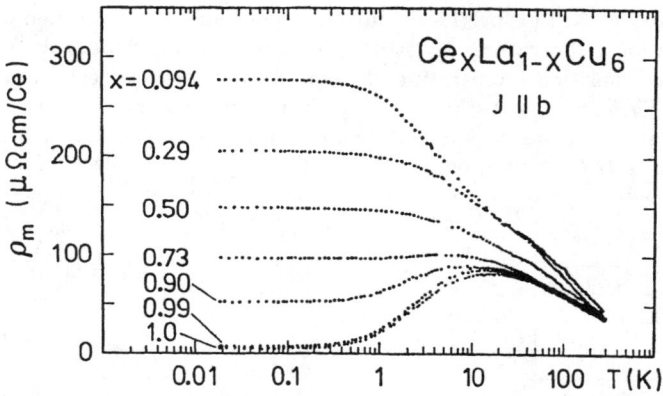

Fig. 9.3. Electrical resistance versus temperature for the Kondo system $Ce_xLa_{1-x}Cu_6$ (Onuki and Komatsubara [9.24])

much reduced in 4f systems due to the localization of the $4f$ wavefunction. Therefore later Kondo effect studies were made mainly with Ce^{3+} impurities. Kondo-like resistivities can also be observed in concentrated Kondo compounds. Typical examples are $CeAl_3$, $CeAl_2$, $CeCu_2Si_2$, $CeCu_6$ and $CeSn_3$. A characteristic property, apart from the resistivity, is the low temperature specific heat C or its Sommerfeld coefficient $\gamma = C/T$, which can have very large values of the order of $1 Jmol^{-1}K^{-2}$, compared to $1 mJmol^{-1}K^{-2}$ for Cu. This very large value, expressed in electron density of states or effective masses m^*/m_0 (up to 200), has led to the name of heavy fermion compounds. In Fig. 9.3 we show typical resistivity curves for the system $Ce_xLa_{1-x}Cu_6$ (Onuki and Komatsubara [9.24]). With $T^* \sim T_K = 10K$ for these compounds, a negative temperature coefficient in temperature, typical for the Kondo effect, has been noted for the dilute $x = 0.094$ for more than two decades. In the concentrated system $x = 1$, a resistivity increase for $T > T^*$ has been observed but for $T < T^*$ the resistance decrease due to the formation of the Kondo lattice with a coherent ground state.

Kondo lattices or heavy fermion systems at low temperatures can be described with the Fermi liquid theory (see reviews cited above). The thermodynamic quantities have temperature dependencies as in the free electron model, but with strongly renormalized parameters. The specific heat is $C = \gamma T$ with γ enhanced by m^*/m, the same being true for the paramagnetic susceptibility so that the Sommerfeld - Wilson ratio R (a dimensionless ratio of magnetic susceptibility χ and γ) remains unchanged, $R = \pi^2 k_B^2 \chi / 3\mu_B^2 \gamma \approx 1$. The electrical resistivity at low temperatures has the form like $\rho = AT^2$ with $A \propto \gamma^2$, therefore also enhanced. This is the Kadowaki-Woods [9.25] relation.

In Table 9.1 we give a compilation of physical properties of mixed valent and heavy fermion compounds. The table is limited to just such materials for which the acoustic properties are discussed.

Table 9.1. Physical properties for mixed valence and heavy fermion compounds. T^* fluctuation temperature, T_Q structural (quadrupolar) transition temperature, T_V valence transition temperature, T_N anti-ferromagnetic (Neel) transition temperature, T_C ferromagnetic (Curie) transition temperature, T_{sc} superconducting transition temperature, γ Sommerfeld spec. heat coefficient, c_B bulk modulus, ν Poisson ratio, Ω Grüneisen parameter

Material	T^*	T_Q	T_V	T_N	T_C	T_{sc}	γ mJmol^{-1}K^{-2}	ρ_0 μΩcm	c_B 10^{11}erg/cm^3	ν	Ω
Ce							13		1.88	0.23	
CeAl$_3$	3						1200		4.6	0.175	-200
CeCu$_6$	5						1600		9.1		$\Omega_a=70\Omega_c>120$
CeRu$_2$Si$_2$	15						385				$\Omega_a=\Omega_c=120$
CeAl$_2$	6			3.8			135		6.9	0.168	50
CeCu$_2$Si$_2$	10			0.7		4.8	750		12.5		$\Omega_a=69\Omega_c=32$
CeAg		15			5.3						
CeB$_6$	7	3.3		2.1			260		16.8	0.033	
CePb$_3$	15			1.1			225				
CeSn$_3$	260						42		5.4		16
CePd$_3$	220						35		10.25	0.312	10
CeBe$_{13}$	340										115
CeNi	140						85				~ 8
Ce$_{.74}$Th$_{.26}$			149					0.1 (150K)		-0.75	
CeRu$_2$					6.3		28		13.0	0.48	
CePd$_2$Al$_3$	20						1000				
EuCu$_2$Si$_2$	>300								1.03(300K)		
SmB$_6$							6.8		9.33(300K)	-0.195	
SmS											
Sm$_{75}$Y$_{25}$S	>300								0.83(300K)	-0.671	
SmPd$_3$				1.4					14.5	0.35	
SmIn$_3$				14.8							
SmSn$_3$		10.9		9.9			100-300		7.0	0.38	
SmTl$_3$				8.6							
SmPb$_3$				5.1					5.87	0.4	
TmSe				3					2.17(300K)	-0.467	
YbAl$_2$							17		4.67(290K)	0.102	
YbInCu$_4$	280		42				50				56
YbAgCu$_4$	100						210				-31
UPt$_3$	80		5	0.5			450	0.5	20.9	0.19-0.5	$\Omega_a=91$ $\Omega_c=32$
UBe$_{13}$	7				0.9		1100	18			$\Omega_c=63$
URu$_2$Si$_2$	70		17	1.2			120	10			$\Omega_a=+57\Omega_c=-15$
UPd$_2$Al$_3$	85		14	1.8			150	3.5			$\Omega=-3.5$

9.2 Electron–Lattice Coupling in Mixed-Valence Systems

As discussed above, magnetic ions with unfilled $3d$ and $4f$ shells are often found in different valence states. The resulting state is said to exhibit interconfigurational or valence fluctuations. Examples are $Fe^{2+} - Fe^{3+}$ ions in spinel compounds which experience inhomogeneous valence or charge fluctuations as discussed in Sect. 7.1. As another example, $Sm^{2+} - Sm^{3+}$ ions in cubic rare-earth compounds experience homogeneous valence fluctuations. Crystal field excitations are not observed for these systems because the fluctuation temperature ($T^* > 300\,K$) for them is rather high (see table 9.1). The crystal field levels are broadened by the valence fluctuations. This is in contrast to heavy fermion systems with rather small $T^* < 100K$ where crystal field excitations can be observed. The case of $5f$ uranium compounds is special. Only in very few U–systems have crystal field excitations been observed (UPd_3, see Sect. 7.2). The larger spatial extent of the 5f wavefunctions should put these compounds into the mixed valence category. In fact, it is often difficult to assign a definite valence state to the U-ions. Nevertheless, some U-compounds show clear heavy fermion properties and, as mentioned above, this issue is still unsettled.

SmS, $Sm_{1-x}Y_xS$

By applying pressure or by alloying, a valence instability can occur in some systems. A famous example is SmS which collapses to an intermediate state under pressure or by alloying with Y, As etc. At ambient pressure, SmS is a semiconductor (black phase with NaCl structure) with Sm^{2+} ions. At room temperature with a pressure of 0.65GPa, it undergoes a first order iso-structural transition to a metallic phase with intermediate valence Sm^{2+}-Sm^{3+} (gold phase) (Jayaraman et al. [9.26]). The p–T phase diagram shows an increasing pressure to lower temperatures ($p \sim 2GPa$ for 4K). Large volume changes can be observed with no change in lattice symmetry, because the trivalent and divalent Sm ions have quite different ionic radii. In very rare cases, $Ce_{0.74}Th_{0.26}$ and $YbInCu_4$, these substances exhibit a valence transition as a function of temperature at ambient pressure (see below).

The electronic and elastic energies must be added in order to describe valence transitions (Hirst [9.27]). The appropriate order parameter for a homogeneous valence transition is the volume strain ε_V (see Sect. 3.2) and the associated soft mode is the bulk modulus c_B. Because of the large volume changes, an anomalous Poisson ratio ν is also expected. The Poisson ratio measures the lateral contraction to a uniaxial pressure, i.e. $\nu = -\varepsilon_\perp/\varepsilon_\parallel$ where $\varepsilon_\perp$, $\varepsilon_\parallel$ are the strains perpendicular and parallel to the applied pressure (see Fig. 9.4 and Sect. 3.3).

For a normal Poisson ratio ($\nu > 0$), there is a negative $\varepsilon_\parallel$ and a positive $\varepsilon_\perp$ (dashed lines in Fig. 9.4) whereas with $\varepsilon_\parallel$ negative and $\varepsilon_\perp$ negative, a $\nu < 0$ results (point dashed curve). It can be seen from Table 9.1, that all the

188   9 Unstable Moment Compounds

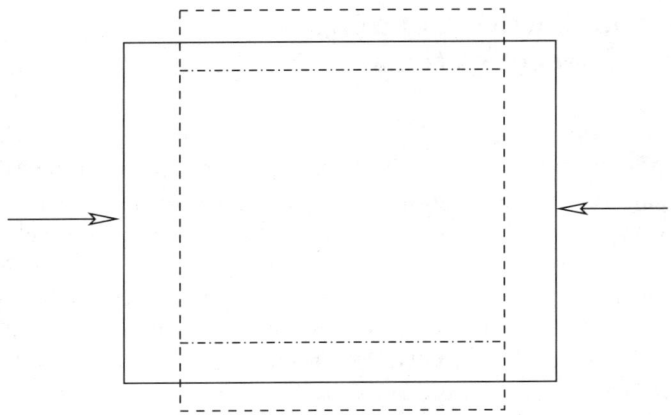

Fig. 9.4. Lateral contraction in mixed valence compounds: dashed lines: normal ν, point-dashed lines: negative ν

heavy fermion compounds have positive Poisson ratios, but several mixed valence compounds, though not all, have negative Poisson ratios. It shows that the soft bulk modulus and the negative Poisson ratio are characteristic for mixed valence compounds. The negative Poisson ratio in mixed valence compounds was first pointed out by Wachter (see Wachter [9.15]).

Figure 9.5 shows the softening of the bulk modulus as a function of x in $Sm_{1-x}Y_xS$ (Penney et al. [9.28]). Here x acts as an internal pressure coordinate. It is seen that at the semiconductor–metal transition for $x \sim 0.2$, c_B is practically zero. For pure SmS the pressure dependent bulk modulus does show a softening and the Poisson ratio shows a strong softening up to 6×10^8Pa, reaching zero for the critical pressure of 6.5×10^8Pa (Hailing et al. [9.29]).

In addition, some phonon modes in the measured phonon dispersion spectra of $Sm_{0.75}Y_{0.25}S$ exhibit anomalies (Mook et al. [9.30]). These are particularly strong for longitudinal phonons in the (111) direction, as shown in Fig. 9.6. The fact that $\omega_{LA}(k) < \omega_{TA}(k)$ for small k is again evidence for a negative Poisson ratio. In the linear part of the dispersion law we have (Chap. 3, Table 3.1) $\rho v_L^2 = 1/3(c_{11} + 2c_{12} + 4c_{44})$ and $\rho v_T^2 = 1/3(c_{11} + c_{44} - c_{12})$. For $v_L < v_T$ follows again $c_{12} < -c_{44}$, i.e. $\nu = c_{12}/(c_{11} + c_{12}) < 0$. Detailed calculations of these phonon dispersion spectra have been performed (Bilz et al. [9.31], Grewe et al. [9.32], Wakabayashi [9.33], Pastor et al. [9.34]).

These experimental results show the strong contrast between Kondo lattice compounds and mixed valence compounds. In the former, there is approximate particle conservation in the heavy fermion band, whereas in the latter we have a transfer of f electrons to the conduction band thereby reducing the screening and making $\nu < 0$. The lateral strain $\varepsilon_\perp$ can be estimated as $\varepsilon_\perp = \delta n_f/(Z - n_f)$ with n_f, δn_f the number and its change under an ex-

9.2 Electron–Lattice Coupling in Mixed-Valence Systems 189

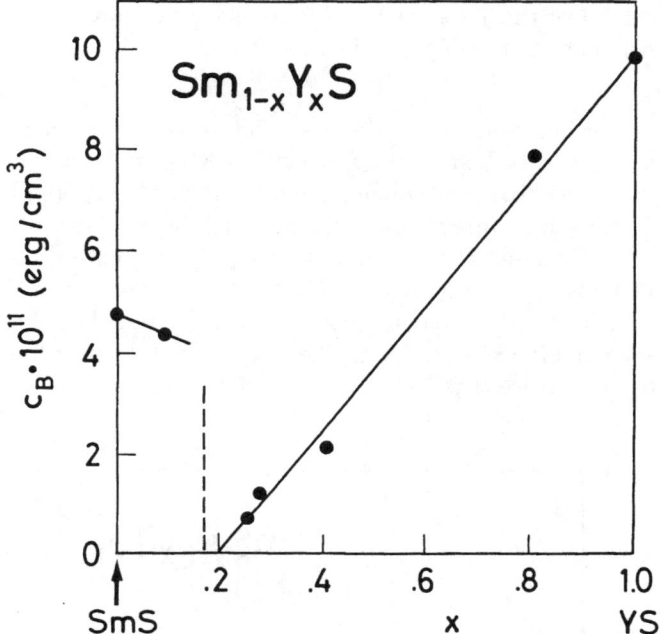

Fig. 9.5. Bulk modulus for $Sm_{1-x}Y_xS$ at 300K as a function of x (Penney et al. [9.28])

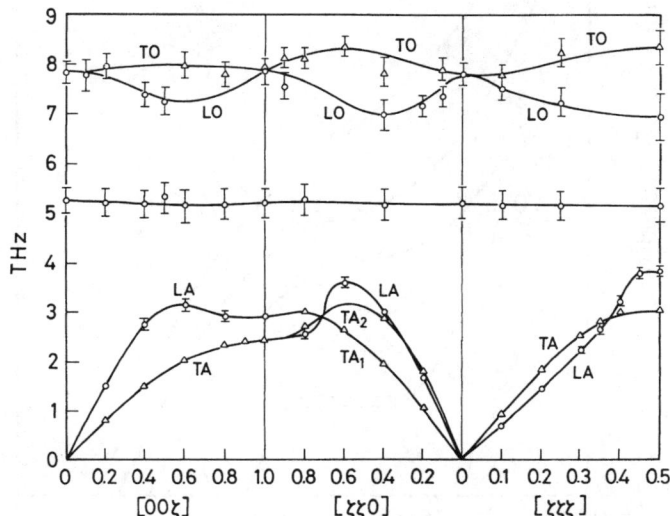

Fig. 9.6. Phonon spectra for the $Sm_{0.75}Y_{0.25}S$ system for various symmetry directions (Mook et al. [9.30])

ternal parameter of the f electron and Z an effective number of unscreened charges at the nucleus of the ion (Lüthi [9.35]).

TmSe$_{1-x}$Te$_x$, Ce$_{1-x}$Th$_x$

Apart from the Sm–compounds discussed above other systems show pronounced mixed valence behaviour. A special case is the cubic mono-chalcogenide TmSe which has the two magnetic configurations 4f^{13}(Tm^{2+}) and 4f^{12}(Tm^{3+}) and which orders anti-ferromagnetically at $T_N \sim 3K$. A great deal of work was put into establishing the $p - V - x$ phase diagram of the alloy system TmSe$_{1-x}$Te$_x$ (Wachter [9.15]). Figure 9.7 shows the $P-V$ phase diagram for this system (Boppart [9.36]). It can be seen that this looks like the well-known liquid-gas transition. The critical point is given by $x_c \sim 0.45$, $p_c \sim 0.8$ GPa. Because of the lack of a symmetry breaking, a critical point is

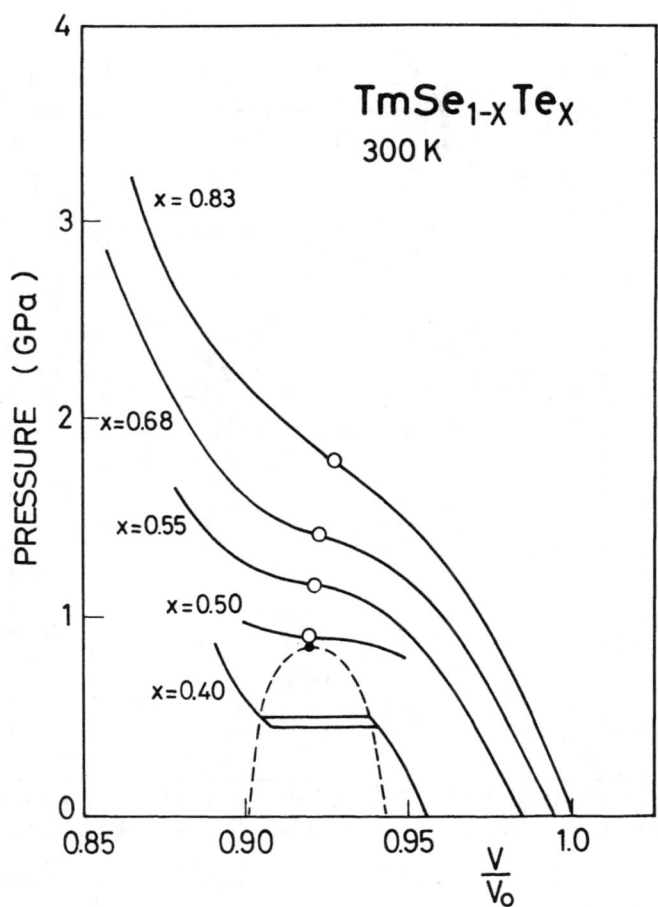

Fig. 9.7. The p-V phase diagram for TmSe$_{1-x}$Te$_x$ system at 300K (Boppart [9.36]). The critical point is indicated by a full dot

expected for this solid state transition. For $x < x_c$ one observes a first order semiconductor–metal transition. The phase transition shows mean field behaviour with classical critical exponents as expected for long-range elastic interactions with a soft bulk modulus at p_c (see Sect. 4.4).

Very similar behaviour to the TmSe–TmTe system has been observed for the $Ce_{1-x}Th_x$ system (Parks and Lawrence [9.37], Lawrence et al. [9.14]) with critical parameters $x_c = 0.265$ and $T_c = 149K$. Again a very small bulk modulus and a negative Poisson ratio have been observed for this $\gamma - \alpha$ iso-structural transition (Wehr et al. [9.38]).

$EuCu_2Si_2$, $YbCu_2Si_2$

Using Brillouin scattering as explained in Sect. 2.7, Mock et al. [9.39] determined the bulk modulus c_B of rare earth compounds with the formula RCu_2Si_2. The compounds with R = La, Nd, Gd, Tb, Er, Y, Ca had a normal c_B, following a straight line with Q/V (Q = valence of R and V = unit cell volume). Compounds with R = Sm or Ce, however, showed slight deviation from this behaviour and compounds with R = Eu or Yb showed a very large deviation, c_B being 30–60% smaller than for normal c_B. These room temperature measurements indicate again an anomalously small bulk modulus.

$YbAl_2$

For the mixed valence Yb compounds the Laves phase $YbAl_2$ has a mixed valency of 2.4 and reduced bulk modulus c_B and Poisson ratio ν (but still $\nu > 0$) in comparison to other rare earths in RAl_2 (Penney et al. [9.40]), see Table 9.1. But the temperature dependence of c_B and ν is weak.

$YbIn_{1-x}Ag_xCu_4$

This system is a widely discussed case nowadays (Sarrao [9.41]). $YbInCu_4$ exhibits an iso-structural transition at $T_V = 42K$ (Felner and Novick [9.42]) at ambient pressure, again without a symmetry change, thereby reducing the valency of the high temperature phase Yb^{3+} to $Yb^{2.8}$ in the low temperature phase (Sarrao [9.41]). The dilution with Ag is used to make this transition more second order. Elastic constants have been measured for this system (Kindler et al. [9.43] and Zherlitsyn et al. [9.44]). Both the bulk modulus c_B and the Poisson ratio ν exhibit softening towards the valence transition. But $c_B > 0$ and $\nu > 0.2$ for all Ag concentrations for this system. The lattice constant changes abruptly through the transition by about 0.14%, i.e. it is larger in the low temperature mixed valence phase (Kindler et al. [9.43], Cornelius et al. [9.45]). In contrast to the Kondo volume collapse model, this would indicate a negative Grüneisen parameter for $YbInCu_4$ (see (9.5) in Sect. 9.3.2 below). But it is positive and only negative for $YbAgCu_4$ (Svechkarev et al. [9.46]) (Table 9.1). The difference in lattice constant behaviour of Ce and Yb–compounds is due to the different f-shell occupation (Müller-Hartmann [9.47]).

With high field ultrasonics, the $B - T$ phase diagram can be traced for this system $YbIn_{1-x}Ag_xCu_4$ as shown in Fig. 9.8a,b. Along the phase boundary, the longitudinal modes exhibit a strong minimum. In the insert we trace

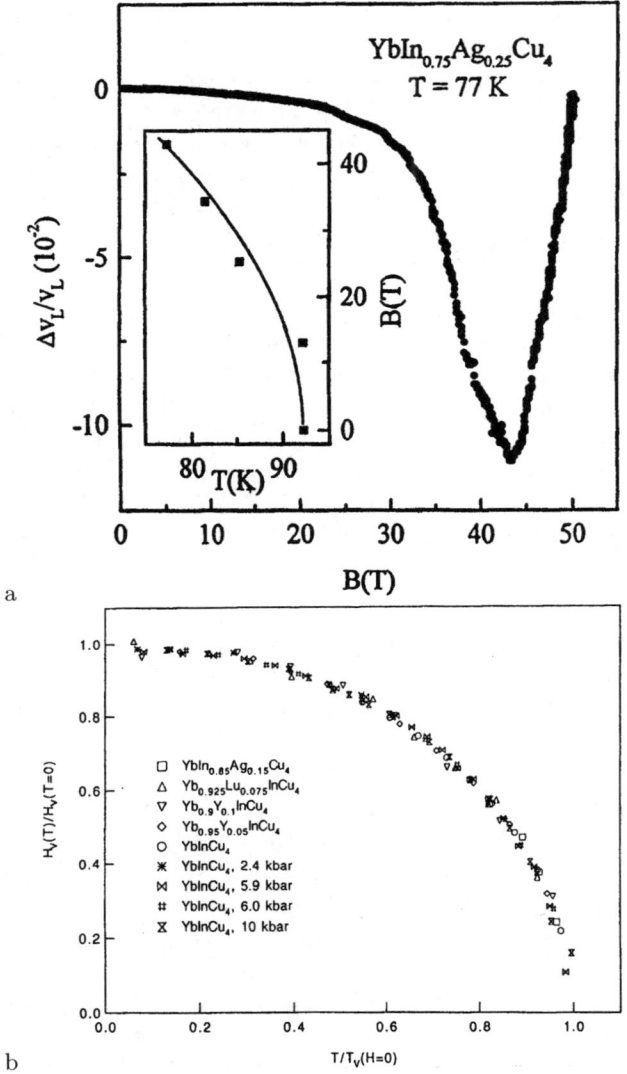

Fig. 9.8. (a) Longitudinal sound velocity for $YbIn_{0.75}Ag_{0.25}Cu_4$ in pulsed fields up to 50T. Insert $B-T$ phase diagram (Zherlitsyn et al. [9.44]), (b) $B-T$ phase diagram for $YbIn_{1-x}M_xCu_4$ with M = Ag, Lu and pressure in reduced units (Immer et al. [9.48])

the B-T diagram from these measurements. It is amazing that all the different compounds $YbIn_{1-x}M_xCu_4$ where M = Ag, Lu and pressure dependent $YbInCu_4$ give a universal $B-T$ phase diagram if plotted on reduced scales: $[B_c(T)/B_c(0)]^2 + [T_c(B)/T_c(0)]^2 = 1$ as shown in Fig. 9.8b (Immer et al. [9.48]). This phase diagram was recently explained with an entropy driven transition for free Yb-ions (Dzero et al. [9.49]). A detailed analysis of the various physical properties was recently given by Goltsev and Bruls [9.50].

Another mixed valence system which exhibits temperature and field induced phase transitions is the Eu–compound $EuNi_2(Si_{1-x}Ge_x)_2$ (Wada et al. [9.51]). Measurements of the differential magnetic susceptibility up to 500 Tesla gave a B–x phase diagram (Platonov et al. [9.52]).

Other Cerium Compounds

$CeSn_3$, $CePd_3$, $CeBe_{13}$, CeNi

There exist a number of Ce–mixed valence compounds ($CeSn_3$, $CePd_3$, $CeBe_{13}$, CeNi) which do not exhibit a complete bulk modulus softening or negative Poisson ratio. Instead they have a minimum in c_B, a maximum in the thermal expansion and the magnetic susceptibility as a function of temperature. In Fig. 9.9a we show linear thermal expansion for $CeSn_3$ and $LaSn_3$ (Takke et al. [9.53]) and Fig. 9.9b,c gives bulk modulus data c_B for $CeSn_3$ and $CeBe_{13}$ (Lüthi et al. [9.55]). The $CeBe_{13}$ data were taken from Lenz et al. [9.54].

These data were interpreted phenomenologically with a scaling Ansatz (Takke et al. [9.53]). We parameterize the electronic Helmholtz free energy density

$$F_e = -kT \sum_i e^{-H/kT} = kTnf\left(\frac{T}{T^\star}\right) \qquad (9.1)$$

with a single energy scale $T^\star$, where n is the number of magnetic ions per unit volume. The change in the bulk modulus is given by $\Delta c_B = \frac{\partial^2 F_e}{\partial \varepsilon_V^2}$ and the change in thermal expansion by $\Delta \beta = -c_B^{-1} \frac{\partial^2 F_e}{\partial T \partial \varepsilon_V}$ and for the change in the electronic specific $\Delta C = -T \frac{\partial^2 F_e}{\partial T^2}$. As the coupling to the lattice, the volume dependence of $T^\star$ is taken in the form of the Grüneisen parameter.

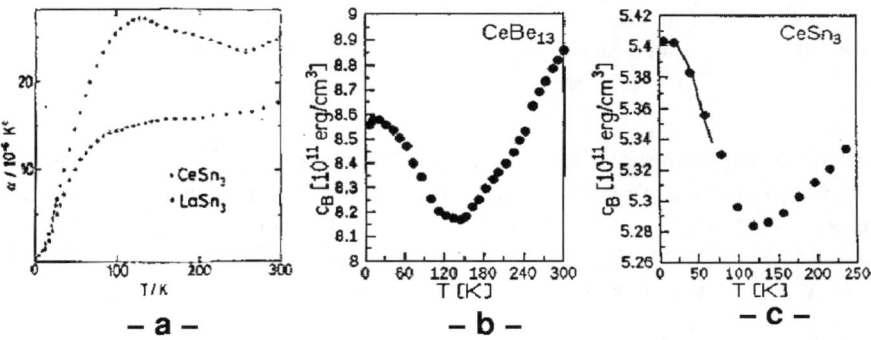

Fig. 9.9. (a) Thermal expansion of $CeSn_3$ and $LaSn_3$ (Takke et al. [9.53]), (b) Bulk modulus c_B as a function of temperature for $CeBe_{13}$ and (c) $CeSn_3$ (Lüthi et al. [9.55], Lenz et al. [9.54])

194   9 Unstable Moment Compounds

$$\Omega = -\frac{\partial \ln T^\star}{\partial \ln V} = -\frac{1}{T^\star} \times \frac{\partial T^\star}{\partial \varepsilon_V} \qquad (9.2)$$

with the neglect of second derivatives $\frac{\partial^2 T^\star}{\partial \varepsilon_V^2}$, the following relations between the thermodynamic derivatives are obtained:

$$\Delta c_B = -\Omega^2 T \Delta C, \quad \Delta c_B = -\Omega^2 T \Delta \beta \quad \text{and} \quad \Delta \beta = \frac{\Omega}{c_B^0} \Delta C. \qquad (9.3)$$

From the third relation follows the usual Grüneisen parameter definition in terms of thermodynamic quantities: $\Omega = \Delta \beta c_B^0/\Delta C$. This gives the identical expression as in (4.23). From Table 9.1 we see that Ω values for mixed valence compounds are of the order of 10. The relations of (9.3) enable us to fit the data in Fig. 9.9. This is shown in Fig. 9.9b,c where we see that c_B changes like $-T^2$ for low temperatures. Likewise the thermal expansion is proportional to the specific heat for $T < T^\star$. In addition, for CeSn$_3$, Δc_B and $\Delta \beta$ are proportional to each other in accordance with the second equation of (9.3). The calculation above is done for isothermal elastic constants. In the next Sect. 9.3, it will be shown that this is justified for these compounds. A more detailed discussion of the electronic Grüneisen parameter will also be given there. A similar analysis has been made by Thompson and Lawrence [9.56].

A similar analysis can also be made for CeNi, an orthorhombic intermediate valence compound. The temperature dependence of various elastic constants has been measured (Bömken et al. [9.57]). The longitudinal modes c_{11}, c_{22}, c_{33} all show a minimum around 130K which can be analysed similarly to the case of CeSn$_3$. The shear waves c_{44} and c_{55} exhibit normal temperature dependence and c_{66} a slight softening below 150K. Likewise inelastic neutron scattering of the phonon branches show appreciably smaller phonon energies for CeNi than for LaNi in the a and b–directions (Clementyev et al. [9.58]). Thermal expansion and magneto-striction show similar anisotropic behaviour like the elastic constants (Creuzet and Gignoux [9.59]). The volume dependent properties were analysed with (9.3), see also Nefedova et al. [9.60]. A detailed analysis, such as that done for anisotropic Kondo lattice substances shown in Sect. 9.3.4, has not been given yet for CeNi.

CeRu$_2$

Another Ce compound, CeRu$_2$, with a Sommerfeld coefficient $\gamma = 28$mJ/Mol K^2 comparable to the ones of CeSn$_3$, CePd$_3$, does not qualify as a mixed valence compound, as seen from Table 9.1. Bulk modulus c_B and Poisson ratio ν have normal values and all physical properties point to a Ce^{4+} ion valency. CeRu$_2$ has, however, intriguing normal and superconducting lattice properties which are discussed in Sects. 10.3, 10.5.

CeNi$_5$

This compound crystallises in the CaCu$_5$ hexagonal structure like PrNi$_5$, which has interesting magneto-elastic properties as discussed in Sects. 5.2.1,

7.2.2 and 7.2.3. In CeNi$_5$, no anomalies in the temperature dependence of the thermal expansion and elastic constants were found (Butler et al. [9.61]). This indicates that a phase transition does not occur down to 4K indicating that the valency of the Ce ion is Ce^{4+} as for CeRu$_2$ discussed above.

Cerium

Elemental Cerium has been investigated ultrasonically at room temperature as a function of pressure (Voronov et al. [9.62]). In polycrystalline samples, the bulk modulus softens somewhat at the $\gamma - \alpha$ transition at 7.5kbar with a volume change of 14%. At the $\alpha - \alpha'$ transition of 51 kbar both longitudinal and transverse modes soften on both sides of the transition, the Poisson ratio stays positive however in the whole pressure range. The equilibrium line of the $\gamma - \alpha$ transition ends in a critical point, characteristic for the mixed valence transition. Alternatively, Cerium has been described as a Kondo lattice substance undergoing the so-called Kondo volume collapse (see Sect. 9.3.2).

Charge Fluctuation Rate

In the foregoing, we have seen that some compounds exhibit strong bulk modulus softening while others experience only effects in the order of % or less. It might be suspected that the lattice structure determines these special lattice dynamical effects. But a satisfactory explanation of these effects has been given by Müller-Hartmann and collaborators by considering the charge fluctuation rate ω_c introduced above in Sect. 9.1 (Kuramoto [9.63], Müller-Hartmann [9.64], Schmidt and Müller-Hartmann [9.65]). Usually ω_c is of the order of acoustical and optical phonon frequencies ω_{ph}. By investigating these phonon modes using Raman scattering, Mock et al. [9.66] and Zirngiebl et al. [9.67] could show that for $\omega_c \gg \omega_{ph}$ the phonon frequencies are unaffected and the bulk modulus change is small. This happens for CePd$_3$ and CeSn$_3$ (Fig. 9.9). For $\omega_c \geq \omega_{ph}$ the phonon frequency of Γ_{1g}-symmetry softens with respect to a trivalent reference compound because the charge relaxation rate can approximately follow the movement of the ions (example CeBe$_{13}$). The bulk modulus softens somewhat more (Fig. 9.9). For $\omega_c \sim \omega_{ph}$ the effects on some phonon modes and bulk modulus are stronger, as observed for SmB$_6$ (Nakamura et al. [9.68]). Finally for $\omega_c < \omega_{ph}$ the phonon frequencies have values according to the valence mixing ratio, but the bulk modulus softens completely. This is the case for SmS, Sm$_{0.75}$Y$_{0.25}$S and TmSe.

These qualitative considerations for $\omega_c \leftrightarrow \omega_{ph}$ for Γ_{1g}-symmetry given above have not been treated theoretically in any detail. For some further discussions of phonons in mixed valence compounds, see e.g. Sherrington and von Molnar [9.69], Khomskii [9.70], Capellmann and Lipinski [9.71].

Finally it should be stressed that negative Poisson ratios are not confined to mixed valence materials. Firstly, in fcc crystals with pressure applied along a [110] direction, the Poisson ratio $\nu_{001} = -\varepsilon_{001}/\varepsilon_{110}$ is positive but $\nu_{11'0} = -\varepsilon_{11'0}/\varepsilon_{110}$ is negative for many elements (Cu, Ag, Au, Pb, Ni, Pd)

(Milstein and Huang [9.72]). Isotropic foams can also have negative Poisson ratios (Lakes [9.73]). Secondly, anisotropic materials can have negative Poisson ratios in certain directions. Negative Poisson ratio materials can have special reflection properties of a (longitudinal) sound wave on a plane. Surface acoustic waves (SAW) also have special propagation characteristics in this case (Lipsett and Beltzer [9.74]).

9.3 Electron–Phonon Coupling in Heavy Fermion Systems

9.3.1 Introduction

In these compounds, introduced at the beginning of this chapter, there is a deformation potential coupling as in ordinary metals (Sect. 8.1) and a magneto-elastic coupling to the CEF states. This latter coupling is operative for $T > T^\star$ where CEF levels show a strain dependence leading to the various thermodynamic effects as discussed in Chap. 5 The deformation potential coupling is the origin of e.g. magneto-acoustic quantum oscillations.

In addition to these contributions, heavy fermion compounds exhibit a special kind of deformation potential coupling for heavy quasiparticle states at temperatures below $T^\star \sim T_K$ (commonly called the Kondo temperature for $4f$ systems). It is the same temperature region where Fermi liquid theory applies (see Sect. 9.1.2). T_K describes the width of the Abrikosov-Suhl resonance T_0 indicated in Fig. 9.1. This special coupling is characterised by an electronic "Grüneisen parameter" Ω. This is a measure for the volume dependence of T_K. Microscopically, it is mainly due to the volume dependence of the sf-hybridization strength V (Thalmeier [9.75], Keller et al. [9.76]). Because the energy scale for this coupling is T_K, it can be expected that elastic anomalies occur for temperatures $\leq T_K$ which are superposed on the magneto-elastic effects that have a temperature scale of the CEF splitting Δ, which is generally an order of magnitude larger than T_K. We then have a breathing mode coupling of the volume strain to the energy states: $d_{\Gamma 1} = \Omega \langle E \rangle$ where $\langle E \rangle$ is a typical energy band width of the order of $T^\star$ (see Fig. 9.1), also called Abrikosov–Suhl resonance and from (8.1)

$$\Omega = -\frac{\partial \ln \langle E \rangle}{\partial \varepsilon} = -\frac{\partial \ln T^\star}{\partial \varepsilon}. \tag{9.4}$$

It will be shown below that this coupling increases the energy gain due to singlet formation. From this discussion, strong effects for longitudinal elastic modes are expected but no effect for the volume conserving shear modes. Fig. 9.10 is a schematic diagram giving the different contributions to the elastic constants.

9.3 Electron–Phonon Coupling in Heavy Fermion Systems

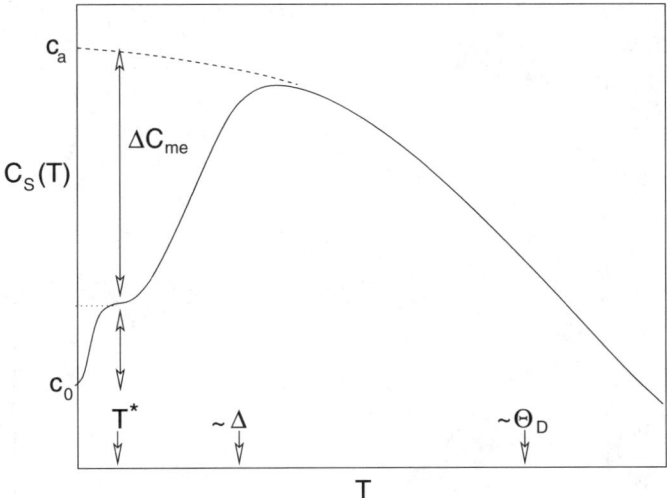

Fig. 9.10. Schematic diagram of adiabatic longitudinal elastic constant for a heavy fermion compound. The different regimes are described in the text

Three different regimes can be distinguished:

1. The background region for high temperatures, as discussed in Sect. 4.2.
2. The crystal field regime where the magneto-elastic coupling is operative in the region of the CEF splittings Δ, as discussed in Sect. 5.2.2.
3. And the low temperature regime ($T \leq T_K$) where the Grüneisen parameter coupling is important.

A good illustration of the foregoing discussion is given by the elastic constants of **CeCu$_6$** (Weber et al. [9.77], Suzuki et al. [9.78]). In Fig. 9.11 we show this case. Clear CEF effects are observed for longitudinal (c_{11}, c_{22} and c_{33}) and shear (c_{44}) waves. Grüneisen parameter effects in the heavy fermion regime ($T \leq 5K$) are observed only for longitudinal modes (c_{11}, c_{33}). In CeCu$_6$ there is an additional structural transition from orthorhombic (D_{2h}^{16}) to monoclinic (C_{2h}^{5}) at about 220K with a soft c_{66} mode (Fig. 9.11, Suzuki et al. [9.78], Weber et al. [9.77]). This softening is very strong with $T_c - T_0 \sim 10^3$K implying a strong strain–order parameter coupling (4.13).

A more detail discussion follows on the heavy fermion region ($T < T^\star$) with large Grüneisen parameters. In compounds where no low temperature structural phase transition occurs, e.g. CeCu$_6$, CeAl$_3$, CeRu$_2$Si$_2$, CeCu$_2$Si$_2$ and UPt$_3$, longitudinal elastic constants show a distinct temperature dependence as shown schematically in Fig. 9.10 and for CeCu$_6$ and CeRu$_2$Si$_2$ later on (Fig. 9.13) and for CeCu$_2$Si$_2$ in Fig. 9.14. We can interpret this temperature dependence using scaling arguments (Yoshizawa et al. [9.79]) as shown below (Sect. 9.3.4). A microscopic theory has been given by Thalmeier [9.75] and Keller et al. [9.76]. It should be noted that for heavy fermion compounds such

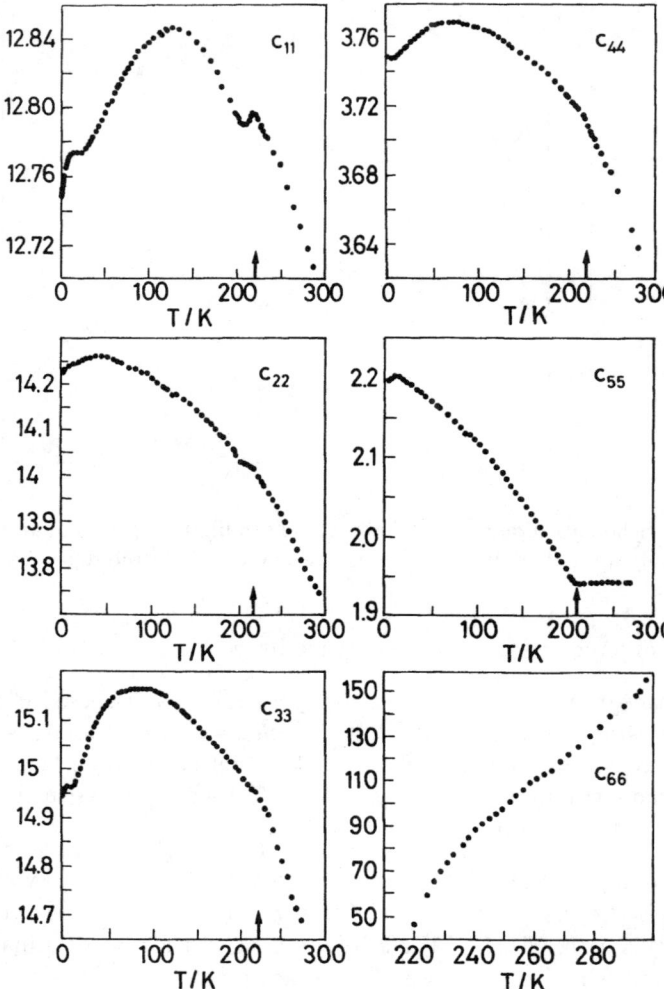

Fig. 9.11. Temperature dependence of elastic constants for CeCu$_6$ in units of 10^{11} erg/cm^3 (Weber et al. [9.77])

as CeAl$_2$, CeB$_6$, CePb$_3$, CeIn$_3$, CeAg and URu$_2$Si$_2$ low temperature magnetic or structural transitions prevent observation of this characteristic temperature dependence (see Table 9.1).

9.3.2 Kondo Volume Collapse

A consequence of the strong Grüneisen parameter coupling is the so-called Kondo volume collapse (Allen and Martin [9.80], Lavagna et al. [9.81], Razafimandimby et al. [9.82]). Here explicit expressions for the Grüneisen parameter and the bulk modulus are given using formulas for the free energy from numerical solutions of the Kondo problem. But one can immediately see that a large positive Ω leads to a shrinking of the volume (volume collapse) and an enhancement of the singlet binding energy (Fulde et al. [9.7]). Taking $T^\star(\varepsilon_v) = T_0(1 - \Omega\varepsilon_v)$, we get at $T = 0$ for the singlet energy density $E_s = -nk_B T^\star + c_B \varepsilon_V^2/2$ and with $\delta E_s/\delta \varepsilon_V = 0$ the volume shrinking ε_V^0 and the singlet binding energy

$$\varepsilon_V^0 = -nk_B T_0 \frac{\Omega}{c_B} \tag{9.5}$$

$$E_s = -nk_B T_0 \left[1 + nk_B T_0 \frac{\Omega^2}{2c_B}\right]. \tag{9.5a}$$

With the exception of polycrystalline CeAl$_3$, all heavy fermion Grüneisen parameters are positive and we have a volume shrinking according to (9.5). In addition, the singlet ground state formation is enhanced through the Grüneisen parameter coupling as shown in (9.5a). The negative Ω for CeAl$_3$ is perhaps a consequence of the polycrystalline nature. Single crystalline CeAl$_3$ becomes magnetic at 1–2K (Lapertot et al. [9.83]).

Table 9.1 quotes Grüneisen parameters for heavy fermion compounds which are of the order of $\Omega \sim 100$. With the values from Table 9.1 for the fluctuation temperature of $T_0 \sim 10$K, for the bulk modulus of $c_B \sim 10^{12}$erg/cm^3 and with $n \sim 10^{22}$ magnetic ions/cm^3 we get from (9.5) a volume strain of $\varepsilon_V \sim -10^{-3}$, a large volume collapse.

The prime example for the Kondo volume collapse is elemental Ce with a quoted volume change ε_V of -14% (see Sect. 9.2). This is then a new interpretation taking Ce rather as a Kondo system than a mixed valence system. Another clear manifestation of a Kondo volume collapse is the temperature dependence of the lattice parameters a and c in CeRu$_2$Si$_2$ (Raymond et al. [9.84]). Below $T^\star \sim 20$K these lattice parameters decrease rapidly. The extrapolation to $T \to 0$K gives a relative volume shrinking of $\varepsilon_V \sim 1.4\ 10^{-3}$ in good agreement with the estimate above. On the other hand, in the case of the mixed valence compound YbInCu$_4$, a volume expansion can be observed (Sarrao [9.41]).

9.3.3 Special Sound Propagation Effects for Large Ω

Using the thermodynamic relation $\left(\frac{dT}{d\varepsilon}\right)_S = \left(\frac{\partial T}{\partial S}\right)_\varepsilon \left(\frac{\partial S}{\partial \varepsilon}\right)_T$ (see Sect. 4.5), with $\left(\frac{\partial S}{\partial \varepsilon_V}\right)_T = -c_B \alpha_T$ and with $C_V = T\left(\frac{\partial S}{\partial T}\right)_V$, gives the following relation used above (Sect. 9.2)

$$\Omega = \alpha_T \frac{c_B}{C_V} = -\left(\frac{\partial \ln T}{\partial \varepsilon_V}\right)_S. \tag{9.6}$$

The large Grüneisen parameter gives rise to special sound wave effects as outlined below:

1. Equation (9.6) says that a strain wave is accompanied by a temperature wave. Substituting $\Omega \sim 100$, $\varepsilon_V \sim 10^{-4}$ and $T = 1K$ leads to a large temperature amplitude of $\Delta T \sim 1\text{mK}$. This has not been measured so far.
2. In the heavy fermion state, sound propagation is adiabatic even at low temperature. For $ql_e \ll 1$ a local temperature can be defined (q is the wave number and l_e is the electron mean free path). Adiabatic sound propagation means that the sound wave period is much smaller than the energy diffusion time (see Sect. 4.5, (4.22)): $\omega\tau \ll (3/2\pi)(v_s/v_F)^2$. This is valid for heavy fermion metals with typical electron relaxation times $\tau \sim 10^{-12}$s and sound velocities of the same order as the Fermi velocity $v_F = \hbar q/m^\star$. The same estimate shows that for normal metals under the same conditions sound propagation is isothermal. Actually it has been argued that in heavy fermion systems one is even in the isolated region, i.e. the sound wave period is much less than the particle diffusion time (Becker and Fulde [9.85]).
3. Another consequence of large Ω is the large difference between the adiabatic c_S and the isothermal c_T. With (4.21) $c_S - c_T = T\beta^2 c_T^2/C_V = \Omega^2 TC_V$, the relative change in the elastic constant is given by

$$\frac{c_S - c_T}{c_T} = \Omega^2 \frac{TC}{c_B}. \tag{9.7}$$

The parameter values used above result in $(c_S - c_T)/c_T \sim 1\%$ even at low temperatures of 1K, whereas for normal metals like Cu at 1K this quantity is 10^{-8}, i.e. the difference between c_S and c_T has no consequences in normal metals.
4. The large difference between c_S and c_T leads to central peak phenomena, i.e. an anomalously large Landau–Placzek ratio which determines the ratio of the quasi-elastic line and the Brillouin lines (Becker and Fulde [9.86], [9.85]). Experimentally this effect has been found in UPt$_3$ (Mock et al. [9.87]) but not yet in Ce compounds.
5. Finally another consequence of large Grüneisen parameters is a possible ultrasonic attenuation mechanism similar to that which is found in fluids (Becker and Fulde [9.86,9.88], Pethick et al [9.89], Bennemann et al. [9.90], Schotte et al. [9.91], Yoshizawa et al. [9.79]). For UPt$_3$ such an attenuation peak has been found (Müller et al. [9.92]), see Fig. 9.12. Around 12K, a longitudinal attenuation peak for c_{11} develops with a ω^2 dependence. This was tentatively interpreted similarly to the attenuation peak discussed for magnetic phase transitions involving a coupling to energy fluctuations (6.6, Sects. 6.1.1, 6.1.2) with thermal diffusion. The result is similar to ultrasonic attenuation in a fluid (Landau–Lifshitz [9.93]):

$$\alpha = \frac{\omega^2}{2\rho v_s}\kappa T \frac{\beta^2}{C_p^2} = \frac{\omega^2}{6\rho v_s^3}\Omega^2 TC_p\tau\left(\frac{v_F}{v_s}\right)^2. \tag{9.8}$$

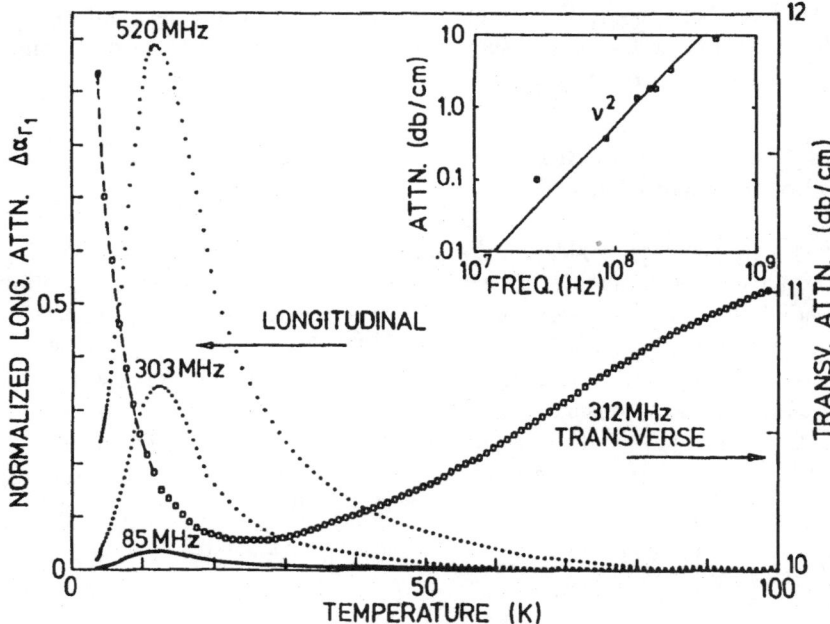

Fig. 9.12. Temperature dependence of the longitudinal ($q \parallel b$) electronic attenuation coefficient in normal UPt$_3$ at various frequencies, showing the 12K fluctuation peak. The peak is absent for transverse sound. Inset shows the quadratic frequency dependence of the peak height (Müller et al. [9.92])

Here κ is the electronic thermal conductivity, β the thermal expansion and Ω the Grüneisen parameter. In (9.8) for κ the gas-kinetic expression $\frac{1}{3}C_p v_F^2 \tau$ was used and for β (9.6). Such an expression gives a similar temperature dependence as observed (see especially Pethick et al. [9.89]) but for 10K and 300 MHz with $\kappa = 15\,\text{mW/Kcm}$, $v_s = 3.9 \times 10^5\,\text{cm/s}$, $\rho = 19.4\,\text{g/cm}^3$ and $\beta_a = 11 \times 10^{-6} K^{-1}$ the calculated α gives 0.06 db/cm instead of the experimental 1.2 db/cm, i.e. a factor of 20 too small. Apparently this effect is not quantitatively understood yet (see also Becker and Fulde [9.88]). Note that the attenuation peak occurs at the same temperature independent of the ultrasonic frequency, eliminating possible dislocation peak phenomena. Similar effects should be observable for other heavy fermion compounds. The thermodynamic properties of UPt$_3$ will be discussed further in Sect. 9.3.4.

Besides the electronic Grüneisen parameter used for heavy fermion compounds, other such parameters can also be defined, e.g. for magnetic transitions $\Omega^m = \mathrm{d}\ln T_m/\mathrm{d}\varepsilon_v$ (Sect. 5.1.3), for superconductors $\Omega^{sl} = \mathrm{d}\ln T_{sl}/\mathrm{d}\varepsilon_v$ (Sect. 10.7.1) or phonon Grüneisen parameters discussed in Sect. 4.2.1 etc. Besides the Ω^T of (9.6) we will also discuss below a Ω^B for meta-magnetic transitions. Anisotropic Grüneisen parameters for heavy fermion systems with tetragonal or hexagonal symmetry Ω_a, Ω_c, besides the usual volume Ω_V, will also be introduced below.

9.3.4 Scaling Approach for the Temperature Dependence of Thermal Quantities

In Sect. 9.2, we introduced the scaling approach for isothermal processes (9.3). Since in heavy fermion materials we often deal with adiabatic elastic constants and adiabatic processes we want to introduce a corresponding scaling calculation (Yoshizawa et al. [9.79]). Equation (9.6) leads to $T = T_0 \exp(-\Omega\varepsilon) = T_0 s$, with s a scaling parameter of the energy. With the density of states $D(E)$ and particle conservation, we have $D(E,\varepsilon)\mathrm{d}E = D(E^\star,0)\mathrm{d}E^\star$ and with $E^\star = E/s$, $\mathrm{d}E^\star = \mathrm{d}E/s$ follows $D(E,\varepsilon) = D(E/s,0)/s$. From this relation we can calculate the thermodynamic potentials:

Internal energy:
$$\begin{cases} U(T,\varepsilon) = \int E f\left(\frac{E}{T}\right) D(E,\varepsilon)\,\mathrm{d}E \\ \text{and}\quad U\left(\frac{T}{s},0\right) = \int E^\star f D(E^\star,0)\,\mathrm{d}E^\star \\ \text{or}\quad U(T,\varepsilon) = sU\left(\frac{T}{s},0\right) \end{cases}$$

Entropy:
$$\begin{cases} S = -k_B n \int \mathrm{d}E\, D(E,\varepsilon)\left[f\ln f + (1-f)\ln(1-f)\right] \\ \text{and} \\ S\left(\frac{T}{s},0\right) = -k_B n \int \mathrm{d}E^\star D(E^\star,0)\left[f\ln f + (1-f)\ln(1-f)\right] \\ \text{or} \\ S(T,\varepsilon) = S\left(\frac{T}{s},0\right) \end{cases}$$

Helmholtz free energy:
$$\begin{cases} F(T,\varepsilon) = U(T,\varepsilon) - TS(T,\varepsilon) \\ \quad = sU\left(\frac{T}{s},0\right) - TS\left(\frac{T}{s},0\right). \end{cases}$$

From these expressions we can derive the thermodynamic quantities:

Specific heat:
$$\begin{cases} C_V = \left(\frac{\partial U}{\partial T}\right)_V = U' = \dfrac{\mathrm{d}U}{\mathrm{d}x}\quad \text{with } x = \dfrac{T}{s} \\ \text{or } C_V = T\left(\dfrac{\partial S}{\partial T}\right)_V = \dfrac{T}{s}S'\quad \text{or}\quad U' = \dfrac{T}{s}S' \end{cases}$$

9.3 Electron–Phonon Coupling in Heavy Fermion Systems

Thermal expansion:
$$\begin{cases} \text{stress} & \sigma = \left(\dfrac{\partial F}{\partial \varepsilon}\right)_T = \left(\dfrac{\partial U}{\partial \varepsilon}\right)_S = s'U\left(\dfrac{T}{s},0\right) \\ \text{because} & S = const \quad \text{implies} \quad \dfrac{T}{s} = const \\ \text{and} & \left(\dfrac{d\sigma}{dT}\right)_\varepsilon = \dfrac{s'}{s}U\left(\dfrac{T}{s},0\right) \\ \text{with} & \left(\dfrac{d\sigma}{dT}\right)_\varepsilon = -\left(\dfrac{\partial \sigma}{\partial \varepsilon}\right)_T \left(\dfrac{\partial \varepsilon}{\partial T}\right)_\sigma = -\alpha_T c_T \\ \text{it follows that} \ \alpha_T & = -\dfrac{s'\,U'}{s\,c_T} \end{cases}$$

Adiabatic elastic constant:
$$\begin{cases} c_S = \left(\dfrac{\partial \sigma}{\partial \varepsilon}\right)_S \\ = \dfrac{\partial}{\partial \varepsilon}\left(s'U\left(\dfrac{T}{s},0\right)\right)_S = s''U\left(\dfrac{T}{s}\right) = \dfrac{s''}{s}U(T,\varepsilon) \end{cases}$$

Isothermal elastic constant:
$$\begin{cases} c_T = \left(\dfrac{\partial \sigma}{\partial \varepsilon}\right)_T \\ = s''U\left(\dfrac{T}{s}\right) - T\dfrac{s'^2}{s^2}U'\left(\dfrac{T}{s}\right) = c_S - \left(\dfrac{s'}{s}\right)^2 C_V T. \end{cases}$$

Likewise (4.21) follows from these scaling laws. With s from above follows $\dfrac{s'}{s} = -\Omega$ and $\dfrac{s''}{s} = \Omega^2$ and

$$\begin{aligned} \text{Thermal expansion coefficient} \quad & \alpha_T = \dfrac{\Omega}{c_T}C_V \\ \text{Adiabatic elastic constant} \quad & c_S = \Omega^2 U(T) \\ \text{Isothermal elastic constant} \quad & c_T = \Omega^2 U - T\Omega^2 C_V \end{aligned} \qquad (9.9)$$

These equations can also be obtained directly using (4.21) and (9.7) together with the scaling procedure of Sect. 9.2 with (9.3). We give some applications of the scaling formulas for Ce and U–compounds.

Cerium Compounds

CeCu$_6$, CeRu$_2$Si$_2$

Figure 9.13 shows the low temperature elastic constants for the heavy fermion compounds CeCu$_6$ and CeRu$_2$Si$_2$ together with a fit for the adiabatic elastic constants (9.9) and from the theory by Thalmeier [9.75]. With $C = \gamma T$ from scaling, we get $c_S = \Omega^2 \gamma T^2/2$ which is valid for the region $T \ll T^*$. It is seen that the fit is very good for the two cases up to $T \sim T^*$. Thalmeier's theory calculates the internal energy beyond the scaling region up to T^* using the so-called slave boson technique (Coleman [9.94]).

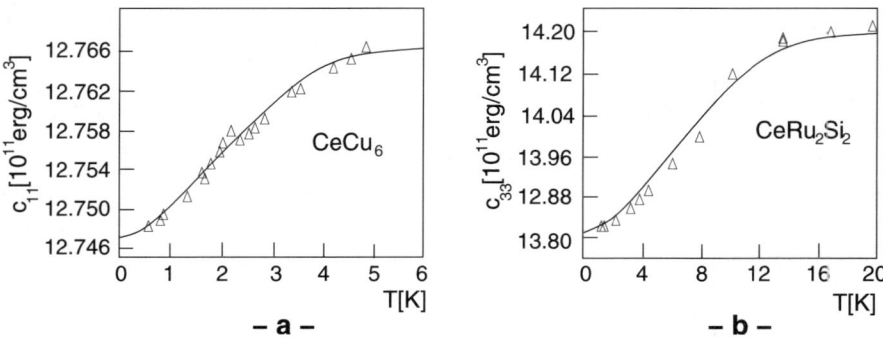

Fig. 9.13. Longitudinal elastic constants in the heavy fermion region for (a) CeCu$_6$, (b) CeRu$_2$Si$_2$, together with a fit as described in the text (Lüthi et al. [9.55])

Actually, both CeCu$_6$ and as well as CeRu$_2$Si$_2$ are not cubic crystals but orthorhombic–monoclinic ($D_{2h}^6 - C_{2h}^5$ for CeCu$_6$) and tetragonal (D_{4h}^{17} for CeRu$_2$Si$_2$). The unit cell of CeRu$_2$Si$_2$ is the same as that of CeCu$_2$Si$_2$ and is shown in Fig. 10.20. The isotropic scaling formulas derived above are therefore not strictly applicable. Anisotropic Grüneisen parameters must be introduced to describe thermal expansion and elastic constants. This can be done conveniently by generalizing (9.9). In the case of tetragonal and hexagonal systems, it is sufficient to introduce (Thalmeier and Lüthi [9.3])

$$\Omega_a = \frac{\partial \ln T^\star}{\partial \varepsilon_{11}}, \qquad \Omega_c = \frac{\partial \ln T^\star}{\partial \varepsilon_{33}}. \tag{9.10}$$

For the elastic constants c_{11} and c_{33} one obtains anisotropic expressions

$$c_{11} = \Omega_a^2 U, \qquad c_{33} = \Omega_c^2 U. \tag{9.11}$$

The aforementioned calculations have been performed with these formulae. In addition, the values $\Omega_{a,c}$ listed in Table 9.1 have been obtained in this way.

CeCu$_2$Si$_2$

This material has the same structure as CeRu$_2$Si$_2$ (space group D_{4h}^{17}, $I4/mm$). The unit cell is shown in Fig. 10.20. CeCu$_2$Si$_2$ exhibits also a T^2 temperature dependence for the longitudinal elastic constant c_{11} and c_{33}. The Grüneisen parameters Ω_a and Ω_c deduced with (9.11) are given in Table 9.1. In Fig. 9.14 we give experimental results for the temperature dependence of c_{11} on a logarithmic scale. It demonstrates CEF effects on elastic constants for $T > 10$K, heavy fermion Grüneisen parameter effects with a typical T^2 dependence for $T_{sc} < T < T^\star \sim 10$K and an anomaly at the superconducting transition temperature $T_{sc} = 0.7$K (see Sect. 10.7.1).

Ce-mono-pnictides

In the cubic CeP, CeAs, CeSb and CeBi with NaCl structure, these low temperature Grüneisen parameter effects discussed above are not observed, because

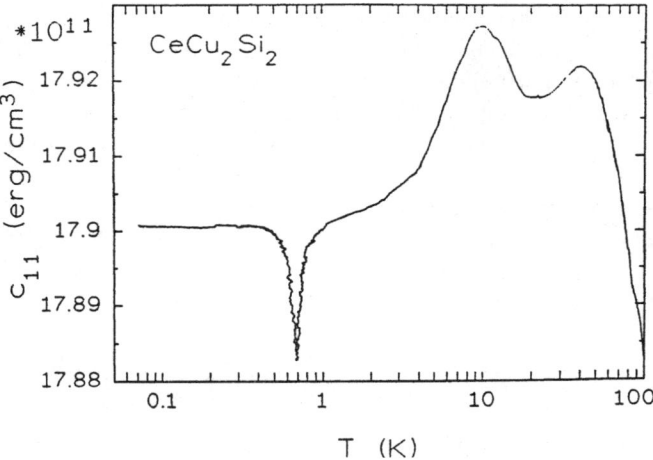

Fig. 9.14. Temperature dependence of the longitudinal c_{11} mode for CeCu$_2$Si$_2$ exhibiting CEF and heavy fermion effects (Bruls et al. [9.95])

of low temperature magnetic phase transitions (T_N(CeBi) = 24K, T_N(CeAs) = 8.2K, T_N(CeSb) = 16K). Instead, crystal field effects from magneto-elastic coupling to the $\Gamma_7 - \Gamma_8$ levels of the Ce^{3+}–ions are observed, as discussed in detail in Sect. 5.2.2. For a reference to elastic constants in Ce-mono-pnictides see Goto et al. [9.96] and references in it. This behaviour of the temperature dependence of the elastic constants is similar to the cases of other cubic Ce–compounds: CeAl$_2$ (Lüthi and Lingner [9.97]), CePb$_3$ (Nikl et al. [9.98]), CeB$_6$ (Lüthi et al. [9.99]) and others discussed above in Sect. 9.3.1.

CeAl$_3$

The first discovered heavy fermion compound CeAl$_3$ (Andres et al. [9.100]) with a large Sommerfeld γ–value (Table 9.1) also does not exhibit typical heavy fermion behaviour in the elastic constants, presumably because of the negative Grüneisen parameter Ω (Table 9.1) and the polycrystalline nature (Niksch et al. [9.101]). As pointed out above, the negative Ω for CeAl$_3$ is not understood (Sect. 9.3.2).

CeNiSn

This orthorhombic compound (ε–TiNiSi type structure) was first considered a Kondo insulator. Later studies showed that the semiconducting properties could be connected with parasitic phases. The purest samples display metallic behaviour. The elastic properties were investigated by Nakamura et al. [9.102] and Holtmeier et al. [9.103]. Both groups displayed anomalous temperature dependence of the longitudinal elastic constants c_{11}, c_{22}, c_{33} in the form of deep minima. These anomalies could be explained with a deformation potential coupling to split bands near the Fermi energy allowing for an energy gap. The idea is that these split bands with high density of states are strongly coupled to the longitudinal strains in contrast to the low density of states near E_F. The shear modes c_{44}, c_{55}, c_{66} did not exhibit any anomalies. Low temperature thermal conductivity experiments could be explained again with an opening of a gap at the Fermi energy (Paschen et al. [9.104]). The field dependence of the longitudinal modes showed a meta-magnetic type of behaviour. This effect we will discuss in Sect. 9.3.5.

The thermal expansion of tetragonal compounds, similarly to the elastic constants and also in the a, c–direction (Lüthi and Ott [9.105], see also Sect. 5.2.1), results in $\alpha_i = \mathrm{d}\varepsilon_{ii}/\mathrm{d}T = \sum s_{ij}\mathrm{d}^2F/\mathrm{d}\varepsilon_{ii}\mathrm{d}T$ which then gives

$$\alpha_a = (s_{11} + s_{12})\Omega_a C + s_{13}\Omega_c C$$
$$\alpha_c = 2s_{13}\Omega_a C + s_{33}\Omega_c C,$$
(9.12)

where the s_{ij} are the elastic compliances and C the specific heat. Because the off-diagonal compliance components can also be negative this can lead to opposite signs in the thermal expansion coefficients as observed in PrNi$_5$ (Lüthi and Ott [9.105]) and UPt$_3$ (see below). Thermal expansion measurements for tetragonal or hexagonal Ce compounds in the normal state have not yet been performed.

Samarium Compounds

There are some medium heavy fermion compounds of Sm with the formula SmX_3, X = Pd, In, Sn, Tl, Pb. They have the cubic AuCu$_3$ structure. As seen from Table 9.1, they all have magnetic transitions in the range 5–15K and Sommerfeld coefficients $\gamma \sim$100–200 mJ/mol K^2. Because of the magnetic and structural transitions, they do not exhibit clear heavy fermion behaviour in their longitudinal elastic constants. But their lattice properties are, of course, completely different from the mixed valent Sm compounds SmB$_6$, SmS. The elastic properties of the SmX_3 compounds were interpreted with the strain–quadrupolar interaction as discussed in Sect. 7.2 (Endoh et al. [9.106]).

Uranium Compounds

UPt$_3$
We discuss first UPt$_3$, a hexagonal close-packed heavy fermion superconductor (space group D_{6h}^4, $P6_3$/mmc) in which many different experiments have been performed. There were also a number of acoustic measurements performed. In Fig. 9.15 we show the crystal structure of UPt$_3$ with two U-atoms per unit cell.

The complete elastic constant tensor of UPt$_3$ has been measured by the pulse echo method by de Visser et al. [9.107] and by the RUS spectroscopy

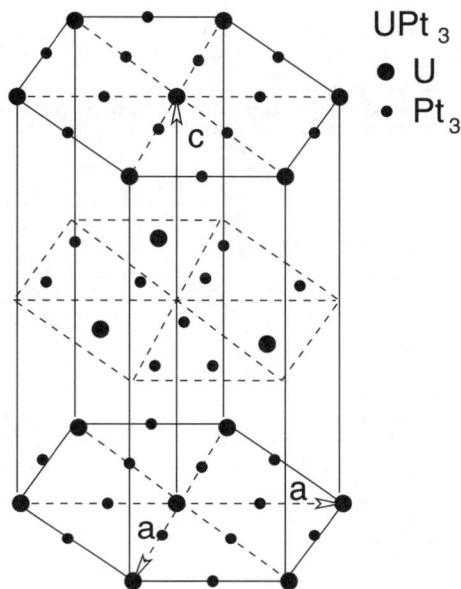

Fig. 9.15. Crystal structure of UPt$_3$ (Lattice constants $a = 5.8$Å, $c = 4.9$Å)

(Sect. 2.2) by Paglione et al. [9.108]. Both methods agree nicely as shown in Table 9.2.

The thermal expansion of UPt$_3$ has characteristic features as shown in Fig. 9.16 (de Visser et al. [9.107]). In contrast to the cubic case of CeSn$_3$ (Fig. 9.9a), it exhibits anisotropy due to the hexagonal symmetry, $\alpha_{a,b}$ and even has a different sign than α_c. This fact does not imply Grüneisen parameters of different signs as usually assumed but negative compliances as just discussed above. For example $s_{13} = -c_{13}/\Delta$ with Δ defined below. Equation (9.12) can be re-written in terms of elastic constants:

$$\alpha_a = [c_{33}\Omega_a - c_{13}\Omega_c]\frac{C}{\Delta}, \alpha_c = [-2c_{13}\Omega_a + (c_{11} + c_{12})\Omega_c]\frac{C}{\Delta} \quad (9.12a)$$

with $\Delta = -2c_{13}^2 + c_{33}(c_{11} + c_{12})$, the same expression used already in Sect. 5.2.1. With the data from Tables 9.2 and 9.3, the values $\alpha_a = 28.53C$ and $\alpha_c = -22.69C$.

The thermal expansion of UPt$_3$ exhibits minima and maxima at about 15K. This implies also a maximum in the electronic specific heat at the same temperature. The low temperature Grüneisen parameters can be conveniently determined solving the equations above (9.12) and (9.12a) for $\Omega_{a,c}$:

$$\begin{aligned} C\Omega_a(T) &= (c_{11} + c_{12})\alpha_a + c_{13}\alpha_c \\ C\Omega_c(T) &= 2c_{13}\alpha_a + c_{33}\alpha_c \,. \end{aligned} \quad (9.10a)$$

Table 9.2. Elastic constant tensor (see App. C) of UPt$_3$ at 300K (de Visser et al. [9.107], Paglione et al. [9.108]) mass density $\rho = 19.4$ g/cm^3

In 10^{12} erg/cm^3	pulse echo method	RUS spectroscopy
c_{11}	3.093	
c_{12}	1.421	1.44
c_{13}	1.70	1.732, 1.695
c_{33}	2.891	2.93
c_{44}	0.372, 0.374	0.379
c_{66}	0.836	0.837

Table 9.3. Physical properties of UPt$_3$ and CeRu$_2$Si$_2$

Material	UPt$_3$	CeRu$_2$Si$_2$
Crystal structure	Hexagonal	Tetragonal
Specific heat γ-values (mJ mol$^{-1}K^{-2}$)	420	380
Fluctuation temperature T^* (K)	25	15
Magnetic easy direction	a-b plane	c-axis
Critical metamagn. field B_c(T) for χ anomaly	20.3	8.07
Soft longitudinal acoustic mode	C_{11}	C_{33}
Thermal Grüneisen parameter Ω_T	$\Omega_a^T = 91$ $\Omega_c^T = 32$	$\Omega_a^T = \Omega_c^T = 117$
Magnetic Grüneisen parameter Ω_B	$\Omega_a^B = 65$	$\Omega_c^B = 115$

Actually the low temperature anisotropic Grüneisen parameters listed in Table 9.1 have been determined with (9.10a). The results in Fig. 9.16 can be modeled with a two–band model, thus giving more fit parameters (Yoshizawa et al. [9.79]).

Figure 9.17 shows the temperature dependence of the main elastic constants for UPt$_3$ (Yoshizawa et al. [9.109]). It is seen that the two longitudinal modes c_{11} and c_{33} exhibit strong anomalies at low temperatures. But also c_{44} has a surprising temperature dependence. The longitudinal modes do not fit the simple adiabatic expressions of (9.11). Contrary to the case of Ce-heavy fermion compounds (Fig. 9.13, 14) the UPt$_3$ longitudinal elastic constants exhibit minima at 30K (c_{11}) and at 15K (c_{33}). This behaviour was qualitatively explained using a two–band model, as for the thermal expansion, (Yoshizawa et al. [9.79]). But another explanation was tried using isothermal elastic constants (Lüthi et al. [9.55]), implying that the estimates in Sect. 9.3.2 are in favour of isothermal conditions in UPt$_3$. This issue is not yet clear. As to the c_{44} mode, inelastic neutron scattering and ultrasonic elastic constant measurements have indicated appreciable changes from zero ($\omega\tau > 1$) to first ($\omega\tau < 1$, hydrodynamic) sound (Yoshizawa et al. [9.109]).

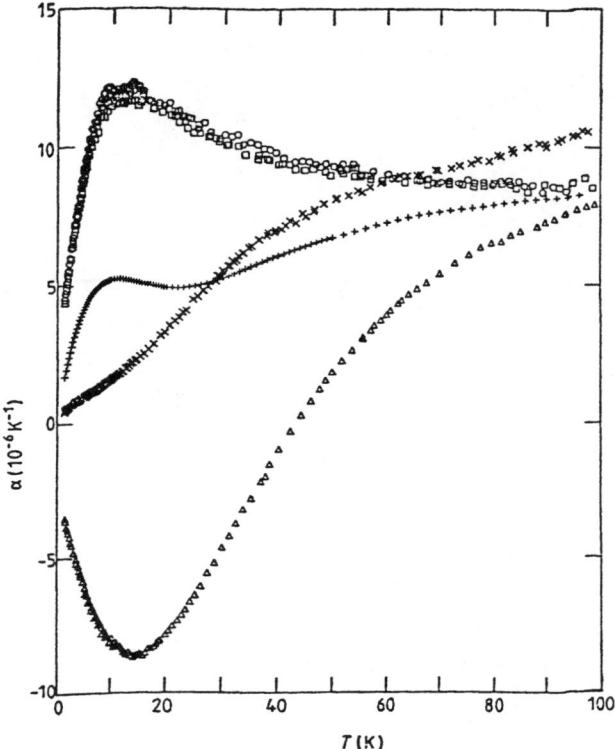

Fig. 9.16. Linear thermal expansion coefficient α_i of UPt_3 along different crystallographic directions (de Visser et al. [9.107]): square a-axis, circle b-axis, triangle c-axis, cross $\alpha_V/3$, x polycrystalline UAl_2

The whole discussion on UPt_3 has indicated that the normal state of this compound has many interesting electron–lattice effects, such as anomalous temperature dependence of thermal expansion and elastic constants, an unexpected attenuation effect and anomalous Landau Placzek ratio. In addition, effects in a magnetic field will also be shown below. Important parameters for UPt_3 and $CeRu_2Si_2$ are listed in Tables 9.1 and 9.3.

URu_2Si_2

Another U-compound exhibiting anomalous temperature dependence of some elastic constants is URu_2Si_2. This body-centered tetragonal compound with the same space group $I4/mmm$ as $CeRu_2Si_2$ has an anti-ferromagnetic (or rather a hidden order parameter) phase transition at $T_N = 17.5K$ and a superconducting transition at $T_s = 1.2K$. In Fig. 9.18 the temperature dependence of some elastic modes of this compound are shown (Wolf et al. [9.110]). With the exception of c_{11}, the modes have a more or less normal temperature dependence. The low temperature Grüneisen parameters, determined

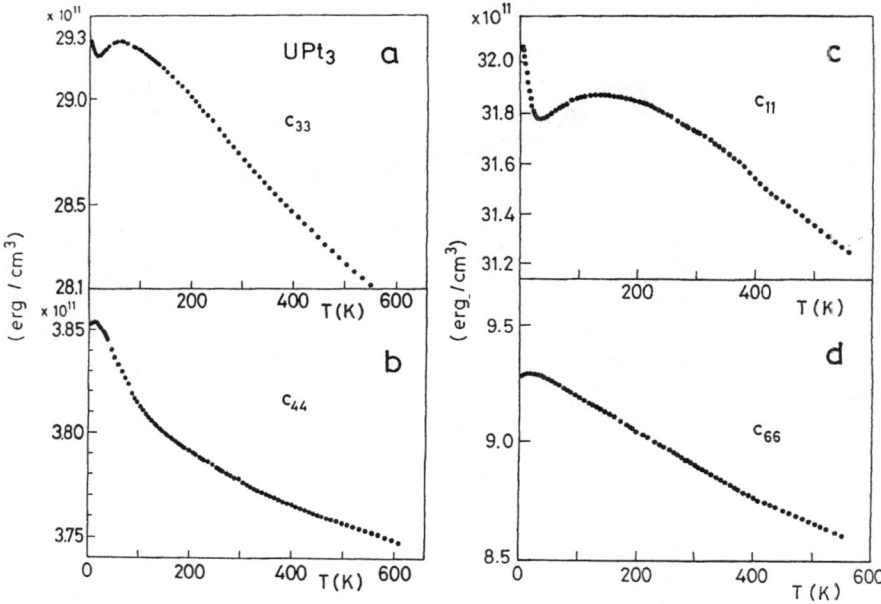

Fig. 9.17. Temperature dependence of the elastic constants c_{11}, c_{33}, c_{44}, c_{66} for UPt_3 from 1 to 600K (Yoshizawa et al. [9.109])

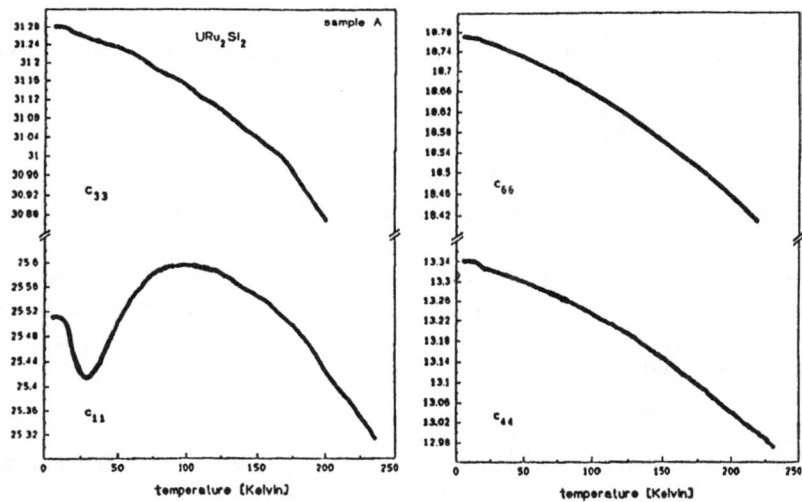

Fig. 9.18. Temperature dependence of the elastic constants c_{11}, c_{33}, c_{44}, c_{66} for URu_2Si_2 (Wolf et al. [9.110])

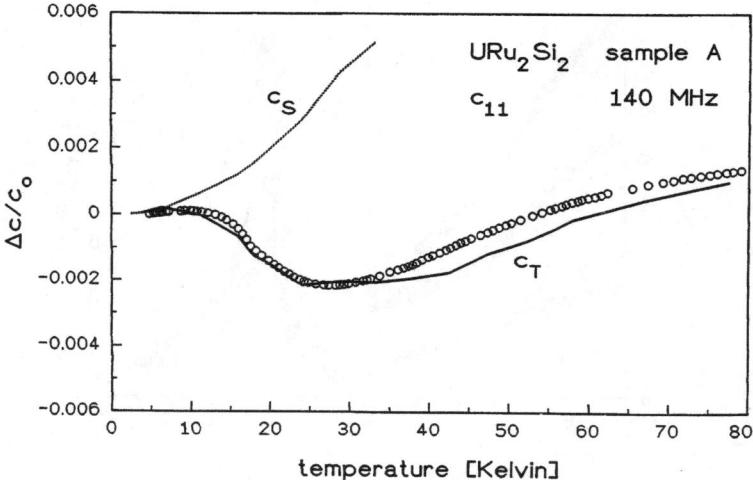

Fig. 9.19. c_{11} mode for URu$_2$Si$_2$ together with calculated fits c_S (dotted line) and c_T (full line) (Lüthi et al. [9.55])

from elastic constant and the pressure dependence of the characteristic temperatures $T^\star$, T_N, T_s, are listed in Table 9.1.

It is seen that only Ω_a is large. The c_{11} mode was interpreted with (9.3) and (9.9). This is shown in Fig. 9.19 where it is seen that the measured c_{11} follows closely the isothermal expression c_T and not the adiabatic c_S.

In Fig. 9.18 and 9.19 the anomalies at T_N are not visible. In the following Fig. 9.20 we give a more detailed view of the low temperature region. We notice anomalies at T_N of the order of 10^{-4} in $\Delta c/c_0$ and strong BCS-like anomalies in the attenuation $\Delta \alpha$ (Wolf et al. [9.111]). It was argued that these anomalies might arise due to the special spin nematic ordering (Lüthi et al. [9.112]). The temperature dependence of the elastic constants in URu$_2$Si$_2$ do not show any evidence for CEF effects, in contrast to recent claims (Santini et al. [9.113]). The nature of the magnetic transition at T_N is not clear yet. The experimental results of Fig. 9.20, especially the attenuation results, point to an opening of a spin or charge–density gap at T_N.

Further experimental results on URu$_2$Si$_2$ will be given in Sect 10.7.

UPd$_2$Al$_3$
For the hexagonal large moment heavy fermion compound UPd$_2$Al$_3$, the temperature dependence of the elastic constants c_{11}, c_{33}, c_{44}, c_{66} have been measured (Lüthi et al. [9.112]). They exhibit CEF -effects for the c_{33} mode (Modler et al. [9.114]) and pronounced anomalies at the magnetic transition temperature $T_N = 14$K. No low temperature heavy fermion behaviour (9.9) has been observed for any of these elastic modes, mainly because of the small Grüneisen parameter (Table 9.1). Also only small anomalies of the order of 10 ppm were measured at the superconducting transition temperature

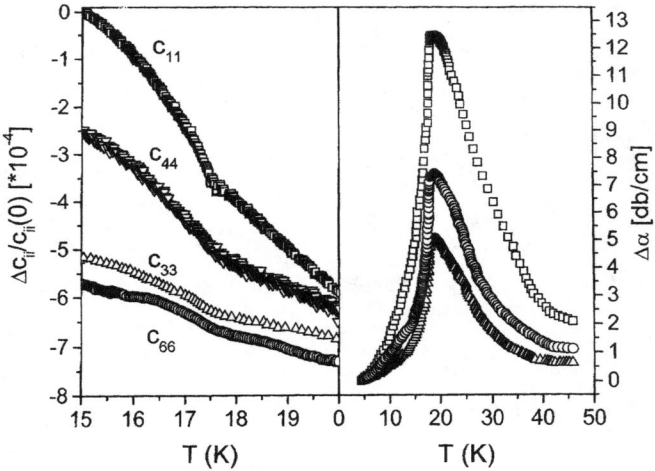

Fig. 9.20. Elastic constants (c_{11}, c_{33}, c_{44}, c_{66}) and ultrasonic attenuation α for c_{11} in URu$_2$Si$_2$ in the vicinity of T_N (Wolf et al. [9.111])

$T_s = 1.8$K. The measurement of the B-T diagram of the anti-ferromagnetic phase agreed with corresponding susceptibility measurements.

Summary

Summarising, the temperature dependence of longitudinal elastic constants in heavy fermion Ce–compounds can be explained with the adiabatic elastic constant formula c_S (9.9), (9.11) whereas the heavy fermion U–compounds are better fitted with the isothermal formula c_T (9.9). This is particularly true for URu$_2$Si$_2$ whereas the case of UPt$_3$ is not clear yet. The scaling procedure, employed in Sect. 9.2 and especially in this Sect. 9.3.4, assumes a single energy scale given by the fluctuation temperature $T^\star \sim T_K$. This implies a constant Grüneisen parameter $\Omega = \partial \ln T^\star / \partial \varepsilon$ for $T < T^\star$. If Ω still increases with decreasing temperature, there is an additional energy scale like the coherence temperature T_{coh}. Such behaviour can be seen in CeCu$_6$ for $T < 1$K (see e.g. Milliken et al. [9.115]). Another interesting feature occurs near a quantum critical point. This will be discussed in Sect. 9.3.6.

Most of the interesting heavy fermion materials (UPt$_3$, URu$_2$Si$_2$, CeCu$_6$, CeRu$_2$Si$_2$, CeCu$_2$Si$_2$) have tetragonal, hexagonal or orthorhombic structure. Therefore the longitudinal elastic constants and the thermal expansion are also anisotropic. It has been shown above how these systems with anisotropic Grüneisen parameters Ω_a and Ω_c can be described.

9.3.5 Sound Wave Effects in Magnetic Fields, Meta-magnetic Transition

Heavy fermions exhibit pronounced effects in high magnetic fields. This is particularly true for substances which have so-called meta-magnetic transitions. This term depicts the simple fact that the differential susceptibility has a maximum at a critical magnetic field B_c and the high field state has an enhanced magnetic moment. With this use of the term meta-magnetism, quite different classes of substances exhibit such effects. Many anti-ferromagnets (e.g. $FeCl_2$, $DyPO_4$ etc.) belong to these materials. For an early review see Stryjewski and Giordano [9.116] and later ones: Lüthi et al. [9.117] and Flouquet et al. [9.118]; this last one especially for $CeRu_2Si_2$.

$CeRu_2Si_2$ & UPt_3

Two nice examples for heavy fermion meta-magnetic behaviour are UPt_3 and $CeRu_2Si_2$, which have already been discussed at length in Sects. 9.3.3 and 9.3.4. Physical properties of these two substances are given in Tables 9.1 and 9.3. The structure of $CeRu_2Si_2$ is the same as for $CeCu_2Si_2$ which is shown in Fig. 10.20. Again the longitudinal modes show the biggest effect. For UPt_3, it is the c_{11} mode propagating in the hexagonal plane, which is also the easy direction of magnetization, and for $CeRu_2Si_2$, it is the c_{33} mode propagating along the easy magnetic c-axis. Figure 9.21 displays the relative velocity changes for these two cases (Kouroudis et al. [9.119]). The meta-magnetic transition is at 20T for UPt_3 and at 8T for $CeRu_2Si_2$.

One observes for the relative elastic constant very deep minima in $\Delta c_{ii}/c_{ii}$ at these transitions of 8% and 30% for UPt_3 and $CeRu_2Si_2$ respectively. In fact there is a strong temperature dependence. For example, at 50 mK the minimum is 50% deep for $CeRu_2Si_2$ (Bruls et al. [9.120]). In order to under-

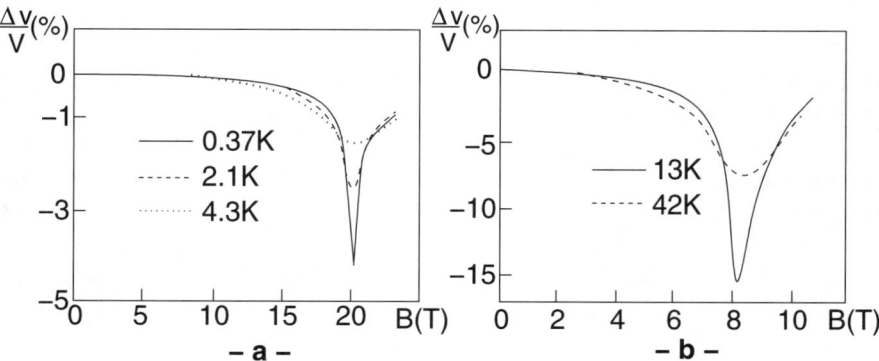

Fig. 9.21. Relative sound velocity for c_{11} mode as a function of magnetic field in (**a**) UPt_3 and (**b**) $CeRu_2Si_2$. All measurements are made for different temperatures as indicated in the figure (Kouroudis et al. [9.119])

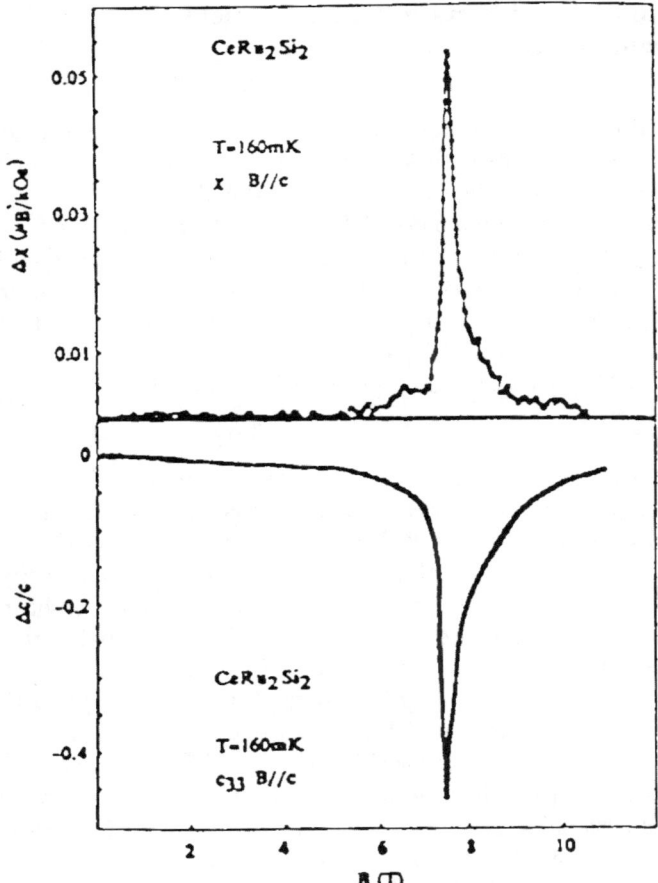

Fig. 9.22. Magnetic susceptibility and c_{33} mode as a function of magnetic field parallel to c-axis in $CeRu_2Si_2$ at 160 mK (Bruls et al. [9.120])

stand these anomalies, Fig. 9.22 compares the sound wave effect for $CeRu_2Si_2$ with the differential magnetic susceptibility (Bruls et al. [9.120]).

The result is that roughly $\Delta c/c_0 \propto -\chi_m$. This result we can phenomenologically understand again with scaling arguments. With the appropriate thermodynamic potential $F = F_0 - \int M dB$ and introducing the electron–lattice coupling by scaling $M(B/B_c)$ with the Grüneisen parameter $\Omega_B = -\partial \ln B_c/\partial \varepsilon_V$ we obtain for the isothermal elastic constant $c = \partial^2 F/\partial \varepsilon^2$

$$c = c_0 - \Omega_B^2 B^2 \chi_m \,, \tag{9.13}$$

where terms with $\partial^2 B_c/\partial \varepsilon^2$ have been neglected. Equation (9.13) can be obtained also from a generalized scaling of the free energy $F(\frac{T}{T^\star}, \frac{B}{B_c})$ by introducing $\Omega_T = -\partial \ln T^\star/\partial \varepsilon_V$ and $\Omega_B = -\partial \ln B_c/\partial \varepsilon_V$ (Thalmeier and

9.3 Electron–Phonon Coupling in Heavy Fermion Systems

Fulde [9.121]). For $T < T^*$ and $\Omega_T = \Omega_B$, (9.13) is again obtained. With $\Omega_T = 117$ and $\Omega_B = 115$, we get a good fit for c_{33} in Fig. 9.21 and 9.13 for CeRu$_2$Si$_2$. Actually, in this case it is also necessary to introduce the anisotropic Grüneisen parameters: $\Omega_B^a = -\partial \ln B_c/\partial \varepsilon_{11}$ and $\Omega_B^c = -\partial \ln B_c/\partial \varepsilon_{33}$ as outlined in Sect. 9.3.4. The various Grüneisen parameters Ω_a and Ω_c are given in Table 9.3.

The equality of Ω_B and Ω_T has been explained by Kaiser and Fulde [9.122]. It was shown that for systems with small Sommerfeld–Wilson ratio R of the order of 1 it follows that $\Omega_B \sim \Omega_T$, as observed for the heavy fermion compounds UPt$_3$ and CeRu$_2$Si$_2$. The Sommerfeld-Wilson ratio is defined as (Sect. 9.1.2)

$$R = \frac{1}{3}\left(\frac{\pi k_B}{\mu_B}\right)^2 \frac{\chi_m}{\gamma}, \tag{9.14}$$

with k_B the Boltzmann constant, μ_B the Bohr magneton, χ_m the magnetic susceptibility and γ the Sommerfeld coefficient of the specific heat. By contrast, transition metal compounds with strong Stoner enhancement have $R > 1$ and $\Omega_B \neq \Omega_T$ because the magnetic susceptibility becomes more volume dependent in this case. In Table 9.3 we see that $\Omega_B \sim \Omega_T$ is fulfilled very well for the two compounds discussed. A microscopic theory for these meta-magnetic transitions including the phonon effects has not been developed yet. For a microscopic theoretical treatment of meta-magnetic transitions (without coupling to phonons) see e.g. Held et al. [9.123], Kuramoto and Kitaoka [9.124].

Using (4.19) relating ultrasonic attenuation α and elastic constant change $\alpha = -\frac{\Delta c}{c^*}\frac{\omega^2 \tau}{2v}$, results in a characteristic relaxation time of spin fluctuations at the meta-magnetic transition (Weber et al. [9.125]). In Fig. 9.23, typical results are shown for α and $\Delta c/c_0$ for different frequencies in UPt$_3$. From these one obtains $\tau \sim 10^{-10}$ s for $T = 0.4$K and $\omega\tau \sim 0.09 \ll 1$. Although the spin fluctuation time τ is much longer than typical electronic relaxation times, $\omega\tau$ is still considerably smaller than 1.

In addition to sound velocity measurements, there are other experiments demonstrating the strong coupling of the heavy fermion state to the lattice. For CeRu$_2$Si$_2$ magneto-strictive studies give very large effects near the meta-magnetic transition of the order of $\Delta a/a \sim 4 \cdot 10^{-4}$ (Mignot et al. [9.126]). Pressure dependent magnetization studies together with the Maxwell relation $(\partial M/\partial p)_{B,T} = (\partial V/\partial B)_{p,T}$ give consistency checks and large volume changes at B_c. A microscopic theory for the meta-magnetic transition involving electronic, magnetic and lattice degrees of freedom has not been established yet. It seems that in substances like CeRu$_2$Si$_2$, we are dealing with a de-localised heavy fermion system for $B < B_c$ and a more localised one for $B > B_c$.

Magneto-striction experiments, together with the one parameter scaling $M = M(B/B_c(p))$, were used to determine thermodynamic quantities at constant volume rather than at constant pressure. If meta-magnetism had nothing to do with magneto-elastic interactions involving volume strains,

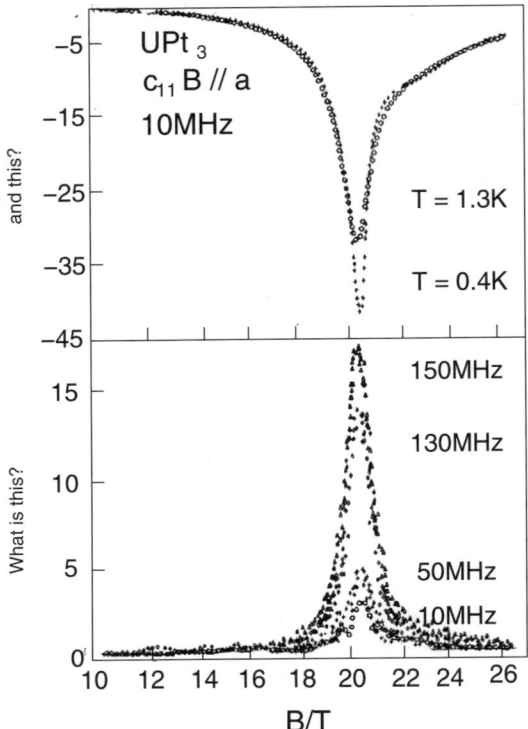

Fig. 9.23. Relative velocity change and attenuation for UPt$_3$ at 1.3 and 0.4K in the frequency range 10–150 MHz (Weber et al. [9.125])

no change in specific heat $C_p \to C_V$ or magnetization $M_p \to M_V$ would be expected. On the other hand, if such a lattice coupling is the primary interaction, the meta-magnetic effect would be absent for constant volume experiments. Experimental results for CeRu$_2$Si$_2$ show rather dramatic changes although a meta-magnetic transition can still be discerned from the data. In Fig. 9.24 magnetization data for M_p and M_V are shown for applied fields at very low temperature in the easy c-axis direction (Matsuhira et al. [9.127]). In the inset the differential susceptibility χ_p and χ_V are also shown. Whereas χ_p exhibits the sharp meta-magnetic peak at 7.8T, χ_V shows a very broad flat maximum at 9.4T. These results indicate that meta-magnetism in heavy fermion compounds is a magnetic phenomenon, but that the coupling to the lattice degrees of freedom cannot be neglected in any quantitative theory.

The CeRu$_2$Si$_2$ compound is also interesting with respect to a possible quantum critical point (Flouquet et al. [9.118]). It will be discussed in Sect. 9.3.6.

The large Grüneisen parameter Ω_B near the meta-magnetic transition makes it possible to generate longitudinal sound electromagnetically. This

9.3 Electron–Phonon Coupling in Heavy Fermion Systems

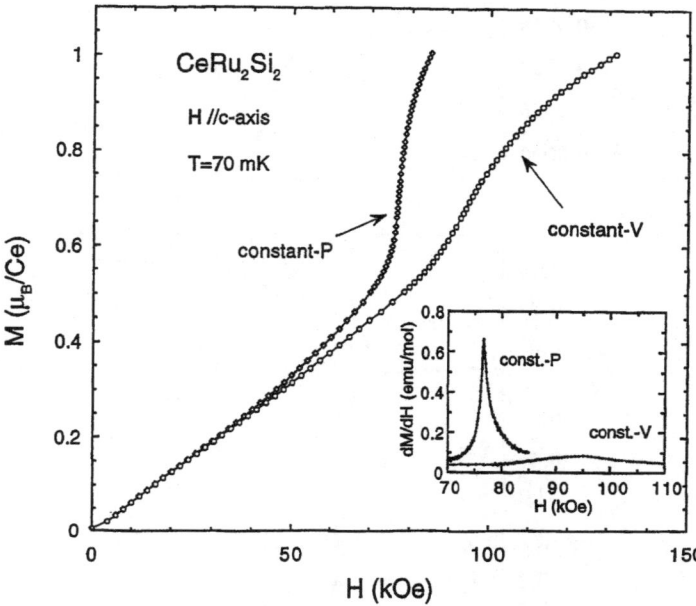

Fig. 9.24. Constant volume and constant pressure magnetization for CeRu$_2$Si$_2$ at 70 mK in magnetic fields applied along the c-axis. Inset gives the differential susceptibility (Matsuhira et al. [9.127])

has been shown in CeRu$_2$Si$_2$ (Hampel et al. [9.128]). The radio frequency field $b(\omega)$ gives rise to a stress $\sigma_{zz} = -\Omega_B B_z \chi_m b(\omega)$.

URu$_2$Si$_2$
In the other heavy fermion U-compound URu$_2$Si$_2$, magnetic field dependent anomalies are also observed. These are shown in Fig. 9.25 for the c_{33} and c_{11} modes at a temperature of 4.2K (Wolf et al. [9.129]). The anomalies occur at field values of 34.5, 36.5 and 39T. The measurements were carried out in pulsed fields up to 50T. Previous magnetization and susceptibility measurements in quasi-static and pulsed fields (de Visser et al. [9.130], Sugiyama et al. [9.131]) gave step function-like behaviour of the magnetization, very similar to the plateau behaviour in low-dimensional spin systems (see Sects. 12.1.2 and 12.3). But URu$_2$Si$_2$ is a 3–dimensional compound with a tiny ordered sublattice magnetic moment of 0.03 m_B/U-atom oriented along the tetragonal c-axis (Broholm et al. [9.132]). In Fig. 9.25 we also show the magnetic susceptibility at 7K in arbitrary units. The step function anomalies in the magnetization, and therefore the susceptibility peaks, are most pronounced at 35T and 39T. For both elastic modes, strong anomalies occur for $T < T_N$. The c_{33}–mode exhibits just one step function-like anomaly at the upper transition at 39T. This is the so-called spin–flop transition. With a strain coupling to the sublattice magnetization of the form $F_{int} = g\varepsilon_{33}(M_{z1}^2 + M_{z2}^2)$, we get a

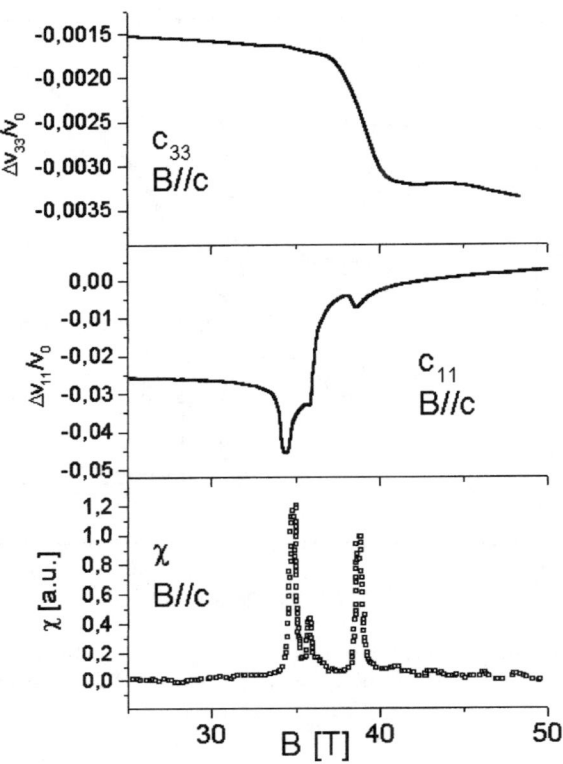

Fig. 9.25. Relative sound velocity measurement for the c_{33} (a) and c_{11} (b) modes in URu_2Si_2 for $B \parallel c$ together with the differential susceptibility. The sound velocity curves are given as a function of field for a temperature of 4.2K (Wolf et al. [9.129])

step function at the spin flop transition. Contrary to the c_{33} mode, we observe large changes at the corresponding transitions for the c_{11} mode (Fig. 9.25). Note the different scale for the two modes c_{11} and c_{33}. The largest change is observed for c_{11} at the 34.5T transition. The field dependence resembles the longitudinal effects at the meta-magnetic transition of UPt_3 and $CeRu_2Si_2$ (Fig. 9.21). Note that the temperature dependence of the c_{11} mode was also anomalous (Figs. 9.18 and 9.19) due to the big Grüneisen parameter. We therefore interpret the transition at 35T as a meta-magnetic transition.

For recent developments on the magnetism of URu_2Si_2 see Jaime et al. [9.133] and the various contributions presented at the SCES Conference 2002. It was found that magneto-caloric effects appear at these high field transitions. For the data of Fig. 9.25, this effect at 4.2K is rather small as seen from a comparison with static measurements (Jaime et al. [9.133]). In the temperature region 5-10K this effect is noticeable. A further ultrasonic study was performed in static and pulsed fields (Suslov et al. [9.134]).

CeCu$_6$, CeNiSn

There are other heavy fermion compounds exhibiting meta-magnetic phenomena in ultrasonic propagation. In CeCu$_6$ with a magnetic field along the easy c-axis, a velocity minimum was observed at $B_c \sim 2.2T$ for very low temperatures of 70mK (Goto et al. [9.135]). Differential susceptibility measurements revealed a maximum around $\sim 2T$ for temperatures below 500mK (Löhneysen et al. [9.136]), typical for a heavy fermion meta-magnetic transition. It was found that short range inter-site magnetic fluctuations are suppressed with a magnetic field of 2T. This indicates again that the meta-magnetic transition leads to a more localised state for $B > B_c$. For CeNiSn, a sound velocity anomaly was observed at low temperatures (80mK) for longitudinal modes c_{11}, c_{33} (Holtmeier et al. [9.103]). It bears the signature of a meta-magnetic-like transition, a broad minimum around $B_c = 8.7T$ for $B \parallel a$–axis. It disappears for $T > 1K$. On the other hand, the temperature dependence of the c_{11} mode exhibits a broad minimum around 8K. This, approximately, corresponds energetically to the B_c value. These meta-magnetic-like magnetic and acoustic anomalies in the heavy fermion substances CeCu$_6$ and CeNiSn are by no means so distinct as can be observed for the materials UPt$_3$ and CeRu$_2$Si$_2$ discussed above.

9.3.6 Non-Fermi Liquid Effects

In recent years, deviations from Fermi liquid theory (see Sect. 9.1.2) are at the center of interest in heavy fermion research. These non-Fermi liquid phenomena can occur near a quantum critical point, e.g. near the critical boundary between the paramagnetic ground state and a magnetic one for $T \to 0$. They can also occur for disordered heavy fermion materials or for systems exhibiting a multichannel Kondo effect. For a review of these phenomena, see Cox and Zawadowski [9.137]. A good example is the alloy system CeCu$_{6-x}$Au$_x$ where for $x \sim 0.1$ a transition from the paramagnetic state ($x < x_c$) to antiferromagnetic long range order ($x > x_c$) occurs (Löhneysen [9.138]). Another case for non-Fermi liquid behaviour occurs for U-ions. Because these ions usually do not have a fixed valency in intermetallic compounds, the interaction of the U^{4+}(5f^2)-ions with conduction electrons leads also to non-Fermi liquid behaviour. This effect is the multichannel Kondo effect mentioned above, in this case named two-channel Kondo effect or quadrupolar Kondo effect. The quadrupole moment of a CEF ground state here is compensated by the conduction electrons in a similar way as the magnetic moment is compensated in conventional Kondo systems (Fulde and Loewenhaupt [9.139], Cox [9.140], Cox and Zawadowski [9.137]). A typical example investigated is U$_{1-x}$Y$_x$Pd$_3$ (Seaman et al. [9.141], Andraka and Tsvelick [9.142]), but see below.

The occurrence of non-Fermi liquid behaviour is manifested by different (mostly logarithmic) dependences on temperature of the different thermodynamic quantities (specific heat and susceptibility) and transport properties like electrical resistivity. For a review see, Cox and Zawadowski [9.137]. The question we want to investigate here is whether sound wave experiments can contribute to the physics of this new ground state.

Elastic Constants

For $CeCu_{6-x}Au_x$ extensive elastic measurements were carried out for $x = 0$ and 0.1 (Finsterbusch et al. [9.143, 9.144]). While for $CeCu_6$ typical heavy fermion Fermi liquid properties were found for the longitudinal elastic modes (Fig. 9.11 and 9.13; Sect. 9.3.2), in $CeCu_{5.9}Au_{0.1}$ the longitudinal modes c_{ii} ($i = 1, 2, 3$) have a linear T dependence for $T < 1K$. The effects measured, however, are very small. For a possible explanation see the next section 9.3.6.

For the case of the quadrupolar Kondo effect, a more detailed answer can be given. $U_{0.2}Y_{0.8}Pd_3$ was compared to UPd_3 (Amara et al. [9.145]). The latter substance (with $U^{4+}5f^2$ and $^3H_4$ ground state), which was discussed in Sect. 7.2.3 and Fig. 7.19, exhibits a quadrupolar phase transition at 7.6K. For polycrystalline $U_{0.2}Y_{0.8}Pd_3$ no signs of an anomalous temperature dependence for longitudinal and transverse waves were observed from 300K to 1.5K. A normal hardening of the elastic constants towards lower temperatures can be observed. In this compound, the U^{4+} ion has the same near environment as the quasi-cubic sites in UPd_3. It has the same coordination number and its first neighbours are located at nearly exactly the same positions. In the framework of the quadrupolar Kondo effect, the quadrupolar susceptibility results in a logarithmic divergence below $T_K = 42K$ for $T \to 0$ and a similar softening for longitudinal (c_L) and transverse (c_T) elastic constants for $T > T_K$ should be observed as in UPd_3. As pointed out above, no softening is observed however. Magnetic field dependence is very small, the effect amounting to less than 1 part in 10^4 at 10T for the relative elastic constant. Because of the absence of any softening for c_L and c_T, one has to conclude therefore that in $U_{0.2}Y_{0.8}Pd_3$ the quadrupolar Kondo effect is not operative.

In the body-centered skutterudite structure (T_h^5) material $PrFe_4P_{12}$, Nakanishi et al. [9.146] come to a different conclusion. With an assumed orbitally degenerate ground state Γ_3, they show that the $c_{11} - c_{12}$ mode shows little softening versus decreasing temperature for magnetic field $B = 0$, but a strong 10% softening in a field of $B = 8T$. They interpret this with a quadrupolar Kondo effect for $B = 0$, which is broken up and no longer operative for $B > 0$. Further experimental and theoretical work is needed to understand the transition from the screened quadrupolar moment at $B = 0$ to a fully developed quadrupolar moment in strong fields.

Magnetic Quantum Critical Point

New interesting phenomena can occur near a quantum critical point. In Fig. 9.26 we give a general phase diagram $T - \delta$ for a heavy fermion sys-

9.3 Electron–Phonon Coupling in Heavy Fermion Systems

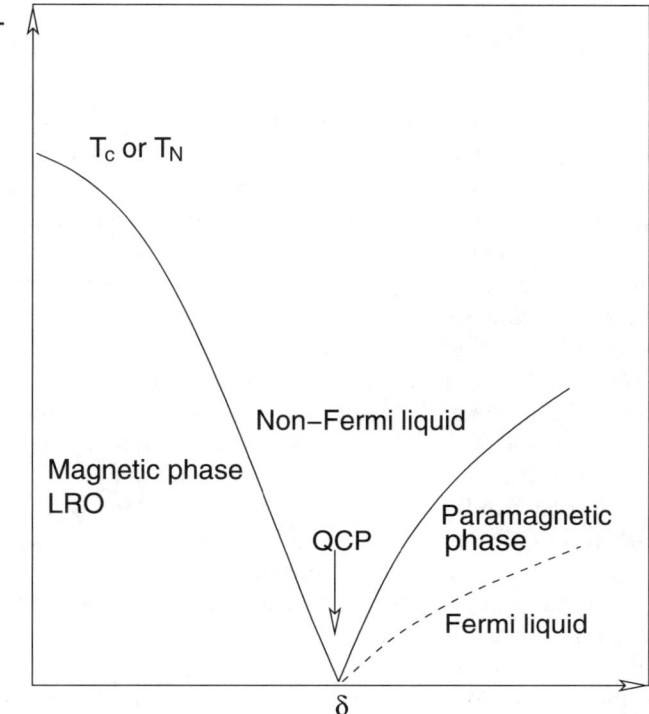

Fig. 9.26. Schematic phase diagram $T-\delta$ of a heavy fermion metal near a quantum critical point. δ is a control parameter (see text) and the various regions near the QCP are indicated

tem indicating magnetic phases and non-magnetic phases with a quantum critical point (QCP) for $T \to 0$. Here δ is a control parameter. δ can be

1. a chemical substitution parameter,
2. the pressure,
3. a magnetic field or
4. the on-site exchange coupling between the f-spin and the conduction electrons.

This picture was originally introduced by Doniach [9.147]. Near QCP, quantum fluctuations prevail and deviations from usual heavy fermion behaviour (Fermi liquid properties) can occur. Recently it has been found that unconventional superconductivity can occur near a QCP (see Sect. 10.7). In addition, it has been argued that at a continuous quantum critical phase transition at $T = 0$, the Grüneisen parameter can diverge in a similar manner as the thermodynamic quantities, such as the specific heat, can diverge near a classical phase transition (Zhu et al. [9.148]). With the magnetic field B as the control parameter, the Grüneisen parameter becomes

$$\Omega_B = -\frac{\left(\frac{\partial S}{\partial B}\right)_T}{T\left(\frac{\partial S}{\partial T}\right)_B} = -\frac{\left(\frac{\partial M}{\partial T}\right)_B}{C_B} = \frac{\left(\frac{\partial T}{\partial B}\right)_S}{T},$$

which is closely related to the magneto-caloric effect (2.9). The analysis gives for Ω_B a scaling expression $\Omega_B \sim r^{-1/z\nu}$ with $r = (B - B_c)/B_c$ or a temperature scale and z the dynamical critical exponent and ν the exponent of the correlation function (see Sect. 6.1.2). Experimentally the thermal expansion coefficient gave a stronger singularity (for $T \to 0$) than the specific heat at very low temperature for $CeNi_2Ge_2$ and $YbRh_2(Si_{0.95}Ge_{0.05})_2$ indicating indeed a diverging Grüneisen parameter (Küchler et al. [9.149]). From these results we would expect also anomalous results for the longitudinal elastic constants towards $T \to 0$, namely a strong softening.

In $CeCu_{6-x}Au_x$, a linear T-dependence of the longitudinal elastic constants for $x = 0.1$ was found as discussed above. Whether this can be interpreted with the scaling theory outlined above is not yet clear.

Another system exhibiting a quantum critical point is $Ce_{1-x}M_xRu_2Si_2$ (Flouquet et al. [9.118]). Substituting La ions for M, giving $Ce_{1-x}La_xRu_2Si_2$, drives the system into a magnetic state with $x \sim 0.07$. In the pure lattice, quantum fluctuations apparently prevent a magnetic phase transition at B_c. The phase change with La alloying means that a negative pressure of a few kbar would give a corresponding phase change for the pure lattice. No ultrasonic experiments have been performed yet in the $Ce_{1-x}La_xRu_2Si_2$ system.

Quantum phase transitions and QCP can also occur in low dimensional spin systems, especially in dimerised spin $\frac{1}{2}$ compounds. Again, external parameters as pressure or magnetic field can drive the system into a magnetic state. Famous examples are $SrCu_2(BO_3)_2$ or $TlCuCl_3$. These cases will be discussed in Sects. 12.2.2, 12.3 and 12.7.

10 Ultrasonics in Superconductors

Ultrasonic attenuation measurements in conventional superconductors played a crucial role in discovering the gap structure and in testing the BCS theory of superconductivity. Later on, this technique and accompanying sound velocity measurements helped in obtaining useful information on quite different types of superconductors, mainly on Type II, such as unconventional and High-Temperature Superconductors. We will survey the field of conventional, unconventional and high temperature superconductivity. After an introduction to superconductivity, we present first a phenomenological treatment of the electron–strain coupling in superconductors (Sect. 10.2). Afterwards we survey elastic properties of different types of superconductors (Sect. 10.3) for $B = 0$. The case of high temperature superconductivity is treated in particular in Sect. 10.4. We then introduce magnetic field-dependent properties of sound propagation, the sound wave interaction with the flux line lattice (Sect. 10.5). Surface wave attenuation in superconductors is treated in Sect. 10.6. Finally we give an overview of sound wave experiments performed in unconventional superconductors in Sect. 10.7.

There are numerous books on superconductivity. Some of them we list here: Schrieffer [10.1], Buckel [10.2], Tinkham [10.3], Tilley and Tilley [10.4], Tsuneto [10.5]. For unconventional superconductors consult Mineev and Samokhin [10.6] and various review articles listed below.

10.1 Introduction

First some introductory remarks on superconductors in general. Many elements, intermetallic compounds and even some oxygen compounds exhibit superconductivity. At the transition temperature T_c, the electrical resistance vanishes very abruptly. The transition temperatures T_c can vary from milli-Kelvin range to more than 100 K. In addition, in the superconducting state, the magnetic field is expelled or cannot penetrate. This is the so-called Meissner-Ochsenfeld effect. In type II superconductors at a field B_{c1} magnetic field starts to penetrate in the form of vortices and the transition to the normal state ends at a field B_{c2}. Fig. 10.1a shows the magnetization for a type I and a type II superconductor. The diamagnetic magnetization rises for a type I superconductor until the critical field B_c, (labeled as B_{c1})

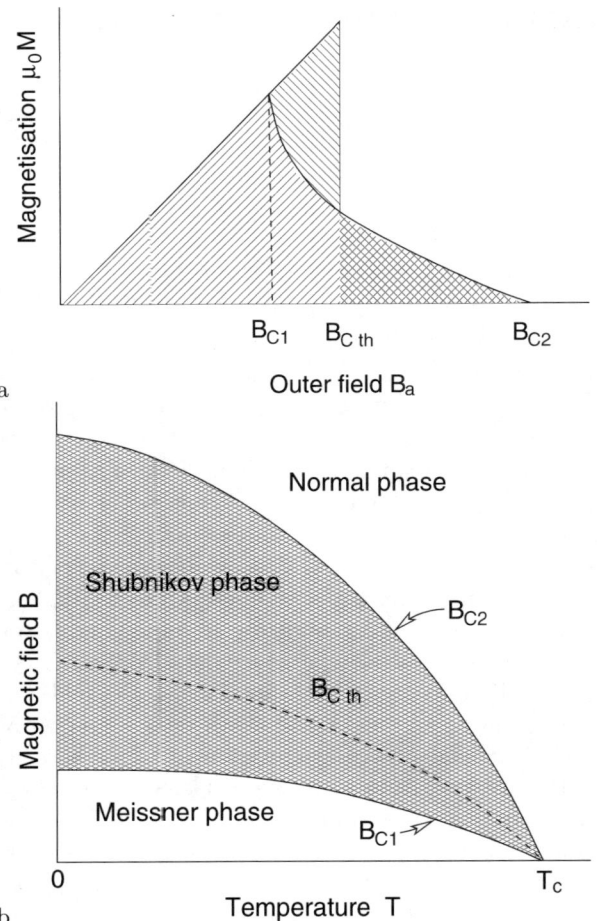

Fig. 10.1. (a) Diamagnetic magnetization for type I and II superconductors (dotted line: type I, full line: type II), (b) B–T phase diagram for a type II superconductor

and falls then abruptly to zero (dotted line). The B–T phase diagram of a type II superconductor with the two phases – "Meissner phase and Shubnikov phase" – are shown in Fig. 10.1b. The thermodynamic field B_{cth} is related to B_{c2} via $B_{c2} = \sqrt{2}\kappa B_{cth}$ with $\kappa = \lambda/\xi$. κ is called the Ginzburg–Landau parameter. Here λ is the penetration depth and ξ the coherence length (Tinkham [10.3]). For type I superconductors $\kappa < 1/\sqrt{2}$ and for type II superconductors $\kappa > 1/\sqrt{2}$.

The underlying mechanism giving rise to superconductivity is, in most cases, the electron–phonon interaction. The forming of Cooper pairs $(\mathbf{k}, -\mathbf{k})$ with this interaction produces an energy gap between the highly correlated ground state and the excited states. This energy gap $\Delta(T)$ is temperature dependent and vanishes at T_c. The spatial extension of the Cooper pairs is the coherence length ξ. It can vary from 10000Å in elemental superconductors to a few Å in oxide superconductors. The force range is therefore mostly much larger than a typical lattice constant. Therefore for most superconductors the Ginzburg criterion (see Sect. 4.4) is fulfilled and a Landau theory or the so-called Ginzburg–Landau theory works well.

The order parameter for s-wave superconductors is $\eta(x) = \langle \psi_\uparrow(x)\psi_\downarrow(x) \rangle$ which is equivalent to the gap parameter $\Delta(x)$. Since these are scalar complex functions $\eta = |\eta| e^{i\Phi} = \Delta e^{i\Phi}$ the Ginzburg–Landau free energy reads, in analogy to (4.9), with $F_n - F_s = \Delta F_{GL}$ and $F_n > F_s$ for $T < T_c$:

$$\Delta F_{GL} = a \, |\eta|^2 + \frac{b}{2} \, |\eta|^4 + \frac{\hbar^2}{2}m \, |\nabla \eta|^2 \; . \tag{10.1}$$

This $B = 0$ free energy is the same as the Landau energy used in Sect. 4.3 with the last term, the gradient term for $\eta(x)$ included. With $a = a_0(T - T_c)$ (where $a_0, b > 0$), then together with $\frac{\partial F_{GL}}{\partial \eta} = 0$ this results in the equilibrium order parameter $|\eta|^2 = -a/b$. Inserted into (10.1) gives

$$\Delta F_{GL} = \frac{a_0^2}{2}b(T - T_c)^2 \; . \tag{10.2}$$

This is the molecular field form of the the free energy density of a superconductor. We can write it as

$$\Delta F_{GL} = \frac{B_c^2}{8\pi} = \Phi_0 \left[\frac{T - T_c}{T_c}\right]^2 \tag{10.2a}$$

with Φ_0 the condensation energy at $T = 0$. With this free energy expression, the strain–electron coupling can be calculated by assuming Φ_0 and T_c to be strain dependent. The specific heat $C = -T\frac{\partial^2 F}{\partial T^2}$ gives $\Delta C = 2\Phi_0 \frac{T}{T_c^2}$ or the specific heat change at T_c

$$\Delta C \, |_{T_c} = C_s - C_n = T_c\frac{a_0^2}{b} = 2\frac{\Phi_0}{T_c} \; . \tag{10.3}$$

Figure 10.2 shows an experimental result for the specific heat for Sn a type I superconductor (Keesom et al. [10.7])). The specific heat jump at T_c is clearly visible. With an isotropic gap the specific heat vanishes exponentially for $T \to 0$.

10.2 Electron–Strain Coupling in Superconductors

Ultrasonic experiments played an important role in the development of superconductivity. From attenuation experiments the gap structure could be determined. In unconventional superconductors, the complicated B–T phase diagram could be calculated. The interaction of vortices with sound waves

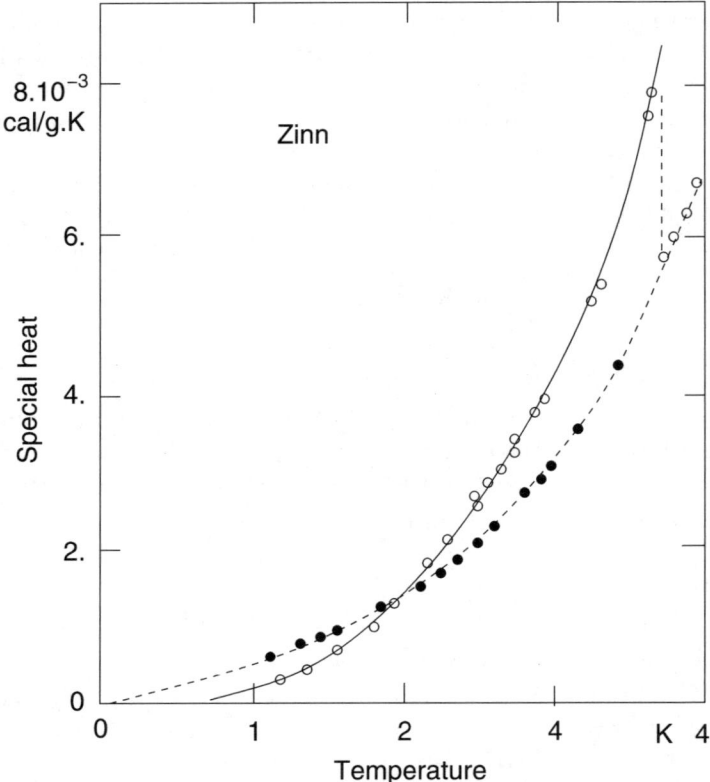

Fig. 10.2. Low temperature specific heat at T_c for Sn as a function of temperature. Open circles: without magnetic field, filled circles: for fields $B > B_c$ (adapted from Keesom and van Laer [10.7])

leads to characteristic sound wave anomalies. These effects and others we will discuss in the following sections. We start here with a molecular field treatment of elastic constants followed by the famous BCS attenuation formula for sound waves.

10.2.1 Elastic Constants

Isothermal elastic constants can be calculated using (10.2) and (10.2a) and assuming a strain dependence of the condensation energy $\Phi_0(\varepsilon)$ and of the critical temperature $T_c(\varepsilon)$. This phenomenological approach has been used first by Testardi [10.8, 10.9].

$$c_s - c_n = \frac{\partial^2 \Delta F}{\partial \varepsilon^2}$$

$$= -\frac{\left(\frac{\partial B_c}{\partial \varepsilon}\right)^2 + B_c \frac{\partial^2 B_c}{\partial \varepsilon^2}}{4\pi}$$

$$= -2\left(\frac{\partial T_c}{\partial \varepsilon}\right)^2 \Phi_0 T \frac{-2T_c + 3T}{T_c^4} + 2\frac{\partial^2 T_c}{\partial \varepsilon^2} \Phi_0 T \frac{T-T_c}{T_c^3} \quad (10.4)$$

$$- \frac{\partial^2 \Phi_0}{\partial \varepsilon^2}\left(\frac{T-T_c}{T_c}\right)^2 + 4\frac{\partial T_c}{\partial \varepsilon}\frac{\partial \Phi_0}{\partial \varepsilon} T \frac{T-T_c}{T_c^3} \ .$$

This formula describes the low frequency isothermal elastic constant for $T \leq T_c$. For conventional superconductors there is no domain formation in the superconducting state. Therefore (10.4) applies to a single domain state. Contact can be made between (10.4) and the strain–order parameter formulas in Chap. 4, (4.12) and (4.14). The first term on the right hand side of (10.4) describes the coupling to the order parameter as given in Chap. 4, (4.12), (4.14). With $F_{int} = h_\Gamma \varepsilon_\Gamma \, |\eta|^2$ at T_c results in the following step function (see (4.14)):

$$\Delta c_\Gamma = -\frac{h_\Gamma^2}{b} \ . \quad (10.5)$$

Likewise the first term of (10.4) results in a step function

$$\Delta c_\Gamma = -2\Phi_0 \frac{\left(\frac{\partial T_c}{\partial \varepsilon}\right)^2}{T_c^2} = -\left(\frac{\partial T_c}{\partial \varepsilon}\right)^2 \frac{\Delta C}{T_c} \quad \text{or} \quad h = \frac{\partial T_c}{\partial \varepsilon a_0} \ .$$

For high symmetry directions, shear waves do not couple because $\frac{\partial B_c}{\partial \varepsilon_{sh}} = \frac{\partial T_c}{\partial \varepsilon_{sh}} = \frac{\partial T_c}{\partial (-\varepsilon_{sh})} = 0$ (Pippard [10.10]). This no longer holds in general for unconventional multi-component order parameters as discussed in Sect. 10.7. The other terms of (10.4) disappear at T_c but dc_Γ/dT has a finite value (Testardi [10.8]). If the next higher term in the strain–order parameter interaction is included: $F_{int} = h_\Gamma \varepsilon_\Gamma \, |\eta_\Gamma|^2 + H_\Gamma \varepsilon_\Gamma^2 \, |\eta|^2$, then an expression like the second term of (10.4) is obtained, i.e. $H_\Gamma \sim \partial^2 T_c/\partial \varepsilon^2$. Applications of (10.4) will be given below (Sects. 10.3–6). A compilation of ultrasonic data for a large variety of superconductors has been given recently (Zherlitsyn et al. [10.11]). Equations (10.1)–(10.5) apply strictly only close to T_c, the region of the validity of Landau theory. Similar expressions as (10.4) can be derived by using the two-fluid model of Gorter and Casimir (for a discussion of the two-fluid model see Tilley and Tilley [10.4]) instead of the Ginzburg–Landau free energy expression.

10.2.2 Ultrasonic Attenuation

Ultrasonic attenuation in superconductors was first studied by Bömmel (see [10.12] and [10.13]). Figure 10.3 displays an attenuation curve of Sn for 10 MHz, both in the normal and superconducting region. It is seen that the attenuation of longitudinal waves in the normal state grows rapidly towards lower temperature due to the increase in the electronic mean free path and

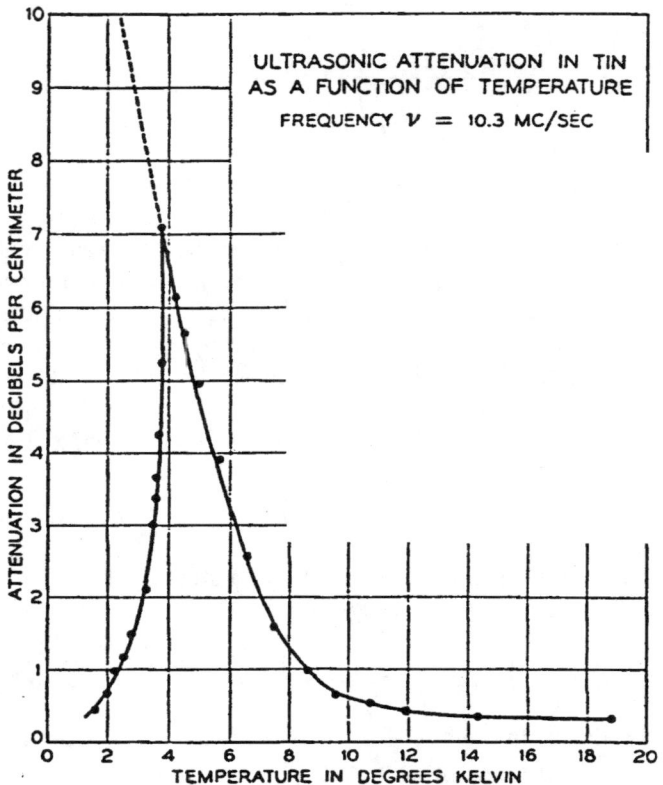

Fig. 10.3. Ultrasonic attenuation in single crystal of tin as a function of temperature for 10.3 MHz longitudinal sound waves (adapted from Boemmel [10.12])

the transition from $ql < 1$ to $ql > 1$ (Sects. 8.2, 8.3). At T_c, the attenuation drops suddenly and decreases continuously for lower temperatures.

In the BCS theory (Bardeen et al. [10.14]) the attenuation coefficient for longitudinal waves for conventional superconductors is written as

$$\frac{\alpha_s}{\alpha_n} = 2f_0(\Delta(T)) = \frac{2}{1+e^{\frac{\Delta}{kT}}}. \tag{10.6}$$

Here Δ is a temperature-dependent isotropic energy gap. The attenuation ratio of the superconducting and normal states becomes, at low temperatures, exponentially small. Close to T_c, α_s has an infinite slope due to the infinite slope of $\Delta(T)$. This is what one observes for isotropic conventional superconductors (Morse [10.15], Levy [10.16]). Equation (10.6) was derived already in the BCS theory using the coherence effects due to the superposition of occupied one-electron states. A derivation of (10.6) is given in the theoretical books listed at the beginning of this chapter (Tinkham [10.3], Tsuneto [10.5]).

The sound attenuation in normal metals has been discussed in Sects. 8.2 and 8.3. As for the normal state, the attenuation in the superconducting state is also proportional to the density of states $N_c(E)$. Due to the gap formation, N_c is strongly modified. In BCS theory it reads

$$N_c^s(E) = N_c^n(0) \frac{E}{\sqrt{(E^2 - \Delta^2)}} \, .$$

Here N_c^n is the normal density of states, $\Delta(T)$ the superconducting energy gap and E is measured from the Fermi level. Similar to the normal state, it is necessary to distinguish between longitudinal and transverse ultrasonic waves and between the region $ql_e < 1$ and $ql_e > 1$ with l_e the electronic mean free path. For longitudinal waves and $ql_e > 1$, a similar expression as for the normal state is obtained with the exception of the coherence terms mentioned above:

$$\alpha_s = \int N_c^{s2}(E) \left\{ 1 - \frac{\Delta^2}{E^2} \right\} [f(E) - f(E')] \, \mathrm{d}E \, .$$

Here the term in brackets $\{\}$ is the coherence factor. It just cancels the N_c^{s2} factor and

$$\alpha_s = -2N_n(0) \int_\Delta^\infty \hbar\omega \frac{\partial f}{\partial E} \mathrm{d}E = 2f(\Delta)\alpha_n \, ,$$

which is precisely (10.6).

The case of transverse waves and $ql_e < 1$ was dealt with by Levy [10.17]. The next Sect. 10.3 deals with conventional superconductors.

10.3 Conventional Superconductors

Ultrasonic attenuation of elemental superconductors have been reviewed in several places (Gottlieb et al. [10.18], Rayne and Jones [10.19], Levy [10.16]). The attenuation obeys usually (10.6) and the velocity anomalies are generally very small. Since for pure metals one is often in the region $ql_e > 1$ (q wavenumber of the sound wave, l_e mean free path of the conduction electrons) the isothermal elastic constant is not defined and therefore velocity changes have not been calculated. This problem was discussed briefly for metals in the normal state in Sect. 8.3.1. But we concentrate here on superconducting compounds investigated by ultrasonic attenuation and velocity changes. These compounds are usually in the limit $ql_e < 1$. We concentrate first on effects without a magnetic field, $B = 0$ properties. The $B \neq 0$ properties are discussed later in Sect. 10.5. The physical parameters of the superconductors discussed below are listed in Tables 10.1 and 10.2.

Table 10.1. Physical properties of A-15 compounds

Material	$T_c(K)$	$T_a(K)$	soft mode
Nb$_3$Sn	18.2	45	$(c_{11}-c_{12})/2$, c_{44}
Nb$_3$Al	18.5		
Nb$_3$Ge	5		
V$_3$Si	17	21	$(c_{11}-c_{12})/2$
V$_3$Ge	6		$(c_{11}-c_{12})/2$
V$_3$Ga	15		

Table 10.2. Physical properties of conventional superconductors presented in Sect. 10.3

Material	$T_c(K)$	$T_a(K)$	Structure	Elastic constants (10^{11} erg/cm^3)			Density g/cm^3
PbMo$_6$S$_8$	13.2		Chevrel	100 K: c_L 9.12	c_T 2.14		5.75
Eu$_{0.6}$ Sn$_{0.4}$ Mo$_6$ S$_8$ Br$_{0.1}$	7.8		Chevrel	100 K: c_L 5.40	c_T 1.61		4.3
HfV$_2$	9	118	Laves C-15	200 K: c_{11} 12.8	c_{12} 10.2	c_{44} 1.7	9.3
CeRu$_2$	6.3		Laves C-15	200 K: c_{11} 13.85	c_{12} 12.72	c_{44} 2.29	10.6
Ba$_{1-x}$K$_x$BiO$_3$	~32		Cubic (Pm3m)				
Yb$_3$Rh$_4$Sn$_{13}$	6.5		ternary stannide	200 K: c_{11} 12.62	c_{12} 6.77	c_{44} 2.05	8.9
Ca$_3$Rh$_4$Sn$_{13}$	7.1		ternary stannide	200 K: c_{11} 13.91	c_{12} 7.33	c_{44} 1.86	8.3
MgB$_2$	40						

A15 Compounds

These compounds are cubic (space group Pm3n) and have the formula A_3B with A a transition metal ion V or Nb and B a metal atom like Si, Ge, Al, Ga or Sn. The B-atoms sit on bcc sites and the A-atoms on the cube faces. These compounds are of considerable technological importance as superconducting magnets with fields up to nearly 20T. In Table 10.1, the main A-15 compounds are listed with some physical properties. The electron–lattice properties are very important. Some of the compounds show a cubic–tetragonal phase transition at T_a, just above the superconducting transition temperature T_c as first observed by Batterman and Barrett [10.20]. Reviews on these compounds are given by Weger and Goldberg [10.21], Testardi [10.9] and Müller [10.22].

A representative picture of the elastic modes for Nb$_3$Sn is shown in Fig. 10.4a (Rehwald [10.23]). Here the individual elastic constants are given as a function of temperature. It is seen that the $c_{11}-c_{12}$ mode softens almost

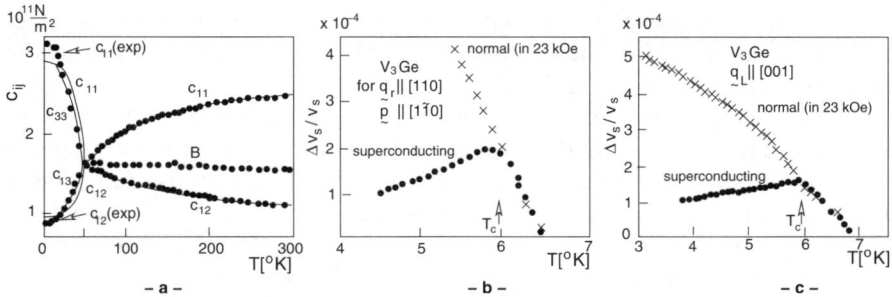

Fig. 10.4. (a) Temperature dependence of the elastic constants of Nb_3Sn (Rehwald [10.23]), (b) temperature dependence of the $c_{11} - c_{12}$ and (c) longitudinal mode near T_c of V_3Ge (Testardi [10.8])

completely down to the structural phase transition of $T_a = 45$ K. But also c_{44} softens considerably. Below this structural transition, the $c_{11} - c_{12}$ mode is strongly damped due to domain wall–stress effects. The elastic constants for $T < T_a$ were determined with the longitudinal (110) mode and the assumption of a constant bulk modulus independent of temperature. The temperature dependence for $T > T_a$ was interpreted with a deformation potential coupling formula (8.8) with $E_F - E_0 = 80$ K, $2d^2 N_0 = 12.1 \times 10^{11}$ erg/cm^3, $c_0 = 9.68 \times 10^{11}$ erg/cm^3. No anomalies were seen at the superconducting transition for $T \leq T_c$ presumably because of strong domain wall–stress effects due to the structural transition. Inelastic neutron scattering experiments on Nb_3Sn show that the softening of the E_g mode extends to about 10% of the Brillouin zone (Axe and Shirane [10.24]).

In V_3Ge clear elastic anomalies are observed at the superconducting transition for the $c_{11} - c_{12}$ mode as shown in Fig. 10.4b (Testardi [10.8], Testardi [10.9]). These have been interpreted by (10.4). No step function is visible at T_c indicating no order parameter–strain coupling for this transverse wave. But the term with the finite temperature derivative clearly explains the data for $T \leq T_c$. Longitudinal waves exhibit similar features.

The softening of some symmetry elastic constants in A-15 compounds for $T > T_a$ was thought to be due to the so-called band Jahn–Teller effect (Labbe-Friedel [10.25]). But more recent theories (Gorkov and Dorokhov [10.26], Bhatt [10.27]) take the 3-dimensional band structure into account. Therefore a softening, due to the electronic redistribution mechanism between strain split and shifted itinerant bands as discussed in Sect. 8.2, (8.8) for volume conserving modes, is possible. The fitting formulas are the same however in both cases as shown in Sect. 8.2.1.

In all models developed for the A-15 compounds, the density of states at the Fermi energy is very large. A magnetic field can have a strong effect leading to field dependent density of states and to effects on the elastic constants and the structural, martensitic phase transition (Dieterich and Fulde [10.28]).

Such an effect was observed in V_3Si with specific heat experiments in magnetic fields of 9T (Maita and Bucher [10.29]). They found a change in the transition temperature of $\Delta T = 0.30$ K.

There are structurally transforming and non-transforming A-15 crystals. In both cases the superconducting transition stops further cubic-tetragonal transformation or the increasing lattice instability. In the transforming crystals the lattice parameters a, c are constant for $T < T_c$ as shown for V_3Si by Batterman and Barrett [10.20]. For non-transforming V_3Si, the soft $(c_{11} - c_{12})/2$ mode shows no more softening below T_c and becomes temperature independent (Testardi [10.8]). This was explained using the Labbé-Friedel model by Dieterich [10.30]. For further details of these interesting effects see Testardi [10.9].

Chevrel Phase Compounds

These are ternary molybdenum sulfide compounds which have the general chemical formula MMo_6S_8 where M stands for a large number of metals. The superconducting properties are mainly determined by the band structure of the Mo_6S_8 cluster, whereas the magnetic behaviour depends on the metal atoms M. Since single crystals of these materials are hardly available most of the work has been carried out on sintered substances. Reviews are available on this class of substances emphasising normal state and superconducting properties (Fischer [10.31], Pena and Sergent [10.32]).

Acoustic measurements have been performed on sintered material of $PbMo_6S_8$ ($T_c = 13.2$ K) and $Eu_{0.6}Sn_{0.4}Mo_6S_8Br_{0.1}$ ($T_c = 7.8$ K) (Wolf et al. [10.33]). In the normal state – for both materials and both longitudinal and transverse modes – a pronounced softening from room temperature down to the superconducting transitions can be observed. The relative changes in the elastic constants amount to 10–20% for both modes and are somewhat smaller than those found in the A-15 compounds. They can be explained quantitatively with the deformation potential formula (8.8). With a bandwidth $\Delta E = E_F - E_0 \sim 70\text{--}90$ K all modes can be fitted rather accurately (Wolf et al. [10.33]).

In Fig. 10.5, the relative elastic constant changes at low temperatures for the longitudinal and transverse modes are shown for the two compounds. The electrical resistivity given also in the figure defines the superconducting transition temperature T_c. It is seen that the strong softening in the normal state stops close to $T_c \pm 1$ K. All the modes increase for $T < T_c$. This can be accounted for with (10.4). It is not the order parameter–strain coupling but rather the other terms involving the second derivative of T_c and the condensation energy Φ and which follow a $T(T - T_c)$ or a $(T - T_c)^2$ law which can explain the temperature dependence. The strain–order parameter coupling is too weak in this case. The full lines from a fit of (10.4) for the longitudinal and transverse modes of $PbMo_6S_8$ shown in Fig. 10.5 give a rather good agreement to the experimental curves. The field-dependent effects of these compounds are discussed in Sect. 10.5.

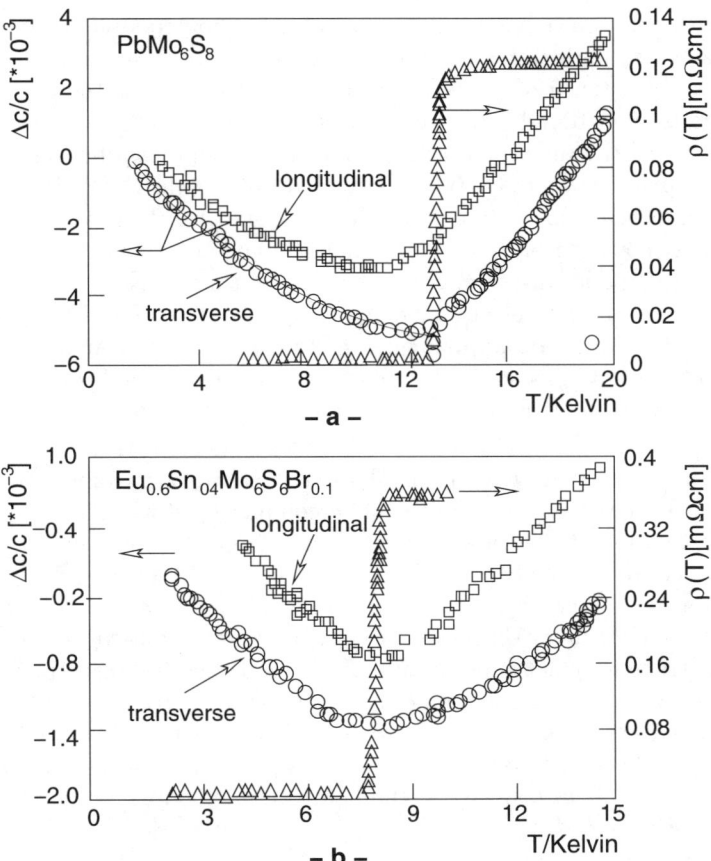

Fig. 10.5. Temperature dependence of the electrical resistance and the relative elastic constant changes $\Delta c/c_0$ for (a) $PbMo_6S_8$ and (b) $Eu_{0.6}Sn_{0.4}Mo_6S_8Br_{0.1}$ each for $T < 20$ K. This temperature dependence for $T > T_c$ and $T < T_c$ can be explained phenomenologically as discussed in the text. Full lines are fits from (3.4) (adapted from Wolf et al. [10.33] and Zherlitsyn et al. [10.11])

HfV_2

This is a Laves phase type C-15 compound like ZrV_2. The structure is shown in Chap. 5, Fig. 5.7. Both compounds, HfV_2 and ZrV_2, have a high density of states at the Fermi energy and a strong influence of spin fluctuations (Keiber et al. [10.34]). HfV_2 is a superconductor with a $T_c = 9$ K. The most interesting property is the temperature dependence of the specific heat in the superconducting state. It does not follow the BCS law but has a power law dependence $C \sim T^3$, similar to heavy fermion superconductors (Lüthi et al. [10.35]), (see Sect. 10.7). But this does not imply that we deal here with unconventional superconductivity. Strong coupling effects can give such a behaviour too (Rainer and Bergmann [10.36]). Therefore HfV_2 can be used to distinguish between strong coupling, anisotropic superconductors and unconventional superconductors. We will discuss this problem again in Sect. 10.7.

But the existence of a specific heat with a power law temperature dependence is immediately followed up by the question on the temperature dependence of the ultrasonic attenuation in the superconducting state. An ultrasonic investigation of HfV$_2$ was carried out (Lüthi et al. [10.35]). First it was found that this compound exhibits a structural phase transition at $T_a = 118$ K. For both polycrystalline material and single crystals, practically all the elastic modes are strongly influenced by this transition. A large softening for the transverse elastic mode c_T (polycrystalline) and for the $c_{11} - c_{12}$ mode (single crystal) on approaching T_a can be observed. At the superconducting transition T_c, a small step function anomaly for c_L is observed. This indicates a sizeable coupling of the strain to the superconducting order parameter (the $\left(\frac{\partial T_c}{\partial \varepsilon}\right)^2$ term in (10.4)). The transverse mode ($c_{11} - c_{12}$) does not exhibit any anomaly at T_c. But one does not find any electronic contribution to the ultrasonic attenuation in the superconducting state. This is similar to the A-15 compounds (see above) but quite different to the case of the heavy fermions like UPt$_3$ (see Sect. 10.7). In both cases, HfV$_2$ and UPt$_3$, one is in the region $ql < 1$, with q the ultrasonic wave vector and l the electronic mean free path. But in UPt$_3$, a noticeable electronic contribution to the ultrasonic attenuation is found because of the high effective mass (high density of states). In HfV$_2$ one has a temperature independent attenuation for $T > T_c$ and even an increasing attenuation for $T < T_c$ for frequencies up to 200 MHz. Also for the thermal conductivity, one finds practically only phonon contributions for $T < T_c$ in contrast to the heavy fermion superconductors (Lüthi et al. [10.35]).

CeRu$_2$

This substance is another Laves phase compound with intriguing properties. It has a superconducting transition at $T_c = 6.3$ K. Ultrasonically, it was investigated by different groups (Goshima et al. [10.37], Wolf et al. [10.38], Yoshizawa et al. [10.39]). In Fig. 10.6, the temperature dependence of the various elastic modes are shown (Wolf et al. [10.38]). It is seen that in the normal state all modes show pronounced softening towards lower temperatures.

The strongest softening is observed for the $c_{11} - c_{12}$ mode. From room temperature to the lowest temperature the softening amounts to more than 50% in the relative elastic constant. CeRu$_2$ is often considered a valence fluctuating material in literature. The elastic constants of Fig. 10.6 show, however, that the Poisson ratio is large and the bulk modulus does not show any trace of softening. Therefore according to the discussion given in Sect. 9.2 CeRu$_2$ has a stable valence ground state with Ce^{4+}. The temperature dependence of the elastic constants has been interpreted correspondingly with a deformation potential coupling and the soft mode has been quantitatively explained as for the Chevrel-phase compound with (8.8) with $\Delta E = 180$ K (Wolf et al. [10.38]). A detailed determination of the deformation potential

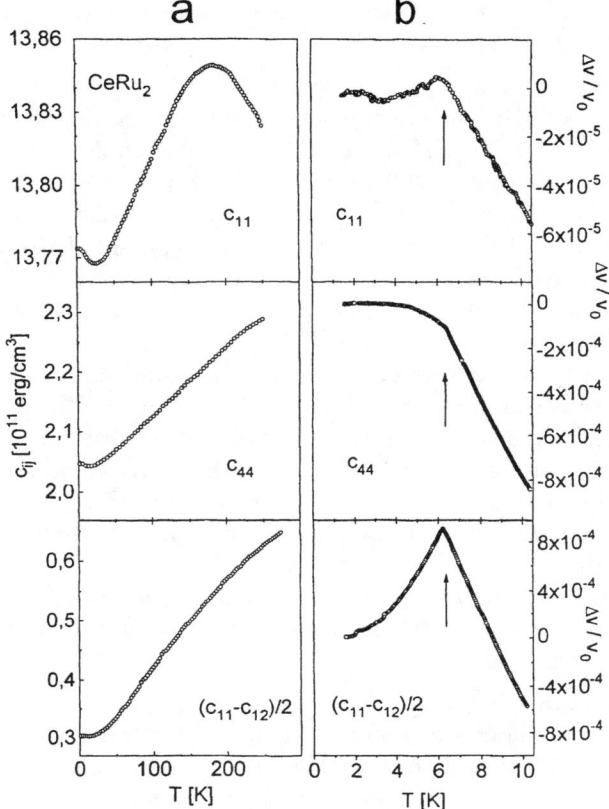

Fig. 10.6. Temperature dependence of the elastic modes c_{11}, c_{44}, $(c_{11} - c_{12})/2$ in CeRu$_2$ (Wolf et al. [10.38]). (a) Temperature range from 250 K to 1 K, (b) below 10 K for B = 0. The superconducting transition is marked by arrows.

coupling constant $d(\Gamma_3)$ shows that a 15% larger d would lead to a cubic–tetragonal phase transition at low temperatures. It is interesting to note that the La$_{1-x}$Ce$_x$Ru$_2$ system exhibits a cubic–tetragonal phase transition for $x < 0.15$ with a transition temperature of 30 K for LaRu$_2$ (Shelton et al. [10.40]). Furthermore, with a pressure of 5.5 GPa, a structural phase transition was induced for CeRu$_2$ around 80 K (Nakama et al. [10.41]). A similar elastic softening was also observed for the isomorphous compounds LaRh$_2$ and CeIr$_2$ (Ozawa et al. [10.42]).

Looking at the elastic modes in the vicinity of T_c (Fig. 10.6b), we can again interpret the temperature dependence in the field-free case for $T < T_c$ with the help of (10.4). No order parameter–strain coupling is observed in this case but the various modes exhibit a temperature dependence due to the higher order strain dependence of the condensation energy and T_c (10.4) (Zherlitsyn et al. [10.11]). The magnetic field-dependent superconducting properties of CeRu$_2$ will be shown in Sect. 10.5.

Ba$_{1-x}$K$_x$BiO$_3$

This copper-free high temperature superconductor has a transition temperature as high as $T_c \cong 32$ K for $x \cong 0.37$. It has cubic symmetry for the potassium concentration $x \geq 0.35$. Superconductivity has, so far, only been observed in the cubic phase (space symmetry group Pm3m), but close to the phase boundary of the cubic and orthorhombic phases. Ultrasonic investigations on this compound have been performed on polycrystalline samples (Levy et al. [10.43]) and single crystals (Zherlitsyn et al. [10.44]). The latter results are discussed briefly below.

An instability of the crystal lattice in the normal state can be observed which leads to a softening of both c_{44} and c_{11} modes. This instability can be clearly observed as a temperature dependent hysteresis for sound velocity and attenuation around 100 K for Ba$_{0.65}$K$_{0.35}$BiO$_3$. X-ray powder diffraction does not reveal any change in cubic structure. This effect is explained with a coupling of the acoustic modes to the an-harmonic oscillations of BiO$_6$ octahedra.

In the superconducting region, small anomalies in velocities (10^{-4} effects) are found and attenuation which can be explained semiquantitatively with (10.4) (Zherlitsyn et al. [10.11]).

Yb$_3$Rh$_4$Sn$_{13}$, Ca$_3$Rh$_4$Sn$_{13}$

In many respects, these compounds have much in common with the Chevrel-structure, Ba$_{1-x}$K$_x$BiO$_3$, CeRu$_2$, and the A-15 compounds, discussed above. In all previous cases one observes a softening of some symmetry elastic constants in the normal state and a pronounced peak effect, as found in many type II superconductors and discussed in detail in Sect. 10.5. The ternary stannide compounds form a phase I (primitive cubic) with two inequivalent sites for the Sn atoms: Sn(1)Yb$_3$Rh$_4$Sn(2)$_{12}$ where Sn(1)Yb$_3$ forms a A-15 type sublattice. These compounds can have a slight disorder of the Yb or Ca and Sn(1) sites, which can also be observed by acoustic means. In these compounds the electron–phonon coupling is much weaker than in say CeRu$_2$. Therefore the detailed analysis of the peak effect and the interaction of the structural and superconducting transition cannot be done as for the previous cases, especially CeRu$_2$. But the temperature dependence of the various elastic modes (Haas et al. [10.45]) in the superconducting region can again be accounted for semiquantitatively with (10.4) (Zherlitsyn et al. [10.11]). Furthermore, an approximate logarithmic dependence of the sound velocity and sound attenuation in the superconducting region is found, similar to superconducting amorphous alloys, as discussed in Chap. 14. Here the slight disorder of the Sn sites mentioned above could be responsible for this.

Other Systems

There are other superconducting systems with interesting physical properties. The acoustic properties of reentrant superconductors of Er$_{1-x}$Ho$_x$Rh$_4$B$_4$ are reviewed by Sun et al [10.46]. The borocarbides with chemical formula e.g. RNi$_2$B$_2$C (R = Y, Er) have not been studied acoustically yet. There is a recent review on the Borocarbides (Müller and Narozhnyi [10.47]). Also the recently, much discussed, new superconductor MgB$_2$ with $T_c = 40$ K has been investigated ultrasonically only for polycrystalline material so far (Ichitsubo et al. [10.48]).

A recent study for single crystals does not reveal any anomalies in the superconducting state (Ichitsubo et al. [10.49]).

10.4 High Temperature Superconductors

Since the discovery of high temperature superconductivity (HTSC) (Bednorz and Müller [10.50]), many acoustic experiments have been performed in these substances. These included first experiments on polycrystalline and sintered materials and later on single crystals. We will discuss here the acoustic experiments involving single crystals. Since no ultrasonic experiments have been performed to test the unconventional d-wave order parameter, we will delay this subject to Sect. 10.7. The magnetic field-dependent effects will be discussed together with the conventional superconductors in the next Sect. 10.5. Single crystal research has been performed in La$_{2-x}$Sr$_x$CuO$_4$ (LSCO), YBa$_2$Cu$_3$O$_7$ (Y–123) and Bi$_2$Sr$_2$CaCu$_2$O$_8$ (Bi–2212). Table 10.3 lists some physical properties of these substances.

La$_{2-x}$(Ba,Sr)$_x$CuO$_4$

Structural transitions are found in this material which are connected with normal state properties and superconductivity. The La-214 compounds with K$_2$NiF$_4$ structure are susceptible to structural transitions from the high temperature tetragonal lattice (THT) to a mid-temperature orthorhombic lattice (OMT) and to a low temperature tetragonal structure (TLT). Such structural changes with analogous accompanying soft acoustic modes were also found in organic perovskite-like layer-structure materials, e.g. (CH$_3$NH$_3$)$_2$MnCl$_4$ (MAMC) (Goto et al. [10.51]), see Sect. 12.6. In La$_2$CuO$_4$, there is a transition at $T_{d1} = 530$ K from tetragonal (space group $I4/mmm$) above to an orthorhombic phase (space group Bmab). Upon substituting Ba, the transition temperature T_{d1} disappears almost linearly with x at $x \sim 0.2$. Interestingly, around $x = 1/8$, there is a low temperature tetragonal phase (space group

Table 10.3. Physical properties of high temperature superconductors

Material		T_c(K)	B_{c1} (T)	B_{c2} (T)	G-.L κ
La$_{2-x}$(Ba,Sr)$_x$CuO$_4$	La cupr.	38		50	50
YBa$_2$Cu$_2$O$_{7-x}$	YBCO	93	0.05	130	85
Bi$_2$Sr$_2$CaCu$_2$O$_{8+x}$	Bi 2212	94		32	100

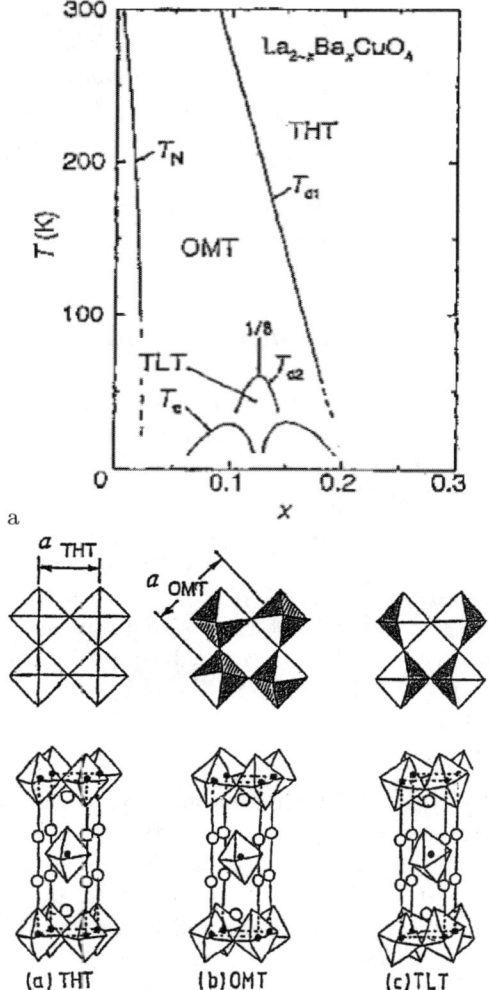

Fig. 10.7. (a) T–x phase diagram for $La_{2-x}Ba_xCuO_4$ with T_N Neel temperature, T_c superconducting transition temperature, T_{d1}, T_{d2} structural transition temperatures, (b) schematic representation of the different crystal structures. Closed and open circles represent Cu and La(Ba) atoms respectively. Oxygen atoms are located at each apex of octahedra drawn by solid lines. The upper figure show the top view of the structures, clarifying how CuO_6 octahedra are tilted (Maeno [10.52])

$P4_2/ncm$) peaked at $T_{d2} \sim 60$ K. For $x = 1/8$, superconductivity disappears. All these facts are presented in figure 10.7. In Fig. 10.7a the $T - x$ phase diagram is shown with the different structural, magnetic and superconducting transitions. In Fig. 10.7b, the different crystal structures are presented in a schematic way (Maeno [10.52]).

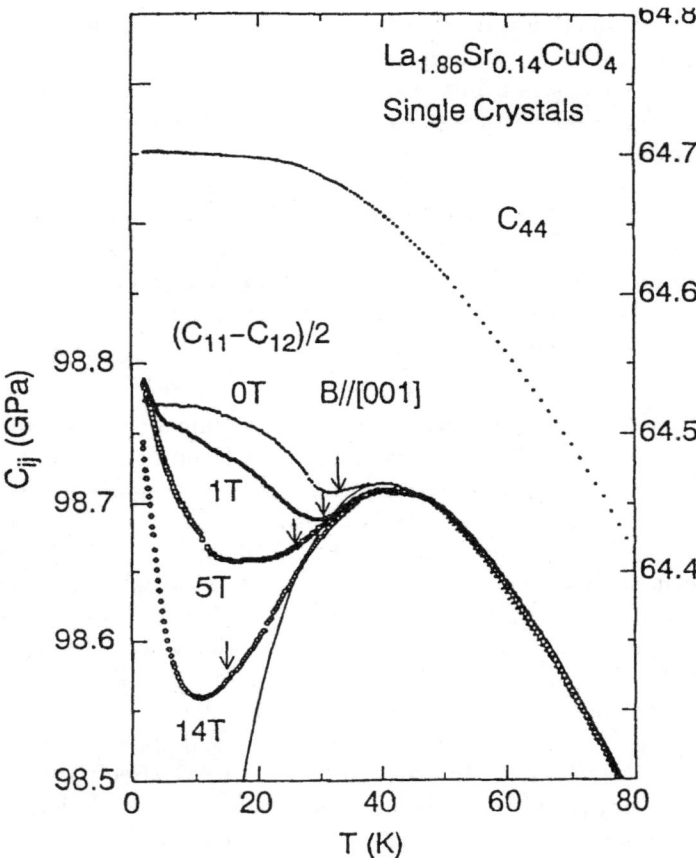

Fig. 10.8. Transverse elastic constants c_{44} and $(c_{11} - c_{12})/2$ in LSCO with different magnetic fields $B \parallel (001)$ (Suzuki et al. [10.54])

Most acoustic measurements were not made in LBCO but rather in $La_{2-x}Sr_xCuO_4$ (Migliori et al. [10.53], Suzuki et al. [10.54]). The former group used the RUS technique (Sect. 2.2.3) to measure the temperature dependence of the elastic modes. In $La_{1.86}Sr_{0.14}CuO_4$, a strong softening of 70% of the c_{66} mode was found at the THT–OMT transition at $T_{d1} \sim 210\,K$ and a step like softening of 10% for the c_{11} mode. Actually the THT–OMT transition involves a soft Brillouin zone phonon mode at the X-point (Birgeneau et al. [10.55]). The softening of the c_{66} mode therefore has to be interpreted as due to strain coupling to fluctuations as discussed in Sect. 4.3.5 and 7.4.3. For LSCO, it was found that a two-dimensional Gaussian fluctuation model with exponents 0.5 for the elastic constant and 1 for the fluctuation time explains the results quantitatively (Migliori et al. [10.56], Nohara et al. [10.57]).

For the same compound, $La_{1.86}Sr_{0.14}CuO_4$, the c_{11}-c_{12} mode exhibits an interesting feature below $50\,K$ as shown in Fig. 10.8. It slightly softens down to the superconducting transition $T_c = 36$ K and hardens below T_c. In strong magnetic fields ($B > B_{c2}$), the softening is more pronounced, indicating that the $\varepsilon_{xx} - \varepsilon_{yy}$ strain in the CuO_2 plane wants to trigger a OMT–TLT structural transition (Suzuki et al. [10.54]). The superconductivity prevents this structural transition apparently.

These results and more detailed structural investigations (Maeno [10.52]) indicate that a Landau-type description of the phase diagram and the thermodynamic properties gives, at best, a qualitative picture. Also the electronic redistribution mechanism between strain split bands (Sect. 8.2, (8.8)) seems not to be responsible for the strong c_{66} softening but rather the structural fluctuation effect discussed above.

$YBa_2Cu_3O_{7-x}$

Several acoustic measurements exist in this high temperature superconductor with $Tc \sim 89$ K. The frequency range extends from 20kHz (vibrating reed technique) up to 1Ghz. In some crystals a small anomaly was found at T_c, in others large hysteretic behaviour was measured for $T < T_c$. All these results are summarised in a review article (Golding [10.58]).

An acoustic experiment, aiming at studying the weakly disturbed vortex system in the vicinity of T_c, is briefly mentioned here. This is in contrast to the experiments discussed in the next Sect. 10.5 involving a strongly-pinned vortex lattice in high magnetic field. The experiment consists of an attenuation measurement with low frequency (3 MHz) longitudinal sound through a 50μm thin plate along the c-axis with magnetic field applied in the ab-plane (Hucho and Levy [10.59]). In order to study a nearly free vortex lattice, a twin–free single crystal was used and the induced vortex motion is parallel to the Cu-O planes. In the attenuation versus field curves, sudden drops of attenuation were observed indicating a drastic reduction of losses in the externally modulated vortex system and an increase of pinning in the vortex system (see Fig. 10.9).

$Bi_2Sr_2CaCu_2O_{8+x}$

In this Bi-2212 crystal, sound waves and surface waves were measured using Brillouin scattering techniques (Boekholt et al. [10.60]) and a complete set of elastic constants was obtained. Two planes (001) and (010) were investigated. Values for c_{11}, c_{12}, c_{13}, c_{33} and c_{44} were obtained at room temperature. The technique used is described in Mock and Güntherodt [10.61] and is briefly mentioned in Sect. 2.7. There is another conventional ultrasonic investigation of this compound by Saint-Paul et al. [10.62]. The temperature dependence of a longitudinal mode in the basal plane and the c_{44} mode were measured. Only the longitudinal mode exhibits a step-like anomaly at T_c whereas the attenuation shows no anomaly.

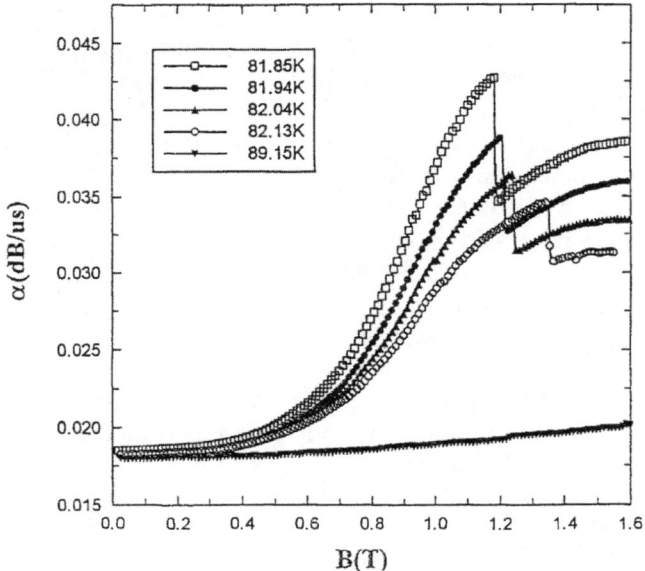

Fig. 10.9. Sound attenuation versus field near T_c in $YBa_2Cu_3O_{7-\delta}$ (Hucho and Levy [10.59])

10.5 Sound Wave Interaction with the Flux Line Lattice

Ultrasonic attenuation and velocity dispersion are good methods to study the flux line lattice in type II superconductors. In the mixed state of type II superconductors, flux lines can interact with the crystal lattice via pinning. Pinning forces prevent the flux line lattice from moving freely. The relative motion of the flux lines with respect to the crystal lattice, or the pinning centers, gives rise to dissipation and hence to velocity dispersion and attenuation effects. The presence of the strongly pinned flux line lattice gives rise to a stiffening of the elastic modes. Especially transverse modes with $B \parallel k$ show large magneto-acoustic effects. Before discussing experiments and models for the vortex–strain interaction, some important facts of the vortex physics in type II superconductors are summarised (Brandt [10.63], Blatter et al. [10.64]).

Each vortex line or flux line carries one quantum of magnetic flux $\Phi_0 = h/2e = 2.07 \times 10^{-7} G cm^2$ caused by the super-currents circulating in this vortex. The magnetic field peaks at the vortex position where the superconducting order parameter vanishes accordingly. For well-separated vortices, the radius of a vortex tube equals the magnetic penetration depth λ and the core radius is larger than the superconducting coherence length ξ. With increasing applied magnetic field, the spacing a_0 of the vortices decreases and the average flux density increases, the flux tubes overlap and the periodic flux density is nearly constant. Finally at the upper critical field $B_{c2} = \Phi_0/2\xi^2$ the vortex cores overlap, the order parameter vanishes and the superconductivity disappears. The flux line lattice (**FLL**) exists when the Ginzburg–Landau parameter $\kappa = \lambda/\xi$ exceeds the value $1/\sqrt{2}$ and when the magnetic field ranges between the lower critical field B_{c1} and B_{c2}. As discussed in Sect. 10.1 for $B < B_{c1}$, the system is in the Meissner phase (Fig. 10.1b).

For elemental type II superconductors like **Nb**, one observes a distinct magnetic field dependence of the ultrasonic attenuation between B_{c1} and B_{c2} (Ikushima et al. [10.65]) which can be roughly explained with a BCS formula (10.6) with a field dependent energy gap due to the volume fraction of the vortices (see e.g. Kudo [10.66]).

For the acoustic experiments in strongly pinned superconductors, the TAFF (thermally assisted flux flow) model is discussed below (Anderson and Kim [10.67], Kes et al. [10.68]). For a detailed discussion see Brandt [10.63]. This model is usually used for the description of the penetration of magnetic flux into a type II superconductor and the temperature dependence of the critical current. One assumption of this model is a strong pinning of the vortices in the sample. The system should be an extreme type II superconductor because the pinning energy U is proportional to the square of the thermodynamic critical field B_c^2 and to the third power of the coherence length ξ^3. The electrical resistivity, in the region where there is a thermally assisted hopping of flux bundles, is given by $\rho = \rho_0 \exp[-U(B)/k_B T]$ (Brandt [10.63]) with ρ_0 constant in this region.

Assuming that the vortex lattice is along the z-axis an elastic strain ε_{zz} has no effect and the c_{33} mode should not couple to the vortices. On the other hand $\varepsilon_{xx} + \varepsilon_{yy} = \Delta f/f$, with f the area perpendicular to z, affects the elastic constant c_{11}. From Hooke's law $T_{xx} = c_{11}\varepsilon_{xx} + c_{12}\varepsilon_{yy} \sim (c_{11}+c_{12})\Delta f/2f$ for elastic isotropy, we notice that the flux lattice compression affects $(c_{11}+c_{12}) \sim (c_{11}-c_{66})$. Finally, c_{44} connected to the shear strain ε_{xz} is affected by a tilting of the vortices. By assuming $Bf = $ const, for these cases gives $c_{11}, c_{44} \sim B^2$.

These acoustic effects were first quantitatively discussed for **NbTi** and high temperature superconductors by Pankert et al. [10.69], Pankert [10.70] and Lemmens et al. [10.71]. By assuming that the vortex motion has relaxational character, they derive dispersion equations for the coupled elastic modes. For transverse modes with $k \parallel B$, the equations read for the relative velocity change $\Delta v_t/v_t$ and the attenuation α:

$$\frac{\Delta v_t}{v_t} = \frac{B^2}{8\pi\rho_m v_t^2} \times \frac{1}{1+\beta^2} \qquad \Delta\alpha = \omega \frac{B^2}{8\pi\rho_m v_t^3} \times \frac{\beta}{1+\beta^2} \qquad (10.7)$$

with ρ_m the mass density and $\beta = (1/4\pi)c^2\rho/v_t^2$ and ρ the electrical resistivity. Note the similarity to the Alpher–Rubin formula (8.9) for the sound wave dispersion in a viscous electron liquid. But contrary to (8.9), the parameter β has no longer any frequency dependence and ρ has here an activated temperature behaviour. Equation (10.7) has all known physical constants and therefore no adjustable parameters. Some examples of these vortex acoustic effects are given below:

High Temperature Superconductors

Sound wave interaction with the flux line lattice was investigated in some detail in ceramic high temperature superconductors. Measurements were performed in Y-123, YBa$_2$Cu$_3$O$_{6.99}$ (Lemmens et al. [10.72]) and in Bi-2223, Bi$_{1.6}$Pb$_{0.4}$Sr$_2$Ca$_2$Cu$_3$O (Lemmens et al. [10.71]) and related substances. Fig. 10.10 gives velocity and attenuation results for Bi-2223 for 5.7 MHz and a magnetic field of 5T along the preferential c-axis together with fits using (10.7). The simple TAFF theory describes the results very well with an activation energy $U \sim 500$K.

Chevrel Phase Compounds

The temperature dependence of the sound velocity in these materials is discussed in Sect. 10.3. The magnetic field dependence of the elastic constants and sound attenuation also show unexpected results. In Fig. 10.11, relative transverse velocity and attenuation data for $B \parallel k$ are given as a function of temperature with the magnetic field as a parameter for the sintered substance Eu$_{0.6}$Sn$_{0.4}$Mo$_6$S$_8$Br$_{0.1}$. Note that the field values are rather high, up to 29T. These experiments were performed in the high field laboratory in Grenoble (Wolf et al. [10.33]). Strong dispersive effects are observed which can be interpreted again with the TAFF model (10.7). The effects scale roughly with B^2 and the velocity gets stiffer with lower temperatures. The full lines are

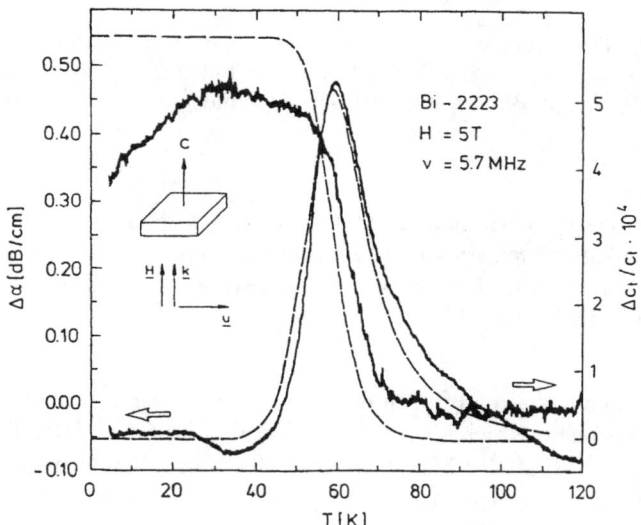

Fig. 10.10. Attenuation and velocity changes for a transverse mode of a textured Bi-2223 material with $B \parallel k \parallel c$. Dotted lines are fits with (10.7). (Lemmens et al. [10.71])

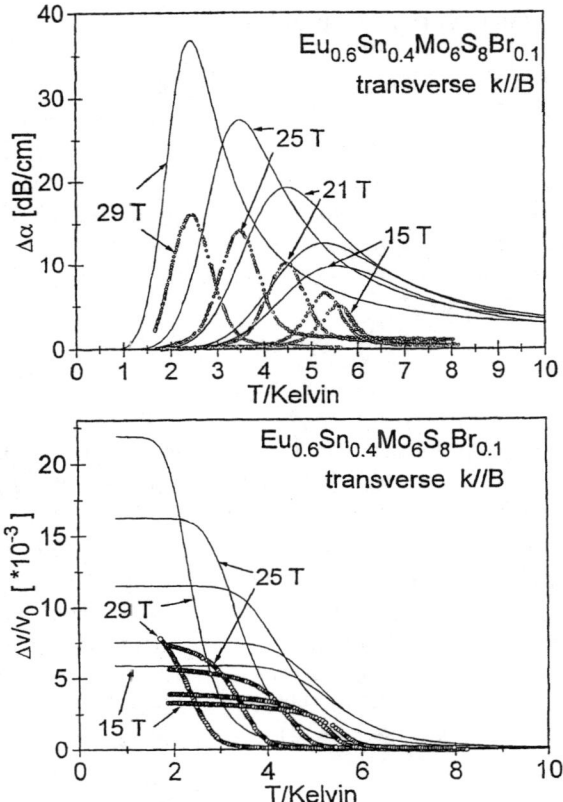

Fig. 10.11. Temperature and magnetic field dependence of the relative velocity and attenuation for $Eu_{0.6}Sn_{0.4}Mo_6S_8Br_{0.1}$. The full lines are fits with the TAFF model (10.7). Longitudinal waves with $k \parallel B$, sound wave frequency 10 MHz (Wolf et al. [10.33])

fits of this model to the experimental curves. Although the fit is far from perfect, the salient features of the model are reproduced. The neglect of a filling-factor correction for this sintered substance accounts probably for the quantitative discrepancy.

$CeRu_2$

This substance has already been described in Sect. 10.3. In a magnetic field, special effects occur which are unique for $CeRu_2$ (Wolf et al. [10.38], Yoshizawa et al. [10.39]). In Fig. 10.12 the field dependence of different modes are shown at 1.5 K together with the magnetization curve, all exhibiting the peak effect. Common to all modes is a linear magnetic field dependence up to $3T$, a hysteretic region around $4T$ and a field independent behaviour for $B > B_{c2}$. In addition the size of the field dependent effects scale with the size of the anomalies in the normal state. The anomalies around $4T$ are due to the so-called

10.5 Sound Wave Interaction with the Flux Line Lattice   245

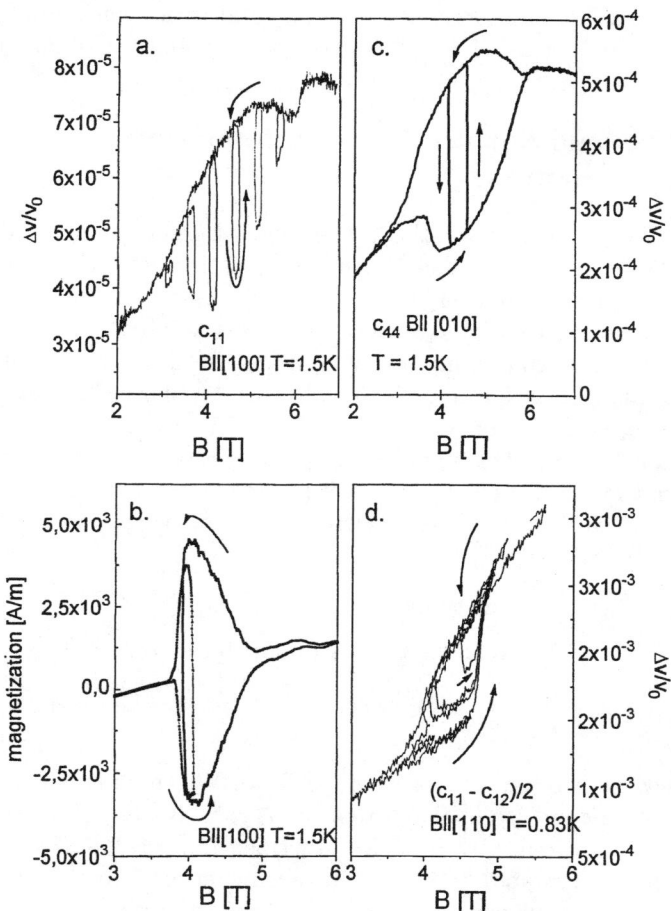

Fig. 10.12. Relative sound velocities of elastic modes for CeRu$_2$ as a function of magnetic field at $T = 1.5$ K. (**a**) c_{11} mode, (**b**) magnetization, (**c**) c_{44} mode, (**d**) $(c_{11} - c_{12})/2$ mode (Wolf et al. [10.38])

peak effect and can be explained using the TAFF model mentioned above. The softening of the Abrikosov flux line lattice near B_{c2} leads to a pinning by weak pinning centers which give rise to an enhanced magnetic irreversibility. One can distinguish between Lorentz–modes, where the particle motion experiences a Lorentz force ($\boldsymbol{u} \perp \boldsymbol{B}$) and non Lorentz–modes ($\boldsymbol{u} \parallel \boldsymbol{B}$), with $\boldsymbol{u}$ the displacement vector. Especially the effects for the Lorentz–modes can be nicely explained with the TAFF model. The hysteretic behaviour can be observed in the magnetization (Fig. 10.12b) and in the different sound wave modes (Fig. 10.12a,c,d). In Fig. 10.12, pictures a) and c) show non-Lorentz modes and d) a Lorentz mode. An experimental review on normal and superconducting properties of $CeRu_2$ is given by Roy and Chaddah [10.73].

10.6 Ultrasonic Surface Wave Attenuation in Superconductors

In the seventies and eighties of the last century a number of ultrasonic attenuation studies were performed using surface acoustic wave techniques (SAW). Thin films can be studied nicely by this technique as well as proximity effects, gap-less superconductivity and surface superconductivity.

Superconductors containing magnetic impurities can exhibit gap-less superconductivity (Abrikosov and Gorkov [10.74]). Similar effects can occur for thin films in a parallel magnetic field (Maki [10.75], Maki and Fulde [10.76]). This case can be studied straightforwardly using SAW. The theory for ultrasonic attenuation in this case is valid with the following conditions (d sample thickness, l_e mean free path, ξ_0 coherence length, λ_L London penetration length) (see Krätzig [10.77]):

$$l_e \ll \xi_0 \qquad \text{(dirty limit)},$$
$$d \ll (l\xi_0)^{\frac{1}{2}} \qquad \text{(constant order parameter across the specimen)},$$
$$d \ll \lambda_{eff} = \lambda_L (\frac{\xi_0}{l_e})^{\frac{1}{2}} \text{ (nearly perfect penetration of the magnetic field)}.$$
(10.8)

The measured attenuation ratio α_s/α_n for SAW can be shown to be equal to the ratio of the longitudinal attenuation α_s^L/α_n^L (Krätzig [10.77]). The experiment is carried out with a device described in Sect. 2.5. The superconducting film is evaporated between the transmitter and receiver on a quartz plate. Extensive experiments with Sn and Pb films of typically $d = 400$Å and $f = 700$ MHz gave the predicted deviations from BCS theory (Sect. 10.2, (10.6)) but the experimentally found pair breaking effect was stronger than theoretically expected (Krätzig [10.77]).

Surface superconductivity may occur if the Ginzburg–Landau parameter $\kappa = \lambda(T)/\xi(T)$ is larger than 0.42 (Saint James and de Gennes [10.78]). In a parallel magnetic field $H_\parallel$ a film, which is already in the normal state in the bulk, may exhibit superconducting surface sheets on both sides of the film. With SAW one can study this effect. In Fig. 10.13 the SAW attenuation for a Pb film is shown. At higher temperature (4.9 K< 7.2 K = T_c), the G–L parameter κ is smaller than 0.42 and no surface superconductivity occurs – only the normal-superconducting transition at $H_{c\parallel}$ as a first order transition is observed. At lower temperature ($T < 4$K) $\kappa > 0.42$. The SAW attenuation shows again an abrupt change at $H_{c\parallel}$ but then followed by a gradual increase until the critical field H_{c3} is reached. In the lower part of the figure, the same experiment is shown with a doped Pb film with a few percent of Bi. In this way, the dirty limit is reached and the phase transition is of second order.

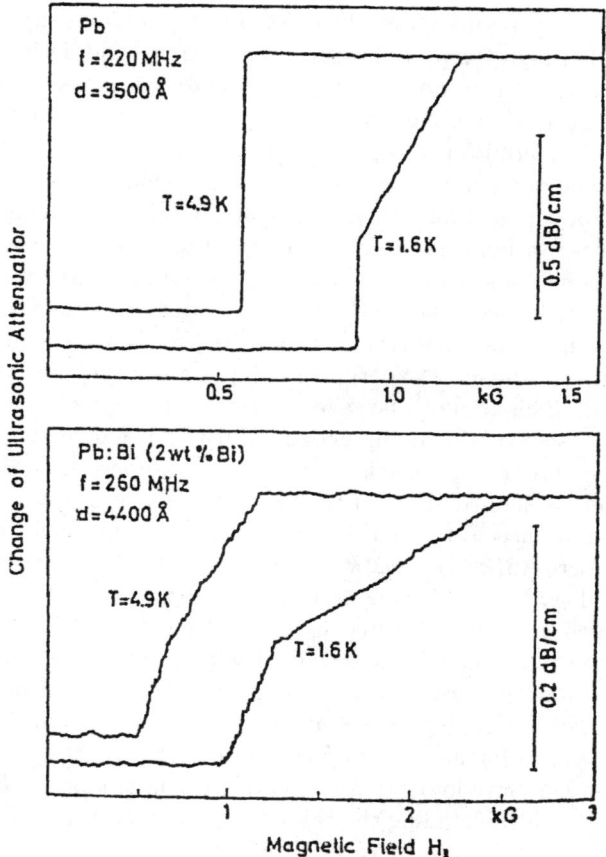

Fig. 10.13. SAW attenuation for a Pb film and a Pb:Bi(2%) film versus the magnetic field $H_\parallel$ (Krätzig [10.77])

Similar SAW experiments in other superconductors were performed: NbN (Fredricksen et al. [10.79]), In Akao [10.80], Nb_3Sn (Fredericksen et al. [10.81]) and the proximity effect was studied in Cu-Pb-Cu sandwiches (Jain and Tilley [10.82]). Finally the experimental verification of superconducting order parameter fluctuations in thin Al films, using SAW attenuation with a frequency of 2GHz, was not successful, possibly because the polycrystalline sample was not clean enough (Robinson et al. [10.83]).

10.7 Unconventional Superconductivity

In the last 20 years, new types of superconductors were discovered which stimulated the field of superconductivity enormously and gave rise to new phenomena and description beyond the BCS theory. We just mention the field of organic superconductors, the class of heavy fermion superconductors, the large field of high temperature superconductivity and derivations from it like Sr_2RuO_4. These new types of superconductors gave rise to speculations on the mechanism of superconductivity and they also gave rise to a new phenomenological description of superconductivity, especially in identifying the symmetry of the order parameter. The following reviews some aspects of this modern field of superconductivity. This will not be a comprehensive review, rather the presentation is limited to the important acoustic experiments performed in this field and to some related key experiments and we will refer to existing review articles. The starting point of unconventional superconductivity was actually the discovery of orbital triplet (p-wave) superfluidity in liquid He^3 where ultrasonic experiments played a key role (see Vollhardt and Wölfle [10.84]). This topic is, however, outside the scope of this book.

What is "unconventional superconductivity"? If gauge symmetry is the only broken symmetry, one speaks of conventional superconductivity. The order parameter, as defined above at the beginning of this chapter, is not invariant if one adds a phase α to each electron, i.e. multiplying ψ_σ by $\exp(i\alpha)$. The order parameter is $\Delta(\boldsymbol{k},T) = |\Delta(\boldsymbol{k},T)| \exp(i\alpha)$ with Δ the energy gap. There is a close analogy to an isotropic ferromagnet with $\boldsymbol{M} = |\boldsymbol{M}| (\cos\alpha, \sin\alpha)$. Here the degeneracy in α is lifted by a small magnetic field. In the superconductor, the phase can be determined with a Josephson experiment. If additional symmetries are broken, one speaks of unconventional superconductivity. These can be, e.g., the point symmetry, the translation symmetry or the time reversal symmetry. In Fig. 10.14 a case is shown where the order parameter can have lower symmetry than the lattice or the Fermi surface (Fulde et al. [10.85]), including the case of high temperature superconductors with $d_{x^2-y^2}$ wave.

The nature of the symmetry breaking field, which makes a compound an unconventional superconductor, will be discussed below for some concrete cases. The order parameter, the Ginzburg–Landau free energy and the strain–order parameter coupling will need to be generalised for this case. This is done in a few special cases.

We will not comment on the pairing mechanism responsible for the superconductivity. In the field of unconventional superconductivity, as defined above, the question on the nature of the pairing mechanism immediately arises. There are many speculations about non-phononic pairing mechanisms. It is very difficult experimentally to decide such questions. Especially ultrasonics cannot contribute to this fundamental aspect of superconductivity. A case of non-phononic pairing mechanism has been recently established: In the heavy fermion system UPd_2Al_3, tunneling experiments and inelastic neutron scattering has given strong evidence for a magnetic exciton exchange (see e.g. Sato et al. [10.86], Thalmeier et al. [10.87]).

Although the symmetry of the unconventional superconducting order parameters is not known with absolute certainty, research of the last few years has shown that for hexagonal UPt_3, it is most probably an even parity d-wave

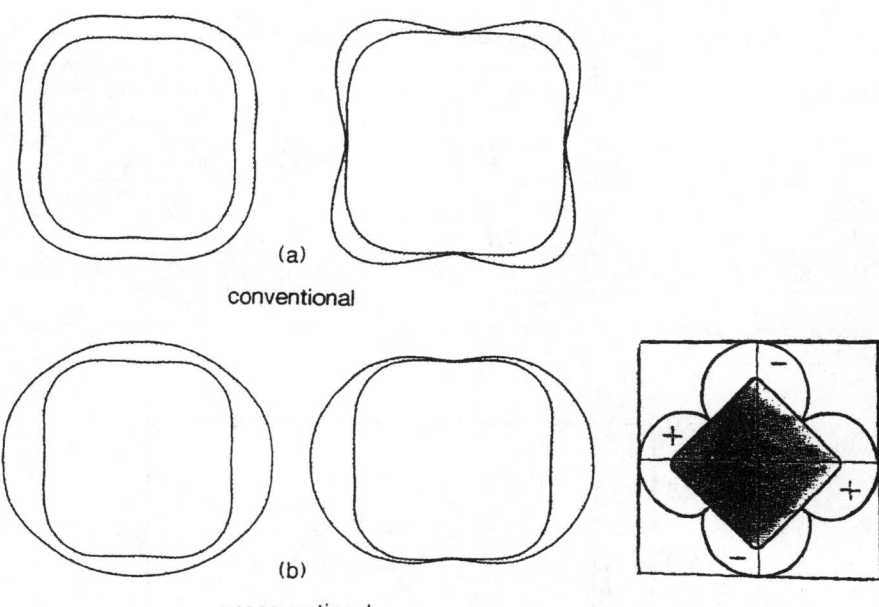

Fig. 10.14. Energy gaps in momentum space for (**a**) conventional and (**b**) unconventional pairing including d-wave pairing from high T_c superconductors (adapted from Fulde et al. [10.85])

singlet order parameter with E_{1g} or E_{2u} symmetry, for tetragonal Sr_2RuO_4 it is an odd parity p-wave order parameter with E_u representation and for high temperature superconductors it is a d-wave singlet order parameter.

The following starts with heavy fermion superconductivity because most of the acoustic work in the field of unconventional superconductivity has been done on these substances. In addition, work is reviewed on Sr_2RuO_4 and organic superconductors. The case of high temperature superconductivity has been dealt with in Sect. 10.4 without discussing any strain–order parameter coupling since relevant experiments do not exist.

10.7.1 Heavy Fermion Superconductivity

The acoustic properties of normal state heavy fermion compounds have been dealt with in Sect. 9.3. A number of these systems exhibit superconductivity – in fact the list of h.f. superconductors is steadily growing. The latest additions to the list are $CeCoCu_5$ (Petrovic et al. [10.88]) and $CeIrIn_5$ (Petrovic et al. [10.89]). Table 10.4 shows some heavy fermion superconductors in which ultrasonic experiments have been performed and which will be discussed below. In this table we list some of their physical properties. There are a number of reviews on heavy fermion superconductivity: Lee et al. [10.90],

Table 10.4. Physical properties of heavy fermion superconductors

Material	T_c	T^*	$\gamma \left(\frac{J}{mol\ K^2}\right)$	$\frac{C_s - C_n}{C_n}$	G–L κ	Superconducting Grüneisen parameters	
CeCu$_2$Si$_2$	0.7	10	0.6	0 - 1.4	≈ 50	$\Omega_a = 60$	$\Omega_c = 1$
UPt$_3$	0.54		0.45	1.6	≈ 23	$\Omega_a = -137$	$\Omega_c = -133$
UBe$_{13}$	0.85		0.75	2.4	≈ 70		
URu$_2$Si$_2$	1.4	17.5(T_N)	0.12	0.6	≈ 105	$|\Omega_a| = 215$	$|\Omega_c| = 63$
UPd$_2$Al$_3$	1.82	14(T_N)	0.15			$\Omega_v = -3.5$	

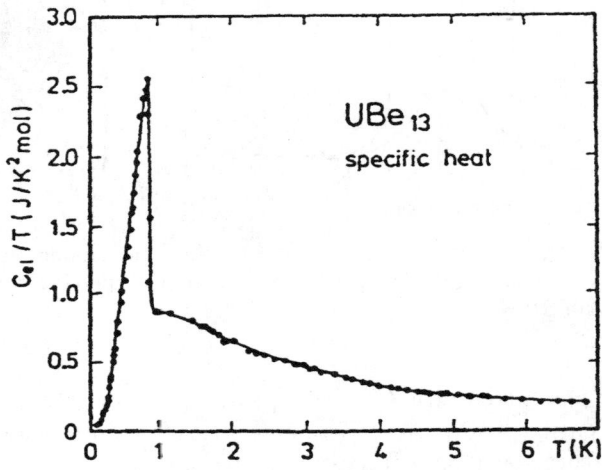

Fig. 10.15. Specific heat in the normal and superconducting state for the heavy fermion UBe$_{13}$ (Ott et al. [10.96])

Gorkov [10.91], Ott [10.92], Fulde et al. [10.85], Sigrist and Ueda [10.93], Grewe and Steglich [10.94], Sarma et al. [10.95].

Heavy fermion superconductivity has several interesting properties. The first one is that compounds with magnetic ions can exhibit superconductivity despite the fact that magnetic ions are strong pair breakers. Apparently in concentrated compounds, they can form a coherent superconducting ground state. Another one is the fact that the heavy quasi particles are responsible for superconductivity. This can be concluded from the specific heat experiments at T_c as shown in Fig. 10.15. The quantity $(C_s - C_n)/C_n$ is of order 1 at T_c (see Table 10.4). The normal specific heat C_n is already much enhanced over normal metals as shown in the γ–values in Table 10.4 and discussed in Sect. 9.1.2. This means that heavy quasi particles also form Cooper pairs. $(C_s - C_n)/\gamma T_c$ is often close to the BCS value 1.4 (Grewe and Steglich [10.94]).

10.7 Unconventional Superconductivity

Table 10.5. Power laws for different physical quantities

Quantity	Nodal point	Nodal line	Gap-less
specific heat C	T^3	T^2	T
NMR relaxation ($1/T_1$)	T^5	T^3	T
Ultrasonic attenuation α	T^3	T^2	
Ultrasonic att. α_T transv.		T^3	
thermal conductivity κ	T^3	T^2	

Yet another interesting property is the power law dependencies of various thermodynamic and transport quantities at low temperatures. For normal superconductors, there is exponential dependence at lower temperature because of the presence of the isotropic gap. The first indication of a power law was the T^3 law in the specific heat of UBe_{13} (Ott et al. [10.96]). Acoustic attenuation also indicates a power law dependence of T^2 or T^3 for longitudinal modes in UPt_3 (Bishop et al. [10.97], Müller et al. [10.98], Thalmeier et al. [10.99]). From such experiments it can be concluded that the gap structure is complicated. This is discussed for some examples. Table 10.5 above lists the various exponents in the power laws that are expected for some thermodynamic and transport properties for $T \ll T_c$. For example a T^2 or T^3 power law in specific heat suggests that the gap vanishes on the Fermi surface along lines or points respectively. The theoretical predictions listed in Table 10.5 have to be taken with care however. Most of them originate from a weak coupling theory with simple density of states considerations, constant relaxation time approximation for the impurity scattering etc. In addition, gap-less superconductivity also gives rise to power laws. As an example of power laws in the specific heat for conventional superconductors, refer to the case of HfV_2 and ZrV_2 (see Sect. 10.3). But from the foregoing remarks, it is not surprising that heavy fermion compounds should exhibit unconventional superconductivity. In the following, some heavy fermion compounds are discussed more closely: UPt_3, URu_2Si_2, UBe_{13}, UPd_2Al_3 and $CeCu_2Si_2$.

For a quantitative discussion of the results presented in the following Figs. 10.16, 10.17 and 10.18, it is first necessary to discuss the symmetry of the order parameter, secondly the nature of the symmetry breaking field and finally the resulting strain–order parameter coupling. This ambitious goal has only been partly achieved for UPt_3 and Sr_2RuO_4. The correct symmetry of the order parameter in the heavy fermion superconductor UPt_3 is still not definitely known. Therefore information on the gap structure, without knowing the order parameter, is still helpful (see Table 10.5). For another approach to determine the nodal structure of the superconducting gap see Sect. 10.7.3.

Table 10.5 is by no means complete. The power law exponents were taken mostly from the review by Sigrist and Ueda [10.93]. They were derived from density of states considerations and are therefore independent of whether we

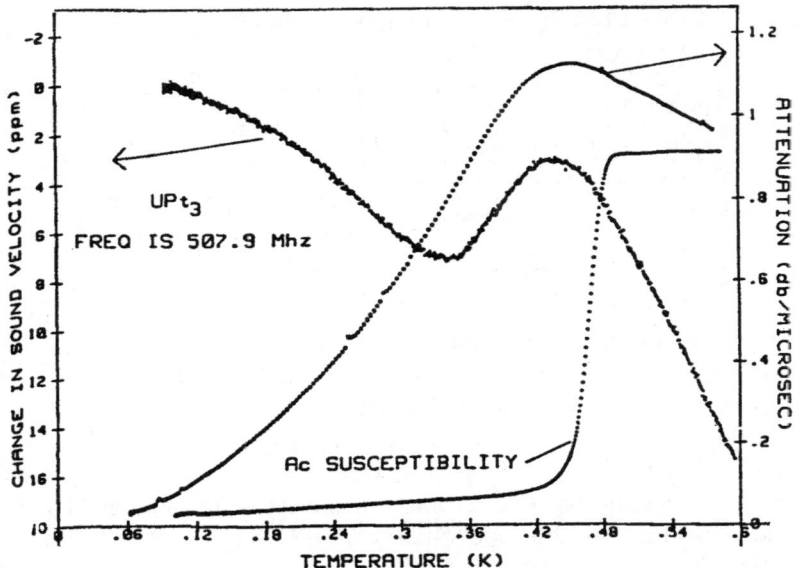

Fig. 10.16. Ultrasonic attenuation, relative longitudinal sound velocity changes and ac-susceptibility for UPt$_3$ (adapted from Bishop et al. [10.97])

deal with an even parity d-wave or an odd parity p-wave order parameter. The case of transverse ultrasonic attenuation α_T has been investigated in more detail by Moreno and Coleman [10.100] for a d-wave scenario calculating the viscosity tensor. The result gives a strong directional dependence with quite different exponents, not included in Table 10.5. Strong gap anisotropies of strong coupling superconductors can also give power laws in thermodynamic and transport quantities if the temperature region is not low enough. A possible case we encountered for HfV$_2$ in Sect. 10.3. Another case will be Sr$_2$RuO$_4$ discussed in Sect. 10.7.2.

UPt$_3$ Strain order parameter coupling

Most of the acoustic experiments for heavy fermion superconductors have been performed in UPt$_3$. The normal state properties have been discussed in Sect. 9.3. We have seen that the normal state acoustic properties can be explained with the large electronic Grüneisen parameter. This coupling will also be of importance in the superconducting state. In Fig. 10.16 ultrasonic attenuation together with velocity change and ac-susceptibility is shown for UPt$_3$ (Bishop et al. [10.97]). As seen, the longitudinal velocity experiences a step function at T_c and the attenuation a power law dependence for $T < T_c$.

The velocity step function (4.4) indicates that there is a coupling of the strain to the square of the order parameter. In fact, in well annealed single crystals, double step functions, as a function of magnetic field, are found

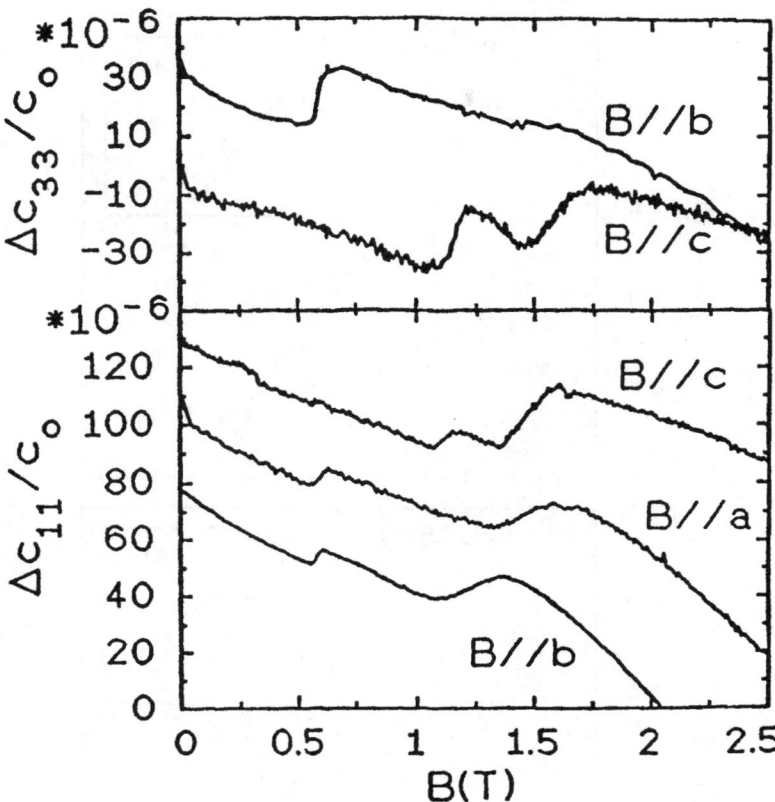

Fig. 10.17. Relative change in the longitudinal c_{11} (lower part) and c_{33} (upper part) elastic constants of UPt$_3$ vs magnetic field at 170 mK (c_{11}) and 50 mK (c_{33}). Indicated in the figure is the orientation of the magnetic field B with respect to the crystal axes (Bruls et al. [10.101])

(Fig. 10.17, Bruls et al. [10.101]). This points to a more complicated $B - T$ phase diagram than for conventional superconductors.

Such an unconventional phase diagram was first determined with specific heat measurements (Fisher et al. [10.102]). Ultrasonic longitudinal velocity measurements give the clearest picture of such B–T phase diagrams (Bruls et al. [10.101], Addenwalla et al. [10.103]) as shown in Fig. 10.18). These data supersede the pioneering ultrasonic attenuation measurements by Müller et al. [10.104]. Shear waves (c_{44} and c_{66}) do not show any velocity anomalies across the phase boundaries (Thalmeier et al. [10.99]). Thermal expansion and magneto-striction measurements (van Dijk et al. [10.105]) and the magneto-caloric effect (Bogenberger et al. [10.106]) have also given the same phase diagram. One can discern three different superconducting phases,

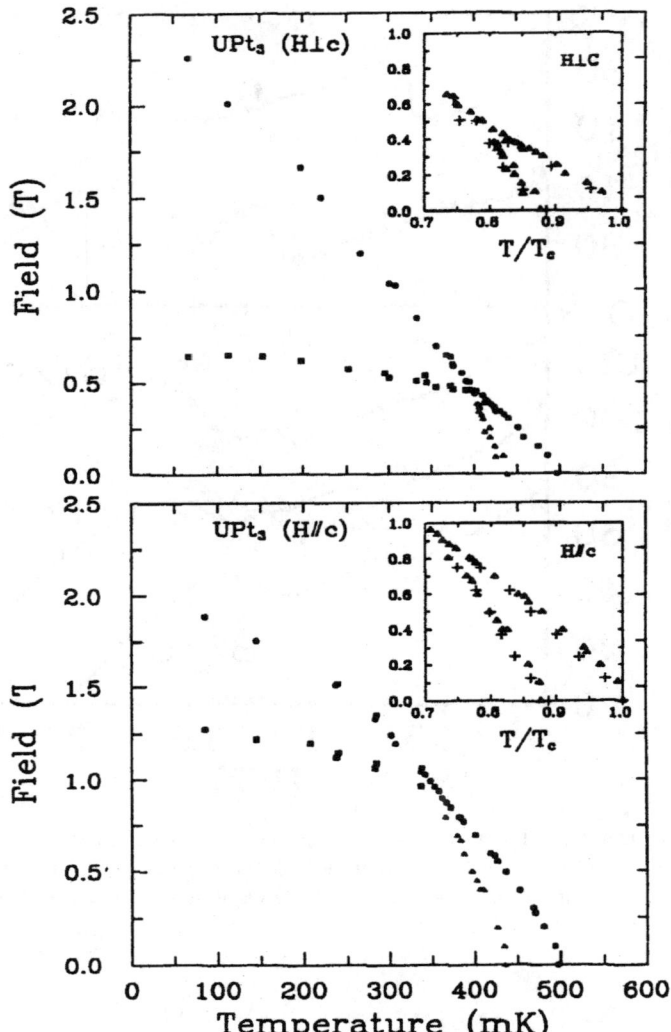

Fig. 10.18. B–T phase diagram of UPt$_3$ in the superconducting state determined with ultrasonic velocity measurements. 3 phases are observed: A between the highest transition temperature T_c^x and the 60 mK lower one T_c^y, B below T_c^y and C the high field phase, see text (Addenwalla et al. [10.103])

labeled A, B and C. These experiments can be analysed with the free energy expressions given below.

According to the program outlined above, the gap function can be written: $\Delta(r_1, r_2, s_1, s_2) = \Delta(r, s_1, s_2) = \varphi(r)\chi_s$ with $r = r_1 - r_2$. For fermion statistics the antisymmetric form is $\Delta(1,2) = -\Delta(2,1)$. Then for singlet pairing $S = 0$, there is an even parity state $\varphi(r) = \varphi(-r)$ like s or d waves.

For triplet pairing $S = 1$, there is an odd parity state $\varphi(\mathbf{r}) = -\varphi(-\mathbf{r})$ like p waves. These considerations are valid for crystals with inversion symmetry and even in case of strong spin–orbit coupling. The order parameter can now be classified with the basis functions of the irreducible representations of the hexagonal D_{6h} point group (Volovik and Gorkov [10.107], Sigrist and Ueda [10.93], Sauls [10.108]). For UPt$_3$ three types of order parameters with symmetries E_{1g}, A_{2u} and E_{2u} were investigated. Using the character table of D_6 (see Table 3.4 in Chap. 3) we get e.g. for a singlet pairing d-wave E_{1g}, the two-dimensional gap function $\Delta(\mathbf{k}) = \eta_x k_x k_z + \eta_y k_y k_z$ as the lowest order basis function. The GL order parameter is then a complex two-component vector $\boldsymbol{\eta} = (\eta_x, \eta_y)$ transforming according to the E_1 representation. Apparently this order parameter leads to an equatorial nodal line (Table 10.5). The E_{2u} representation also gives a two-component order parameter and the strain–order parameter coupling has the same form as for the E_{1g} representation. Therefore it is not possible to distinguish from ultrasonics between the two representations. Apparently it is still an unsettled question which representation best fits all results (see below).

In UPt$_3$ the symmetry breaking field seems to come from the antiferromagnetic order in the basal plane measured by neutron scattering (Aeppli et al. [10.109], Frings et al. [10.110]). As pointed out by Joynt [10.111] this leads to a coupling energy of the form $F_c = \gamma m^2(|\eta_x|^2 - |\eta_y|^2)$ which gives a symmetry lowering from hexagonal to orthorhombic. In Fig. 10.19 the antiferromagnetic order together with the temperature dependence of the Bragg peak intensity is given (Aeppli et al. [10.109]). The magnetic moment is very small (0.02 m_B). The Landau theory can be constructed from second and third order invariants (Volovik and Gorkov [10.107]): Together with F_c from above it reads for, e.g., E_{1g}:

$$F_L = \alpha \left(|\eta_x|^2 + |\eta_y^2| \right) + \frac{\beta_s}{2} \left(|\eta_x|^4 + |\eta_y|^4 \right) + \\ (\beta_1 + \beta_2 \cos(2\Phi)) |\eta_x|^2 |\eta_y|^2 + \gamma m^2 \left(|\eta_x|^2 - |\eta_y|^2 \right). \tag{10.9}$$

Here Φ is the intrinsic phase, $\alpha = \alpha_0(T - T_0)$ and the β_i are the fourth order parameters with $\beta_s = \beta_1 + \beta_2 + \beta_3$. For D_{6h}, $\beta_3 = 0$. Equation (10.9) leads to the desired splitting of T_c:

$$T_c^x = T_0 + \frac{\gamma m^2}{\alpha_0} \qquad T_c^y = T_0 - \frac{\gamma m^2}{\alpha_0} \times \frac{\beta_1}{\beta_2}. \tag{10.10}$$

For the order parameter we get for the (1,0) phase (A-phase) ($T_c^y < T < T_c^x$) $\eta_x^2 = \alpha_0 \frac{T_c^x - T}{\beta_1 + \beta_2}$ and for the (1,i) phase (B-phase) ($T < T_c^y$) $\eta_x^2 = \alpha_0 \frac{T_c - T}{2\beta_1}$, $\eta_y^2 = \alpha_0 \frac{T_c^y - T}{2\beta_1}$. We do not discuss the B–T phase diagram quantitatively because this would require the introduction of the full Ginzburg–Landau free energy with field dependent gradient terms. This has been done for the various symmetries including E_{1g} (Joynt and Taillefer [10.112], Sauls [10.108], Sigrist and Ueda [10.93], Machida et al. [10.113]).

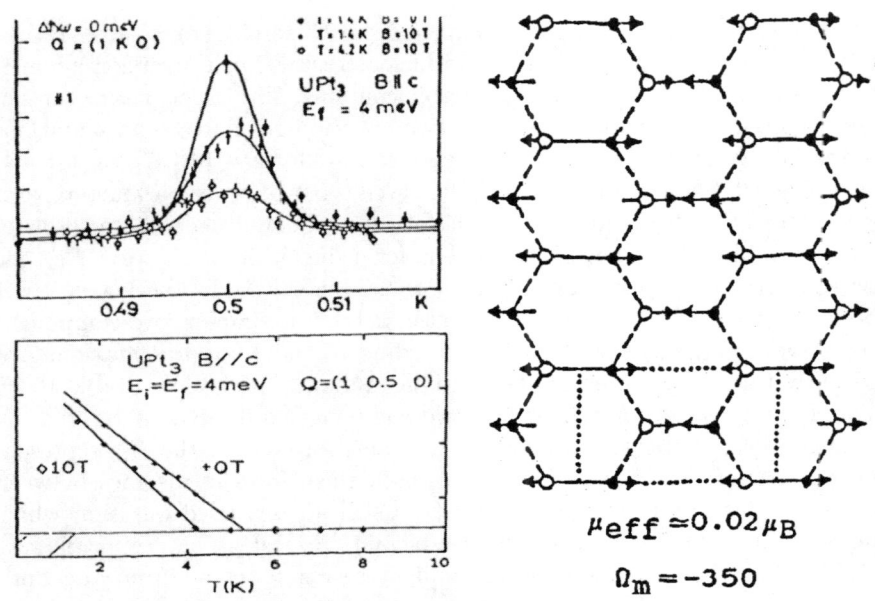

Fig. 10.19. Anti-ferromagnetic structure in UPt$_3$: (a) Elastic scan of the AF-peak and temperature dependence of the AF peak, (b) AF structure (adapted from Aeppli et al. [10.109])

Strain order parameter coupling is now discussed. Already in the normal state, a strong coupling of the strain to the heavy electron states can be observed, characterised by a very strong enhancement of the coupling strength (Grüneisen parameters) of longitudinal modes (see Sect. 9.3) (Thalmeier et al. [10.99]). We give again the strain–order parameter coupling F_{sp} for the E_{1g} symmetry. With the Landau free energy F_L from (10.9) we get analogously F_{sp} for longitudinal and shear strains

$$F_{sp} = [g_V(\varepsilon_{xx} + \varepsilon_{yy}) + g_{V'}\varepsilon_{zz}](|\eta_x|^2 + |\eta_y|^2) + \\ g_{66}(\varepsilon_{xx} - \varepsilon_{yy})(|\eta_x|^2 - |\eta_y|^2) + g_{66}\varepsilon_{xy}(\eta_x\eta_y + \eta_x\eta_y).$$
(10.9a)

With $F = F_c + F_L + F_{sp}$ one can calculate the elastic constant steps as before (Thalmeier et al. [10.99]). Note that the c_{44} mode ($\varepsilon_{xz}, \varepsilon_{yz}$) does not enter F_{sp} and no step is expected. For c_{66}, it is assumed that $|g_{66}| \ll |g_V|, |g_{V'}|$. Introducing superconducting Grüneisen parameters

$$\Omega_c = -\frac{\left(\frac{\partial \ln T_c}{\partial \varepsilon_{zz}}\right)}{T_c} = \frac{g_{V'}}{\alpha_0 T_c} \quad \text{and} \quad \Omega_{a,b} = -\frac{\left(\frac{\partial T_c}{\partial \varepsilon_{xx}}\right)}{T_c} = \frac{g_V}{\alpha_0 T_c}$$

one can express the elastic constant changes at the T_c^x transition as Ehrenfest relations

$$\Delta_1 c_{11} = -T_c^x \Omega_a^2 \Delta_1 C$$
$$\Delta_1 c_{33} = -T_c^x \Omega_c^2 \Delta_1 C \tag{10.9b}$$

with $\Delta_1 C$ the specific heat step at the T_c^x transition. Analogous expressions can be obtained for the second phase transition with $\Delta_2 c_{ii}$ and $\Delta_2 C$ (Thalmeier et al. [10.99]). It turns out that the superconducting Grüneisen parameters are of the same order of magnitude but opposite in sign to the ones in the normal state (table (10.4) and table 9.1). We want to emphasize that strain order parameter coupling can give an adequate description of the elastic constant changes for UPt$_3$ but the symmetry of the order parameter cannot be determined uniquely as discussed above. Because of the even–symmetry of the strains ε_{ij}, (10.9a) is valid for E_{1g} and E_{2u} representations.

UPt$_3$ Ultrasonic Attenuation
Another important acoustic measurement in UPt$_3$ is the acoustic attenuation for the different elastic modes. For longitudinal waves a T^3-law is found in the temperature region 0.06–0.42 K for both the c_{11} and c_{33} modes (Bishop et al. [10.97], Müller et al. [10.98], Thalmeier et al. [10.114]) as also shown in Fig. 10.16. According to Table 10.5 this would indicate vanishing gaps at points. A calculation with a d-wave order parameter, similar to the E_{1g} discussed above, gives a qualitative agreement with these data (Monien et al. [10.115]). Around the superconducting transition, the attenuation is also anomalous (Müller et al. [10.98], Schenstrom et al. [10.116]). It exhibits a pronounced fluctuation peak just at T_c. This was tentatively explained again with formula (9.8) with the specific heat anomaly at T_c. Because $\omega\tau \ll 1$ in UPt$_3$ and for acoustic frequencies ~ 100 MHz, one is in the hydrodynamic regime and a coupling to collective order parameter modes is not possible.

Of particular importance is the attenuation of the shear waves α_{44} and α_{66}. The elastic constants c_{44} and c_{66} do not exhibit any anomaly as a function of temperature (10.4), especially no order–parameter–strain coupling (10.9a). But for the corresponding attenuations, different temperature dependences are found (Shivaram et al. [10.117], Thalmeier et al. [10.114], Ellman et al. [10.118]). For $\alpha_{44}(q \parallel a, e \parallel c)$, a T^3 law is found and for $\alpha_{66}(q \parallel a, e \parallel b)$ it is a T law. These results are a clear demonstration of the gap anisotropy in UPt$_3$ in addition to the unconventionality due to the power law dependencies. The measurements by Ellmann et al. [10.118] could even give the different T_c^x and T_c^y, i.e. they could detect the A-phase with the attenuation α_{66}. These experiments could be quantitatively explained with an E_{1g} or E_{2u} order parameter representation leading to point nodes and basal plane line nodes (Graf et al. [10.119]). In this theory, appropriate to heavy fermion compounds, the hydrodynamic limit $\omega\tau \ll 1$ applies and the viscosity tensor is evaluated in a similar way to the previously mentioned method (Moreno and Coleman [10.100]).

Furthermore uniaxial pressure effects give additional evidence for the phase diagram, the tetra-critical point and the symmetry of the order parameter. Such ultrasonic experiments have been carried out by Boukhny et al. [10.120]. All these experiments (ultrasonic attenuation, thermal conductivity and static properties) give a nodal structure of the gap compatible with an E_{1g} or E_{2u} order parameter. A comprehensive review on the superconducting phases in UPt$_3$ has recently appeared (Joynt and Taillefer [10.112]).

CeCu$_2$Si$_2$

This was the first heavy fermion superconductor found (Steglich et al. [10.121]). Single crystal growth is very complicated due to peritectic solidification processes and it has taken many years to get a large-enough crystal of sufficient quality to perform ultrasonic measurements (Sun et al. [10.122], Nüttgens et al. [10.123]). One serious problem was the loss of Cu during the crystal growth. Annealing of the crystal in a Cu atmosphere helped to solve it. In Fig. 10.20 we show the crystal structure of CeCu$_2$Si$_2$. It crystallises in the tetragonal ThCr$_2$Si$_2$ structure (space group D_{4h}^{17}). The lattice constants are for $a, b = 4.10$Å and for $c = 9.93$Å. The lattice constant a is also the smallest Ce-Ce distance. Figure 10.20 shows that the Ce ions are arranged in planes separated by Cu and Si layers.

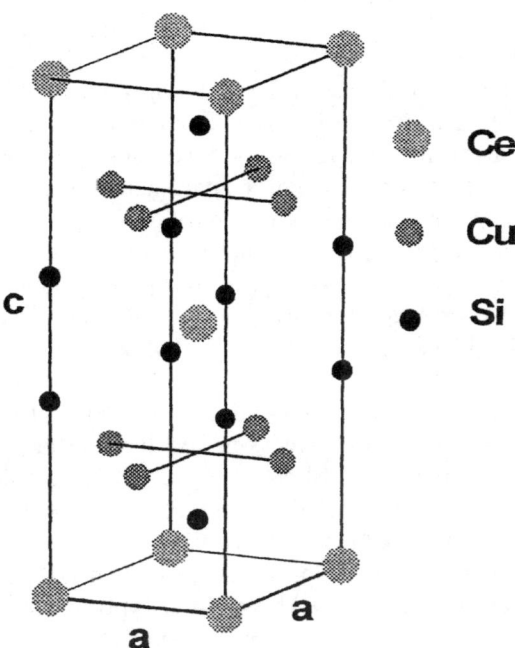

Fig. 10.20. Crystal structure of CeCu$_2$Si$_2$

10.7 Unconventional Superconductivity   259

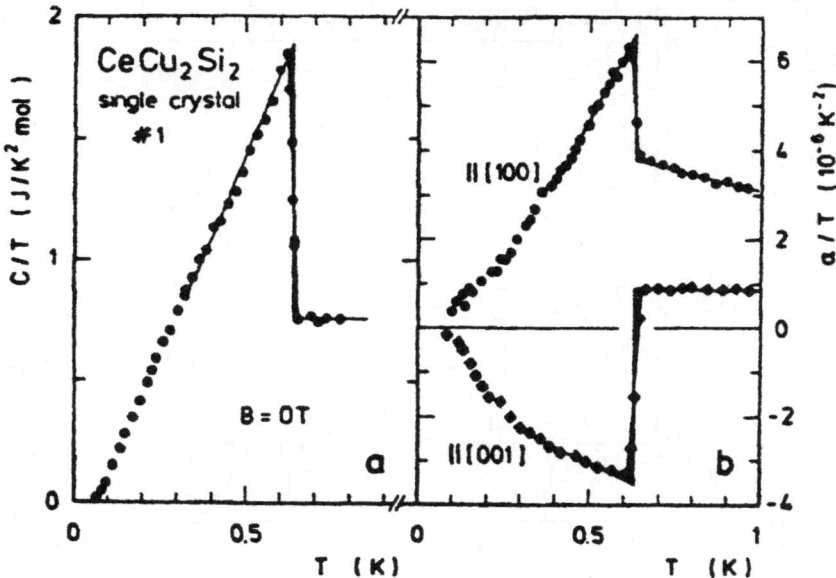

Fig. 10.21. Temperature dependence of the Specific heat as C/T and linear thermal expansions $\alpha_{a,c}/T$ for CeCu$_2$Si$_2$ (Lang et al. [10.124])

Specific heat and linear thermal expansion along the a and c axis are shown for the low temperature range in Fig. 10.21 (Lang et al. [10.124]). For both thermodynamic quantities, C/T and $\alpha_{a,c}/T$ a power law temperature dependence can be observed in contrast to a BCS temperature dependence. In addition a pronounced change at T_c is observed, indicating again that the heavy quasi-particles are responsible for the superconductivity. The thermal expansion has a different temperature dependence for the a and c direction, as noted already for the normal state behaviour in tetragonal and hexagonal crystals (Chaps. 5 and 9, Fig. 9.16). Using (9.12), (9.12a), one can determine from these experiments superconducting Grüneisen parameters or the pressure dependencies of T_c (Lang et al. [10.124]) see Table 10.4.

In the normal state of CeCu$_2$Si$_2$, CEF and Grüneisen parameter effects are observed in the temperature dependence of the elastic constants – as shown in Fig. 9.14 of Chap. 9 (Bruls et al. [10.125]) and in high magnetic fields magneto-acoustic quantum oscillations (Wolf et al. [10.126]). These were discussed briefly in Sects. 8.3.2 and 9.1.

CeCu$_2$Si$_2$ exhibits an interesting low temperature B–T phase diagram (Fig. 10.22a,b). The superconducting phase is embedded in the so-called A–phase. At higher magnetic fields there exists another phase called B–phase. In Fig. 10.22a, the different phases are shown and in Fig. 10.22b an isothermal line for $\Delta c_{11}/c_{11}$ is given for $T = 0.5$ K. From such curves the phase diagram was constructed. It exhibits typical signatures of a strain–order parameter

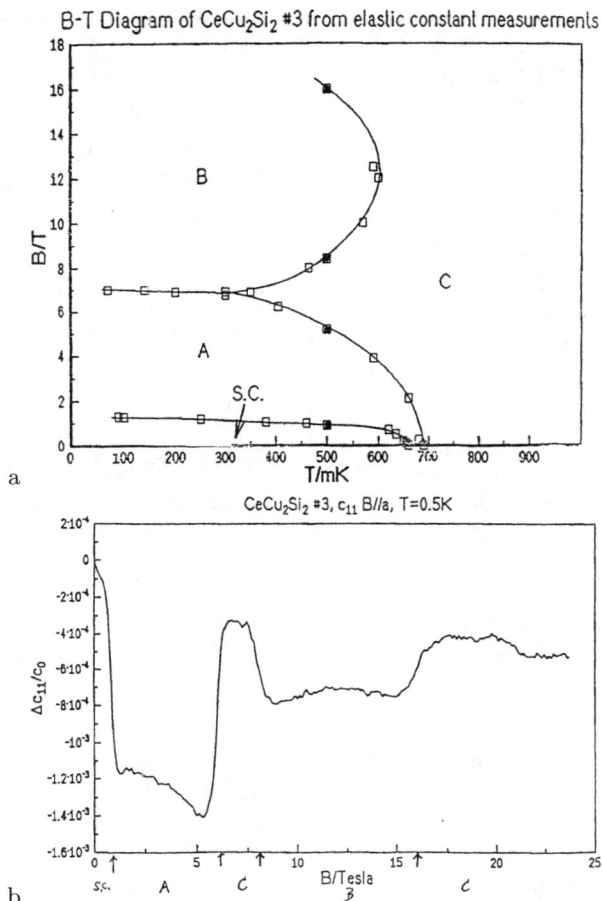

Fig. 10.22. (a) B-T phase diagram for CeCu$_2$Si$_2$ (Bruls et al. [10.127]), (b) isotherm of relative elastic constant change for c_{11} mode at $T = 0.5$ K up to 26 T

coupling as discussed before (10.5). Going from a high symmetry phase to a low symmetry phase, the elastic constant change is negative, e.g. $C \to B(16T)$ and for the reverse process positive for $B \to C(8T)$ and again negative for $C \to A(6T)$. The positive step for $A \to S.C.$ is due to an expulsion of the A phase by the superconducting phase. This could be realized by analysing the magnetic field dependence of the longitudinal elastic constant and by analysing a thermodynamic cycle in the B–T phase (Wolf [10.126], Bruls et al. [10.127]).

It should be stressed that CeCu$_2$Si$_2$ single crystals exist which do not show an A–phase. Consequently the elastic constant change from the normal to the superconducting state is negative as observed by Finsterbusch [10.128]. The parameters which determine crystals with the A–phase, with reduced

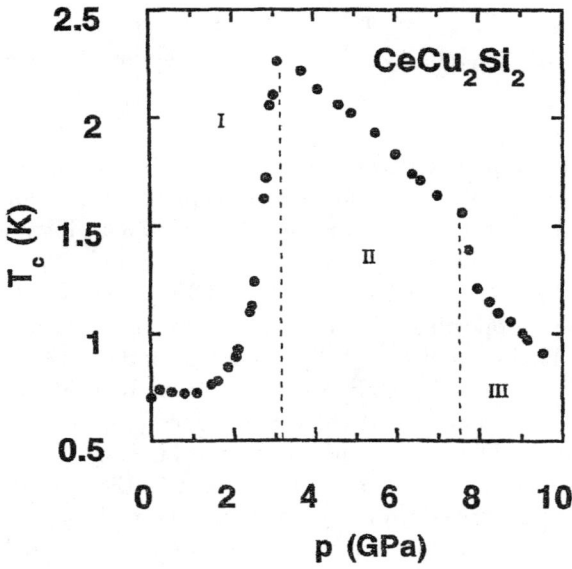

Fig. 10.23. CeCu$_2$Si$_2$: $p - T_c$ phase diagram (Thomas et al. [10.130])

A–phase and with no A–phase are not yet known for crystal growth. In addition the true nature of the A–phase is not yet known, it has magnetic and structural properties. Recently elastic neutron scattering experiments could determine the A phase as an incommensurate spin density wave state (Stockert et al. [10.129]).

Application of hydrostatic pressure up to 9GPa gives a $p - T_c$ phase diagram which exhibits roughly a constant T_c up to 2Gpa and then a sharp increase of T_c to more than 2 K for 3.5 GPa followed by a linear decrease up to 7.6GPa and then a sharp decrease (Thomas et al. [10.130]). This phase diagram is shown in Fig. 10.23. The possible A–phase is suppressed already at very low pressure. The complicated $p - T_c$ diagram, with the three regions I, II, III indicated in the figure, suggests that different mechanisms are operative in these regions.

The difference of the superconducting phase in CeCu$_2$Si$_2$ to the ones in UPt$_3$ and URu$_2$Si$_2$ (to be discussed below) is obvious. In the latter two substances, the superconducting phase is embedded in a anti-ferromagnetic phase which provides the symmetry breaking field and seems to be stable although the magnetic moments are extremely small. In CeCu$_2$Si$_2$, the A–phase seems to be unstable by entering the superconducting phase or is not present at all in certain crystals. Therefore a strain–order parameter analysis as made for UPt$_3$ is not feasible. A formal Landau description of the interplay of the A–phase (order parameter $OP\chi$) and the superconducting phase ($OP\eta$) can be given. With the free energy $F = \alpha \, |\eta|^2 + \beta \, |\eta|^4 + a \, |\chi|^2 + b \, |\chi|^4 + g \, |\eta|^2|\chi|^2$ we introduce a repulsive term for the coupling between the two phases which gives a diminution of $|\chi|$ for $T \leq T_c$ if $T_A > T_c$. But a detailed microscopic description of the superconductivity and the different phases for CeCu$_2$Si$_2$ is lacking.

URu$_2$Si$_2$

This tetragonal compound has an anti-ferromagnetic transition at 17.5 K with a very small magnetic moment and a superconducting transition at 1.2 K. The normal state properties were presented in Sect. 9.3.4. They exhibit strong Grüneisen parameter effects especially for the c_{11} mode and anomalous behaviour for c_{11} and c_{33} near T_N. No sign of CEF effects were observed. The superconducting properties were investigated by means of ultrasound (Wolf et al. [10.131], Thalmeier et al. [10.99]). Again, as in the normal state, only the c_{11} mode exhibits effects, namely a steplike anomaly at T_c of $\Delta c/c_0 \sim 10^{-4}$ (Fig. 10.24a). The anomalies of the other modes (c_{33}, c_{44} and c_{66}) amount to less than 10^{-5}. From these measurements and analogous susceptibility measurements, a B–T phase diagram as shown in Fig. 10.24b can again be constructed. In the case of URu$_2$Si$_2$ the elastic anomalies and the B–T phase diagram do not give any hint for an unconventional superconductivity. Note that in Fig. 10.24b the difference of c_{11} and ac-susceptibility for sample B points to inferior crystal quality; the superconducting transitions are less steep than for crystal A shown in Fig. 10.24a.

UBe$_{13}$

An acoustic investigation for longitudinal high frequency sound of 0.9–2.4 GHz has been carried out in this material. In Fig. 10.25 sound velocity and attenuation are shown for 1.7 GHz (Golding et al. [10.132]). The relative sound velocity exhibits a pronounced step function of 0.25×10^{-4} size at $T_c = 0.86$ K. This is in accordance with a strain coupling to the order parameter as discussed above (10.5). On the other hand, the attenuation exhibits a power law behaviour for low temperatures, $\alpha \sim T^2$, and a pronounced attenuation maximum just below T_c. The power law dependence could indicate a nodal line structure of the gap function (Table 10.5). The attenuation peak follows an ω^2 law which indicates that this is in the hydrodynamic regime. This follows also on application of (4.19). The velocity change and the attenuation peak of Fig. 10.25 give a relaxation time $\tau \sim 8 \times 10^{-12}$s which leads to $\omega\tau < 0.08$. This conclusion is similar to the case of UPt$_3$ where also $\omega\tau < 1$. A coupling of the strain wave to a collective mode is therefore improbable.

An acoustic investigation of the system U$_{0.98}$Th$_{0.02}$Be$_{13}$ has also been made (Batlogg et al. [10.133]). This system has a complicated phase diagram with two phase transitions $T_{c1} \sim 425$ mK and $T_{c2} \sim 360$ mK, the former being a bulk superconducting transition and the latter still being under discussion.

UPd$_2$Al$_3$

This compound was briefly discussed in Sect. 9.3.4. The small electronic Grüneisen parameter makes all acoustic effects very small, also the ones in the superconducting state. Therefore we cannot add more to this topic. But it should be mentioned again that in this compound inelastic neutron scattering

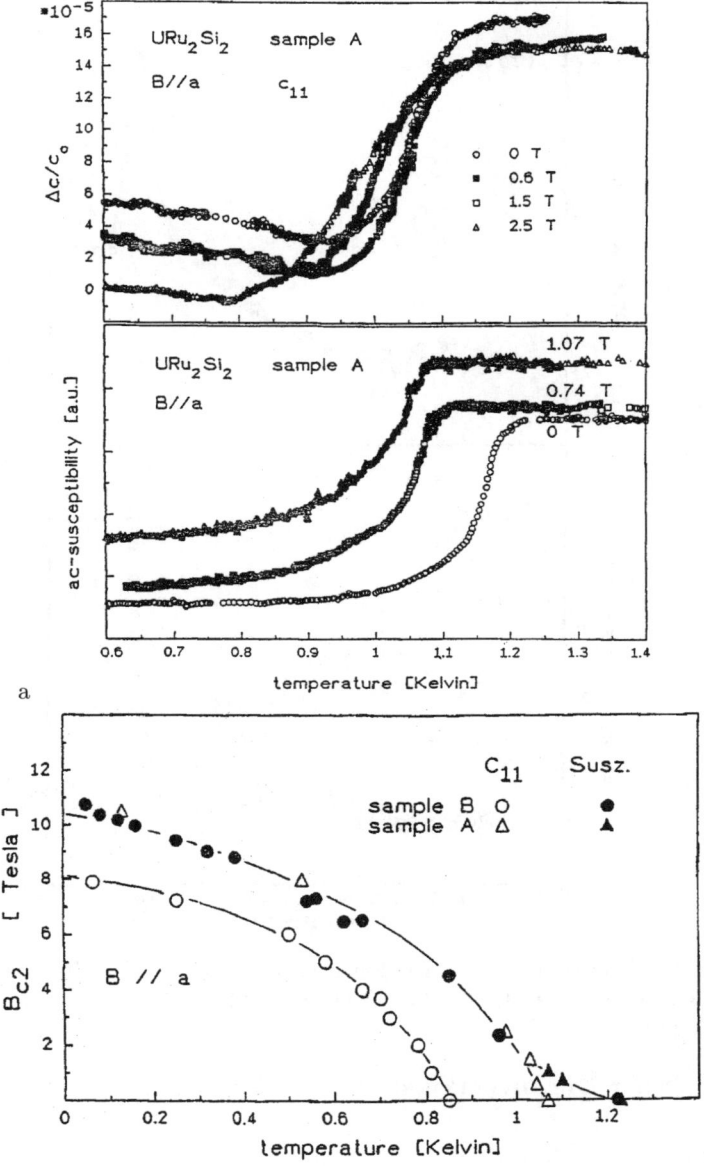

Fig. 10.24. (a) Temperature dependence of the c_{11}-mode and ac-susceptibility for URu_2Si_2, (b) resulting B–T phase diagram for URu_2Si_2 (Wolf et al. [10.131])

and tunneling experiments have established a non-phononic mechanism for superconductivity, a so-called magnetic exciton exchange (Sato et al. [10.86], Thalmeier et al. [10.87]).

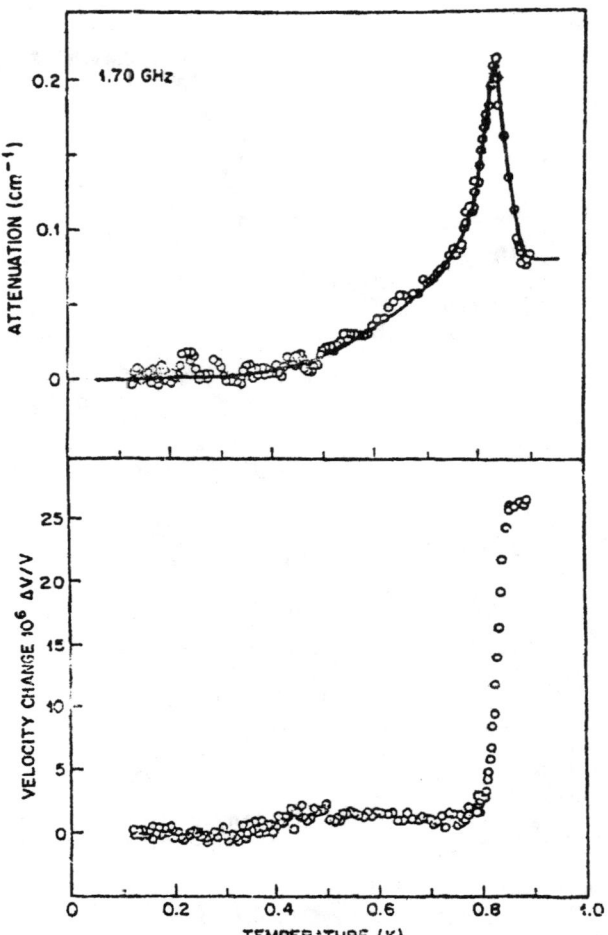

Fig. 10.25. Acoustic attenuation and relative sound velocity for UBe_{13} as a function of temperature at 1.7 GHz (Golding et al. [10.132])

10.7.2 Other Unconventional Superconductors

Sr_2RuO_4

The layered tetragonal perovskite compound Sr_2RuO_4 exhibits superconductivity at $T_c \sim 1.4\,\mathrm{K}$ (Maeno et al. [10.52]). It has the same crystallographic structure as the high temperature superconductors, but in contrast to them the alternating RuO_2 planes are responsible for the strongly correlated two-dimensional electronic states. The absence of a Hebel-Slichter coherence peak in NQR, power law dependencies for specific heat and NQR, muon spin relaxation studies and other experiments, point to an unconventional nature of superconductivity as suggested theoretically (Rice and Sigrist [10.134], Sigrist et al. [10.135]). Unlike the High Temperature Superconductors, Sr_2RuO_4 is very sensitive to doping (Mackenzie et al. [10.136]). This fact and similarities to the correlated quantum liquid He^3 (Vollhardt and Wölfle [10.84]), especially its A–phase, led to the prediction of a p-wave pairing spin triplet superconductor. This state breaks time reversal symmetry and has a two-dimensional complex order parameter. It belongs to the E_u representation of the tetragonal crystal point group D_{4h}. For a recent review on this substance see Maeno et al. [10.137].

Several acoustic studies have been performed in this compound. In Sect. 8.3.2 and Fig. 8.10 the normal state acoustic properties have been discussed, especially magneto-acoustic quantum oscillations (MAQO). It was found that the c_{33} mode has the largest effect. This is somewhat different for the superconducting phase. Here c_{11} and c_{33} have relatively weak temperature dependences, but $(c_{11} - c_{12})/2$ as a B_{1g} symmetry mode exhibits a significant temperature dependence which can be interpreted with (10.4) using the second and third term (Matsui et al. [10.138]). This is shown in Fig. 10.26 together with specific heat and ultrasonic attenuation data. The latter quantity is interesting because it can be fitted to a power law in temperature $\sim T^2 - T^3$ dependent on the temperature region, with T^2 for the low temperature range. According to Table 10.5, this might suggest a nodal line order parameter. The peak in $d\alpha/dT$ just below T_c is common to shear wave attenuation in superconductors (see Gavenda [10.139]). More recent experiments for longitudinal and transverse modes with propagation and polarization in the tetragonal plane give similar power law dependencies (Lupien et al. [10.140]). All these experiments would exclude a node-less p-wave state. But strong anisotropies of the energy gap could simulate such temperature dependencies. For further developments see also Sect. 10.7.3.

One interesting point of the experiment by Matsui et al. [10.138] is the symmetry analysis for the different modes. From the acoustic de Haas van Alphen experiments (see Sect. 8.3.2), it can be concluded that only A_{1g} strains couple strongly to the Fermi surface sheets α and β, but not the B_{1g} strains ($c_{11} - c_{12}$ mode). This latter mode gives the strongest superconducting effect as discussed above and a coupling to the large γ surface. A recent ultrasonic study of Sr_2RuO_4 (Okuda et al. [10.141]) confirms previous experiments and gives a detailed fit with a molecular field equation similar to (10.4).

In contrast to UPt_3 and UBe_{13}, clear additional attenuation anomaly in the vicinity of T_c has not been observed (Matsui et al. [10.138] and Lupien et al. [10.140]). According to a Landau–Khalatnikov type analysis (Sigrist [10.142]) one could expect a coupling to the order parameter relaxation mode in the limit $\omega\tau \sim 1$ (see Sect. 4.3). The order parameter strain coupling reads, similar to (10.9a) for UPt_3, (Sigrist [10.142])

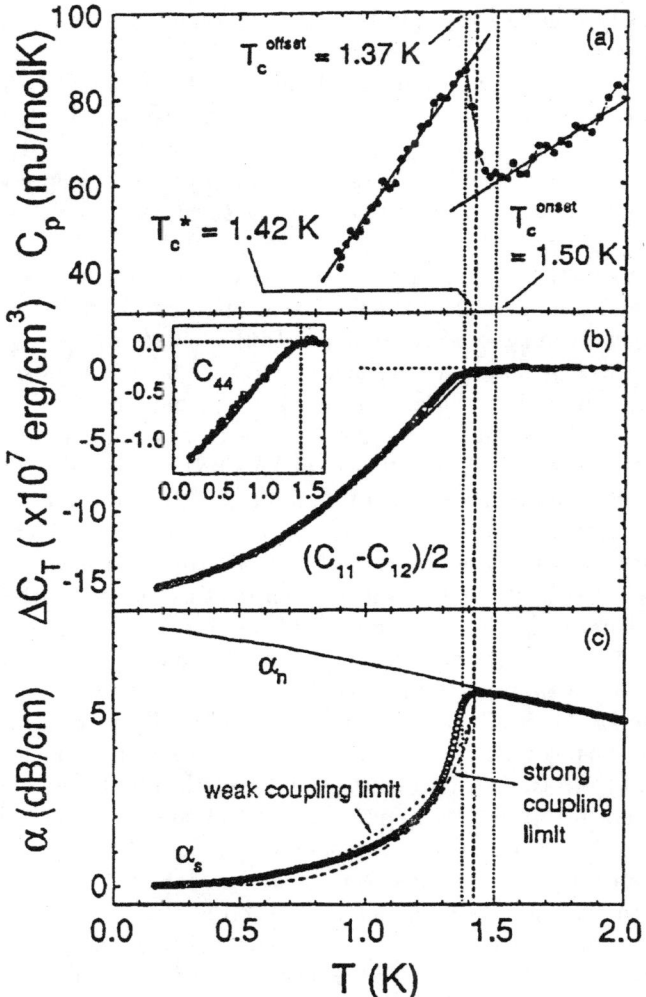

Fig. 10.26. Temperature dependence of specific heat (a), elastic constant (b) and attenuation for $(c_{11} - c_{12})/2$ (c) in Sr_2RuO_4 (Matsui et al. [10.138])

$$F_{sp} = [g_V(\varepsilon_{xx} + \varepsilon_{yy}) + g_{V'}\varepsilon_{zz}] |\eta|^2 + \\ g_{B2g}\varepsilon_{xy}(\eta_x^\star\eta_y + \eta_x\eta_y^\star) + \\ g_{B1g}(\varepsilon_{xx} - \varepsilon_{yy})(|\eta_x|^2 - |\eta_y|^2). \quad (10.11)$$

The elastic constant changes and the attenuation can be calculated from the complete free energy density expressions, together with the Landau–Khalatnikov formulation. The expressions for the elastic constant changes Δc_{ij} for this unconventional superconductor are slightly more complicated than the ones given in (10.4), valid for a conventional superconductor. As discussed above, first order effects, involving $\partial T_c/\partial \varepsilon_{ij}$, a strain–order parameter coupling, have not yet been observed in Sr_2RuO_4 with the possible exception of the c_{11} mode (Okuda et al. [10.141]).

Organic Superconductors

Earlier work on organic superconductors has been reviewed by Jerome and Schulz [10.143]. The first organic superconducting compound, $(TMTSF)_2PF_6$, was discovered in 1980. Recent work has concentrated on quasi-two-dimensional BEDT-TTF based organic superconductors. The transition temperature of $(BEDT-TTF)_2ReO_4$ was $T_c = 12.8\,K$ (Kini et al. [10.144]). In these materials, free carriers are created by charge-transfer processes with the electron–donor molecule D. Since the pressure–temperature phase diagram consists of anti-ferromagnetic, superconducting and metal insulator phases, it is suspected that unconventional superconductivity is being observed. But this has not been definitely proven, see recent reviews by Lang [10.145], Lang and Müller [10.146]) and Lefebvre et al. [10.147]. There is one acoustic experiment which deals with this class of superconductors (Yoshizawa et al. [10.148], Simizu et al. [10.149]). In κ-$(BEDT-TTF)_2Cu(NCS)_2$, longitudinal sound velocities in the frequency range of 20–96 MHz exhibited at $T_c = 8.9\,K$ a step-like function anomaly of $\Delta v/v \sim 5.6 \times 10^{-4}$, as expected for a strain–order parameter coupling of (10.5). The B–T phase diagram from this experiment does not give any indication for unconventional superconductivity. More recently, the system κ-$(BEDT-TTF)_2Cu[N(CN)_2]Cl$ was investigated with ultrasound (Fournier et al. [10.150]). Apart from several superconducting transitions and a metal-insulator transition a compressibility divergence at the critical point of the Mott transition was observed.

10.7.3 Other Methods to Probe the Energy Gap Structure

In recent years, new experiments have been devised in order to learn more about the gap structure of the unconventional superconductors. One of these experiments, close to the case of ultrasonics, is a thermal conductivity study. The experiment is also a directional probe – like ultrasonic attenuation – sensitive to the relative orientation within the thermal flow, the magnetic field and the nodal directions of the order parameter. Therefore angular dependent thermal conductivity can give detailed information on the gap structure. The thermal conductivity is due to the unpaired quasi-particles for $T < T_c$. Experiments have been carried out in recent years for UPt_3 (Lussier et al. [10.151]), for Sr_2RuO_4 (Tanatar et al. [10.152], Izawa et al. [10.153]), for UPd_2Al_3 (Chiao et al. [10.154]), for $CeCoIn_5$ (Izawa et al. [10.155]) and for high temperature superconductors (Chiao et al. [10.156]). Relevant theoretical papers on this subject are Vekhter et al. [10.157], Thalmeier and Maki [10.158]. These experiments seem to confirm the results for the order parameter symmetries deduced from the acoustic experiments.

10.7.4 Summary

The previous sections have shown that the field of unconventional superconductivity is rapidly expanding. Starting with the heavy fermion superconductor $CeCu_2Si_2$, there are now quite different materials exhibiting unconventional superconductivity. Apart from the Uranium compounds UPt_3, UBe_{13}, URu_2Si_2 and UPd_2Al_3, the new Cerium materials $CeCoCu_5$ and $CeIrIn_5$ have been recently discovered. They belong to the class of heavy fermions exhibiting superconductivity near a quantum critical point (see Sect. 9.3.6). The latest addition is the cubic skutterudite compound $PrOs_4Sb_{12}$ (Bauer et al. [10.159]) which is the first Pr-based heavy fermion superconductor. Apparently it has a non-trivial superconducting B–T phase diagram as discussed already by various theory groups (Ichioka et al. [10.160], Goryo [10.161], Miyake et al. [10.162], Izawa et al. [10.163]). In addition to the unconventional superconducting B–T phase diagram (limited by $B < 2.3T$, $T < 1.85\,\mathrm{K}$), so far it exhibits a field-induced phase for $B > 4T$ of unknown origin.

Recent ultrasonic experiments revealed a strong softening of the symmetry elastic constants c_{44} and $(c_{11} - c_{12})/2$ in the normal state towards the superconducting transition temperature $Tc = 1.85\,\mathrm{K}$ (Goto et al. [10.164]). These experiments could not determine the ground state of the Pr^{3+}-ion in this skutterudite compound (Γ_1 or Γ_3) because of strong dispersion effects due to the rattling motion of the off-center Pr-ions. Note that other higher order multipole-strain effects were observed also exclusively in Pr-compounds (Sect. 7.2.2). For other off-center effects on elastic constants see Sect. 14.1.

In many cases the unconventional superconductivity could be inferred from the structure and symmetry of the order parameter as investigated for, e.g., in UPt_3 or Sr_2RuO_4. In other systems, such as UPd_2Al_3, a non-phononic superconducting mechanism, as discussed at the beginning of Sect. 10.7, could be extracted.

Ultrasonic studies have helped a great deal to elucidate the intriguing properties of unconventional superconductivity. Examples discussed in previous sections are the investigation of:

1. The evasive A-phase and the superconducting phase in $CeCu_2Si_2$.
2. The strain–order parameter coupling and the detailed B–T–p phase diagram established in UPt_3.
3. Ultrasonic attenuation experiments in UPt_3 for longitudinal and shear waves and accompanying theories for $\omega\tau \ll 1$ giving strong evidence for the order parameter symmetry.
4. The anomalous temperature and magnetic field dependence of elastic modes in Sr_2RuO_4 especially its large anisotropy.
5. The temperature dependent and high field properties of elastic modes in URu_2Si_2.

An analogous ultrasonic investigation of the other recently discovered heavy fermion superconductors would be interesting, especially the ones close to a quantum critical point because they seem to have especially strong electron-lattice couplings (see Sect. 9.3.6).

An ultrasonic investigation for $PrOs_4Sb_{12}$ in the superconducting state would also be interesting. The experiment by Goto et al. [10.164] suggests that quadrupolar fluctuation could give a mechanism for superconductivity in this case. One would expect that the strain–order parameter coupling would be likewise strong.

11 Coupling to Collective Excitations

In this chapter, we study the coupling of sound waves to collective excitations in solids and the various effects which result from it. As collective excitations we consider plasmons, helicons, polaritons and spinwaves. We will first introduce each collective mode and then describe the coupling to sound waves and the coupled dispersion spectra. We will then show special effects due to this coupling and discuss relevant experiments.

The case of solitons is not discussed here. Recently, phonon-solitons were experimentally investigated using high power laser pulses (Hao and Maris [11.1]). Since the spinwave equations are nonlinear, spinwave–solitons can be observed already at low power levels (see Demokritov et al. [11.2]). In the framework of the magnetic Sine-Gordon system the phonon-soliton interaction was studied theoretically (Schöbinger and Jelitto [11.3]).

In Sect. 11.1, we discuss first plasmons, helicons and their interaction with sound waves. Sect. 11.2 presents spinwaves and their interaction with phonons, the so-called magneto-elastic waves. Experiments in ferrimagnets and anti-ferromagnets are also presented.

11.1 Plasmons and Helicons

11.1.1 Dielectric Tensor

We first consider the dielectric tensor for a conductive medium. The free electron approximation is used and the dielectric tensor with classical equations of motion are derived. The static magnetic field is taken in the z-direction and an isotropic medium is used. The equation of motion reads

$$m\frac{d\bm{v}}{dt} = e\bm{E} + \frac{e}{c}[\bm{v} \times \bm{B}] - m\frac{\bm{v}}{\tau}. \tag{11.1}$$

Here, the Newton equation of motion gives the forces due to the electric field $\bm{E}$, the magnetic field $\bm{B}$ (Lorentz force) and a damping term characterised by a relaxation time τ. With $\bm{v} = \bm{v}_0 \exp(-i\omega t)$, the cyclotron frequency $\omega_c = eB/mc$ and the current density $\bm{j} = ne\bm{v} = \sigma\bm{E}$ with n the number of carriers per unit volume, the conductivity tensor (σ_{ij}) is:

11 Coupling to Collective Excitations

$$\sigma_{xx} = \sigma_{yy} = \frac{ne^2}{m}\left(-i\omega + \frac{1}{\tau}\right)\frac{1}{\left(-i\omega + \frac{1}{\tau}\right)^2 + \omega_c^2}$$

$$\sigma_{xy} = -\sigma_{yx} = \frac{ne^2}{m}\omega_c \frac{1}{\left(-i\omega + \frac{1}{\tau}\right)^2 + \omega_c^2} \quad (11.2)$$

$$\sigma_{zz} = \frac{ne^2}{m}\frac{1}{-i\omega + \frac{1}{\tau}},$$

and the dielectric tensor $\varepsilon = \varepsilon_\infty + 4\pi i \sigma/\omega$ has the matrix form

$$\varepsilon = \begin{pmatrix} \varepsilon_1 & i\varepsilon_2 & \\ -i\varepsilon_2 & \varepsilon_1 & \\ & & \varepsilon_3 \end{pmatrix}.$$

With the screened plasma frequency $\Omega_p^2 = 4\pi n e^2/m\varepsilon_\infty$ and ε_∞ the dielectric constant of the medium, the tensor components in the limit $\omega\tau \gg 1$ are given by:

$$\varepsilon_1 = \varepsilon_{xx} = \varepsilon_{yy} = \varepsilon_\infty\left[1 + \frac{\Omega_p^2}{\omega_c^2 - \omega^2}\right]$$

$$\varepsilon_3 = \varepsilon_{zz} = \varepsilon_\infty\left[1 - \frac{\Omega_p^2}{\omega^2}\right] \quad (11.3)$$

$$\varepsilon_2 = \varepsilon_{xy} = \varepsilon_\infty \Omega_p^2 \frac{\omega_c}{\omega(\omega_c^2 - \omega^2)}.$$

For $B = 0$, the non-diagonal components are zero and the diagonal components are $\varepsilon_1 = \varepsilon_3 = \varepsilon$ with the well-known result that $\varepsilon < 0$ for $\omega < \Omega_p$ and $\varepsilon > 0$ for $\omega > \Omega_p$. Electromagnetic waves propagate only for $\omega > \Omega$ but not for $\omega < \Omega$ (see below).

Using the Maxwell equations $\mathrm{rot}\,\boldsymbol{E} = -1/c\,\frac{d\boldsymbol{B}}{dt}$ and $\mathrm{rot}\,\boldsymbol{H} = 4\pi/c\,\boldsymbol{j} + 1/c\,\frac{d\boldsymbol{D}}{dt}$ with $\boldsymbol{B} = \mu\boldsymbol{H}$ we get $\mathrm{rot}\,\mathrm{rot}\,\boldsymbol{E} = -\mu/c\left[4\pi/c\,\frac{d\boldsymbol{j}}{dt} + 1/c\,\frac{d^2\boldsymbol{D}}{dt^2}\right]$ and with $\mu \sim 1$ and $\boldsymbol{D} = \varepsilon_\infty \boldsymbol{E}$, this results in

$$-\boldsymbol{k}(\boldsymbol{k}\cdot\boldsymbol{E}) + k^2\boldsymbol{E} = \frac{\omega^2}{c^2}\varepsilon\boldsymbol{E} \quad \text{with} \quad \varepsilon = \varepsilon_\infty + 4\pi i\frac{\sigma}{\omega}. \quad (11.4)$$

Depending on which geometry is chosen, we can obtain the plasma polariton or the helicon excitations as discussed in the following.

11.1.2 Plasma Polariton

Choosing the geometry $\boldsymbol{B} = (0,0,B)$ and $\boldsymbol{E} = \boldsymbol{E}_0\exp[i(k_x x - \omega t)]$, (11.4) gives

$$k_x^2 = \frac{\omega^2}{c^2}\varepsilon_V \quad \text{with} \quad \varepsilon_V = \varepsilon_1 - \frac{\varepsilon_2^2}{\varepsilon_1}. \quad (11.5)$$

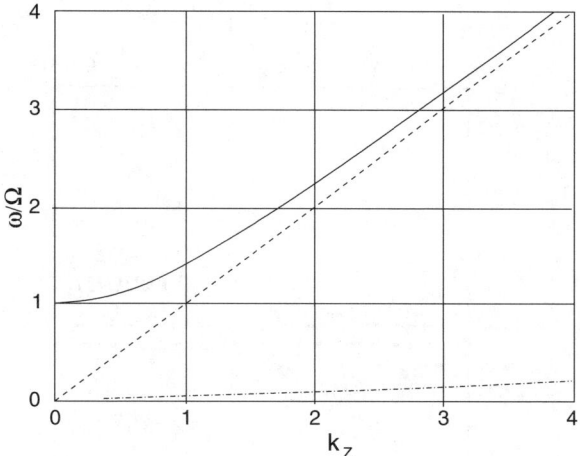

Fig. 11.1. Dispersion spectra for collective excitations of volume plasma–polaritons, ω/Ω vs k_z. Full line polariton dispersion $\omega/\Omega = \sqrt{(1 + k_z^2 c^2/\varepsilon \Omega^2)}$, dashed lines electromagnetic wave $\omega/\Omega = ck_z/(\varepsilon\Omega)$ and sound wave dispersion $\omega = v_s k_z$. For clarity, the ratio v_s/c is exaggerated

This is the dispersion equation of the volume plasma polaritons. For $B = 0$ we get

$$k^2 = \left(\frac{\omega^2}{c^2}\right)\varepsilon_1 = \frac{\omega^2}{c^2}\varepsilon_\infty \left(1 - \frac{\Omega_p^2}{\omega^2}\right). \quad (11.6)$$

For $\omega < \Omega_p$, the wave vector k is imaginary (no propagating wave) and for $\omega > \Omega_p$ there is a propagating wave which for large ω approaches the electromagnetic dispersion $k = \omega/(c\sqrt{\varepsilon_\infty})$. In Fig. 11.1 we show the dispersion spectrum $\omega(k)$ for the volume plasma-polariton. Here the damping is neglected ($\omega\tau \gg 1$). The magnetic field dependence for this geometry shall not concern us. The electromagnetic wave and the sound wave dispersion, together with the polariton mode, is indicated in the figure. We see from it that with acoustic waves, $\omega = v_s k$, we do not have any coupling to the plasma polaritons.

It is possible, however, to have a coupling of plasma waves or plasma polaritons with optical phonons. This is shown in Fig. 11.2 for the case of n-type GaAs (Mooradian et al. [11.4]). Plotted are the undisturbed longitudinal and transverse optical phonon frequencies ω_{LO}, ω_{TO} and the unperturbed plasma frequency together with the plasma–polariton branches as a function of the square root of carrier concentration.

Because long wavelength acoustic phonons do not couple to these plasma polaritons, this topic will not be further discussed.

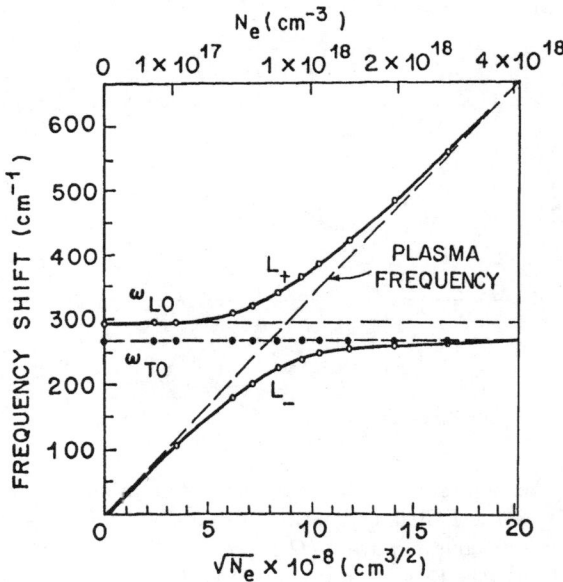

Fig. 11.2. Frequency shifts measured in light scattering spectra of coupled plasmon LO-phonon modes of n-type GaAs (Mooradian et al. [11.4])

11.1.3 Helicons and Alfven Waves

We take the geometry $\boldsymbol{k} = (0, 0, k_z)$ and $\boldsymbol{B} = (0, 0, B_z)$, i.e. $\boldsymbol{B} \parallel \boldsymbol{k}$. Then with $E_\pm = E_x \pm i E_y$ and again $\omega \tau \gg 1$, (11.4) has solutions for E_z and $E_\pm$:

$$E_z : \omega = \Omega_p$$
$$E_+ : k_z^2 = \left(\frac{\omega^2}{c^2}\right) \varepsilon_\infty \left[1 - \frac{\Omega_p^2}{\omega}(\omega - \omega_c)\right] \quad (11.7)$$
$$E_- : k_z^2 = \left(\frac{\omega^2}{c^2}\right) \varepsilon_\infty \left[1 - \frac{\Omega_p^2}{\omega}(\omega + \omega_c)\right].$$

For $\omega < \omega_c < \Omega_p$, the E_+ mode is a propagating mode, but for $\omega_c < \omega < \Omega_p$ the E_+ is a non-propagating mode. For $\omega > \Omega_p > \omega_c$, the mode is again propagating. Likewise for $\omega < \omega_c < \Omega_p$ and for $\omega_c < \omega < \Omega_p$ the E_- mode is non-propagating. The propagating E_+ mode is shown in Fig. 11.3a. For very small frequencies $\omega \ll \omega_c$ the dispersion relation for the helicon mode is

$$\omega = \omega_c k_z^2 \frac{c^2}{\varepsilon_\infty} \Omega_p^2. \quad (11.8)$$

As an electromagnetic propagating wave in a metallic medium, this $\omega(k_z)$ is rather unusual. The phase velocity is very small and will be discussed below.

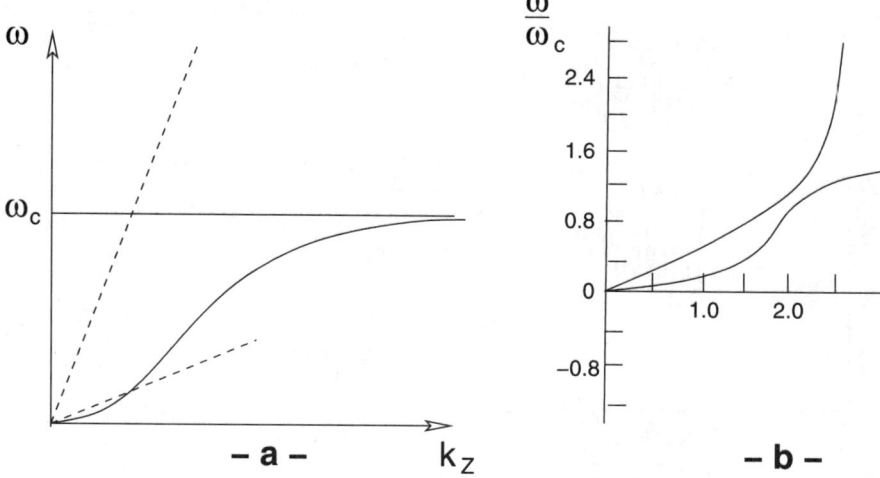

Fig. 11.3. (a) Helicon mode: Full line is the helicon dispersion spectrum (11.7). Also indicated are ω_c (light solid line), the electromagnetic spectrum $\omega = cq$ and the sound dispersion $\omega = v_s k$ (both dotted lines), (b) coupled helicon-phonon dispersion (see text)

The dispersion spectrum has the same structure as for free electrons. The helicon modes were discovered in the metal sodium (Bowers et al [11.5]) and studied subsequently in In, Al and Cu (Cotti et al. [11.6]). Many experiments in semiconductors were performed in the microwave region. For a review of these latter experiments, see e.g. Brazis et al. [11.7].

For compensated metals, with the number of electrons equal to the number of holes ($n_e = n_h$), analogous electromagnetic waves exist — the so-called Alfven waves. For the solid state they were discovered in the semimetal Bi (Buchsbaum and Galt [11.8]) and thoroughly analysed for a good metal Ga by Bezuglyi et al. [11.9]. The dispersion law for Alfven waves is $\omega = v_A k$, which is quite different to the case of helicons (11.8). The wave velocity is quite large, for Ga at $T < 4\,\text{K}$ $v_A \sim 10^7 \text{cm/s}$ which is only a factor of 7 smaller than the Fermi velocity v_F. In addition, v_A has a strong magnetic field dependence. Since the geometry for Alfven waves is the same as for helicons, $\boldsymbol{B} \parallel \boldsymbol{k}$, the Alfven wave–phonon interaction is small because the dispersion relation is the same for both Alfven and sound waves but with different velocities $v_S < v_A$. This is unlike the case of helicon–phonon interaction discussed next in Sect. 11.1.4. Therefore, the Alfven waves will not be further discussed.

11.1.4 Helicon–Phonon Interaction

For small frequencies we get a resonant helicon–phonon interaction at the intersection of the helicon dispersion law, which is proportional to k^2, and the

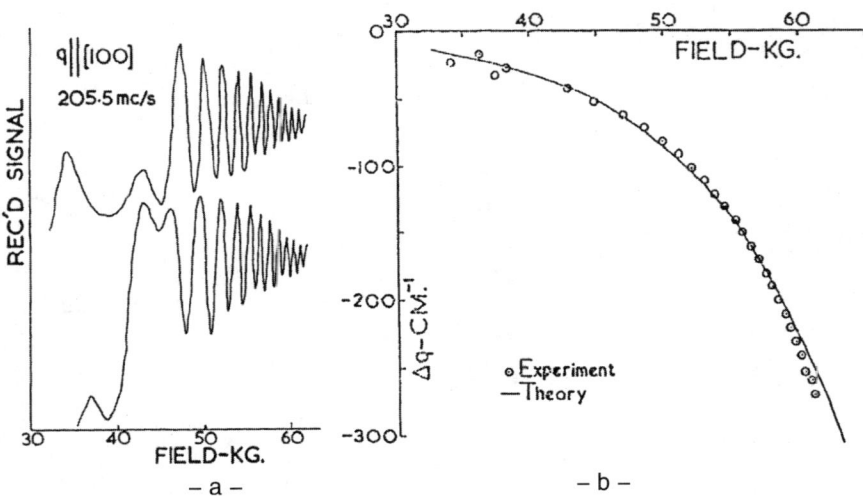

Fig. 11.4. Helicon–phonon interaction in K. (a) Longitudinal field dependence for 205.5 MHz shear waves, (b) field variation of the difference in Δq of the left and right handed circularly polarized sound waves (Blaney [11.12]).

linear sound wave dispersion, see Fig. 11.3b. This means that the conditions $\omega_{ph} = \omega_h$ and $k_{ph} = k_h$ must be fulfilled, where the indices ph stands for phonon and h for helicon. This gives with (11.8): $v_s k_{ph} = \omega_c k_h^2 c^2 / \varepsilon_\infty \Omega_p^2$ or solving for the magnetic field B_c

$$B_c = v_s^2 \frac{4\pi n e}{\omega} c . \tag{11.9}$$

With $v_s \sim 10^5 \text{cm/s}$, $\omega/2\pi = 100 \text{ MHz}$ and $n = 10^{22} \text{cm}^{-3}$ we get $B_c \sim 3.2 \text{T}$. Helicon–phonon resonance was first studied with rf-oscillators and pick up coils at 20–30 MHz in potassium (Grimes and Buchsbaum [11.10]). Later on, it was investigated using ultrasonics (Rosenman [11.11], Blaney [11.12]). Since the transverse sound waves have electromagnetic fields which can interact with the helicon waves, this results in a coupled dispersion spectrum. It reads (Quinn and Rodriguez [11.13], Langenberg and Bok [11.14]):

$$\left(k_z^2 - \Omega_p^2 \frac{\omega}{\omega_c} c^2\right)\left(k_z^2 - \frac{\omega^2}{v_s^2}\right) = n\, m\, \omega \omega_c \frac{k_z^2}{\rho v_s^2} . \tag{11.10}$$

Here, circularly polarized shear waves couple to the helicon wave with the same polarization, thus giving the coupled mode spectrum as indicated in Fig. 11.3b. Equation (11.10) has the usual form of a resonant interaction between coupled modes $(\omega^2 - \omega_{ph}^2)(\omega^2 - \omega_h^2) = f(\omega, \omega_c, k)$ where ω_{ph}, ω_h are the sound wave and helicon frequencies respectively.

Figure 11.4 shows ultrasonic experimental data for potassium (K) exhibiting this coupling effect (Blaney [11.12]). Figure 11.4a gives the amplitude modulation of ultrasonic waves using linearly polarised AC–transducers. As in the Faraday effect, the linearly polarized wave splits into two oppositely circularly, polarized waves. One of them couples to the helicon mode, the other not. Therefore one has a continuous rotation of the polarization direction. As in the Faraday effect, it is possible to plot $\Delta k = k_+ - k_-$ as given in Fig. 11.4b (see (13.10)). The critical field B_c (11.9) in this experiment was estimated to be $B_c \sim 7\mathrm{T}$, which was not quite reached. The agreement between theory (11.10) and experiment (Fig. 11.3b) is nearly perfect.

We have given a somewhat simplified version of plasmons, helicons and their interaction with phonons. The dielectric tensor was written down in the free electron approximation. More sophisticated treatments can be found in the cited literature and in Abrikosov [11.15], Quinn [11.16]. Helicon–phonon effects are discussed in greater detail by Gudkov and Gavenda [11.17].

The analogue of helicon waves in the case of compensated metals, the so-called Alven waves as discussed above in Sect. 11.1.3 is not covered, because of the reasons given there and because to our knowledge no ultrasonic study has been made for the geometry $\boldsymbol{B} \parallel \boldsymbol{k}$ of these waves. In addition, similar to the case of plasma–polaritons, the case of optical phonon–polaritons is not discussed further since it is likewise irrelevant for ultrasonic waves. Polaritons are met again in the form of surface polaritons when discussing the symmetry aspects of non-reciprocal effects (Sect. 13.5).

11.2 Magneto-elastic Waves

The interaction of spinwaves and phonons results in new collective modes, the magneto-elastic waves. They can occur in ferromagnets, ferrimagnets and anti-ferromagnets. For our purposes, we are only interested in effects with sound waves. Therefore, the excitations are only treated for long wavelengths and small wave vectors and thus use a continuum approach. A treatment, however, will be given of spinwaves with dipole interaction included. First ferromagnets and ferrimagnets will be covered and later on anti-ferromagnets (Sect. 11.2.7). Only insulators are dealt with.

11.2.1 Ferromagnetic Spinwaves with Dipolar Interaction

The well-known theory by Herring and Kittel [11.18] is presented here. It takes in the framework of a continuum approach of the equation of motion for the magnetization

$$\frac{\mathrm{d}\boldsymbol{M}}{\mathrm{d}t} = \alpha[\boldsymbol{M} \times \nabla^2 \boldsymbol{M}] + \gamma[\boldsymbol{M} \times \boldsymbol{H}], \qquad (11.11)$$

with $\boldsymbol{H} = \boldsymbol{H}_0 + \boldsymbol{H}_d$ and $\boldsymbol{H}_0$ the externally applied field and $\boldsymbol{H}_d$ the demagnetising field. α is proportional to the exchange constant J (5.6) and the exchange torque term is a development up to second order in $M(r)$. γ is the magneto-mechanical factor $\gamma = g\mu_B/\hbar$ (see App. A). With $M = M_0 \exp i(kr - \omega t)$ and $H = (0, 0, H)$ (11.11) results in

$$\omega = \alpha M_0^z k^2 + \gamma H \quad \text{with} \quad M_0 = (M_x, M_y, 0). \tag{11.12}$$

In this expression, the demagnetising field is only included statically in H with $H_d = -NM_0$. Introducing the demagnetising field, also dynamically, due to dipolar interaction with the oscillating M results in

$$\frac{dM}{dt} = \alpha \left[M \times \nabla^2 M \right] + \gamma [M \times H] + \gamma [M \times h],$$

with $M = M_z + m$ so that $\text{rot}\, h = 0$ and $\text{div}\, b = \text{div}(h + 4\pi m) = 0$. Since H and M_z are constant in space we have the linearised system of equations

$$\begin{aligned}
\frac{dm_x}{dt} &= -\alpha M_z \nabla^2 m_y + \gamma H_z m_y - \gamma M_z h_y \\
\frac{dm_y}{dt} &= \alpha M_z \nabla^2 m_x - \gamma H_z m_x + \gamma M_z h_x \\
[\nabla \times h] &= 0 \\
\nabla(h + 4\pi m) &= 0.
\end{aligned} \tag{11.13}$$

With the input $m = m_0 \exp i(kr - \omega t)$ and $h = h_0 \exp i(kr - \omega t)$, we get for the dipole field $h = -4\pi(m \cdot k)k/k^2$. With $k_x^2 + k_y^2 = k^2 \sin^2 \Theta$ where Θ is the angle between k and the z-axis, the dispersion spectrum reads

$$\omega^2 = (\gamma H_z + \alpha M_z k^2)(\gamma H_z + \alpha M_z k^2 + 4\pi \gamma M_z \sin^2 \Theta). \tag{11.14}$$

This dispersion spectrum is shown in Fig. 11.5. In contrast to (11.12). there is a spinwave band spanned by the limiting values for $k \perp M_z$ ($\Theta = \pi/2$) and $k \parallel M_z$ ($\Theta = 0$). Ferromagnetic resonance experiments probe this spectrum for $k = 0$. Depending on whether these resonance experiments are performed with a disk and the field in the plane of the disk or perpendicular to the disk, a $k = 0$ magnon can be exited at the upper or lower end of the $k = 0$ gap. Also, the resonance linewidth depends strongly on this dipolar spinwave band since the $k = 0$ mode can be degenerate with the $k \neq 0$ spinwaves. For more details on this interesting field of ferromagnetic or ferrimagnetic resonance and relaxation we refer to the specialized literature such as Clogston et al. [11.19], Sparks [11.20], White [11.21].

This spinwave band presented in Fig. 11.5 is not only valid for ferromagnets, but also for ferrimagnets. In this latter case, there is in addition an optical spinwave mode which gives rise to the exchange resonance in the mm wave region. This does not concern us here when dealing with ultrasound $\leq$Ghz. Most of the experiments with the long wavelength spinwaves in insulating substances have been carried out in Yttrium Iron Garnet Garnet ($Y_3Fe_5O_{12}$) abbreviated as YIG. We will discuss this substance and some of the relevant experiments below in Sect. 11.2.5.

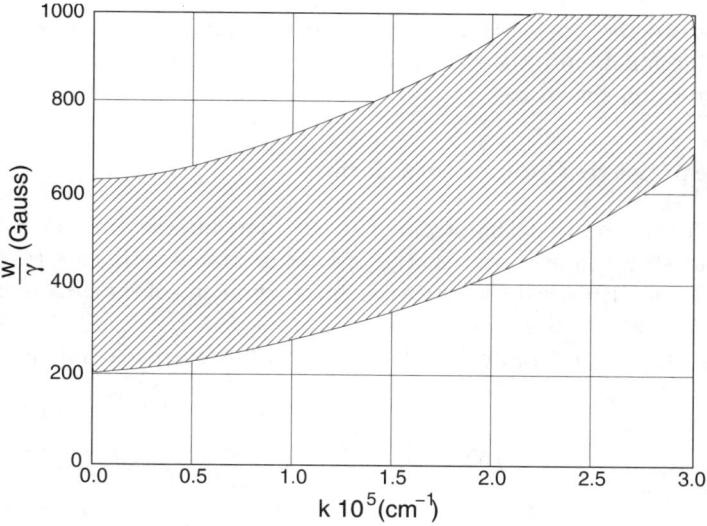

Fig. 11.5. Spinwave band for a ferromagnet with dipolar interaction, (11.14). The parameters are taken for YIG at room temperature: $M = 139$G, exchange $D = \alpha M/\gamma = 5.2 \times 10^{-9}$Gcm^2

11.2.2 Magneto-static Modes

For small wave vectors and wavelengths of the size of the sample, the mode spectrum depends on the sample shape. These are the magneto-static modes, often called Walker modes (Walker [11.22]). These modes can be excited in ferromagnetic resonance with inhomogeneous fields. In (11.11), the exchange torque can be neglected resulting only in $\mathrm{d}\boldsymbol{M}/\mathrm{d}t = \gamma[\boldsymbol{M} \times \boldsymbol{H}]$. With the static field along the z-axis, the linearized equation for m_x, m_y give for $\boldsymbol{B} = \boldsymbol{H} + 4\pi\boldsymbol{M}$ and $\boldsymbol{H} = H_z + \boldsymbol{h}$

$$\boldsymbol{B} = \begin{pmatrix} \mu & \kappa & \\ -\kappa & \mu & \\ & & 1 \end{pmatrix} \boldsymbol{H},$$

with $\mu = 1 - 4\pi\gamma^2 M_z H_z/\{\omega^2 - (\gamma H_z)^2\}$ and $\kappa = 4\pi\gamma M z i\omega/\{\omega^2 - (\gamma H_z)^2\}$.
Introducing the potential Φ with $\boldsymbol{h} = \mathrm{grad}\Phi$ gives

$$\mu\left[\frac{\partial^2 \Phi}{\partial x^2} + \frac{\partial^2 \Phi}{\partial y^2}\right] + \frac{\partial^2 \Phi}{\partial z^2} = 0$$

and $\Delta\Phi = 0$ for the region outside the sample. The eigenvalues of these equations give the magneto-static resonances for the various sample geometries. For larger $\boldsymbol{k}$, they merge into the spinwave branches. The magneto-static region extends for small k to about $k \leq 10^2$cm^{-1} which is shown in Fig. 11.5

very close to the origin. It goes over into the spinwave region without exchange and finally into the spinwave region with exchange which exhibits the $\omega(k)$ dispersion shown in Fig. 11.5.

11.2.3 Spinwave–Phonon Interaction

Section 5.1 introduced the strain–magnetic ion interaction, especially the single ion–strain interaction. In the ordered ferromagnetic or ferrimagnetic region, the quadrupole operators with the components of the bulk magnetization can be expressed as, e.g., $O_{xy} = n(J_x J_y + J_y J_x)/2 = M_x M_y$. Therefore the magneto-elastic free energy density contribution reads thus for a cubic crystal

$$F_{me} = \frac{b_1}{M_0^2}(M_x^2 \varepsilon_{xx} + M_y^2 \varepsilon_{yy} + M_z^2 \varepsilon_{zz}) \qquad (11.15)$$
$$+ \frac{b_2}{M_0^2}(M_x M_y \varepsilon_{xy} + M_x M_z \varepsilon_{xz} + M_y M_z \varepsilon_{yz})$$

with b_i the magneto-elastic coupling constants. For $M_z \cong M_0$, the saturation magnetization, we get the linearized magnon–phonon interaction term

$$F'_{me} = \frac{b_2}{M_0}(M_x \varepsilon_{xz} + M_y \varepsilon_{yz}) . \qquad (11.15a)$$

And with the effective field $\boldsymbol{H}_{eff} = -\mathrm{d}F'_{me}/\mathrm{d}\boldsymbol{M}$ and the effective stress $T_{ij} = \mathrm{d}F'_{me}/\mathrm{d}\varepsilon_{ij}$, results in the coupled equations of motion for the spinwaves and phonons. Instead of the general case, we treat two special cases with a) $M_z \parallel \boldsymbol{k}$ and b) $M_z \perp \boldsymbol{k}$.

a) Case of $k_z =$ propagation direction ($\boldsymbol{k} \parallel M_z$)
With $\varepsilon_{xz} = \frac{1}{2}\frac{\partial u_x}{\partial z}$ and $\varepsilon_{yz} = \frac{1}{2}\frac{\partial u_y}{\partial z}$ the equations of motion read with (11.13) and (3.6)

$$\begin{aligned} \frac{\mathrm{d}M_x}{\mathrm{d}t} &= \gamma H_z M_y + \gamma \frac{b_2}{2}\frac{\partial u_y}{\partial z} \\ \frac{\mathrm{d}M_y}{\mathrm{d}t} &= -\gamma H_z M_x - \gamma \frac{b_2}{2}\frac{\partial u_x}{\partial z} \\ \frac{\mathrm{d}^2 u_x}{\mathrm{d}t^2} &= v_t^2 \frac{\partial^2 u_x}{\partial z^2} + \frac{b_2}{\rho M_0}\frac{\partial M_x}{\partial z} \\ \frac{\mathrm{d}^2 u_y}{\mathrm{d}t^2} &= v_t^2 \frac{\partial^2 u_y}{\partial z^2} + \frac{b_2}{\rho M_0}\frac{\partial M_y}{\partial z} \end{aligned} \qquad (11.16)$$

and the longitudinal mode u_z does not couple. The exchange term has been neglected. The system of equations can be decoupled by introducing the circular magnetization $M^\pm = M_x \pm M_y$ and the circular displacement components $u^\pm = u_x \pm u_y$:

11.2 Magneto-elastic Waves

$$\frac{dM^+}{dt} = -i\gamma H_z M^+ - i\gamma b_2 \frac{\partial u^+}{\partial z}$$
$$\frac{dM^-}{dt} = i\gamma H_z M^- + i\gamma b_2 \frac{\partial u^-}{\partial z}$$
$$\frac{d^2 u^+}{dt^2} = v_t^2 \frac{\partial^2 u^+}{\partial z^2} + \frac{b_2}{\rho M_0} \frac{\partial M^+}{\partial z} \quad (11.16a)$$
$$\frac{d^2 u^-}{dt^2} = v_t^2 \frac{\partial^2 u^-}{\partial z^2} + \frac{b_2}{\rho M_0} \frac{\partial M^-}{\partial z}.$$

With plane wave expressions for $M^\pm$ and $u^\pm$, the coupled magneto-elastic dispersion equation are given by

$$(\omega \pm \omega_m)(\omega^2 - v_t^2 k^2) \pm s v_t^2 k^2 = 0, \quad (11.17)$$

with $\omega_m = \gamma H_z + \alpha M_z k^2$ and the coupling constant $s = \frac{\gamma b_2^2}{2} \rho M_0 v_t^2$. ω can be solved by

$$\omega = \frac{\omega_m + v_t k_z}{2} \pm \left[\frac{(\omega_m - v_t k)^2}{4} + s v_t^2 \frac{k_z^2}{2} \omega_m \right]^{1/2}. \quad (11.17a)$$

For this geometry, the dipolar contributions disappear ($\Theta = 0$), it is the so-called Faraday geometry (see Sect. 13.3). In Fig. 11.6 we show the spinwave, phonon and magneto-elastic wave branches using equations (11.14) and (11.17).

b) Case of k_x = propagation direction ($M_z \perp \mathbf{k}$)
In this case the equations of motion read

$$\frac{dM_x}{dt} = \gamma H_z M_y$$
$$\frac{dM_y}{dt} = -\gamma (H_z + 4\pi M_z) M_x - \frac{\gamma b_2}{2} \frac{\partial u_z}{\partial x}$$
$$\frac{d^2 u_x}{dt^2} = v_l^2 \frac{\partial^2 u_x}{\partial t^2} \quad (11.18)$$
$$\frac{d^2 u_y}{dt^2} = v_t^2 \frac{\partial^2 u_y}{\partial t^2}$$
$$\frac{d^2 u_z}{dt^2} = v_t^2 \frac{\partial^2 u_z}{\partial x^2} + \frac{b_2}{\rho M_0} \frac{\partial M_x}{\partial x}.$$

In this case only the displacement component u_z couples to the spinwaves. The medium is birefringent. The dispersion spectrum reads

$$(\omega^2 - v_t^2 k_x^2)(\omega^2 - \omega_m^2) = \gamma H_z s v_t^2 k_x^2 \quad (11.18a)$$

with $\omega_m^2 = (\gamma H_z + \alpha k_x^2)(\gamma H_z + \alpha k_x^2 + 4\pi \gamma M_0)$.
In this case the dipolar contribution is noticeable with $\Theta = \pi/2$. It is the so-called Voigt–Cotton–Mouton geometry which will be discussed later in Sect. 13.3.

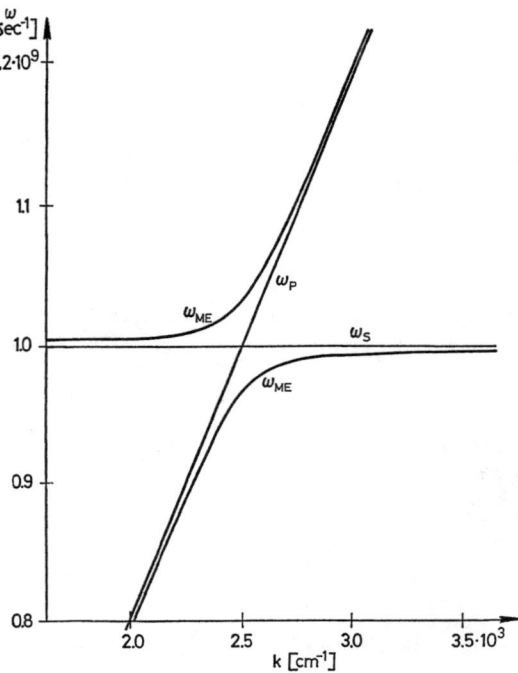

Fig. 11.6. Magneto-elastic dispersion spectrum ω_{ME} versus k (11.17a) together with the spinwave mode ω_S and the acoustic mode ω_P. The parameters are taken for YIG at room temperature: $M = 139G$, $\gamma = 1.76 \times 10^7$ $(Gs)^{-1}$, $b_2 = 6.96 \times 10^6 \text{erg/cm}^3$, $\rho = 5.17 \text{g/cm}^3$, $c_{44} = 7.64 \times 10^{11} \text{erg/cm}^3$, $H_z = 200G$

These coupled spinwave–phonon modes were first investigated theoretically by Kittel [11.23] and by Vlasov and Ishmukhametov [11.24]. The derivation of the coupled spinwave–phonon interaction, as well as the spinwave modes of (11.12) and (11.14) did not include any magnetic anisotropy terms. The derivation holds for isotropic ferromagnets. Corresponding equations, analog to (11.12) and (11.14) including magnetic anisotropy of the easy axis or easy plane type for ferromagnets, anti-ferromagnets, rare earth metals etc. can be found, e.g., in "Spinwaves and magnetic excitations" by Borovik-Romanov and Sinha [11.25].

11.2.4 Magneto-elastic Gap

In cases of strong magneto-elastic coupling constants, there can be a further modification of the equations (11.17a) and (11.18a) above. The magneto-elastic energy, as in (11.15), leads to static equilibrium states for m_0 and for the strain ε_{ij}^0. In this case the dynamical treatment of the coupled equations of motion involve magnetization and strain oscillations around m_0 and ε^0, i.e. for $\Delta m = m - m_0$ and $\Delta \varepsilon_{ij} = \varepsilon_{ij} - \varepsilon_{ij}^0$. Therefore there will be renormalization

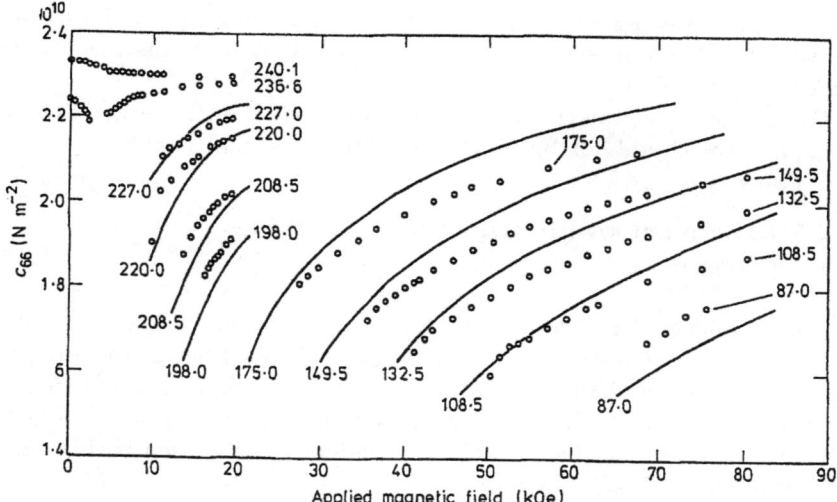

Fig. 11.7. c_{66} in Tb at various temperatures determined from the ultrasonic waves propagating parallel to the field applied along the hard a axis. The full curves show the calculated soft-mode behaviour of c_{66} for the various temperatures. A few examples of the field dependence of c_{66} observed above $T_N = 230\,\mathrm{K}$ are also given (Jensen and Palmer [11.28]).

terms involving m_0 and ε^0. For the uniform spinwave mode in a ferromagnet, a magneto-elastic contribution for the spin gap frequency $\omega_{me} = \gamma b^2/m_0 c$ is obtained. For YIG, with the parameters quoted in Fig. 11.6, $\omega_{me} = 6 \times 10^7 s^{-1}$ which is negligible compared to the spinwave gap of $10^9 s^{-1}$ shown in this figure. The same is true for the orthoferrites like ErFeO$_3$ near the reorientation with magneto-elastic coupling constants of the same order as YIG, as discussed in Sect. 6.2. On the other hand, the giant magneto-striction compound TbFe$_2$, discussed in Sects. 5.4 and 13.3.1, has $b \sim 10^8 \mathrm{erg/cm}^3$ which leads to observable magneto-elastic spinwave gaps. Such experiments have not been performed so far in TbFe$_2$.

The most interesting effects have been observed for the heavy rare earth metal Tb. Apart from an anti-ferromagnetic transition at $T_N = 230\,\mathrm{K}$, Tb has a further transition from a helical spin arrangement to a basal plane ferromagnet at $T_C = 220\,\mathrm{K}$. The strong magneto-striction leads to a magneto-elastic spinwave gap which has been observed with inelastic neutron scattering (Nielsen et al. [11.26]). For a theoretical treatment of these effects see e.g. Chow and Keffer [11.27].

The basal plane elastic constant c_{66}, with strain ε_{xy}, is also strongly affected in this case of strong magneto-elastic interaction. For magnetic fields applied in the hard direction a-axis, the c_{66} mode softens towards the critical field H_c. This field is needed to change the magnetization from the easy b-axis to the hard a-axis. Figure 11.7 shows an experiment and the calculation for this softening effect in the ferromagnetic phase of Tb (Jensen and Palmer [11.28]). Strong ultrasonic attenuation near H_c prevents the observation of the full softening.

For reviews concerning the magneto-elastic gap see e.g. Turov and Shavrov [11.29] and the cited book by Borovik-Romanov and Sinha [11.25].

11.2.5 Experiments with Magneto-elastic Waves in a Ferrimagnet

Some experiments on magneto-elastic waves will be shown in Sect. 13.3. These are the magneto-acoustic analoga of magneto-optical effects, the so-called Faraday and Cotton–Mouton–Voigt effects. Here we show another aspect of magneto-elastic waves, the coherent excitation of spinwaves and magneto-elastic waves (Eshbach [11.30]). These experiments have been performed in the ferrimagnet $Y_3Fe_5O_{12}$ (YIG).

In the cubic garnet lattice $Y_3Fe_5O_{12}$ (YIG), the unit cell contains 8 molecules. In Fig. 11.8, the garnet structure is shown (Pauthenet [11.31]).

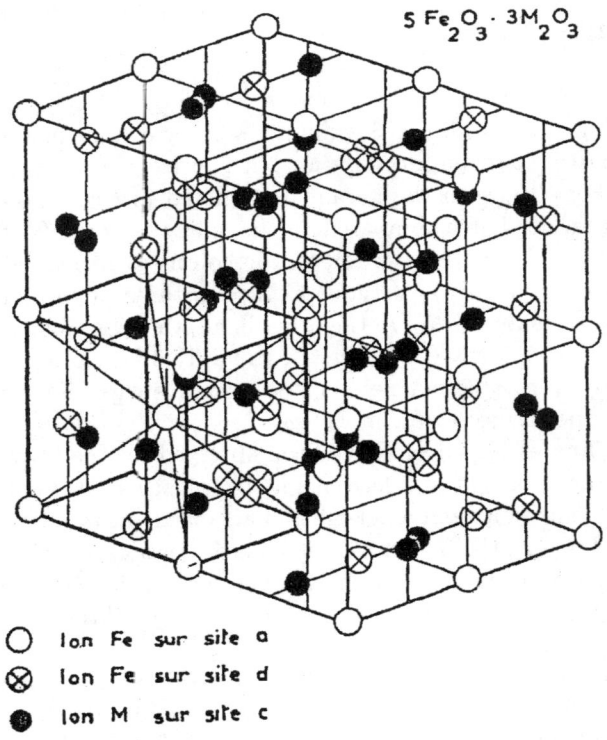

Fig. 11.8. Structure of the cubic garnet $Y_3Fe_5O_{12}$. Open circles Fe^{3+} on a sites, x Fe^{3+} on d-sites, closed circles Y ion on c sites (Pauthenet [11.31])

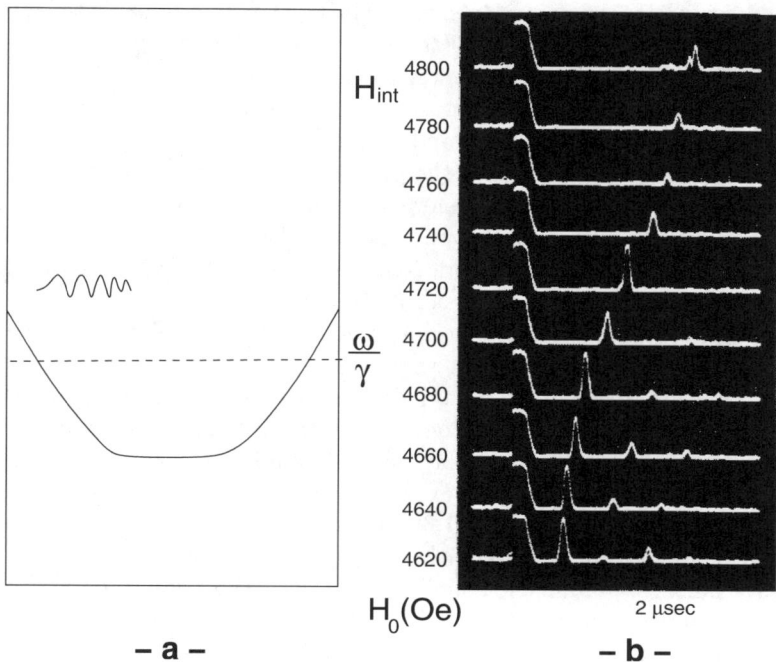

Fig. 11.9. (a) Effective magnetic field across the disk with applied microwave frequency ω/γ. Indicated is also the propagating decaying spinwave at the crossover field, (b) oscilloscope traces showing 9420 Mc magneto-elastic wave echoes at various applied field values (Eshbach [11.30])

The Fe^{3+} ions are situated on 24 tetrahedral d-sites and 16 octahedral a sites. The Y^{3+} sit on 24 dodecahedron c-sites. With the strongest exchange J_{ad} between the two a- and d-sites, there are two sublattice collinear ferrimagnet with a Curie temperature of $T_C = 570$ K. This crystal has small magnetocrystalline anisotropy and a small ferromagnetic resonance linewidth making it very suitable for dynamical experiment with spinwaves (Sparks [11.20]) as shown below and in Sect. 13.3.

In the spinwave experiment, the excitation mechanism of spinwaves is via dipolar coupling in nonuniform magnetic fields due to demagnetizing effects (Schlömann [11.32], [11.33]). In a disk with magnetic field perpendicular to the plane of the disk, the demagnetizing field is approximately constant $4\pi M$ in the inner part of the disk and it diminishes towards the edge down to approximately $2\pi M$. Thus the net internal field $H_{int} = H - N(x)M$ is a function of the radial distance from the disk center and is stronger near the edge. The spinwave has a dispersion law given by (11.14) with $\Theta = \pi/2$. With $\omega/2\pi \sim 9$ GHz, microwave radiation spinwaves are excited at the crossover of ω/γ and the internal field (see Fig. 11.9a). The spinwaves are accelerated towards the disk center and de-accelerated on approaching the opposite edge where they radiate the detection signal. The transit time signal is the integral $T = \int ds/v_g$ with v_g the group velocity. It depends sensitively on the applied field (Fig. 11.9b). The experimental realisation is due to Eshbach [11.30], [11.34]. Best agreement for the transit time is achieved by taking the magneto-elastic wave spectrum (Fig. 11.6 and (11.18a)).

This experiment is similar to the propagation of a particle in a box. The internal field gives the potential energy, the difference between ω/γ and $\bar{H}_{int}$ gives the kinetic energy and, beyond the crossing point, the decaying wavefunction couples to the dipole field. Near the cross over of the spinwave spectrum and the sound wave dispersion both spectra are changed to magneto-elastic waves (Fig. 11.6) which changes the time T–integral given above radically. This is the first experiment which demonstrates spinwave propagation directly. A short review of this and similar experiments is given by Damon and van de Vaart [11.35]. Technological applications of these magneto-static–magneto-elastic waves as, e.g. delay lines, have not yet been realized.

Similar spinwave propagation experiments were recently performed in confined geometries (Jorzick et al. [11.36], Demokritov et al. [11.2]). In micrometer size, magnetic thin permalloy ($Ni_{80}Fe_{20}$) films of different geometries, the spinwave pulses were launched similarly to the method described above, but the detection was achieved with Brillouin scattering.

11.2.6 High Power Level Effects

Another way to study magneto-elastic waves experimentally would be by applying high power level ferromagnetic resonance. In this method, a distinction is made between transverse field pumping (Suhl [11.37]) and parallel pumping (Schlömann et al. [11.38]). In the latter method, a critical field h_c can be observed above which large power absorption occurs. In linear theory (11.13) with $h_{rf} \parallel H_z$, no absorption occurs. However with increasing power in a disc shaped sample with x-z plane, the magnetization performs not a circular path but an undulating one. For one revolution of M there are two revolutions of M_z. This leads to a distinct microwave power absorption threshold at h_c, given by the balance of power input and dissipation. Physically spinwave pairs are created perpendicular to the z-axis. For more details of this effect see Sparks [11.20]. If these spinwaves are in the crossover region of the sound wave branch (see Fig. 11.6), magneto-elastic waves could be studied in this way. A theoretical study has been made for the transverse pumping case (Schlömann [11.39]) but no experimental study has been performed yet.

11.2.7 Sound Wave Experiments in Anti-ferromagnets

The analogous effects for magneto-elastic interactions in anti-ferromagnets are discussed below. Melcher and Bolef [11.40] have investigated the elastic and magneto-elastic properties of cubic RbMnF$_3$. We discussed above the critical sound wave properties of this substance near T_N in Sect. 6.1. A similar

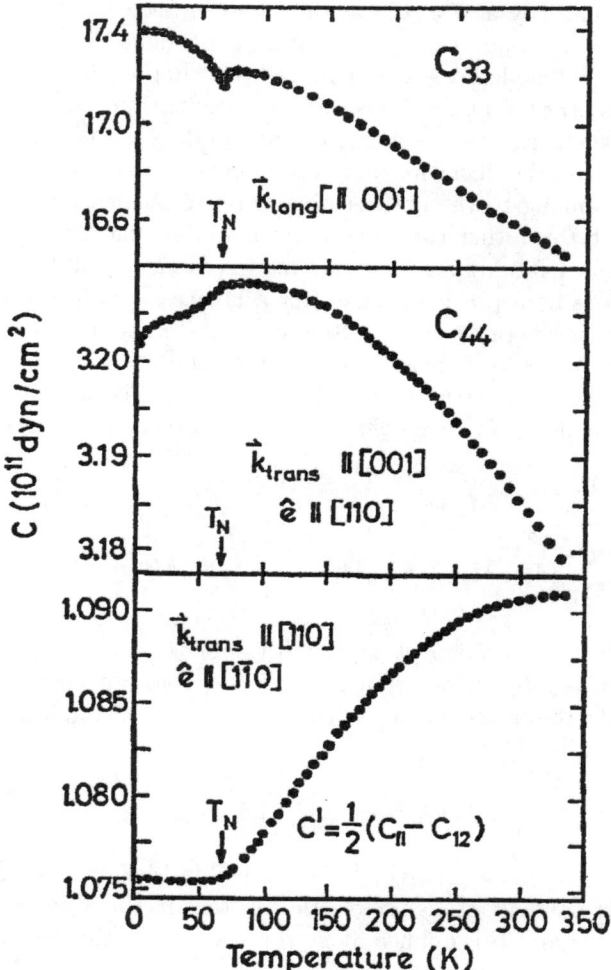

Fig. 11.10. Temperature dependence of the elastic constants for MnF_2 (Melcher [11.41])

expression for the magneto-elastic energy in cubic crystals, as used for the ferrimagnet (11.15) but now with coupling terms to the sublattice magnetization, can describe the measured effects.

Another interesting anti-ferromagnet is the tetragonal compound MnF_2. Again ultrasonic effects at the magnetic phase transition, $T_N = 67.3\,\text{K}$, were discussed in Sect. 6.1. Elastic constant measurements as a function of temperature and magnetic field were performed by Melcher [11.41, 11.42]. In Fig. 11.10 the temperature dependence of the elastic modes c_{33}, c_{44} and $(c_{11} - c_{12})/2$ are given (Melcher [11.41]). As expected from the coupling constant discussion in Sects. 5.1.3 and 6.1, the longitudinal mode exhibits a pronounced anomaly at T_N. It was discussed in Sect. 6.1.3. The transverse mode c_{44} shows an anomalous temperature dependence for $T \leq T_N$. Finally, the $(c_{11} - c_{12})/2$ mode gives a pronounced softening in the paramagnetic region which is arrested at T_N. This softening has nothing to do with the Neel phase transition, the volume magneto-striction does not couple to this mode in first order and the linear magneto-elastic interaction discussed below does not give a coupling to the anti-ferromagnetic resonance modes. It is interesting to note that other rutile structure materials like ZnF$_2$ and NiF$_2$ also exhibit similar softening for the same mode (Rimai [11.43]). This softening could therefore be a precursor effect for a lattice instability. The c_{44} mode also shows some less pronounced softening at low temperatures for ZnF$_2$, NiF$_2$ (Rimai [11.43]) and for MgF$_2$ (Rimai et al. [11.44]).

For uniaxial symmetry, with z the symmetry axis, the magneto-elastic interaction, analogously to the ferromagnetic case (11.15), reads

$$F_{me} = \frac{B}{M_0^2}[(M_{1x}M_{1z} + M_{2x}M_{2z})\varepsilon_{xz} + (M_{1y}M_{1z} + M_{2y}M_{2z})\varepsilon_{yz}] + \frac{D}{M_0^2}[(M_{1x}M_{2z} + M_{2x}M_{1z})\varepsilon_{xz} + (M_{1y}M_{2z} + M_{1z}M_{2y})\varepsilon_{yz}]$$
(11.19)

with M_{ij} the jth component of the ith sublattice ($j = x, y, z$ and $i = 1, 2$). From this interaction follows that c_{33} should not couple to the anti-ferromagnetic resonance modes whereas c_{44} gives a coupling of the form (Melcher [11.42])

$$c_{44} = c_{44}^0 - 2\frac{b^2}{K}\left[\frac{B_c^2}{B_c^2 - B^2}\right].$$
(11.20)

Here $b = B - D$, $B_c = (2B_E B_A)^{1/2} = 9.305$T in MnF$_2$ at 4K, $B_A = K/M_0$ is the anisotropy field, c_{44}^0 is the elastic constant in the absence of magneto-elastic coupling and B is applied along the easy axis of this anti-ferromagnet. Such a behaviour of the elastic mode c_{44} as a function of applied field B is observed as shown later in Sect. 13.2, Fig. 13.3. Equation (11.20) is valid for $B < B_c$ the so-called spin flop field. At B_c the elastic modes exhibit a sharp singularity in the form of a dip for c_{44} and c_{33} (Melcher [11.42]). In addition, the attenuation exhibits also a sharp spike at B_c (Shapira [11.45]).

With these techniques one can map the B–T phase diagram of an uniaxial anti-ferromagnet. This has been done for MnF$_2$ and is shown in Fig. 11.11. Applying the field along the easy c-axis at low enough temperatures can drive the spin system into a canted spin flop state. The gain in Zeeman energy outweighs the loss in exchange and anisotropy energy. The figure shows that the spin flop state is realized close to T_N.

Two mechanisms for this resonance effect at the phase boundary have been suggested. As the anti-ferromagnetic resonance mode softens towards the spin flop, a resonant spinwave–phonon coupling can occur. This effect has been considered by Tani [11.46]. Another possible mechanism is the domain

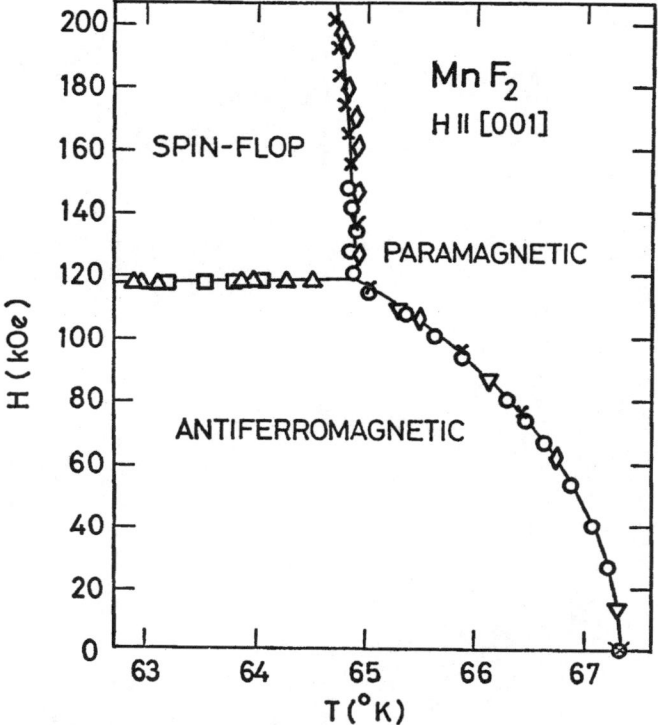

Fig. 11.11. Magnetic phase diagram of MnF_2 for $B \parallel (001)$. Data points from ultrasonic attenuation measurements (Shapira [11.45])

wall–stress effect. Note that the anti-ferromagnetic–spin flop transition is of first order. The sound attenuation in the flopped phase is rather high. Similar phase boundary measurements have been performed also in FeF_2, CoF_2, Cr_2O_3, α-Fe_2O_3, Cr, EuTe and $GdAlO_4$ (Shapira [11.45]).

Such experiments, and related thermal expansion and susceptibility experiments, can be used to investigate multi-critical points (bi, tri and tetracritical) (Fisher [11.47], Rohrer [11.48]) and the field dependence of $T_N(B)$. Such experiments have been performed on MnF_2, Cr_2O_3, $RbMnF_3$ (Shapira and Becerra [11.49]).

Another interesting experiment was carried out in the easy-plane antiferromagnet $FeBO_3$ (Svistov et al. [11.50]). Applying microwave ultrasound ($\approx$ 1GHz) changes the magnetization due to magneto-elastic interaction. These changes can be measured with a SQUID magneto-meter.

Further sound wave experiments in anti-ferromagnets will be discussed in Sect. 13.2 (rotationally invariant magneto-elastic effects) and in Sect 13.3.2 (acoustical Faraday effect).

12 Ultrasonics in Low Dimensional Spin and Electronic Peierls-Systems

In recent decades, the study of low dimensional spin systems has grown to be a very important branch of "solid state physics". This came about partly due to the discovery of "high temperature superconductivity". In these substances, the two-dimensional Cu planes, with Cu^{2+} spins $\frac{1}{2}$ and one-dimensional spin chains play a prominent role. But the interest has also arisen due to several theoretical and experimental discoveries, such as the Haldane conjecture, spin dimerization, magnetization plateaus etc. Moreover, there are conjectures predicting superconductivity through hole doping in such systems (for a review see Dagotto and Rice [12.1]). In addition, low dimensional electronic systems, so-called Peierls systems, were of interest in the fifties and sixties of the last century: especially the prediction of the Kohn anomaly triggered experiments in low dimensional electronic systems. The clearest system found in this respect was KCP (see Sect. 12.5).

In this chapter, the sound wave–spin excitation mechanism occurring in such materials will be discussed together with the various ultrasonic experiments performed in this field. Some of the experiments, relevant to this chapter, have been described already in previous chapters. Examples are the charge ordering phenomena which can occur in low dimensional spin systems and the spin–Peierls effect (Sect. 7.1). In addition, ultrasonic experiments for a Peierls system are described for layer structure materials and for a case of Bose–Einstein condensation of magnons. The discussion on surface acoustic wave experiments in the two-dimensional electron gas, exhibiting the quantum Hall effect, is postponed until Sect. 13.6.

In spite of the three dimensionality of the crystal lattice, materials can be synthesized which have spin chains in them with large exchange interactions within the chain and very small exchanges between the well-separated chains. This behaviour occurs especially for organic materials. As an example, this is demonstrated for the Ni dinuclear oxalato(ox)-bridged compound whose spin chain for the $Ni^{2+}(S=1)$ ion is clearly seen in Fig. 12.1 (Narumi et al. [12.2]). The chemical formula is abbreviated to [Ni_2(Medpt)$_2$(μ-ox)(H_2O)]. Ni^{2+} pairs, coupled by oxalate bridges, are linked through the hydrogen bond and the interactions perpendicular to the a-axis are usually so weak that this compound can be regarded as an $S = 1$ dimer spin–pair compound. We will show and discuss the magnetization curve of this compound later in Fig. 12.8.

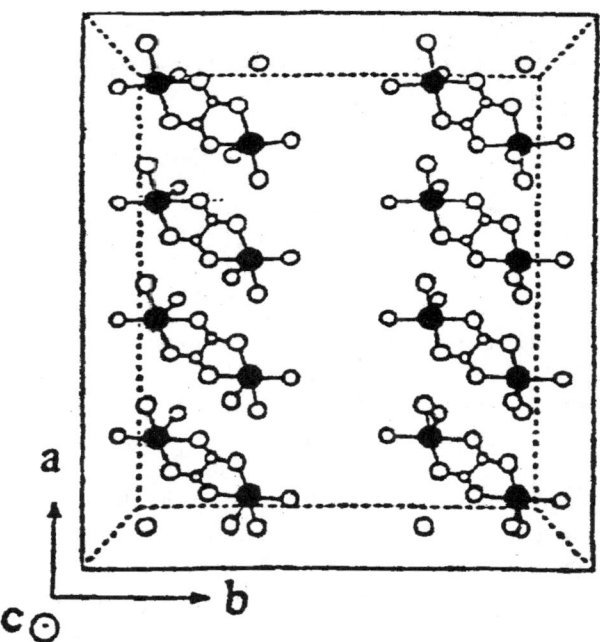

Fig. 12.1. Schematic view of [Ni$_2$(Medpt)$_2$(μ-ox)(H$_2$O)$_2$](ClO$_4$)$_2$2H$_2$O. The shaded circles represent Ni^{2+} ion sites. The coupling of Ni^{2+} pairs via oxalate bridges is shown (Narumi et al. [12.2])

The magnetic properties of some important low dimensional spin substances (Sect. 12.1) are discussed below, followed by a presentation of some ultrasonic effects, both as a function of temperature (Sect. 12.2) and magnetic field (Sect. 12.3) for such systems. The latter parameter, taken to very high values (50 Tesla), enables the investigation of interesting new effects. In Sect. 12.4, recent new experiments in thermal conductivity for such low dimensional spin systems are discussed. Section 12.5 investigates briefly the Peierls and spin–Peierls effect with some representative compounds like KCP and the an-organic spin–Peierls compound CuGeO$_3$. In Sect. 12.6, a two-dimensional electronic layered system is discussed. Finally, in Sect. 12.7, the compound TlCuCl$_3$ is covered which is claimed to have a Bose–Einstein condensation of magnons. Reviews covering the older literature are e.g. de Jongh and Miedema [12.3] and Mikeska and Steiner [12.4].

12.1 Magnetic Properties of Low Dimensional Spin Systems

12.1.1 Uniform Chain

The physics of low dimensional spin systems started early in 1931 when Bethe determined the wavefunction of the anti-ferromagnetically coupled spin $\frac{1}{2}$ chain (Bethe [12.5]). Even today, this problem is still under consideration, both experimentally and theoretically. Due to quantum fluctuations, the ground state of a uniform anti-ferromagnetic spin $\frac{1}{2}$ chain is disordered, even at $T = 0$. The calculated isotropic magnetic susceptibility of such a chain is shown in Fig. 12.2. It has a maximum at $T_{max} = 0.641J$ (Eggert et al. [12.6], Eggert [12.7], Klümper [12.8]) with J the exchange constant between nearest neighbor spins. An experimental result for this effect has been given by Takagi et al. [12.9].

The ground state of the spin $\frac{1}{2}$ chain is a singlet. The dispersion of the triplet excitations (spinon pairs) consists of a continuum between the energies

$$E_1 = \frac{\pi}{2} J \, |\sin q| \quad \text{and} \quad E_2 = \pi J \, |\sin(q/2)| \,. \tag{12.1}$$

There is no energy gap between ground state and excited states. Therefore, due to quantum fluctuations, even at $T = 0$, continuum states can be occupied leading to a finite susceptibility. The magnetization starts, for the same reason, with a finite slope in a characteristic way till saturation (Fig. 12.3).

A nice physical realization of the magnetization for a homogeneous spin $\frac{1}{2}$ chain has been given for CuPzN by Hammar et al. [12.10]. Figure 12.4 shows results for a new Cu(II) coordination–polymer, an analogous chain system. This substance, Cu(II)-bispyrazoldihydroxybenzene (abbreviated as CuCCP)

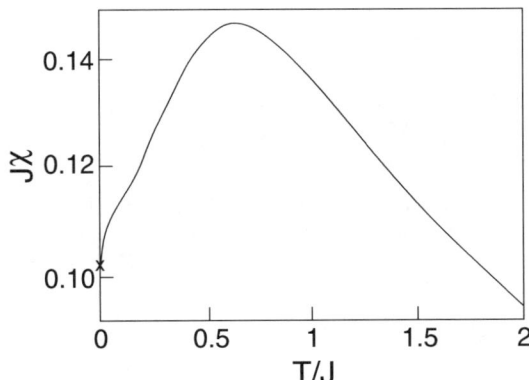

Fig. 12.2. Calculated magnetic susceptibility of an isotropic spin $\frac{1}{2}$ chain: $J\chi$ versus T/J. The symbol x indicates $J\chi(T=0)$ (Eggert [12.7])

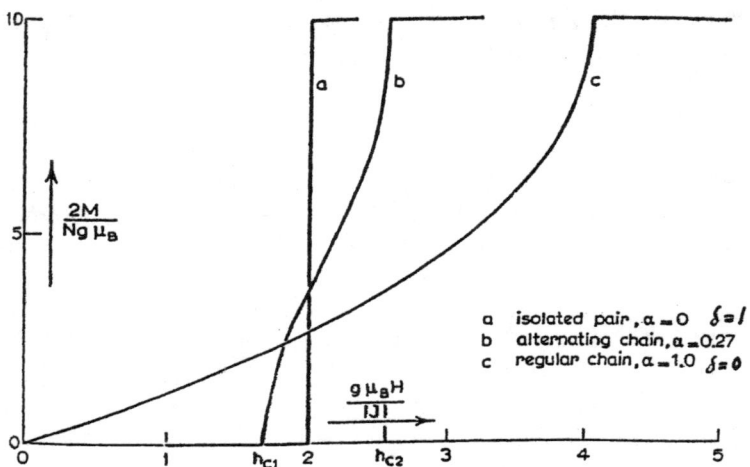

Fig. 12.3. Magnetization per site versus magnetic field at $T = 0$ for different cases (see text). (**a**) isolated pair $\alpha = 0$, $\delta = 1$, (**b**) alternating chain $\alpha = 0.27$, (**c**) regular chain $\alpha = 1$, $\delta = 0$

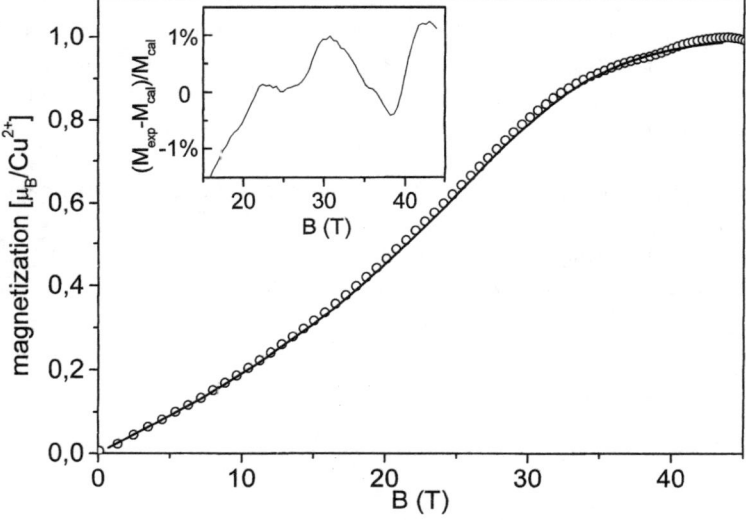

Fig. 12.4. Measured magnetization of CuCCP as a function of magnetic field for $T = 4.2\,\mathrm{K}$ together with a numerical fit. The inset shows the deviation from experimental and calculated magnetization for $T = 6.5\,\mathrm{K}$ (adapted from Wolf et al. [12.12])

has triclinic symmetry with one formula unit per unit cell. It is seen that the measured magnetization follows closely the theoretical one (Fabricius et al. [12.11]) with an exchange constant $J = 20\,\mathrm{K}$ and a corresponding saturation field of ≈ 30T (Wolf et al. [12.12]). There is some disagreement between theory and experiment only close to saturation. Note that the measurement was performed at 4.2 K in pulsed fields and the calculated curve was taken at 6.5 K due to the magneto-caloric effect (see Sect. 2.4). The deviation between experiment and theory amounts to about 1% as shown in the inset of Fig. 12.4.

12.1.2 Dimerized Chains

There are very few ideal uniform spin chain systems with isotropic exchange interactions. The majority of them are dimerized. In the uniform chain, there is only one nearest neighbour isotropic exchange constant J. For alternating exchange constants the situation is quite different. This alternating spin chain can be described by the following Hamiltonian:

$$H = J_\alpha \sum_i (S_{2i-1} \cdot S_{2i} + \alpha S_{2i} \cdot S_{2i+1})$$
$$= J_\delta \sum_i ((1+\delta) S_{2i-1} \cdot S_{2i} + (1-\delta) S_{2i} \cdot S_{2i+1}). \tag{12.2}$$

Here $J_\delta = (1+\alpha) J_\alpha/2$ and $\delta = (1-\alpha)/(1+\alpha)$. If $\alpha = 1$, $\delta = 0$, then this results in the uniform spin chain discussed above whereas $\alpha = 0$, $\delta = 1$ is the case of isolated dimers. Through the alternation, an energy gap arises between the singlet ground state and the excited triplet. With increasing alternation (smaller α) the dispersion of (12.1) decreases and the energy gaps at $q = 0$ and π increase. Likewise, the magnetic susceptibility for low temperature has the form (Bulaevskii [12.13]) $\chi = a(\alpha)/T * \exp(-J_\alpha \Delta(\alpha)/T)$ where a and Δ are α–dependent parameters. Unlike in Fig. 12.2, the susceptibility goes strongly to 0 for $T \to 0$. Also the magnetization has a different behaviour. Figure 12.3 shows the magnetization ($T = 0$) for an isolated dimer ($\alpha = 0$), for a uniform chain ($\alpha = 1$) and an alternating chain ($0 < \alpha < 1$). The uniform chain has been calculated by Eggert [12.7]. The isolated dimer has a singlet–triplet excitation gap Δ. If the lowering of the lower Zeeman branch of the triplet is equal to this gap, the magnetization attains its saturation value. The magnetization of the alternating chain lies between these two cases.

Spin dimerization can also occur through frustration although there is a qualitative difference to the alternating chain. The ground state in frustrated chains breaks translational invariance although the Hamiltonian does not. If one considers exchange between nearest neighbours (nn) and next nearest neighbours (nnn) with exchange constants J and J' respectively, results in spontaneous spin dimerization if $J'/J > 0.241$ (Haldane [12.14], Eggert [12.7]).

What are the mechanisms to generate dimerized chains? There are several possibilities: there can be a dimerization due to the crystal structure. Typical examples are Vanadyle pyro phosphate (($VO)_2P_2O_7$) or the two dimensional

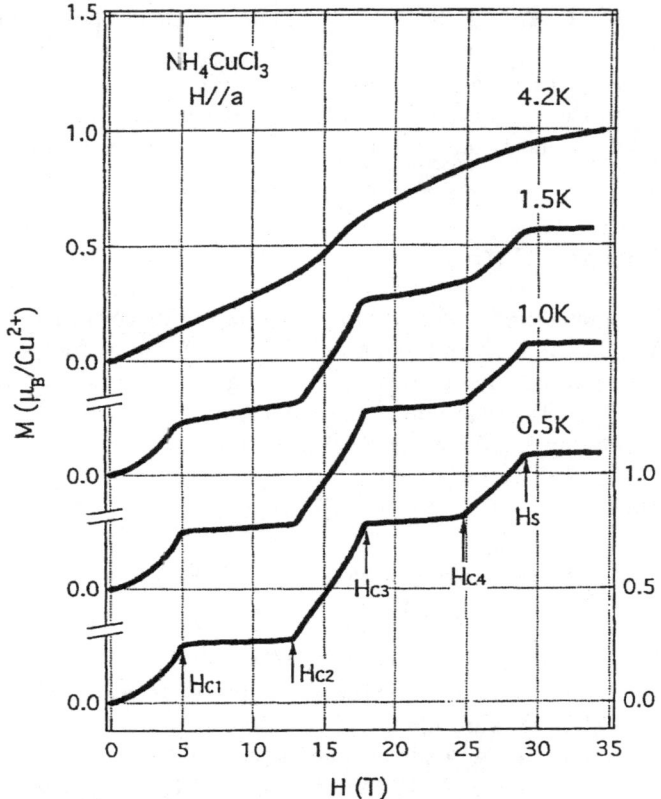

Fig. 12.5. Magnetization plateaus in NH_4CuCl_3 for different temperatures (Shiramura et al. [12.15])

dimerized system Strontium Copper Borate ($SrCu_2(BO_3)_2$). Another possibility is the spin–Peierls effect, a dimerization due to the exchange striction mechanism (see Sects. 5.1.3, 12.5). A famous example for this is $CuGeO_3$. Further still is the possibility of a spin dimerization by frustration, as mentioned above. Finally, charge ordering as discussed in Sect. 7.1 can give a dimerization.

Dimerized spin chains and spin lattices can lead to quite unexpected magnetization behaviour. The different cases illustrated in Fig. 12.3 do not exhaust all possibilities. Dimerized spin chains and spin ladders can give rise to magnetization plateaus at low temperatures. A typical example for this effect is shown in Fig. 12.5 for NH_4CuCl_3 (Shiramura et al [12.15]).

Pronounced plateaus for $M/M_0 = \frac{1}{4}, \frac{3}{4}$ and 1 are observed at low temperatures where M_0 is the saturation magnetization. For increasing temperatures, the effect disappears rapidly (in Fig. 12.5, this happens for $T \geq 4.2\,\mathrm{K}$). Theories exist that give a necessary condition for the existence of such plateaus and

12.1 Magnetic Properties of Low Dimensional Spin Systems

energy gaps in the one-dimensional spin chain case (Oshikawa et al. [12.16]) and generalised to higher dimensions by Oshikawa [12.17]. This condition reads:

$$n(S - \sigma) = \text{integer} . \tag{12.3}$$

Here n is the number of magnetic ions with spin S in a magnetic unit cell. For values of $\sigma = M/$magnetic ion, which, for the right hand side of (12.3), gives an integer, an energy gap and a plateau are expected. For non-integer values the system should be gap-less. Here are some examples for (12.3):

1. The alternating chain ($n = 2$) with $S = \frac{1}{2}$ results in plateaus and gaps for $\sigma = 0$ at $B = 0$ and for $\sigma = \frac{1}{2}$, the saturation magnetization.
2. For a uniform chain ($n = 1$), only the saturation plateau $\sigma = \frac{1}{2}$ occurs. For $B = 0$ the system is gap-less.
3. For $S = 1$, the uniform spin chain, where $B = 0$, $\sigma = 0$ and $nS = 1$, i.e. a spin gap, the so-called Haldane gap occurs.
4. For the case of NH_4CuCl_3 with 2 coupled chains, with $n = 8$ and $S = \frac{1}{2}$, plateaus for $M/M_0 = \frac{1}{4}, \frac{1}{2}, \frac{3}{4}$ and 1 occur. Experimentally the plateau at $\frac{1}{2}$ is missing.
5. A triple chain $n = 3$ and $S = \frac{1}{2}$, results in $3 \times (\frac{1}{2} - \frac{1}{6}) = 1$ with a plateau at $M/M_0 = \frac{1}{3}$ and for $\sigma = S$, gives $M/M_0 = 1$. $CsCuCl_3$ behaves like a triple chain, at least for the magnetization perpendicular to the c-axis.

How is it possible to explain physically the existence of such plateaus and spin gaps? The simplest model is the strongly dimerized ($\alpha = 0$, $\delta = 1$) spin 1 chain, which is realized in nature by the compound $Ni_2(Medpt)_2(\mu$-ox$)(H_2O)_2$ introduced above in Fig. 12.1. On considering such an isolated anti-ferromagnetically coupled $S = 1$ dimer, we have a singlet ground state E_S and a triplet E_t and quintuplet E_Q as excited states.

With the Hamiltonian $H = J\mathbf{S}_1\mathbf{S}_2 - B(S_1^z + S_2^z)$, the energies are $E_S = -2J$, $E_t = -J - B$ and $E_Q = J - 2B$. These energies are plotted in Fig. 12.6a. It is seen that the singlet remains as the ground state until $B_{c1} = J$, the triplet ($S = 1$) from B_{c1} till $B_{c2} = 2J$, followed by the saturation quintuplet state ($S = 2$). The magnetization for $T = 0$ as $M = -dE/dB$ is shown in Fig. 12.6b. There are plateaus for $M/M_0 = 0, \frac{1}{2}$ and 1 as expected. For the following, the lowest excited states for this isolated dimer system is of interest. These can be read off from Fig. 12.6a and they are presented in Fig. 12.6c. It is seen that the magnetic excitations have a soft mode behaviour. The energies tend to zero at the change of the plateaus. Within the plateaus, there is an energy gap so the plateau can form. At the change from one plateau to the next there is a soft magnetic mode. The dotted line indicates the small sound wave frequency of typically 100 MHz, which cuts the soft magnetic excitation at the plateau changes. Therefore the softening of the magnetic excitations can be detected by a resonant interaction with the sound waves as shown later on (see Sect. 12.3).

What we have done for an isolated $S = 1$ dimer can be extended numerically to alternating chains ($0 < \alpha < 1$) and also for $S = \frac{1}{2}$ chains. This

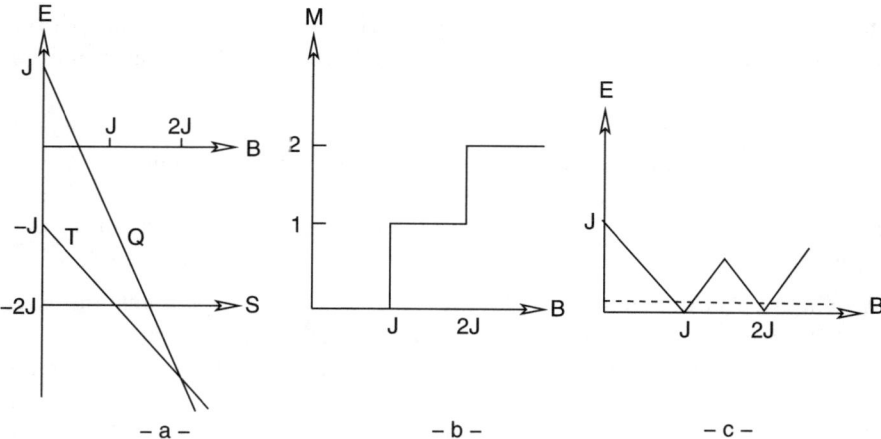

Fig. 12.6. The isolated spin 1 dimer: (a) Energies for the singlet, triplet and quintuplet states for an isolated spin 1 dimer, (b) magnetization for the isolated spin 1 dimer, (c) lowest excited states for the isolated spin 1 dimer

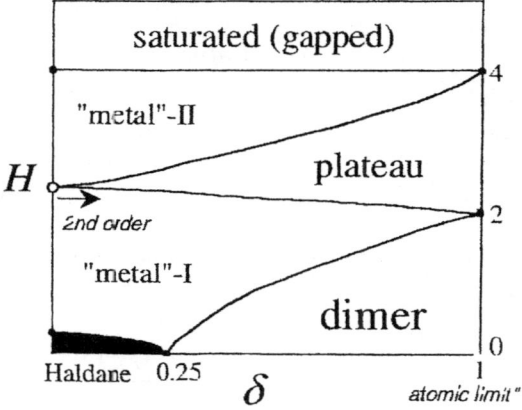

Fig. 12.7. Phase diagram for an alternating $S=1$ chain as a function of alternation parameter δ (Totsuka [12.18])

has been done for many different cases (see e.g. Totsuka [12.18]). A typical example of what can be expected is shown in Fig. 12.7 (Totsuka [12.18]). The limit $\alpha = 0(\delta = 1)$ has been treated above. For $\alpha = 1(\delta = 0)$, the system has an energy gap because we deal here with a $S = 1$ system which shows a Haldane gap (Haldane [12.19]) until $\delta = 0.25$. This is also a consequence of (12.3) as discussed above. For general α, δ, the important difference to the previous case ($\alpha = 0$, $\delta = 1$) is the fact that the magnetization does not jump from one plateau to the next at B_c as shown in Fig. 12.6, but that it passes a certain field region as seen e.g. for NH_4CuCl_3 in Fig. 12.5.

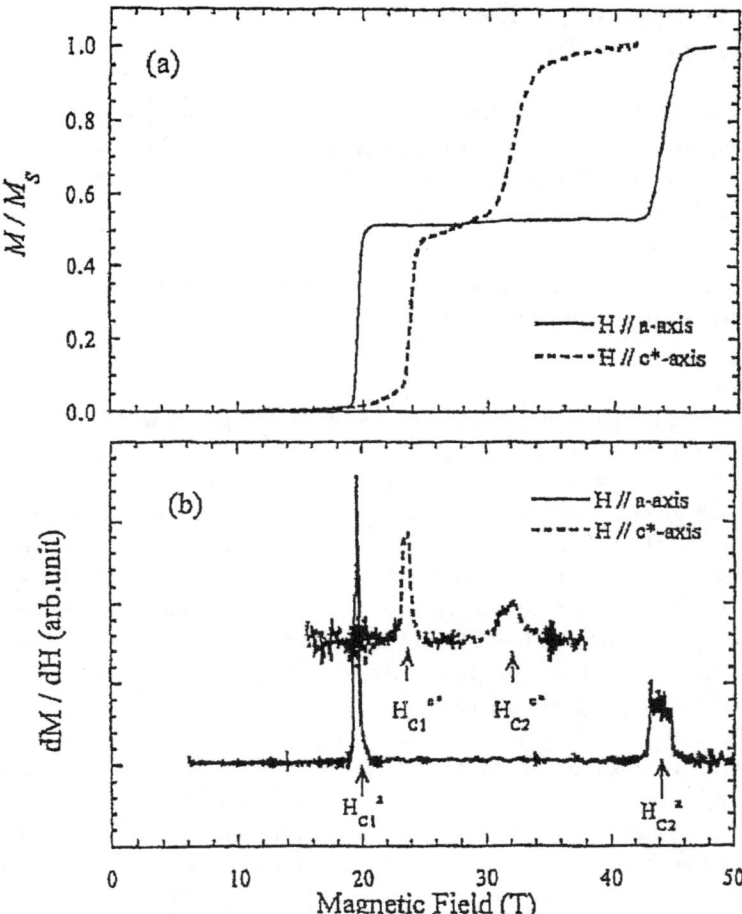

Fig. 12.8. (a) High field Magnetization versus magnetic field for the spin 1 dimerized substance $Ni_2(Medpt)_2(\mu\text{-ox})(H_2O)_2$. The field is applied along the a and c^*-axis (Narumi et al. [12.2]), (b) differential susceptibility dM/dH. Arrows show peak positions corresponding to the transition fields

The case of an isolated spin 1 dimer is realised to a high degree for the organic substance $Ni_2(Medpt)_2(\mu\text{-ox})(H_2O)_2$ whose structure is shown in Fig. 12.1. As discussed before, this structure gives isolated pairs of Ni^{2+} ($S = 1$). The magnetization is shown in Fig. 12.8 where the similarity with Fig. 12.6b is evident for $B \parallel a$. Therefore this substance is a good representative for studying ground state and excited states for a strongly dimerized system as discussed above (Fig. 12.6).

12.2 Temperature Dependence of Elastic Constants in Low Dimensional Spin Systems

In the following, we discuss first the temperature dependence of elastic modes and their attenuation for low-dimensional spin systems. Later, in Sect. 12.3, we show results of magnetic field-dependent sound wave modes with attenuation.

12.2.1 Temperature Dependence of Elastic Constants in Quasi One-Dimensional Spin Systems

CsNiCl$_3$

The first quasi low-dimensional spin system investigated with ultrasonics was CsNiCl$_3$ (Almond and Rayne [12.20], Trudeau et al [12.21]). CsNiCl$_3$ belongs to the family of ABX$_3$ compounds like CsCuCl$_3$ (Sect. 7.2.1). It is hexagonal (space group $P6_3/mmc$) and shows easy axis anti-ferromagnetism with $T_N = 4.4$ K. The Ni^{2+} ion ($S = 1$) has an orbital singlet A_2 ground state. Therefore a one-dimensional anti-ferromagnetic Heisenberg Hamiltonian description with a small magnetic anisotropy should describe the magnetic data adequately. The neutron scattering data were mostly interpreted assuming the Haldane conjecture (existence of a spin gap for $S = 1$ one-dimensional systems) to work (Kakurai et al. [12.22], Affleck [12.23]), as discussed in Sect. 12.1.

Figure 12.9 shows longitudinal acoustic velocity and attenuation for hexagonal c-axis propagation (c_{33} mode) as a function of temperature (Trudeau et al. [12.21]). It exhibits a velocity minimum around 40 K and an attenuation maximum at around 27 K. These anomalies can be interpreted with a theory based on exchange striction spin–phonon coupling (Sect. 5.1.3) for a one-dimensional spin chain (Fivez et al. [12.24], Fivez [12.25]). It predicts an attenuation maximum and velocity minimum at the same temperature $T \approx |J| = 32$ K. Therefore such a single chain theory can interpret the ultrasonic results of Fig. 12.9 at best semiquantitatively. The inter-chain exchange in this substance is more than an order of magnitude smaller than the intra-chain exchange.

CsNiF$_3$

This material is another ABX$_3$ compound with a quasi-one-dimensional spin system of hexagonal symmetry with easy basal plane behaviour. The exchange interaction is ferromagnetic within the chains. Anti-ferromagnetic coupling between the chains leads to a Néel transition $T_N = 2.7$ K. In this compound, magnetic soliton-like excitations have been discussed by Mikeska [12.26].

Young's modulus E was measured with a vibrating reed technique (see Sect. 2.2.6) in the frequency range of $1 - 4$ kHz. $E_\parallel$ parallel to the chain axis c exhibits strong softening from room temperature down to about 30 K

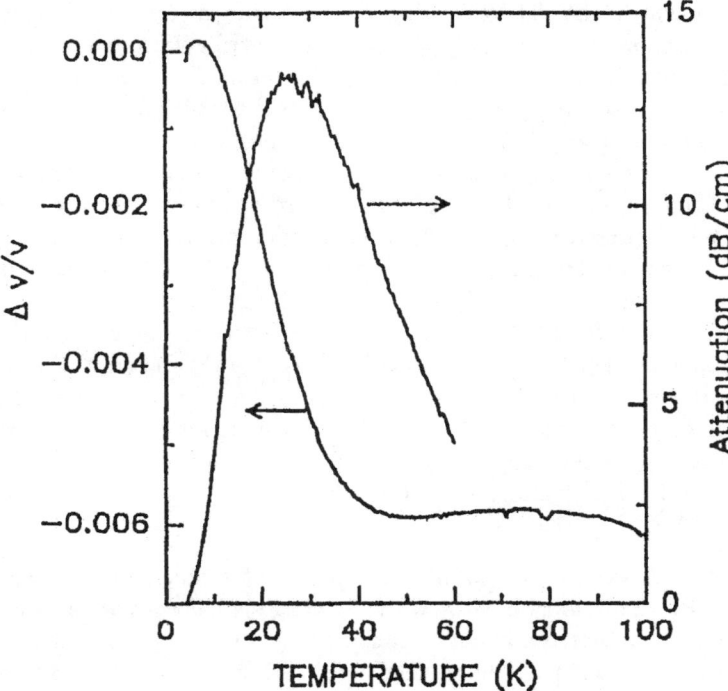

Fig. 12.9. Velocity and attenuation for the c_{33} mode in CsNiCl$_3$ (Trudeau et al. [12.21])

with a further small anomaly below T_N (Barmatz et al. [12.27]). The overall softening of $E_\parallel$ is about 45%. Perpendicular to the chain axis $E_\perp$ exhibits a hardening of about 11%. On the other hand Brillouin scattering data for the c_{11} and c_{33} modes did not reveal any temperature-dependent softening but rather a normal hardening (Käräjämäki et al. [12.28]). Since the c_{33} mode corresponds approximately to the $E_\parallel$ module, this effect could be due to large dispersion. According to (4.18) and (4.18a), the $E_\parallel$ module is in the thermodynamic $\omega\tau \ll 1$ region whereas in the Brillouin scattering for $\omega\tau \gg 1$ (τ = spin fluctuation time), the background elastic constant can be observed. Similar behaviour has been encountered before in Sects. 6.1.3 and 7.2.2.

NH$_4$CuCl$_3$

This system is another ABX$_3$ compound which has been investigated ultrasonically (Schmidt et al. [12.29]). This compound belongs, together with KCuCl$_3$ and TlCuCl$_3$, to the monoclinic structure class ($P2_1/c$) and has been reviewed by Tanaka et al. [12.30]. NH$_4$CuCl$_3$ has an anti-ferromagnetic transition at $T_N = 1.3$ K (Kurniawan et al. [12.31]). The structure consists of double chains of edge sharing CuCl$_6$ complexes which run parallel to the short a-axis. There are several exchange paths and the magnetic structure and its excitation are not yet known. At the NH$_4$-order-disorder transition at $T_a = 70$ K both modes c_{22} and c_{66} show some anomaly. T_a denotes the transition where the ammonium molecules NH$_4$ start to order their orientation. For a general discussion of elastic properties at an order–disorder phase transition, see Appendix G. In the magnetic susceptibility, no anomaly is found at T_a for NH$_4$CuCl$_3$. The c_{22} mode exhibits a pronounced elastic constant dip of $\sim 0.5\%$ and the c_{66} mode a broad minimum. But the temperature dependence of the c_{ij} cannot be easily interpreted as for CsNiCl$_3$. This order–disorder transition at 70 K lowers the crystal structure such that there are three different Cu-dimers A, B and C which, in the b–c plane, have one A and one C dimers and two B dimers (Matsumoto [12.32]). This has important consequences for the magnetic properties which are quite surprising as already shown in Fig. 12.5 for the magnetization. It will be fully discussed in the next Sect. 12.3. The other two monoclinic crystals KCuCl$_3$ and TlCuCl$_3$ have no order–disorder phase transition. Consequently they have no inequivalent dimers and also do not show plateaus in their magnetization curves.

CsCuCl$_3$

Still another ABX$_3$ compound investigated by ultrasonic means is the hexagonal CsCuCl$_3$. This substance shows at $T_a = 421$ K a cooperative Jahn–Teller effect and an anti-ferromagnetic transition at $T_N = 10.5$ K. Therefore the acoustic properties of this substance have already been discussed in Sect. 7.2.1 (see Fig. 7.11). Low temperature acoustic properties have been measured in high magnetic field (Wolf et al. [12.33, 12.34]), see below Sect. 12.3.

TMMC

Tetramethylamoniummangantrichlorid (CH$_3$)$_4$NMnCl$_3$ (TMMC) is a quasi one-dimensional planar spin substance with Mn^{2+}-ions. It has an antiferromagnetic transition at a very low temperature of $T_N = 0.835$ K and a typical one-dimensional behaviour in the magnetic susceptibility for $T > T_N$ with a maximum around 50 K. In addition, this substance exhibits structural phase transitions at $T_a = 389$ K and 126 K. The three phases belong to the following space groups: $P6_3/mmc$ – Phase I' for $T > 389$ K; hexagonal $P6_3/m$ – Phase I for $126 K < T < 389$ K; monocl. – Phase 2 for $T < 126$ K.

Elastic constant measurements for this system produced the following results (Braud et al. [12.35], Finsterbusch [12.36]) see Fig. 12.10:

c_{66}-mode: It exhibits strong softening from 280 K down to 160 K of about 35%

c_{44}-mode: It exhibits a 12% softening in the same temperature interval, somewhat less than c_{66}.

c_{33}-mode: It could be measured down to He temperature. It has a step function behaviour at the 126 K phase transition with a steep maximum around 210 K.

c_{11}-mode: It exhibits a broad maximum around 200 K and a step function behaviour at the 389 K transition

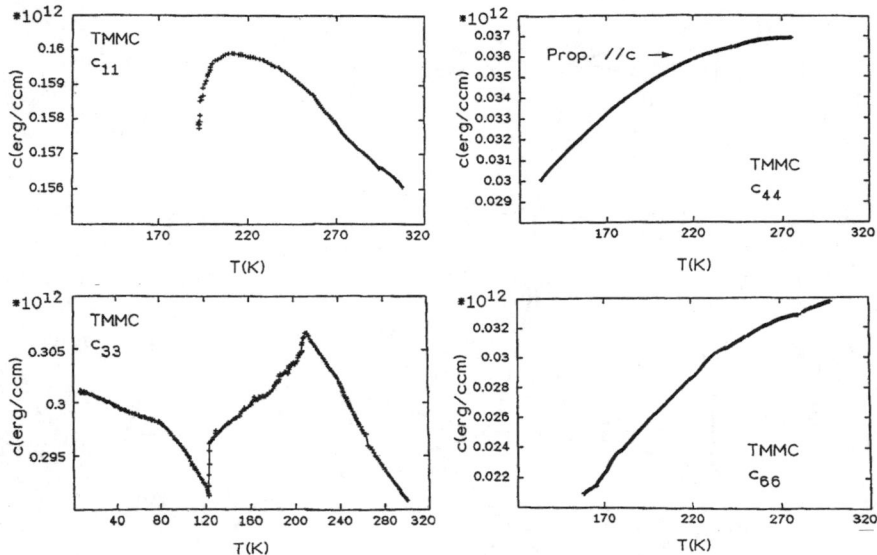

Fig. 12.10. Temperature dependence of the elastic modes c_{11}, c_{33}, c_{44}, c_{66} for TMMC (Finsterbusch [12.36])

With the exception of c_{33}, all modes disappear below 126 K due to strong domain wall–stress effects. The 126 K structural transition is apparently due to the partial ordering of tetramethylamonium groups (TMA). The dynamic of these groups could be responsible for the sharp maximum in c_{33} at 210 K. A bilinear strain–order parameter coupling is only given for the c_{66} mode – with the strain ε_{xy} the order parameter – but not for the c_{44} mode. Therefore c_{66} should be the soft mode for this transition and the c_{44} temperature dependence would be fluctuation dominated (see Sect. 7.4.3).

12.2.2 Case of Two-Dimensional Dimer Spin Systems

Two materials are discussed here:

1. $SrCu_2(BO_3)_2$ is the ideal two-dimensional spin system exhibiting magnetization plateaus and other pronounced effects and
2. K_2NiF_4 which has typical two-dimensional spin properties although it shows three-dimensional ordering already at $T_N = 97.2$ K.

$SrCu_2(BO_3)_2$

Introduction

The tetragonal (D_{2d}^{11}) two dimensional spin dimer compound $SrCu_2(BO_3)_2$ is one of the most interesting low-dimensional spin system. The lattice constants

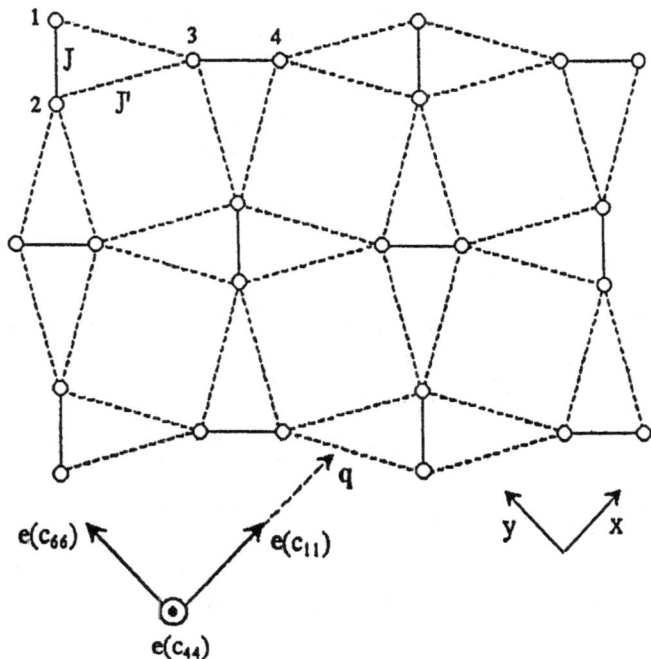

Fig. 12.11. SrCu$_2$(BO$_3$)$_2$: Schematic view of the Cu^{2+}-dimers in the CuBO$_3$ plane. Open circles are Cu^{2+} ions. Exchange interactions J, J' are indicated. Propagation direction q and polarization vector e for the different elastic modes are also given

are $a = b = 8.995$Å and $c = 6.649$Å. It consists of alternately stacked CuBO$_3$ and Sr layers. In Fig. 12.11 we show a Cu-O plane where the Cu^{2+} spin $\frac{1}{2}$ dimers have an exchange J (full lines) and are orthogonally coupled with a smaller exchange J' (dotted lines). This spin–dimer arrangement is equivalent to the so-called Shastry–Sutherland model (Shastri and Sutherland [12.37]). In addition in the same figure we have added the coordinate system and the propagation and polarization directions of the different elastic modes.

The ground state of this spin $\frac{1}{2}$ system is known exactly (Miyahara and Ueda [12.38], [12.39], Müller-Hartmann et al. [12.40], Totsuka et al. [12.41]) because it can be mapped on to the Shastry–Sutherland model. This compound does not order down to 0.5 K but it is close to a quantum critical point with $J'/J \approx 0.68$. This means that this system would become antiferromagnetic for $J'/J \geq 0.7$. For a discussion of quantum critical points see Sect. 9.3.6.

Besides the spectacular magnetic field properties, discussed in the next section, it shows also very interesting temperature dependences of the various thermodynamic functions (specific heat, thermal expansion, magnetic susceptibility and elastic constants). The thermodynamic derivatives of this dimer system can be discussed similarly to the single ion thermodynamic properties

of crystal field split rare-earth ions covered in Sect. 5.2. We will see that a
RPA molecular field treatment accounts quantitatively for the magnetic and
strain susceptibility effects.

It has recently been found (Choi et al. [12.42] and references therein) that
SrCu$_2$(BO$_3$)$_2$ undergoes a structural transition at $T_a = 395$ K. The tetragonal
space groups are for the high temperature phase $I4/mcm$ and for the low
temperature phase $I\bar{4}2m$. This is caused by interlayer interactions. Since the
sound propagation directions reported below are within the ab-layer, the
effect of these couplings and of the resulting phase transition should be small
on the elastic effects.

Magnetic Susceptibility and Elastic Constants

We start with the discussion of magnetic susceptibility and elastic constants.
These quantities are shown in Fig. 12.12 (Zherlitsyn et al. [12.43], Wolf et
al. [12.44]).

It is seen in Fig. 12.12 that magnetic susceptibility χ_m and elastic constants c_{ij} exhibit pronounced maxima and minima as a function of temperature, typical for low-dimensional spin systems. High temperature series expansions have been employed to describe the magnetic susceptibility
(Weihong et al. [12.46]). These phenomena can be described quite successfully using the generalised susceptibility approach of Sect. 5.3 (Zherlitsyn et
al. [12.43], Wolf et al. [12.44]). If one introduces with χ_m^0, the single dimer
magnetic or strain susceptibility, then the full magnetic susceptibility or the

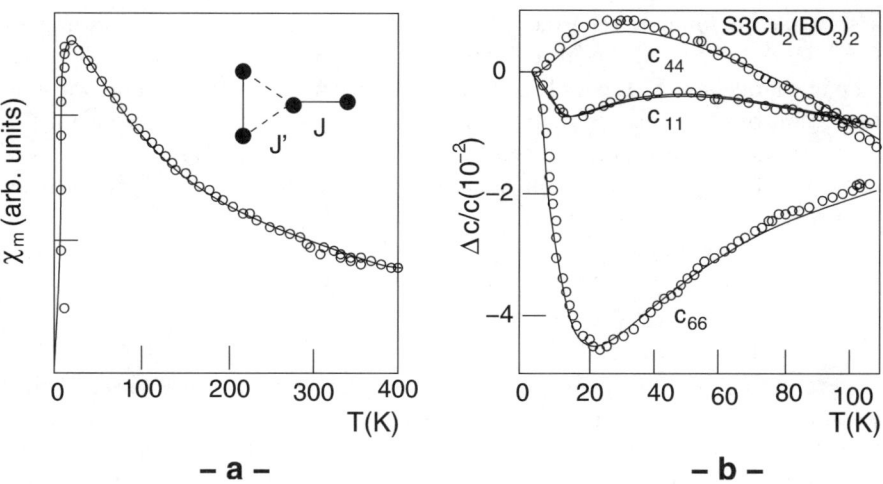

Fig. 12.12. SrCu$_2$(BO$_3$)$_2$: (**a**) Temperature dependence of the magnetic susceptibility: open circles experiment from Kageyama et al. [12.45], solid line fit as described
in text, (**b**) temperature dependence of the elastic constants: Open symbols are from
the experiment and solid lines are fits as described in the text (Wolf et al. [12.44])

elastic constant follows from (5.17) and (5.21). For example for the magnetic susceptibility, we obtain

$$\chi_m = \frac{\chi_m^0}{1 - j\chi_m^0}, \qquad (12.4)$$

where j is the full exchange interaction between the orthogonal dimers shown in Fig. 12.11. According to this figure $j = 4J'$. The single dimer susceptibility χ_m^0 can easily be calculated by considering the splitting of a dimer into a singlet ground state and an excited triplet state at energy Δ. The free energy for this system gives $F = -k_B T \ln Z$ with the partition function $Z = 1 + 3\exp(-\Delta/k_B T)$. For $\chi_m^0 = -\mathrm{d}^2 F/\mathrm{d}B^2$ where $B \to 0$ this gives

$$\chi_m^0 = 2\frac{e^{-\Delta/k_B T}}{k_B T Z}. \qquad (12.5)$$

The fit of this calculation (12.4), (12.5) is shown in Fig. 12.12a as a full line. The agreement is remarkable. The following parameters were used for this fit: the singlet-triplet splitting $\Delta = 30\,\mathrm{K}$ was taken from previous susceptibility calculations and inelastic neutron scattering (Kageyama et al. [12.47]). It is not the expected value $J = 100\,\mathrm{K}$ but already a renormalized value. The inter-dimer exchange $j = -273\,\mathrm{K}$ turns out to be $4J'$ as expected.

A calculation of the elastic constants follows the same line (Wolf et al. [12.44]). First it is necessary to discuss the spin–phonon coupling. As in the case of $CsNiCl_3$ shown above, it is the exchange striction coupling (Sect. 5.1.3) which is responsible for the anomalous sound wave effects. This follows from a discussion of longitudinal and transverse waves in this substance (see below). Furthermore the single ion magneto-elastic coupling for Cu^{2+} ($S = \frac{1}{2}$) is small because of the vanishing quadrupole matrix elements. The exchange striction coupling was also used for the critical sound wave attenuation phenomena occurring at magnetic phase transitions (Sect. 6.1). It is due to a modulation of the exchange integral by a sound wave. As discussed in Sect. 5.1.3, a spin–sound wave coupling occurs (5.10) which is reproduced for convenience here again:

$$E_{exs} = \sum_{i,\delta} \left(\frac{\mathrm{d}J}{\mathrm{d}\boldsymbol{\delta}} \cdot \boldsymbol{e}_q\right)(\boldsymbol{q} \cdot \boldsymbol{\delta})(\boldsymbol{S}_i \cdot \boldsymbol{S}_{i+\delta})e^{i(\boldsymbol{q}\cdot\boldsymbol{R}_i - \omega t)}. \qquad (12.6)$$

From this equation, the polarization dependence of the different sound wave modes can be deduced. Usually (12.6) says that shear waves propagating along symmetry directions do not couple with this mechanism whereas longitudinal waves do. However the geometry in Fig. 12.11 shows that the c_{66} mode, with propagation and polarization in the plane of the dimers can couple with the spin system (propagation vector $\boldsymbol{q}$ and polarization vector $\boldsymbol{e}$ make an angle of $45°$ with the dimer axis) whereas the c_{44} mode, with polarization vector perpendicular to the plane, can couple only in higher order.

12.2 Elastic Properties of Low Dimensional Spin Systems

Table 12.1. Elastic constants c_{ij} at $T = 1.5\,\text{K}$ and exchange striction coupling constants $|\partial\Delta/\partial\varepsilon|$ and K for SrCu$_2$(BO$_3$)$_2$ (Wolf et al. [12.44])

c_{ij}	c^0_{ij} (10^{11} erg/cm^3)	K (K)	$\left|\frac{\partial\Delta}{\partial\varepsilon}\right|$	$\left|\frac{\partial\Delta'}{\partial\varepsilon}\right|$	$\frac{\partial^2\Delta}{\partial\varepsilon^2}$
c_{11}	25.8	23	970		
c_{66}	8.2	-19	1300	3770	
c_{44}	0.96				1680

The generalised strain susceptibility formula (5.21) is taken again to calculate the temperature dependence of the elastic constants.

$$c_\Gamma = \frac{d^2 F}{d\varepsilon_\Gamma^2} = c^0_\Gamma - g^2 N \chi_{str} \tag{12.7}$$

with $\chi_{str} = \frac{\chi^0_{str}}{1 - K\chi^0_{str}}$.

K is the strength of the ($q = 0$) dimer–dimer interaction and χ^0_{str} is the strain susceptibility of a single dimer. N is the density of dimers. (12.7) is the strain analogue of the magnetic susceptibility (12.4). We start with the c_{11} mode. Using the free energy for a single dimer from the calculation of the magnetic susceptibility, gives

$$\chi^0_{str} = \frac{e^{-\Delta/k_B T}}{k_B T Z^2}\ .$$

The fit with the parameters $\Delta = 30\,\text{K}$ (as for the magnetic susceptibility) and $g = |\partial\Delta/\partial\varepsilon| = 970\,\text{K}$ gives a very good fit (see Fig. 12.12 and Table 12.1). Note that the maximum of the magnetic susceptibility at 18 K and the minimum of the c_{11} mode at 13 K is correctly described by this simple RPA-molecular field theory. Next we turn to the c_{66} mode which gives a much stronger anomaly than c_{11} and a minimum at 21 K. For the correct description of this shear mode, it is necessary to take a spin phonon coupling to the same triplet state at 30 K but, in addition, also to the next higher triplet state $\Delta' = 58\,\text{K}$, as determined from inelastic neutron scattering (Kageyama et al. [12.47]). Therefore the partition function is

$$Z = 1 + 3e^{-\Delta/k_B T} + 3e^{-\Delta'/k_B T}$$

and

$$\chi^0_{str} = \left(\frac{k_B T}{Z^2}\right)\left\{Z\frac{d^2 Z}{d\varepsilon^2} - \left(\frac{dZ}{d\varepsilon}\right)^2\right\}\ .$$

Using (12.7) with $\Gamma = 66$ in Fig. 12.12, again gives a very good agreement with the experiment. The coupling constants are now larger: $|\partial\Delta/\partial\varepsilon| = 1300\,\text{K}$ and $|\partial\Delta'/\partial\varepsilon| = 3770\,\text{K}$ (Table 12.1). The dimer–dimer coupling K turns out to be ferrodistortive for the c_{11} mode and anti-ferro distortive for the c_{66} mode (Table 12.1). Finally for the c_{44} mode, it is known from the discussion of the polarization dependence above that this mode couples only in higher order in the strain. Therefore instead of a coupling $(1/J)\mathrm{d}J/\mathrm{d}\varepsilon = (1/\Delta)\mathrm{d}\Delta/\mathrm{d}\varepsilon$ there is a higher order coupling constant $(1/J)\mathrm{d}^2 J/\mathrm{d}\varepsilon^2 = (1/\Delta)\mathrm{d}^2\Delta/\mathrm{d}\varepsilon^2$ and for the c_{44} mode we get

$$c_{44} = c_{44}^0 + 3N\frac{\mathrm{d}^2\Delta}{\mathrm{d}\varepsilon^2}\frac{e^{-\Delta/k_B T}}{Z}. \tag{12.8}$$

This expression gives again a good fit to the experimental data with a value of the coupling constant $\mathrm{d}^2\Delta/\mathrm{d}\varepsilon^2 = 1680\,\mathrm{K}$ (see Fig. 12.12). In summary, it can be concluded that for $SrCu_2(BO_3)_2$, the exchange striction coupling describes the temperature dependence of the elastic constants very well.

Specific Heat and Thermal Expansion

The specific heat of $SrCu_2(BO_3)_2$ was measured by Kageyama et al. [12.48]. The temperature dependence exhibits a Schottky–type maximum at 8 K. Taking independent dimers with a singlet–triplet splitting of 30 K from above gives a maximum at 10.5 K. A calculation using the transfer matrix method gives a better agreement with the experiment (Miyahara and Ueda [12.39]).

The thermal expansion α was measured recently in the tetragonal plane (Lang et al. [12.49]). It is shown for the low temperature region in Fig. 12.12. For the direction in the tetragonal plane it is negative (expansion) with a minimum at 8 K, coinciding with the maximum in the specific heat. Taking again an independent dimer model gives a proportionality between specific heat C and thermal expansion α: $\alpha = \frac{\partial \ln \Delta}{\partial \varepsilon}\frac{C}{c_{11}}$. In the c–direction there is likewise a positive maximum at 8 K, which arises due to the special negative compliances $s_{13} = -c_{13}/\Delta'$ with $\Delta' = c_{33}(c_{11}+c_{12}) - 2c_{13}^2$ in tetragonal and hexagonal materials. For details, see the case of $PrNi_5$ in Sect. 5.2.1 and UPt_3 in Sect. 9.3.4. The values of α are smaller than 20×10^{-6} therefore any length corrections to the elastic constants are negligible.

The coupling constant can be determined from α and C by using (9.12). Since the full compliance tensor or elastic constant tensor is unknown, only an order of magnitude estimate is possible. With $\alpha_a \approx \frac{\partial \ln \Delta}{\partial \varepsilon}\frac{C}{c_{11}}$, we get from Fig. 12.12 for $T = 8\,\mathrm{K}$, $\alpha_a = -20 \times 10^{-6}\,\mathrm{K}^{-1}$. With $C = 6400\,\mathrm{mJ/KMol}$, $c_{11} = 25.8 \times 10^{11}\,\mathrm{erg/cm^3}$, molecular weight $M = 232.3\mathrm{g}$, density $\rho = 4.1\mathrm{g/cm^3}$ and $n = 0.74 \times 10^{22}\,\mathrm{cm^{-3}}$, we get $\frac{\partial \ln \Delta}{\partial \varepsilon} = -65.5$. From elastic constant data (Table 12.1), we get $|\frac{\partial \ln \Delta}{\partial \varepsilon}| = 32$ within a factor of 2 for this rough estimate.

K_2NiF_4

This quasi two-dimensional spin system orders anti-ferromagnetically at $T_N = 97.2\,\mathrm{K}$. It has a layered tetragonal perovskite structure with $Ni^{2+}(S = 1)$ and an orbital singlet. In fact this structure is the same as the one for high temperature cuprate superconductors $La_{2-x}Sr_xCuO_4$ discussed in Sect. 10.4. K_2NiF_4 does not qualify for a Haldane gap system because of the large T_N

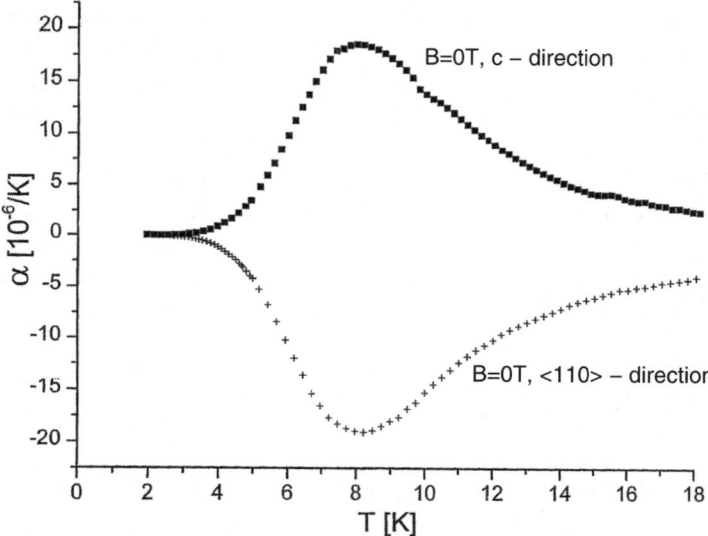

Fig. 12.13. Thermal expansion of $SrCu_2(BO_3)_2$ at low temperatures along the c-axis and in the tetragonal plane (Lang et al. [12.49])

and because of the dimensionality. Despite the rather high T_N, it nevertheless exhibits typically low-dimensional effects in neutron scattering, specific heat and magnetic susceptibility. The broad maximum of the magnetic susceptibility lies around 200 K. These various physical properties are reviewed by de Jongh and Miedema [12.3].

Sound propagation experiments do not reveal typical two-dimensional behaviour like $SrCu_2(BO_3)_2$ discussed above. In fact acoustic anomalies are only observed for longitudinal sound propagation in the tetragonal plane in the vicinity of T_N (Gorodetsky et al. [12.50]). However, in contrast to the three-dimensional case treated in Chap. 6, sound velocity and sound attenuation effects are much reduced for K_2NiF_4. The sound velocity anomaly at T_N amounts to $\Delta v/v \sim -7 \times 10^{-5}$ compared to 10^{-3} effects in three dimensional materials (see Sect. 6.1.3 and Fig. 6.4). Also the sound attenuation α at T_N only becomes noticeable for frequencies above 500 MHz. These small effects are not due to small exchange striction coupling constants $\partial J/\partial \varepsilon_a$ but rather due to small critical correlation effects. On the other hand, shear waves propagating in a, c directions and longitudinal waves propagating in c direction show negligible critical effects around T_N.

12.3 Magnetic Field Effects

Magnetic field effects on ultrasonic waves in low dimensional spin systems are very important for studying the ground state and excited states of these systems. In the last few years, a number of such investigations has been carried out, especially in very high magnetic fields with a pulsed field facility which can reach fields up to $50T$ (Wolf et al [12.51]). It should be stressed that ultrasonics is one of a few spectroscopies which can be used in pulsed fields. In Sect. 2.4, the measuring techniques for pulsed field ultrasonic experiments are described. Ultrasonic velocity changes and attenuation can be measured with the so-called quadrature method with high accuracy. This method has been applied to a number of problems (not only low-dimensional spin compounds) but also in the mixed valence compound $YbIn_{1-x}Ag_xCu_4$ (Zherlitsyn et al. [12.52]), see Sect. 9.2, in the heavy fermion compound URu_2Si_2 (Wolf et al. [12.53]) see Sect. 9.3.3 and in EuB_6 (Zherlitsyn et al. [12.54]). Here we show some effects of pulsed field ultrasonics for low dimensional spin systems: CuCCP, $(VO)_2P_2O_7$, $SrCu_2(BO_3)_2$ and NH_4CuCl_3.

CuCCP

This Cu-polymer exhibits clear signatures of a one dimensional homogeneous spin chain as demonstrated for the magnetization in Fig. 12.4 and for the temperature-dependent magnetic susceptibility with a maximum at $T = 12\,K$ (Wolf et al. [12.12]). It was possible to propagate sound of $50\,MHz$ through a hard-pressed powder of this material. Fig. 12.14 displays relative elastic constant results for magnetic fields up to 40T. A pronounced minimum is observed in the longitudinal elastic constant at the onset of magnetization saturation.

This feature can be described with a thermodynamic model used before for $(VO)_2P_2O_7$ (Wolf et al. [12.34]). We take a free energy approach as outlined in Sect. 4.3 using the Landau free energy density $F_L = a(m/m_0)^2 + b(m/m_0)^4 - Bm$. For the magneto-elastic interaction we take $F_{me} = G\,n\,\varepsilon_{xx}(m_x/m_0)^2$ for a longitudinal mode and we get with the total free energy density $F = \frac{1}{2}c_L^0\varepsilon_{xx}^2 + F_L + F_{me}$ from $dF/dm = 0$ a $dm/d\varepsilon = -2G\,n\,m/m_0^2\chi_m$ and for the elastic constant $c_L = d^2F/d\varepsilon_{xx}^2$

$$c_L = c_L^0 - 4G^2n^2\left(\frac{m_x^2}{m_0^3}\right)\chi_m(B)\,. \quad (12.9)$$

Here χ_m is the differential magnetic susceptibility, n is the number of magnetic ions per unit volume and m/m_0 is the reduced magnetization. Note that in ordinary Landau theory with temperature exponents $\frac{1}{2}$ for m and 1 for χ_m (see Sect. 4.3.2), a step function behaviour at T_c occurs for the elastic constant as discussed before in Sect. 4.3.4 for a quadratic order parameter coupling. In the present case, however, the experimental values for $m(B)$ and the differential susceptibility $\chi(B)$ are taken resulting in a reasonable description of the field dependent elastic constant as shown in Fig. 12.14. But as seen from the figure caption to Fig. 12.14, it is necessary to take temperature changes due to the magneto-caloric effect into account for $T < 7\,K$ similar to the case of magnetization (Fig. 12.4).

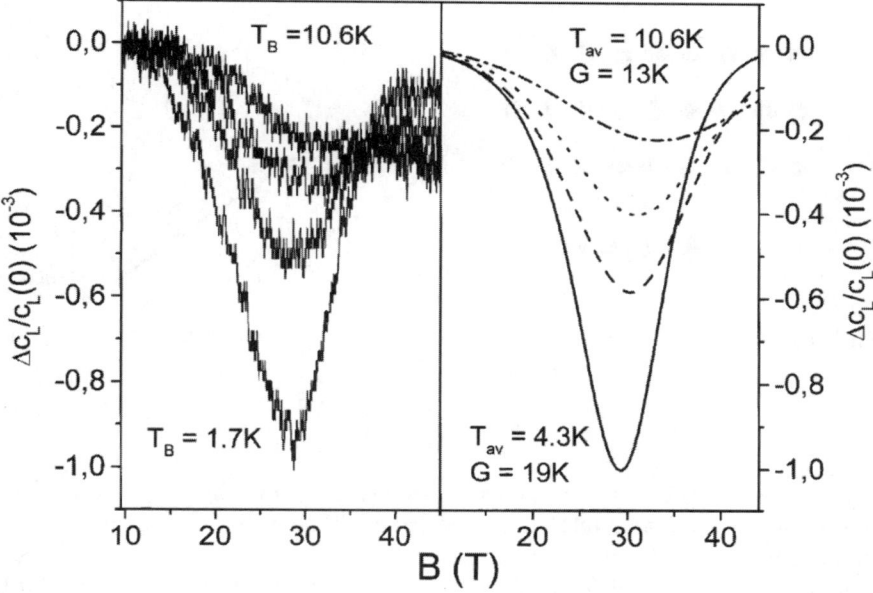

Fig. 12.14. Elastic constant results for CuCCP for 50 MHz longitudinal sound as a function of magnetic field for different bath temperatures $T_B = 1.7$ K, 4.2 K, 7.3 K and 10.6 K. The second figure shows a calculated fit for average temperatures (due to the magneto-caloric effect) $T_{av} = 4.3$ K, 6.5 K, 7.5 K and 10.6 K (see text) (Wolf et al. [12.12])

(VO)$_2$P$_2$O$_7$

In this case of vanadium phosphorous oxide, we deal with an instructive example where we can study the main physics of dimerised systems, namely the interaction of sound waves with low lying magnetic excitations. Vanadyle–pyrophosphate (VO)$_2$P$_2$O$_7$ is a nice example for this. It can exist in two modifications, an ambient pressure and in a high pressure modification (Saito et al. [12.55]). The majority of experiments has been carried out so far in the low pressure modification. An encyclopedic survey of this compound is given by Johnston et al. [12.56]. A schematic description of the structure is given in Fig. 12.15a. In this orthorhombic substance, inelastic neutron scattering revealed (Garrett et al. [12.57]) that we are dealing with V^{4+}($S = \frac{1}{2}$) chains along the b-axis with alternating exchange interactions. (Note that the nomenclature of the axis here is from Garrett et al. [12.57] and not from Saito et al. [12.55]). The strongest exchange is across a PO$_4$ bridge. There are two such chains in adjacent ab–planes for the ambient pressure grown crystal. The interaction between the chains is relatively weak (Uhrig and Normand [12.58,12.59]). The strongest exchange and the biggest dispersion is along the b–axis. Inelastic neutron scattering (Garrett et al. [12.57]), magnetic susceptibility and ESR (Prokofiev et al. [12.60]) and high field magnetization

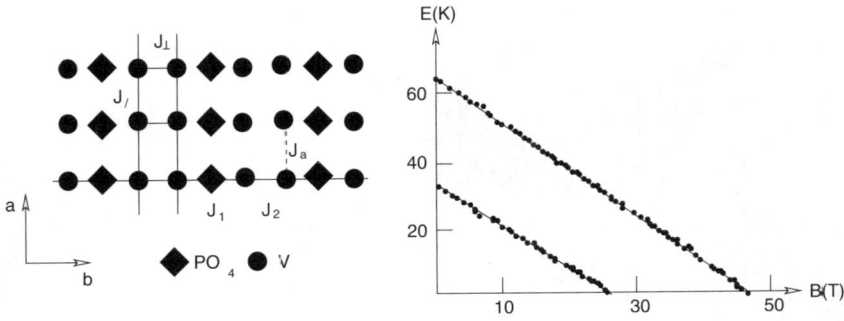

Fig. 12.15. (a) Schematic view of $(VO)_2P_2O_7$ in the ab–plane, (b) excitation energies $E_{exc}(B)$ for the two chains

(Narumi et al. [12.61]), revealed two energy gaps for the two inequivalent chains of ca. 35 K and 68 K at the Γ–point. The ESR and magnetization experiments gave the two lowest branches of the excited triplet states as a function of magnetic field as seen in Fig. 12.15b (Prokofiev et al. [12.60]). They go to zero at 26T and 47T respectively.

Now, it is clear that sound waves of 143 MHz cross these triplet excitations very close to the intersection with the singlet and, if symmetry allowed, make an interaction with them. Figure 12.16 shows an experimental result with longitudinal waves and magnetic field along the a–axis (Wolf et al. [12.34]). One clearly notices a similar resonance minimum as for the case of CuCCP for the sound velocity at 27T which is most pronounced at the lowest temperature (1.6 K) and disappears rapidly with increasing temperature. The effect has disappeared completely at 11 K. In addition, the minima shift to higher fields with increasing temperature. A precursor of the resonance with the higher triplet mode can also be noted.

One can explain these features again with the same simple thermodynamic model as used above for CuCCP (Wolf et al. [12.34]). The inset to Fig. 12.16 gives the result of this calculation for the lower triplet mode again using (12.9). In this case, unlike CuCCP, m/m_0 and χ are calculated using the singlet–triplet model. Note that at 1.6 K, the resonance has already shifted from 26T to 28T. This is the reason that the higher resonance for $B < 50T$ is not observed. The temperature shift of the resonance is also reproduced and the effect disappears also for $T > 11$ K. So the salient features are nicely reproduced with this simple model. The high field side of the resonance is not accurately reproduced by this model. This thermodynamic interpretation implies, as above for CuCCP, $\omega\tau \ll 1$ with τ the relaxation time of the spin excitations (see Sect. 4.3.6). A combination of $\Delta c/c$ from Fig. 12.16 and the ultrasonic attenuation $\alpha(B)$ (Lüthi et al. [12.62]) together with (4.19), gives a $\omega\tau \sim 0.1$. Therefore we do not have a resonant interaction between the sound wave and the magnetic excitation as in $SrCu_2(BO_3)_2$.

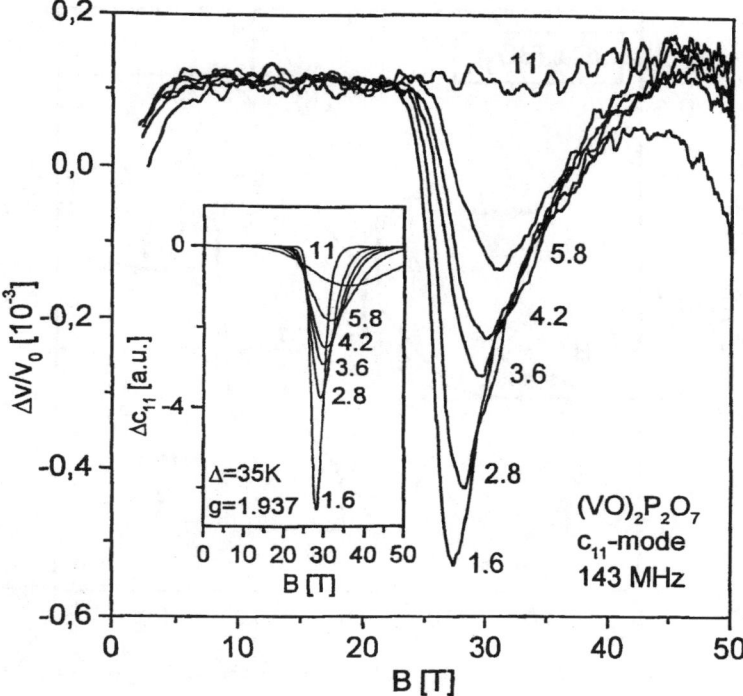

Fig. 12.16. Field dependence of the sound velocity of c_{11} in $(VO)_2P_2O_7$, frequency 143 MHz; inset: model calculation (Wolf [12.34])

SrCu$_2$(BO$_3$)$_2$

This two-dimensional spin dimer system also has interesting magnetic properties, in addition to the thermal properties shown in Figs. 12.12 and 12.13. As a function of magnetic field, it exhibits plateaus at fractional values of the saturation magnetization $m/m_0 = \frac{1}{8}, \frac{1}{4}, \frac{1}{3}$ (Kageyama et al. [12.45], Onizuka et al. [12.63]). Higher lying plateaus should occur for $B > 60T$ but could not be reached so far. Fig. 12.17 shows such a magnetization measurement (Onizuka et al. [12.63]). Similar to Fig. 12.5 for NH$_4$CuCl$_3$, well-developed plateaus, especially at low temperatures (0.08 K) in SrCu$_2$(BO$_3$)$_2$, can be observed. The small difference in magnetization for the different magnetic field directions ($B \parallel c$, $B \perp c$) can be explained with the different g–factors ($g_c \neq g_\perp$ (Nojiri et al. [12.64]).

The next Fig. 12.18 shows the relative change in elastic constants $\Delta c/c$ as a function of magnetic field up to 50T for the modes c_{11}, c_{44} and c_{66}. These are the same modes as already shown in Fig. 12.12 for the temperature dependence. The polarization and propagation directions are shown in Fig. 12.11. They behave quite differently in the magnetic field. While the c_{44} mode has a very weak field dependence (only small steps of the order of

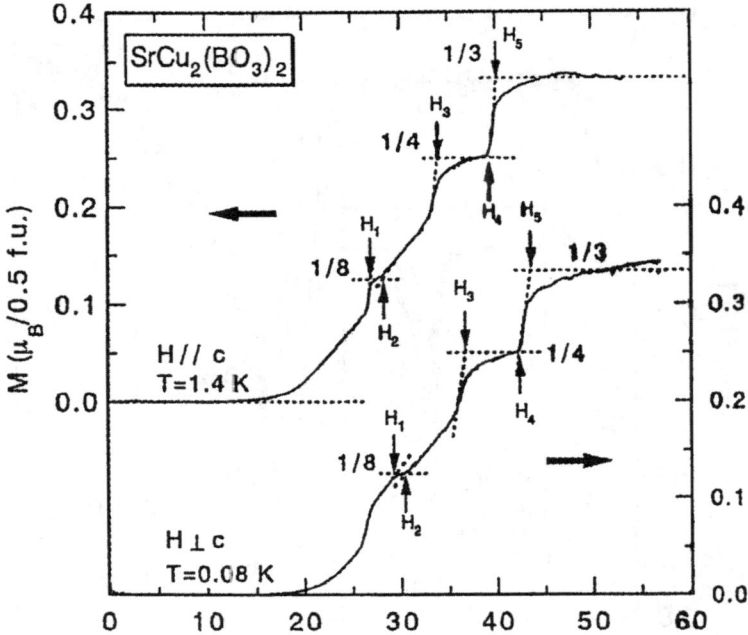

Fig. 12.17. Magnetization for $B \parallel c$ and $B \parallel a$ in SrCu$_2$(BO$_3$)$_2$ for $T = 0.08$ K and 1.4 K (Onizuka et al. [12.63])

10^{-3} at each plateau), the c_{11} mode exhibits sharp features at the onset of the $\frac{1}{8}$ plateau and somewhat less pronounced at the $\frac{1}{4}$ plateaus and a small effect at the $\frac{1}{3}$ plateau. The largest effect by far, however, is shown by the c_{66} mode. It has an overall softening of 25% by the $\frac{1}{4}$ and the $\frac{1}{3}$ plateaus and a somewhat smaller effect at the $\frac{1}{8}$ plateau.

These field dependent effects can be interpreted nicely with the same formalism as for the temperature dependence (Sect. 12.2). We first notice that the large minima in the relative elastic constant for the c_{66} mode occur at 36,3T and 42.9T. These are the field values where the magnetization (Fig. 12.17) changes to the $\frac{1}{4}$ and $\frac{1}{3}$ plateaus respectively. The analogy to the resonant interaction between the sound wave mode and the magnetic excitation is evident. The smaller minima in the steep slope at 26.5T and 32.4T have to be related with the $\frac{1}{8}$ plateau. The minima for the longitudinal mode c_{11} (which is shown enlarged in Fig. 12.19a,b together with the attenuation) are just inverse to the case of c_{66}. The biggest minimum occurs at 26.7T just at the $\frac{1}{8}$ plateau with a broad shoulder at 25.3T. The minima before both the $\frac{1}{4}$ and the $\frac{1}{3}$ plateau at 36.2T and 43.4T respectively, are distinctly smaller. Figure 12.19a,b shows the temperature dependence of these plateau effects for the c_{11} mode together with an ultrasonic attenuation curve

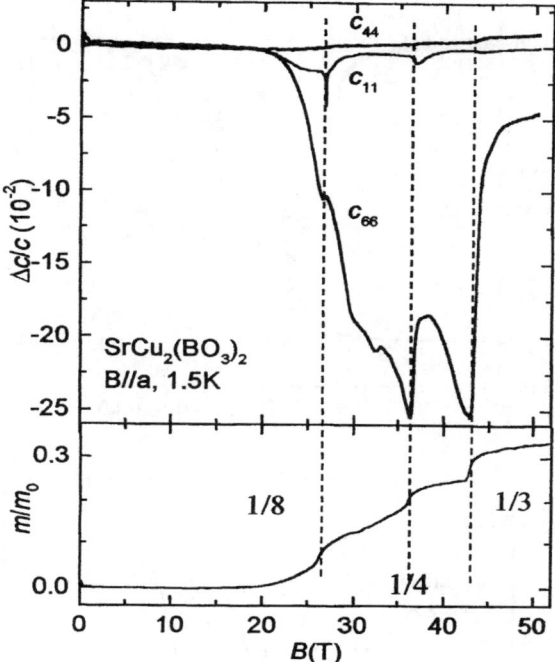

Fig. 12.18. SrCu$_2$(BO$_3$)$_2$: Magnetic field dependence of the elastic constants c_{11}, c_{44} and c_{66} for $T = 1.5$ K (Wolf et al. [12.44]). Also shown is the magnetization for the same field range (Onizuka et al. [12.63])

(Lüthi et al. [12.65]). The anomaly of c_{11} at 26.7T is so sharp that it cannot completely be resolved with the pulsed magnetic field technique.

The smallness of the magnetic field effects for the c_{44} mode can be similarly explained as in the case of the temperature dependence (Sect. 12.2). The strain for this mode does not lie in the a–b plane and couples therefore only in higher order to the spin system.

As mentioned above, the strong effects for the c_{11} and especially for the c_{66} mode can be explained with the resonant interaction of the acoustic mode ω_a and the magnetic excitation ω_m. Instead of a thermodynamic argument, as for (12.9) and $\omega\tau < 1$, we use coupled equations of motion, similar to the case of spin wave–phonon interaction (Chap. 11). They give close to the intersection

$$(\omega - \omega_a)(\omega - \omega_m) = g^2. \qquad (12.10)$$

The coupling g can be frequency and k dependent. For a realistic description, damping terms for the modes $\omega_m + i/\tau$ had to be introduced. A quantitative theory for such a resonant interaction does not exist yet.

Finally, the question why different plateaus couple with different strengths to the various elastic modes needs to be addressed (see the different magnetic

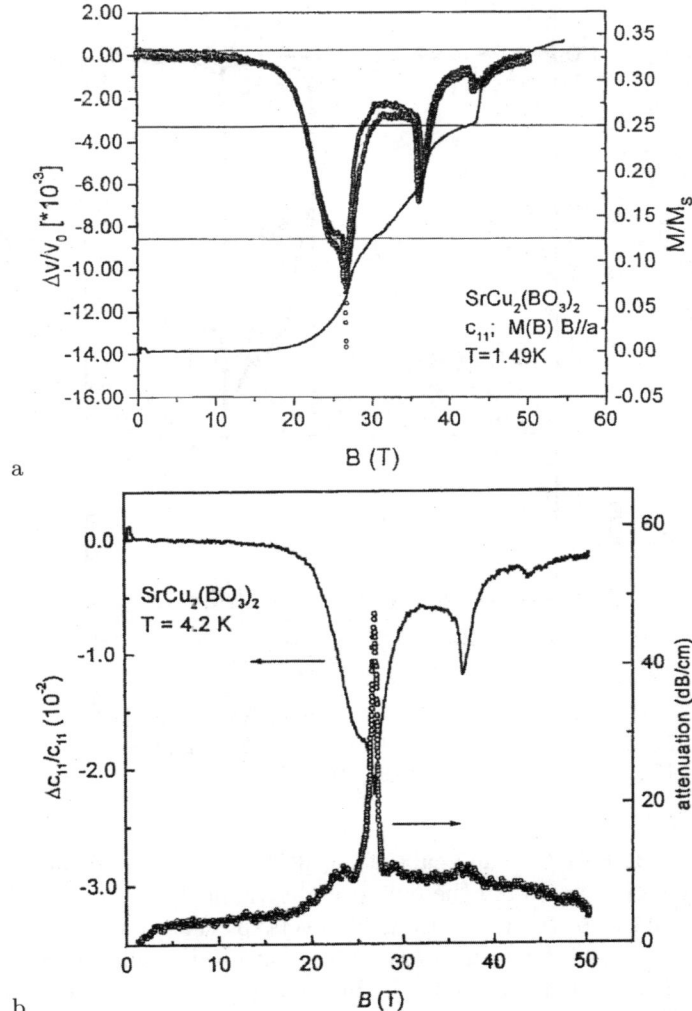

Fig. 12.19. (a) SrCu$_2$(BO$_3$)$_2$: Magnetic field and temperature dependence of the plateau effect for the c_{11} mode, (b) SrCu$_2$(BO$_3$)$_2$: Ultrasonic attenuation as a function of magnetic field for the c_{11} mode at 86.1 MHz (Lüthi et al. [12.65])

field behaviour for the c_{11} and c_{66} modes in Figs. 12.18, 12.19a,b). For this, the symmetries of the elastic modes in the point symmetry group D_{2d} appropriate for SrCu$_2$(BO$_3$)$_2$ are taken into consideration. c_{11} transforms as A_1 and c_{66} as B_2. Since these modes couple to the soft magnetic modes of the same symmetry and since the order parameter of a given plateau must have the same symmetry as the soft mode, it can be concluded that the order parameters of the plateaus $\frac{1}{4}$ and $\frac{1}{3}$ must have a B_{2g} symmetry while the $\frac{1}{8}$ plateau could have A_{1g} or B_{2g} symmetry. First calculations of the condensed triplet distributions in these plateaus give some evidence for this symmetry assignment (Miyahara and Ueda [12.39]).

For the $\frac{1}{8}$ plateau, NMR experiments have been performed recently in static high magnetic field (Kodama et al. [12.66]). It was found that the beginning of the plateau occurs with a discontinuous jump in the magnetisation, similar to the c_{11} change in Fig. 12.19a,b, indicating a first order transition. Furthermore a magnetic superstructure was found which can be explained with displacements due to local strains generated by the same exchange striction mechanism (Miyahara et al. [12.67]) as used in this chapter for the elastic constant anomalies.

NH_4CuCl_3

Figure 12.5 displays the magnetization versus magnetic field for different temperatures. The magnetization exhibits clear plateaus at $m/m_0 = \frac{1}{4}$ and $\frac{3}{4}$ but not at $\frac{1}{2}$ as expected from (12.3). This has been explained tentatively by Matsumoto [12.32] using the inequivalent three dimers discussed in Sect. 12.2.1. With each dimer contributing $\frac{1}{4}$ per unit cell, the magnetization arises due to the splitting of the triplet in the magnetic field. The lowest branch of the triplet gives rise to the magnetization when it crosses the singlet state. The A dimer gives $\frac{1}{4}$ of the magnetization, the two B dimers add another $\frac{1}{2}$ which leads to $\frac{3}{4}$ of the total magnetization, and finally the C dimer brings the system into saturation. The ESR experiments give clear indications of a soft mode at the beginning and end of each plateau (Kurniawan et al. [12.68], Schmidt et al. [12.29]) as expected. Again the model of Matsumoto can account for the different ESR branches. Comparing Figs. 12.5 and 12.20, the ESR branches originating at $H_{c1} = 5T$ and $H_{c3} = 18T$ are triplet–singlet excitations whereas the ones ending at $H_{c2} = 13T$ and $H_{c4} = 25T$ are singlet–triplet excitations. The acoustic waves interact again with these soft magnetic modes. In Fig. 12.20 we give experimental results for the c_{66} mode at 1.5 K (Schmidt et al. [12.29]). We give the sound wave results together with the magnetization and with the magnetic susceptibility (from Shiramura et al. [12.15]) and low frequency ESR results. It is seen that the acoustic result has essentially the same field dependence as the magnetic susceptibility i.e. $\Delta c \sim -\chi_m$. This is not quite the same as for the case of $(VO)_2P_2O_7$ (12.9). A detailed Ginzburg–Landau theory for the different phases given recently by Matsumoto and Sigrist [12.69] can explain various features of the measurements of Fig. 12.20.

$CsCuCl_3$

This system was introduced previously in Sect. 7.2.1. Also in this substance magnetization plateaus were found at $\frac{1}{3}$ of the saturation magnetization (Nojiri et al. [12.70]) for fields in the hexagonal plane at around 31T. High-field ultrasonic measurements were performed in this substance at low temperature up to 50T (Wolf et al. [12.33,12.34]). The c_{11} mode exhibits large sound velocity and sound attenuation anomalies in the region 30-33 T accompanied by some hysteresis. The step function-like anomaly in the elastic constant can

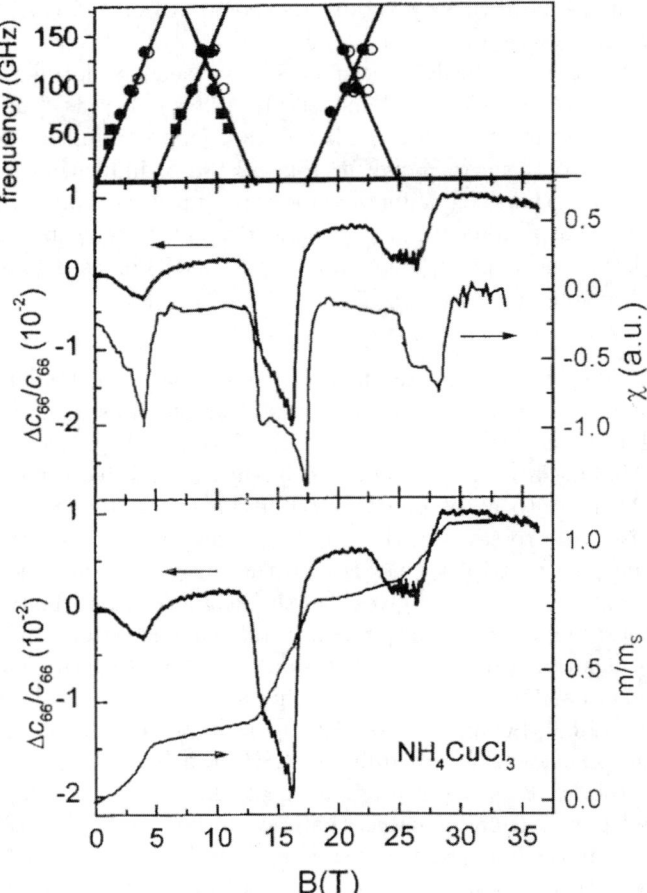

Fig. 12.20. NH_4CuCl_3: (a) ESR results for frequencies < 150 GHZ, indicating the relation between the plateau fields and the sound wave anomalies, (b) relative elastic constant change c_{66} as a function of magnetic field ($\boldsymbol{B} \parallel a$) up to 37T in NH_4CuCl_3 together with differential susceptibility, (c) same elastic constant together with magnetization. The ultrasonic frequency is 49 MHz. The measurements have been taken at 1.5K (Schmidt et al. [12.29])

be explained with a magneto-elastic coupling of the ε_{xx} strain with the triangular spin lattice (Wolf et al. [12.34]). There is an additional anomaly seen in the field region 10–15T, which is due to the recently found incommensurate phase region IC3 (Stüsser et al. [12.71]).

In summary, low dimensional spin systems give a rich variety of interesting physical effects which do not show up in higher dimensions. They can be studied particularly well in very high magnetic fields such as pulsed fields up to 50–60T. These experiments constitute a nice application of the technique

12.4 Thermal Conductivity in Low Dimensional Spin Systems

of pulsed field ultrasonics. Detailed microscopic theories for these high-field ultrasonic effects are still lacking.

12.4 Thermal Conductivity in Low Dimensional Spin Systems

Ultrasonic waves in low dimensional spin systems exhibit strong anisotropies in the sound velocities for different propagation directions. This was shown in NaV$_2$O$_5$ for the c_{55} and c_{66} modes (Fig. 7.7) and in SrCu$_2$(BO$_3$)$_2$ for the c_{44} and c_{66} modes (Fig. 12.12). Thus in thermal conductivity, due to the spin–phonon coupling, similar anisotropies for these substances should be expected. Although such experiments have just begun to be performed, the compounds studied so far have given quite unexpected results. These we review briefly in the following. Results of thermal conductivity experiments in NaV$_2$O$_5$ (Vasil'ev et al. [12.72]), in SrCu$_2$(BO$_3$)$_2$ (Hofmann et al. [12.73], Kudo et al. [12.74]) and in Sr$_{14}$Cu$_{24}$O$_{41}$ (Sologubenko et al. [12.75], Hess et al. [12.76]) are shown below. In all these substances, a two-peak structure is found in the thermal conductivity as a function of temperature. The interpretation of these phenomena, however, is different in the various substances.

NaV$_2$O$_5$

This substance was discussed in Sect. 7.1. It exhibits a charge ordering phenomenon at $T_c = 34$ K, the result of which is shown schematically in Fig. 7.5. For $T > T_c$, the average valence is 4.5. In Fig. 12.21, the thermal conductivity κ is shown together with the magnetic susceptibility χ (Vasil'ev et al. [12.72]). κ exhibits a broad maximum at 70 K and a steep rise just below T_c with a subsequent maximum at ≈ 15 K. The authors tentatively explain their results with a phonon thermal conductivity for $T > T_c$ which decreases on approaching T_c due to the enhancement of phonon scattering. The onset of charge ordering at T_c apparently switches off this extra scattering and leads to a huge increase of κ. The measured sound velocity anomalies in this substance (Schwenk et al. [12.77]) (see Figs. 7.6, 7.7) can be substantial, especially for the c_{66} mode for $T \geq T_c$. This could explain at least part of the decrease of κ in this temperature region. A quantitative analysis of the results of Fig. 12.21 has not been given yet.

SrCu$_2$(BO$_3$)$_2$

Thermal conductivity κ in this compound has been measured by two groups with qualitatively similar, but quantitatively somewhat differing results (Hofmann et al. [12.73], Kudo et al. [12.74]). The more complete results, both experimentally and theoretically, (Hofmann et al. [12.73]) are shown in Fig. 12.22. In zero field, both κ_a and κ_c show pronounced double-peak structures as a function of temperature. For both directions, the low-T maximum is drastically suppressed by a magnetic field. These results are explained with a purely phononic thermal conductivity. The double peak structure and its

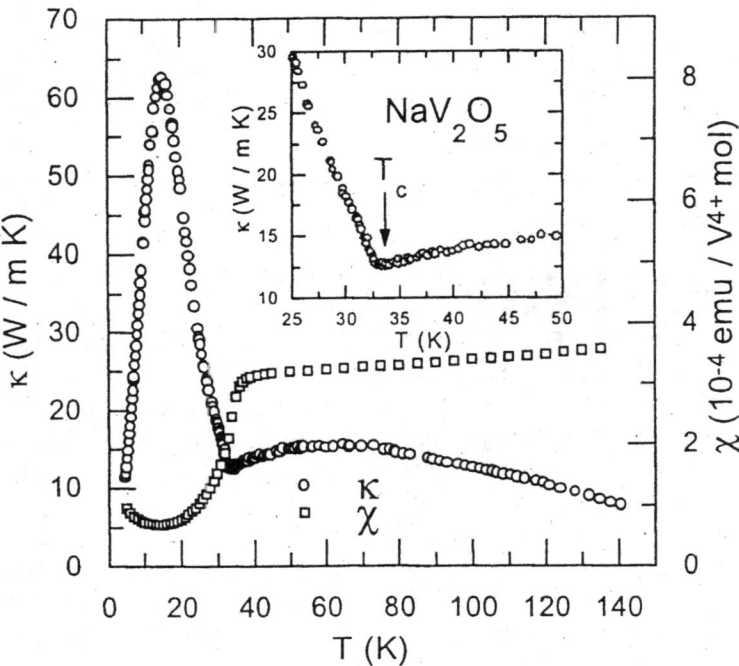

Fig. 12.21. Thermal conductivity κ and magnetic susceptibility χ as a function of temperature for NaV$_2$O$_5$. Inset: κ near charge order transition T_c (Vasil'ev et al.1998)

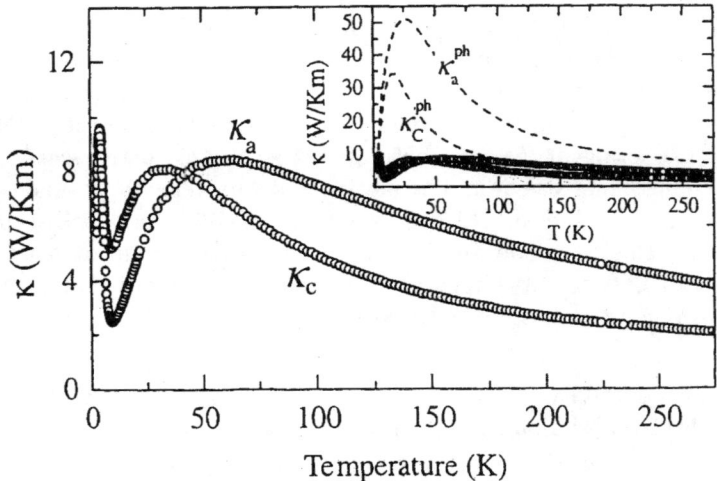

Fig. 12.22. Thermal conductivity of SrCu$_2$(BO$_3$)$_2$ along (κ_a) and perpendicular (κ_c) to the magnetic planes (Hofmann et al. [12.73]). Inset shows conventional phononic κ_a, κ_c without magnetic scattering

field dependence is explained by a strong damping of the phonon heat current due to resonant scattering of phonons by magnetic excitations. The low temperature thermal conductivity, with the maxima around 4.5 K, is suppressed by a magnetic field. This excludes magnon transport because the lowest excitations are triplet excitations with a decreasing branch in the magnetic field (see previous Sect. 12.3). The higher temperature thermal conductivities, with maxima at ~ 60 K for κ_a and ~ 30 K for κ_c, are field independent. The triplet excitation has practically no dispersion as determined from inelastic neutron scattering (Kageyama et al. [12.47]), hence vanishing magnon group velocity. The magnon thermal conductivity is again negligible. The phonon thermal conductivity mechanism with strong scattering on magnetic excitations gives a quantitative explanation. The minima of κ_a and κ_c occur at the maxima of C/T with C the specific heat. The latter is the temperature derivative of the magnetic entropy and thus directly related to the number of magnetic excitations which serve as scatterers for the phonons.

$Sr_{14}Cu_{24}O_{41}$

The crystal structure of this compound contains two, quasi one-dimensional magnetic subsystems along the c–axis as shown in Fig. 12.23. One subsystem is a sheet-like arrangement of Cu_2O_3 two-leg ladders where the Cu^{2+} $S = \frac{1}{2}$ spins are strongly coupled via a Cu-O-Cu super-exchange, while the other subsystem is an array of CuO_2 $S = \frac{1}{2}$ spin chains with weak magnetic interactions. While even the stoichiometric compound $Sr_{14}Cu_{24}O_{41}$ is hole doped, the holes are predominantly located in the chains. This can be deduced from a simple valence consideration (see Fig. 12.23): The Sr-plane gives +14; the ladder plane $(7 \times 2 \ Cu^{2+} + 7 \times 3 \ O^{2-}) = -14$; the next Sr-plane again +14 and the chains $(6Cu^{3+} + 4Cu^{2+} + 2 \times 10 \ O^{2-}) = -14$. This leads to an

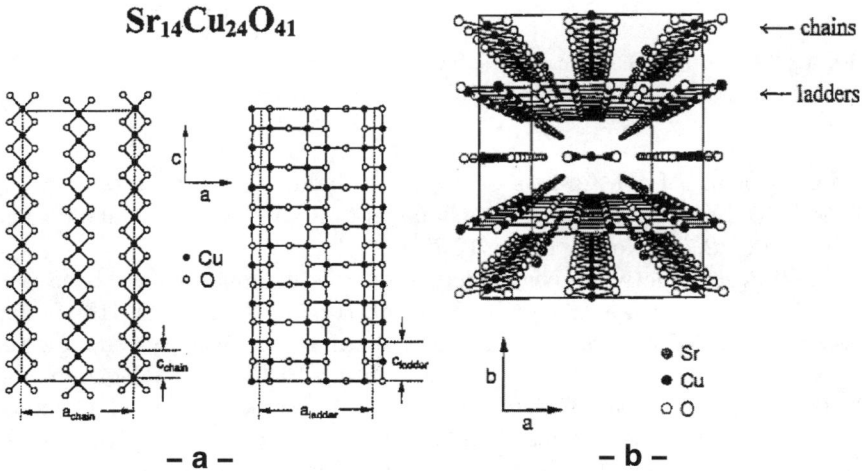

Fig. 12.23. Schematic representation of the Structure of $Sr_{14}Cu_{24}O_{41}$

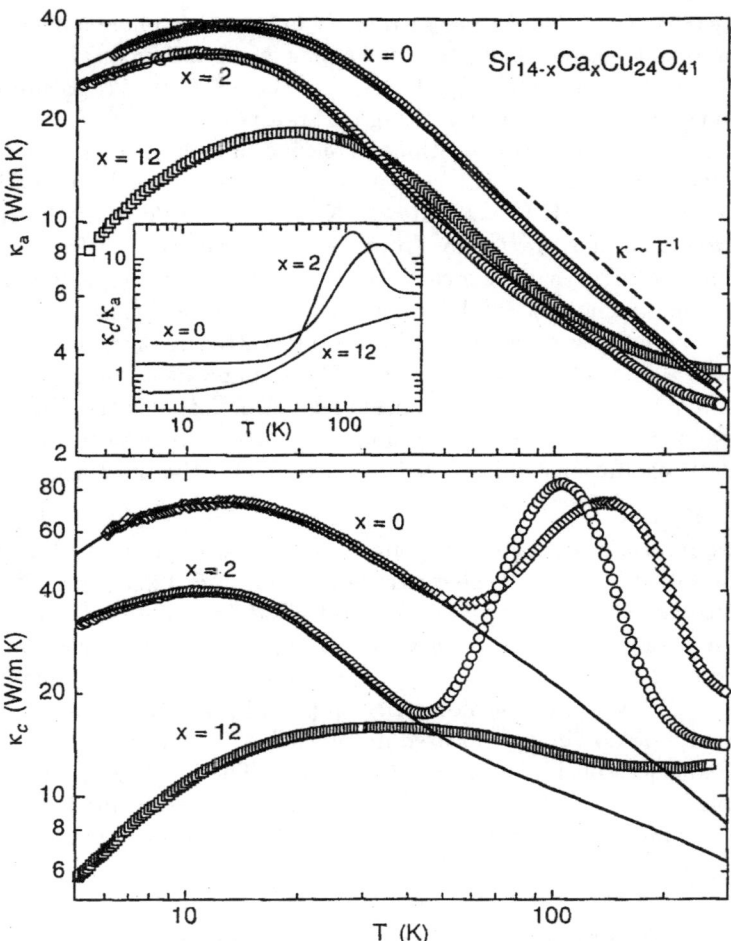

Fig. 12.24. Thermal conductivity κ for a and c directions for $Sr_{14-x}Ca_xCu_{24}O_{41}$ with $x = 0, 2$ and 12 (Sologubenko et al. [12.75])

average valence of the Cu-ions of 2.25. The ladders have a large spin gap of $\Delta \sim 400$ K and the gap in the chains is only $\Delta \sim 114$ K (Matsuda and Katsumata [12.78], Koenig et al. [12.79]).

Thermal conductivity measurements on this compound have been performed by three groups (Sologubenko et al. [12.75], Kudo et al. [12.80] and Hess et al. [12.76]). Figure 12.24 shows thermal conductivity data for the a and c–direction for $Sr_{14-x}Ca_xCu_{24}O_{41}$ with $x = 0, 2, 12$ (Sologubenko et al. [12.75]). Again a two-maxima function was measured with the lower maximum due to phonon thermal conductivity. The upper maximum around 140 K, occurring only for κ_c but not for κ_a or κ_b, however, seems to be due to magnetic excitations propagating along the spin ladders.

The occurrence of transport of magnetic excitations in these compounds is surprising. So far, for magnetic materials, magnon transport has been only thoroughly investigated in the ferrimagnet $Y_3Fe_5O_{12}$ (YIG) whose properties we have discussed in Sect. 11.2. Spinwave transport could only be observed at low temperature (Douglas [12.81], Lüthi [12.82], Slack and Oliver [12.83]). The scattering processes for spinwaves in ferrimagnets are manifold and one needs a very good crystal with very small ferromagnetic resonance linewidth as found in YIG. In addition, a sizable magnon–phonon interaction is needed in order to change from phonon to magnon thermal conductivity at the emitter and vice versa at the detector (Sanders and Walton [12.84]). On the other hand, in low dimensional substances like $Sr_{14}Cu_{24}O_{41}$ with an appreciable energy gap of hundreds of degrees the only scatterers are imperfections of the crystal. One expects an activated behaviour for κ_c along the ladders and chains, (observed by Hess et al. [12.76]) and no magnetic heat conduction κ_a and κ_b. In $Ca_9La_5Cu_{24}O_{41}$ the magnetic excitation thermal conductivity is even bigger (Hess et al. [12.76]). This has apparently to do with the fact that in this crystal the hole doping in the ladder is much smaller.

Summary

We have seen that the two maxima structure of the thermal conductivity in these three examples have a different physical origin. In the first two cases, they are due to phonon heat transport, but with different scattering mechanisms. By far the most surprising effect is in the telephone compound (and related ones) $Sr_{14}Cu_{24}O_{41}$ where magnetic heat transport has been identified. More experimental and theoretical information is needed to understand the surprisingly large mean free path of the magnetic excitations in this case.

Apart from these three substances, similar experiments have been performed in low dimensional spin systems. Examples are $CuGeO_3$ (Takeya et al. [12.85], Vasil'ev et al. [12.72]), or $KCuF_3$ (Miike and Hirakawa [12.86]) and Yb_4As_3 (Köppen et al. [12.87]). But we will not discuss these compounds here any further. A review on "energy transport in one-dimensional spin systems" has recently appeared (Sologubenko and Ott [12.88]).

12.5 Peierls and Spin Peierls Effects

One-dimensional metals are unstable towards structural deformations. This is the so-called Peierls effect (Peierls [12.89]). Periodic ion displacements give rise to new Brillouin zone and if the Fermi energy lies within the created energy gaps at the new zones, the energy of the system is lowered. So for a one-dimensional metal with one conduction electron per atom there is a structural transition with a dimer formation, i.e. every second ion is displaced and the new Brillouin zone vectors change from $k_{BZ} = \pm\pi/a$ to $k_{BZ} = \pm\pi/2a$. This holds for non-interacting one electron states. Furthermore, for a

324   12 Ultrasonics in Low Dimensional Spin Systems

one-dimensional metal, the dielectric constant (Lindhard function) exhibits a divergence leading also to a structural anomaly at $q = 2k_F$ (with k_F the Fermi wave vector). Here the screened Coulomb interaction of the electron gas is taken into account. A half-filled band leads equivalently to a Peierls distortion. Two systems exhibiting the Peierls effect are discussed: KCP and (TaSe$_4$)$_2$I.

There is an analogous effect, called Spin–Peierls effect. The exchange striction coupling, discussed in Sect. 5.1.3, gives rise to a structural instability for one-dimensional spin systems. Apart from organic Cu-compounds, there is one anorganic compound CuGeO$_3$, which exhibits such an effect. This will be briefly discussed below.

There is an interesting book on one-dimensional metals (Roth [12.90]). It emphasises the material aspects and many physical properties of a wide variety of substances, with a strong emphasis on organic polymers.

K$_2$Pt(CN)$_4$Br$_{0.3}$·3H$_2$O (KCP)

This mixed valence platinum compound is a quasi one-dimensional conductor and represents a prototype of a Peierls system (not to be confused with a spin–Peierls compound). For one-dimensional metallic systems, the dielectric constant shows a divergence for $q = 2k_F$, the Fermi surface diameter, due to the screening of the electron gas as explained above. The effect of this instability on the phonon spectrum is called the Kohn effect (Kohn [12.91]). For one-dimensional metals like KCP, there is a giant Kohn anomaly, but the phonon softening of $2k_F$ phonons is not complete, there is no three-dimensional long range order down to low temperatures (Renker et al. [12.92, 12.93]). We mention this system here because the temperature dependence of the elastic constants has been interpreted with a pseudo spin–lattice model (Kurihara et al. [12.94]). For a review of Peierls structural transitions see Bulaevskii [12.95].

The elastic constants for KCP have been measured by Doi et al. [12.96]. It was found that the c_{66} mode exhibits strong softening of 50% from 200 K down to 100 K. By extrapolation, a structural transition at about 60 K is expected but was not found. The c_{44} mode shows smaller softening but a pronounced attenuation below 100 K.

These data can be explained semiquantitatively by modeling the thermal rotations of some water molecules by pseudo-spins. By introducing $S = 1$ Ising spins for four water molecules and by taking a spin lattice coupling of the form $H_{SL} = g\varepsilon_{xy}(S_1^2 - S_2^2 + S_3^2 - S_4^2)$, one obtains a $c_{66}(T)$ with a minimum at 90 K and strong attenuation around that temperature. With this coupling, the strain induces a polarization in the pseudo spin system, which in turn leads to the elastic constant renormalization (Kurihara et al. [12.94]).

(TaSe$_4$)$_2$I

This tetragonal compound (space group $I422$) represents another Peierls system with a transition temperature of $T_p \approx 260$ K. The chain axis is the c-axis. This transition is a semiconductor–metal transition with the electrical resistance showing activated behaviour below T_p. In this substance, sound velocity, sound attenuation and thermal expansion experiments have been performed (Saint-Paul [12.97, 12.98]). From the measured elastic modes c_{11}, c_{33}, c_{44}, c_{66}, the c_{44} mode exhibits strong softening on approaching T_p accompanied by strong attenuation around T_p. There is an anisotropy of the two c_{44} modes with $q \parallel c$, $u \perp c$ and $q \perp c$, $u \parallel c$ respectively with the latter exhibiting fluctuation contributions near T_p.

CuGeO$_3$

Similar to the Peierls effect in one-dimensional metallic systems like KPC, discussed above, a spin–Peierls effect can occur due to spin–strain coupling of the exchange striction type (Sect. 5.1.3). Theories predict a phonon softening at the spin dimerization phase transition (Pytte [12.99]). CuGeO$_3$ is a quasi one-dimensional spin $\frac{1}{2}$ system which undergoes a structural transition at $T_{SP} = 14.5$ K. This is the only an-organic Spin–Peierls system found so far. But it does not exhibit a soft phonon mode as predicted. For a discussion of the temperature dependence of symmetry elastic constants for this low dimensional spin–Peierls compound we refer to Sect. 7.1. There CuGeO$_3$ and the charge order compounds NaV$_2$O$_5$, Yb$_4$As$_3$ are treated and compared together. Especially the related shear elastic modes of CuGeO$_3$ and NaV$_2$O$_5$ are shown together and compared (see Fig. 7.7). Unlike NaV$_2$O$_5$, the Spin–Peierls compound CuGeO$_3$ does not exhibit any substantial elastic constant softening at the Spin–Peierls transition T_{SP}. The shear modes c_{44}, c_{55} and c_{66} give only small anomalies at T_{sp} of the order of $\Delta v/v \approx 10^{-4}$ in the relative sound velocities for $T = 300$ K (Ecolivet et al. [12.100], Schwenk et al. [12.77]). For the longitudinal modes c_{11}, c_{22} and c_{33}, step function-like behaviour at T_{sp} of the order of $10^{-4} - 10^{-3}$ can be observed (Saint-Paul et al. [12.101]) that is in agreement with a quadratic strain–order parameter coupling (Sect. 4.3.4). The absolute elastic constants of CuGeO$_3$, with a mass density of $\rho = 5.1$ g/cm^3 for $T = 300$ K, are (Ecolivet et al. [12.100]): in 10^{11} erg/cm^3, $c_{11} = 6.4$, $c_{22} = 3.76$, $c_{33} = 3.173$, $c_{44} = 3.53$, $c_{55} = 3.53$, $c_{66} = 1.84$ $c_{12} = 3.21$, $c_{13} = 4.69$, $c_{23} = 2.27$.

The lattice dimerization, which occurs below T_{SP}, gives rise to a magnetic spin singlet ground state separated from the excited triplet. Application of a magnetic field suppresses the quantum fluctuations and suppresses finally the singlet–triplet energy gap. The B–T phase diagram consists of a low field dimerized spin liquid, followed by an incommensurate phase at higher fields. The phase diagram has been mapped with different experimental techniques. Ultrasonic experiments have been performed by Saint-Paul et al. [12.101] and Poirier et al. [12.102]. Substituting Zn for Cu in CuGeO$_3$ by an amount of 2% changes the phase diagram radically. A low temperature anti-ferromagnetic phase is created by such a substitution. Again the phase diagram could be mapped for this case by ultrasonic experiments (Saint-Paul et al. [12.103]).

12.6 Perovskite-Type Layer-Structure Materials

Another type of quasi low-dimensional materials are the perovskite-type layer structures with the chemical formula $(C_nH_{2n+1}NH_3)_2MCl_4$ where $n = 1, 2, 3$ and $M =$ Mn, Fe, Cd, Cu. They exhibit a quasi two-dimensional character in their physical properties and have a variety of low temperature structural phase transitions. They also show quasi two-dimensional magnetic properties for the Mn compounds at low temperatures. In this respect they are related to the other quasi two-dimensional perovskite layer compound K_2NiF_4 which has been discussed in Sect. 12.2.2. There is also the structural similarity to the high temperature superconductors and its parent compound La_2CuO_4 discussed in Sect. 10.4.

Acoustic measurements have been carried so far out on the following $(C_nH_{2n+1}NH_3)_2MCl_4$ compounds: For $n = 1$ with $M =$ Cd and Mn, abbreviated MACdC and MAMC by Goto et al. [12.104, 12.105]) and with $M =$ Fe it is MAFeC by Goto et al. [12.106]; for $n = 2$ with $M =$ Mn by Zherlitsyn et al. [12.107] and with $M =$ Fe by Suzuki et al. [12.108]. In addition, Brillouin scattering experiments were performed for $(CH_3NH_3)_2FeCl_4$ by Yoshihara et al. [12.109] and for $n = 2$ $(C_2H_5NH_3)_2FeCl_4$ by Yoshihara et al. [12.110]. These systems exhibit order–disorder type of structural phase transition starting from a tetragonal high temperature phase (THT), changing to an orthorhombic type phase (ORT), then to a low temperature tetragonal phase (TLT) and finally to a monoclinic phase (MLT).

MAMC

As an example we take MAMC: The corresponding phase transition temperatures are with space group nomenclature:

THT	– 392K –	ORT	– 255K –	TLT	– 94K –	MLT
$I4/mmm$		$Abma$		$P4_2/ncm$		$P2_1/b$

Various elastic modes show anomalies at the different transition temperatures and the c_{66} mode with the ε_{xy} strain in the tetragonal plane exhibits strong softening in the two tetragonal phases THT and TLT, as shown in Fig. 12.25. Note that the c_{66} mode softens practically completely in the TLT phase approaching the phase transition temperature from below.

The phase transitions arise from orientation and reorientation of the molecular (CH_3NH_3) groups and a corresponding order–disorder model, describing the sequence of the phase transitions, has been developed by Blinc et al. [12.111]. The soft acoustic mode c_{66} in the TLT phase arises due to a bilinear strain–order parameter coupling and can be quantitatively fitted with (4.13) (Goto et al. [12.105]), see also Ishibashi and Suzuki [12.112], Geick and Lüthi [12.113], Couzi [12.114]. For a similar order–disorder model with order parameter–strain coupling see also Appendix G.

12.6 Perovskite-Type Layer-Structure Materials

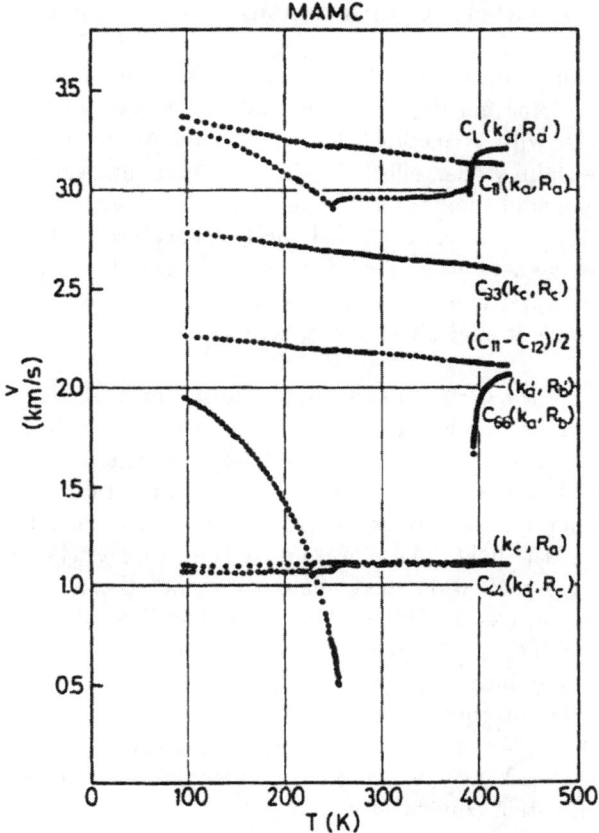

Fig. 12.25. Sound velocities of MAMC in the THT, ORT and TLT phases (Goto et al. [12.105])

MAFeC

Especially the TLT–ORT phase transition was investigated in detail by Goto et al. [12.106] for $(CH_3NH_3)_2FeCl_4$. The c_{66} mode was measured as a function of temperature with hydrostatic pressure as a further parameter. A pressure of 6 kbar gave large changes of ≈ 20 K for the change in the transition temperature. In addition, the transition of THT to ORT can be described by the XY model, involving an order parameter with two spin components. Critical dynamics procedures were applied to the various modes and the specific heat, exhibiting a λ-type anomaly at this THT–ORT phase transition. Brillouin scattering experiments (Yoshihara et al. [12.109]) for the same system gave very similar results as the ultrasonic experiments (5 MHz), namely a strong softening for the c_{66} mode in the TLT region, indicating that $\omega\tau < 1$ even in the 30 GHz region.

12.7 Bose–Einstein Condensation of Magnons in TlCuCl$_3$

The AMX_3 compounds, with A, M metallic ions and X a halogen, exhibit a wide range of fascinating magnetic and structural effects. The hexagonal compounds, with A monovalent cations and M divalent $3d$ ions, show a cooperative Jahn–Teller effect (CsCuCl$_3$ treated in Sect. 7.2.1) or quasi one-dimensional spin chains (CsNiCl$_3$ in Sect. 12.2.1). Another class form the monoclinic TlCuCl$_3$, NH$_4$CuCl$_3$ and KCuCl$_3$ (space group $P2_1/c$). Of these, NH$_4$CuCl$_3$ is a magnetization plateau substance treated before in Sects. 12.2.1 and 12.3. These three compounds form double spin chains, the structure having double chains of edge-sharing octahedra CuX$_6$ along the a-axis (Tanaka et al. [12.30]).

The spin $\frac{1}{2}$ of the Cu^{2+} ions in TlCuCl$_3$ form dimer singlet ground states, with excited triplet states that have an excitation gap of $\approx 7.5\,\mathrm{K}$. With an external field of this magnitude, this singlet–triplet gap collapses and long range staggered magnetic order perpendicular to the magnetic field sets in. This is a field-induced, canted anti-ferromagnet with a magnetization component parallel to the field. Therefore, unlike in ferromagnets where magnons diminish the magnetization, here magnons create the magnetization. The magnetic anisotropy is negligible. The B–T phase boundary can be written $B - B_c \approx T^\Phi$ with $\Phi = 2.2$ and $B_c = 5.61\,\mathrm{T}$. Such a quantum spin system can be mapped onto an interacting Bose gas (Nikuni et al. [12.115]) and the phase transition can be interpreted as a Bose–Einstein condensation. For $B > B_c$, a gap-less soft magnon mode with a linear dispersion law has been observed (Ch. Rüegg et al. [12.116]), which is apparently the so-called Goldstone mode for the Bose–Einstein condensate.

A recent sound attenuation study at low frequencies of 5 MHz found a hysteretic behaviour at the $T_c - B_c$ boundary, indicating a first order phase transition (Sherman et al. [12.117]). They found exchange striction coupling constants of the same order of magnitude as for other low-dimensional spin systems discussed above. Further sound wave experiments could help in elucidating the spinwave–phonon interaction.

12.8 Conclusion

In recent years, the field of low-dimensional physics has become very important. Apart from the quantum Hall effect physics (which is dealt with in Sect. 13.6), materials research has found many interesting compounds exhibiting typical low-dimensional phenomena for electronic systems (Peierls effect) and spin systems (spin Peierls effect, Haldane conjecture, plateau substances etc.) as pointed out in the beginning of this chapter.

Ultrasonic studies for the electronic properties of low-dimensional systems were confined more or less to determine the temperature dependence of elastic constants in Peierls systems. On the other hand, for low-dimensional spin systems sound wave experiments could determine the electron–phonon coupling mechanisms responsible for the temperature dependence of the elastic constants and the thermal expansion. In addition, they could be used to investigate the plateau substances in more detail by using pulsed high magnetic fields. Haldane systems have not been investigated so far with ultrasonics. But first acoustic experiments on Bose condensation in magnetic systems have been carried out.

13 Symmetry Effects with Sound Waves

Because of the tensor properties of elastic strains and elastic constants, a wide variety of experiments exhibiting symmetry properties can be performed. These range from symmetry effects in magnetized materials, to rotationally invariant magneto-elastic interactions, to non-reciprocal effects using surface acoustic waves and to the magneto-acoustic analoga of magneto-optical effects such as optical activity, Faraday effect, Cotton-Mouton effect, Kerr effect. One special symmetry effect, using the time reversal symmetry, is being explored with low frequency ultrasound in quite different media (see e.g. Fink [13.1]). An echo experiment using time reversal symmetry was dealt with in the discussion of structural phase transitions (Sect. 7.4.3), where an experiment using phonon echoes was discussed. Another short discussion of time reversal symmetry in phonon propagation was given in Sect. 2.3.

The various effects mentioned above will be described. Some of these properties have been discussed already in Chap. 11. The symmetry aspects of elastic strains, based on group theory, have been treated already in Sect. 3.2. Use will also be made of them here. The magneto-acoustic analogue of some magneto-optic effects has been discussed thoroughly in a recent book by Gudkov and Gavenda [13.2].

First, the symmetry breaking due to a magnetic field as seen in the field dependence of the elastic constants (Sect. 13.1) is discussed. An analogous symmetry breaking through a structural phase transition is mentioned also for a clear-cut case ($DyVO_4$). This is followed by a survey on rotationally invariant magneto-elastic effects (Sect. 13.2). Additionally, a number of magneto-acoustic effects which are analoga of magneto-optic effects (Sect. 13.3) are given followed by a short section on acoustical activity (Sect. 13.4). Non-reciprocal effects, as observed with acoustic surface waves (Sect. 13.5), are discussed next. Finally, we discuss surface acoustic wave experiments for systems exhibiting the integral and fractional quantum Hall effect (Sect. 13.6). With these latter experiments, a component of the conductivity tensor for the two-dimensional electron gas can be measured. At low enough temperature and with high enough SAW frequencies one observes unexpected features which can apparently be traced to new kinds of electronic states in this system.

13.1 Magnetic Field Induced Symmetry Breaking

Consider a system of magnetic ions in a paramagnetic state. As an example, the cubic Laves phase compound (space group Fd3m) CeAl$_2$ is given for $T > T_N = 3.8$ K. The Ce^{3+} ion ($J = 5/2$) splits in the CEF into a ground state doublet Γ_7 and a quartet Γ_8 at 90 K above. Although this compound has typical features of a dense Kondo compound (Bredl et al. [13.3]) and the CEF excitations are anomalous (Loewenhaupt et al. [13.4]), this does not influence the symmetry considerations. Various physical properties of this substance have been discussed in detail in Sect. 5.4. The structure of CeAl$_2$ is shown in Fig. 5.7.

Now consider the shear elastic constant c_{44} with propagation along a principal cubic axis and polarization perpendicular to it along another cubic axis. Taking a magnetic field B along the y-axis (010), gives three different geometries for the c_{44} propagation modes:

1. $\boldsymbol{q} \parallel (001)$, $\boldsymbol{e} \parallel (100)$ i.e. (q_z, e_x)
2. $\boldsymbol{q} \parallel (001)$, $\boldsymbol{e} \parallel (010)$ i.e. (q_z, e_y)
3. $\boldsymbol{q} \parallel (010)$, $\boldsymbol{e} \parallel (001)$ i.e. (q_y, e_z).

In zero magnetic field, the sound velocities in these three geometries are the same. Magnetic field dependencies of these three modes are shown in Fig. 13.1 (Lüthi and Lingner [13.5]). Of these three geometries, one and two are first considered. The small but significant difference between two and three are the subject of the next Sect. 13.2.

Notice in Fig. 13.1 a large difference between the c_{66}–like mode (q_z, e_x) and the c_{44}–like mode (q_z, e_y). The magnetic field along the (010)–axis makes the crystal elastically tetragonal. These magnetic field dependent modes, with the magneto-elastic interaction, can be calculated (Sect. 5.1.2). From the temperature dependence of the elastic constants (Lüthi and Lingner [13.5]), it is shown that the c_{44} mode (with Γ_5 symmetry, see Sect. 3.2) exhibits strong softening which can be analysed with the magneto-elastic single ion Hamiltonian (5.8):

$$H_{me} = g_3 \sum_i [\varepsilon_{xy} O_{xy,i} + \varepsilon_{xz} O_{xz,i} + \varepsilon_{yz} O_{yz,i}] \tag{13.1}$$

with $O_{ij} = J_i J_j + J_j J_i$ the corresponding quadrupolar operators. The fit to the experimental temperature dependence gives a very large magneto-elastic coupling constant $\mid g_3 \mid = 270$ K/ion. The physical significance of this large coupling constant for this Laves phase structure material has been discussed earlier in Sect. 5.4. Calculating the strain susceptibility in the presence of a magnetic field for the two geometries (q_z, e_x) and (q_z, e_y) with (5.16c) gives the full lines in Fig. 13.1. The salient features of the experiment are clearly recovered, the magnetic field makes the c_{44} mode in the plane perpendicular to B to a c_{66} mode and the magnetic field dependences of the two modes (c_{44} and c_{66}) have opposite signs. The non-perfect agreement of the strain

13.1 Magnetic Field Induced Symmetry Breaking

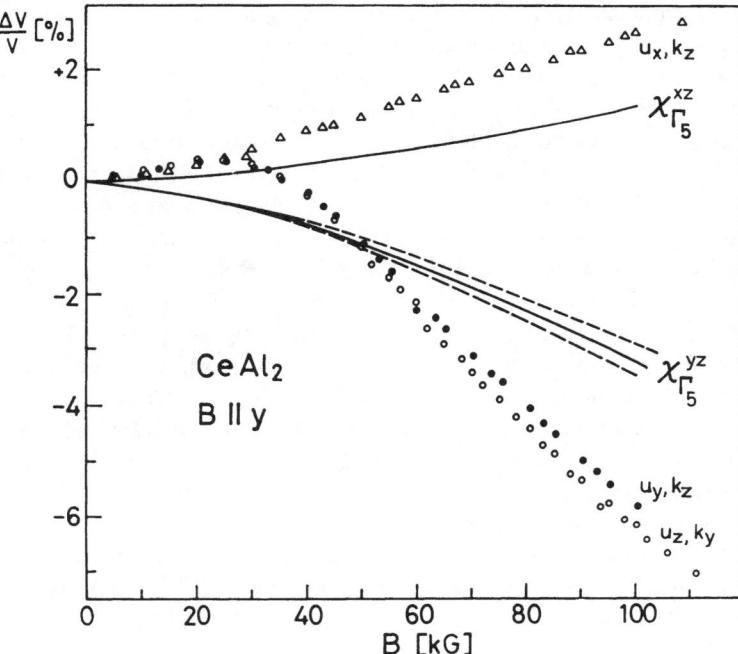

Fig. 13.1. Magnetic field dependence of the three different c_{44} modes at $T = 4.3\,\text{K}$ for CeAl$_2$ with $\boldsymbol{B}$ along the y-axis. Open circles $(q_y e_z)$–mode, closed circle (q_z, e_y)–mode, triangle (q_z, e_x)–mode. The full lines are calculations for (q_z, e_x) and (q_z, e_y)–modes. Broken lines correspond to splitting of (q_z, e_y) and (q_y, e_z)–modes due to rotationally invariant magneto-elastic interaction (see text), sound wave frequencies 10, 50, 90 MHz (Lüthi and Lingner [13.5])

susceptibility fit to the measured elastic modes in Fig. 13.1 is probably due to the vicinity of the magnetic phase transition. The experiment was carried out at 4.3 K.

Note that for the geometry considered, both acoustic modes propagate in the x-direction. Therefore any correction due to magneto-striction affects both modes in the same way. Since the ultrasonic effects are of the order of several % (Fig. 13.1) and the magneto-striction effects in CeAl$_2$ of the order of 10^{-4} (Fawcett et al. [13.6]), these corrections are completely negligible.

The effect on the elastic constant is due to the field-induced quadrupolar matrix element. It leads to a symmetry breaking effect in which the degeneracy of high symmetry elastic modes (e.g. transverse modes along a cubic axis) is lifted by the magneto-elastic interaction. The result of this symmetry breaking for the modes (q_z, e_x) and (q_z, e_y) leads immediately to magneto-acoustic birefringence (Voigt effect). This we will discuss, together with other magneto-acoustic effects in Sect. 13.3. Another ultrasonic investigation of CeAl$_2$ for the c_{44}–mode with $\boldsymbol{k} \parallel \boldsymbol{B}$ has been given with similar results to ours by Roth et al. [13.7]. The B–T phase diagram of CeAl$_2$ was determined with thermal expansion and ultrasonic experiments by Schefzyk et al. [13.8].

There are other systems which exhibit similar effects. A cubic ferromagnet shows a similar elastic constant dependence. Care has to be taken of domains in this case. For fields larger than the saturation field, the ferromagnetic crystal is no longer strictly cubic but tetragonal (or trigonal) due to magneto-strictive effects. Likewise, the c_{66} and c_{44} modes will be different in a single magnetic domain crystal. Such effects have been observed for ferromagnetic GdZn (Rouchy et al. [13.9]) and CoPt (Rouchy et al. [13.10]). By far the strongest effect of this kind is observed in RFe$_2$ crystals (with R rare earth ion), which exhibit giant magneto-striction effects (see Sect. 5.4). In TbFe$_2$, Tb$_{0.3}$Dy$_{0.7}$Fe$_2$, single crystals, poly-crystals or amorphous material, strong magnetic field-induced symmetry effects are observed (Cullen et al. [13.11]). The degeneracy of e.g. elastic c_{44} modes is broken and relative velocity differences of up to 26% have been observed in comparison to $< 1\%$ in another cubic ferrimagnet, magnetite Fe$_3$O$_4$ (Moran and Lüthi [13.12]). The change of the elastic properties related to the symmetry lowering due to the magnetisation M is called the morphic effect (Mason [13.13], de Lacheisserie [13.14]). It has to be described with higher order magneto-elastic contributions (see also Sect. 13.2).

ΔE Effect

For magnetic fields smaller than the saturation field, however, domain wall–stress effects play an important role. This is the ΔE effect (Becker and Döring [13.15], Bozorth [13.16]). For example, in Ni the c_{44} mode in the ferromagnetic state exhibits strong softening by lowering the temperature for $T < T_c$. This softening can be eliminated with a magnetic field sufficient to magnetise the sample into a single domain state. Similar effects have been observed for the elastic moduli E and G. The detailed interaction between a strain wave and a domain wall has not yet been studied. The field dependence of the magnetization for a soft ferromagnet with small magnetic anisotropy is governed first by domain wall displacement followed by domain rotation. This can be described quantitatively with the so-called Néel phase model (Néel [13.17]). But the domain wall–stress effects for these cases have been investigated theoretically only for metallic ferromagnets (Simon [13.18]). Qualitatively, it is possible to argue that an applied stress influences the magnetization direction, which gives rise to an additional magneto-strictive strain. Therefore the E module is lowered. The normal E module is the one in the fully saturated magnetic state. With (3.12b), assuming that the compressibility K is not influenced by the magnetization process, it can be concluded that beside the E module, the shear module G is also influenced: $\Delta(1/G) = 3\Delta(1/E)$. For further details see Becker and Doering [13.15].

A similar related effect to the symmetry breaking for ferromagnets – a structural symmetry breaking – can be observed at a quadrupolar phase transition in DyVO$_4$ (Gorodetsky et al. [13.19]). This case was treated in some detail in Sect. 7.2.2. A transverse c_{44} mode propagating along the z axis splits in a single domain state into c_{44} and c_{55} due to the tetragonal–orthorhombic phase transition. The appropriate free energy term coupling of the strain ε_{ij} to the order parameter η reads $F_{int} = c_{44}^0(\varepsilon_{yz}^2 + \varepsilon_{xz}^2) + K(\eta/2)(\varepsilon_{yz}^2 - \varepsilon_{xz}^2)$ leading to the symmetry breaking elastic constant $c_{44} = c_{44}^0 + K\eta$ and $c_{55} = c_{44}^0 - K\eta$ at the phase transition $T_a = 14\,\mathrm{K}$ (Gorodetsky et al. [13.19]). With a magnetic field of $\sim 0.6\mathrm{T}$, a single structural domain state in this case can be obtained in order to observe the effect more clearly.

13.2 Rotationally Invariant Magneto-elastic Effects

Looking again at Fig. 13.1, a small splitting in the magnetic field of the two c_{44} modes (q_y, e_z) and (q_z, e_y) can be seen. Within elasticity theory, these modes should be equivalent, belonging to the strain ε_{yz}. This magnetic field dependent effect is due to rotational invariance effects. This effect can be qualitatively explained with Fig. 13.2 (note that the axes are labeled x, z rather than y, z).

The rotational part ω_{ij} has opposite signs for the (q_x, e_z) and (q_z, e_x) modes ($\omega_{ij} = -\omega_{ji}$). Contributions to the rotational part of the deformation tensor (3.3), (3.4b) have been neglected so far. The meaning of the rotational part is illustrated in this figure for the the case of the transverse mode c_{44} in a cubic crystal. For 4f, ions this rotational interaction results from the effects

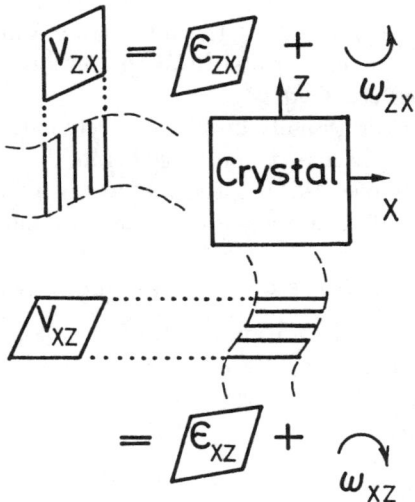

Fig. 13.2. Contributions to the total deformation tensor v_{kl} for two transverse c_{44} modes propagating along a cubic axis. The rotational part ω_{kl} has opposite sign for the (q_z, e_x) and (q_x, e_z) modes ($\omega_{zx} = -\omega_{xz}$).

of lattice rotations on the $4f$ shell. It vanishes if the wave function of the $4f$ electrons can freely follow the rotation of its surrounding. In a magnetic field the (induced) quadrupole moments are pinned and cannot rotate freely with the crystal, therefore the rotational contributions in the interaction should also influence the elasticity.

Rotational invariance in elasticity theory was discussed before by Brown [13.20] and Tiersten [13.21]. The first discussion with an experimental verification using sound waves was given by Eastman [13.22] for $Y_3Fe_5O_{12}$ and Melcher [13.23, 13.24] for MnF_2. Subsequent theories for this effect are from Dohm and Fulde [13.25], Goto et al. [13.26], Jensen [13.27].

Rotational invariance in a magneto-mechanical system was investigated for the first time in the famous Einstein–de Haas experiment (Einstein and de Haas [13.28]). In this experiment, the resulting rotation of a freely suspended magnetic wire, such as iron, due to the reversal of the direction of the magnetization, is measured. Angular momentum conservation forces the rod to rotate if the magnetization is changed. Actually, Einstein took the angular momentum conservation for granted and considered the effect as an indication for the existence of Ampere molecular currents. Rotational invariance requires that the interaction of a magnetic ion, having angular momentum $\boldsymbol{J}$, with the strained and rotated crystal is equal to the interaction of the purely strained crystal with the ion whose angular momentum $\boldsymbol{J}$ has been rotated in the reverse sense. For $B = 0$, neglect of rotational contributions has no consequences because in the long wavelength limit, the rotational tensor describes homogeneous rotations of the crystal which cannot influence the elastic constants. The complete theory of rotational effects for sound propagation (Dohm and Fulde [13.25]) shows that it is necessary to include all terms up to second order in ω_{ij} of the interaction Hamiltonian. A restriction to linear rotational terms would lead to spurious effects in the elastic constants at zero magnetic field.

The calculation of rotationally invariant effects is formulated just for the case of the c_{44} modes discussed above. The full theory can be studied in the cited references above. We use the method of Goto et al. [13.26] and start with the rotational invariance principle as formulated just above for the Hamiltonian:

$$H(\boldsymbol{J}, \varepsilon, \omega) = H(\boldsymbol{D}^{-1}\boldsymbol{J}, \varepsilon) \ . \tag{13.2}$$

Here $\boldsymbol{D}$ is the matrix for a rotation around the y-axis by an angle ω_{zx}. Its components are $D_{xx} = D_{zz} = 1$ and $D_{zx} = -D_{xz} = \omega_{zx}$. The effect of this rotation can be described by a unitary operator $R(\omega_{zx}) = \exp(i\omega_{zx}J_y)$ which acts on the CEF states. Then

$$H(\boldsymbol{J}, \varepsilon, \omega = R^{-1}H(\boldsymbol{J}, \varepsilon)R = R^{-1}(H_{CEF} + H_{str})R \ . \tag{13.3}$$

Here $\boldsymbol{H}_{str} = -g_{\Gamma 5}\sum_i \varepsilon_{zx}O_{zxi}$ is the single ion magneto-elastic Hamiltonian used previously (13.1). It is the first order strain contribution used before (5.7), (5.8). Expansion of (13.3) up to second order in the rotation leads to a total magneto-elastic Hamiltonian $H_{me} = H_{str} + H_{rot}$ with

$$H_{rot} = -\sum_I [\omega_{zx}\Lambda_{zx} + g_{\Gamma 5}\varepsilon_{zx}\omega_{zx}\Omega_{zx} + \lambda_{zx}\omega_{zx}^2], \tag{13.4}$$

where

$$\Lambda_{zx} = i[J_y, H_{CEF}] = 20B_4[(J_z^2 - J_x^2)(J_xJ_z + J_zJ_x) + (J_xJ_z + J_zJ_x)(J_z^2 - J_x^2)]$$
$$\Omega_{zx} = i[J_y, O_{zx}] = 2(J_x^2 - J_z^2). \tag{13.5}$$

Note that the new magneto-elastic Hamiltonian H_{rot} does not create new coupling constants, all terms are determined by the strain coupling $g_{\Gamma 5}$ and the CEF parameter B_4. Including B_6 would enhance the complexity of the formulas enormously. The effect of H_{me} on the elastic constants $c_{44}(q_x, e_z)$ and $c_{44}(q_z, e_x)$ can be calculated via the strain susceptibility with the free energy in the same way as before (Sect. 5.2) (Wang et al. [13.29]). The last term in H_{rot}, together with H_{str}, renormalizes both $c_{44}(q_x, e_z)$ and $c_{44}(q_z, e_x)$ by the same amount. But the terms in H_{rot}, which are linear in $\omega_{zx} = -\omega_{xz}$, lead to a correction equal in magnitude but opposite in sign for both modes. This is due to the opposite sense of rotation with respect to the field direction in both modes (see Fig. 13.2). The difference is given by

$$\frac{c_{44}(zx) - c_{44}(xz)}{c_{44}^0} = \frac{N}{8c_{44}^0}[\chi_-(B,T) - \chi_+(B,T) + 4g_{\Gamma 5}\langle\Omega_{zx}\rangle], \tag{13.6}$$

where $\chi_\pm$ are quadrupolar susceptibilities of the operators $O_\pm = g_{\Gamma 5}O_{zx} \pm \Lambda_{zx}$, respectively and $\langle\Omega_{zx}\rangle$ denotes the thermal average. For zero field $\chi_+ = \chi_-$ and $\langle\Omega_{zx}\rangle = 0$, leading to $c_{44}(q_x, e_z) = c_{44}(q_z, e_x)$ as expected.

Sound propagation experiments testing rotational invariance in this way were first performed by Melcher [13.23, 13.24]) in tetragonal MnF$_2$ at low temperatures in the anti-ferromagnetic state. He used the same geometry as in Fig. 13.2 for the tetragonal geometry. Fig. 13.3 shows the experiment with the two modes $c_{44}(q_z)$ and $c_{44}(q_{110})$.

The $c_{44}(q_{001})$ has a field dependence like $h = B_c^2/(B_c^2 - B^2)$ with $B_c = 9.3$T the spin flop field for the uniaxial anti-ferromagnet (see Chap. 11 (11.20)). Note that anti-ferromagnetic domains should not disturb this result seriously. The effect is of order of 10^{-5} in $\Delta v/v$. Similar experiments were subsequently performed in RVO$_4$ (with R = Nd, Dy, Tb. Tm) by Bonsall and Melcher [13.30] and for HoVO$_4$ by Goto et al. [13.26], in paramagnetic TmSb by Wang and Lüthi [13.29], in CeAl$_2$ by Lüthi and Lingner [13.5] and in ferromagnetic CuPt by Rouchy et al. [13.9]. c_{44} is given for CeAl$_2$ in Fig. 13.1 together with a calculation as outlined in (13.6). In addition to the symmetry effect discussed in Sect. 13.1 ($c_{66} \neq c_{44}$), effect due to rotational invariance can be observed. The calculated splitting for the two modes agree nicely with the observed ones although the absolute values of the strain susceptibilities are somewhat different due to the vicinity of the magnetic phase transition.

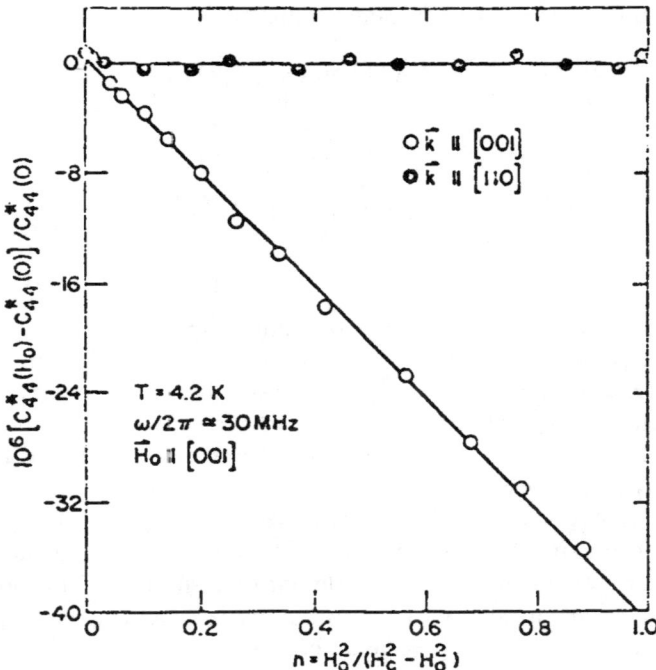

Fig. 13.3. The change in the effective elastic constant c_{44}^* for two propagation directions as a function of the magnetic field dependent parameters $h = H_0^2/(H_c^2 - H_0^2)$; $H_c = (2H_E H_A)^{1/2} \cong 93 kOe$ (Melcher [13.23])

A thorough study was carried out for TmSb, a paramagnetic cubic substance with NaCl structure. In Sect. 5.2 we introduced this substance by showing the CEF level scheme in Fig. 5.3, specific heat and thermal expansion in Fig. 5.4 and the temperature dependence of magnetic susceptibility and elastic constants in Fig. 5.5. All these quantities are CEF dominated and quantitatively understood. The CEF excitations and parameters have been measured (Birgeneau et al [13.31]) and no dispersion of the CEF excitation has been found. Therefore a single ion picture for Tm^{3+} is valid in this case. This means that all parameters for testing rotational invariance in TmSb (Fig. 13.4) are known and no adjustable parameters are left.

Figure 13.4 shows relative sound velocity measurements for the c_{44} mode at $T = 2$ K (Wang and Lüthi [13.29]). The effect is relatively large – of the order of 1%. In contrast to the case of $CeAl_2$ discussed above (Fig. 13.1), the agreement for TmSb for the absolute values and the relative splitting of the c_{44} modes (13.6) is quite good. Other higher order contributions (from B_6 terms of the CEF, second order magneto-elastic contributions, magneto-elastic coupling to $J = 4$ operators etc.) are negligible in this case. More recently, Jensen included magnetic dipolar contribution and made further

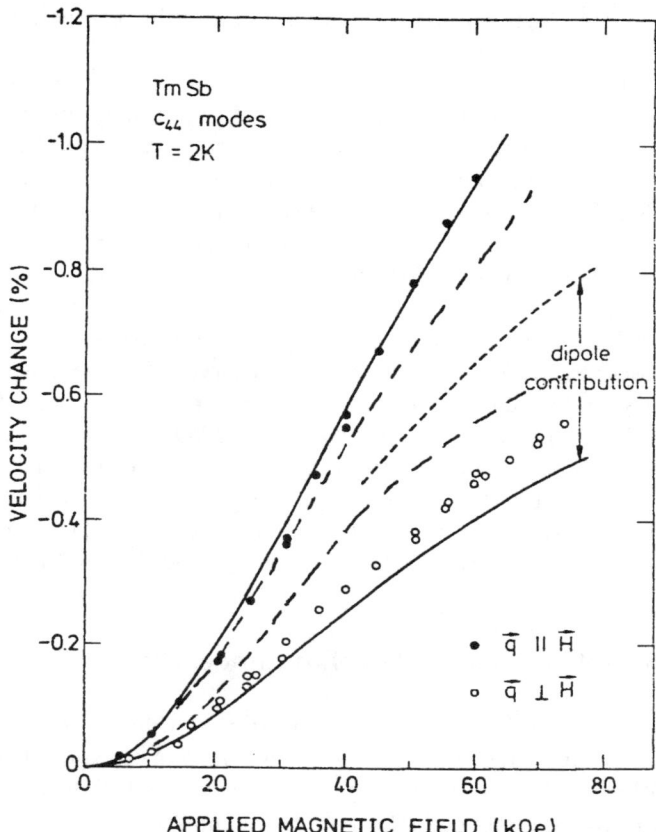

Fig. 13.4. Relative velocity changes of the c_{44} mode in TmSb at 2 K as a function of applied field along (100). Broken lines with rotationally invariant theory (Dohm and Fulde [13.25]), full lines with additional dipolar contributions (Jensen [13.27]). Experimental results from Wang and Lüthi [13.29]

improvements on the agreement with the experiment (see Fig. 13.4). The reason for this effect is the further anisotropy this dipolar interaction introduces as a directional dependence for sound waves.

Analogous effects have been observed for the $(c_{11} - c_{12})/2$ mode in TmSb (Wang and Lüthi [13.29]). In this case magneto-striction corrections as length changes to the sound velocity are important (Jensen [13.27]). The effect of higher order strain contributions is difficult to estimate quantitatively (Dohm [13.32]).

In ferromagnetic materials, the symmetry of the crystal is reduced for $T < T_C$, similar to the effect of a magnetic field in the paramagnetic case discussed above (Sect. 13.1). The effect on the elastic constants due to the magnetization orientation with respect to the crystallographic axes, is the so-called morphic effect. It is also taken into account in the rotationally invariant magneto-elastic theories (Rouchy et al. [13.33]).

Finally, it might be asked whether such rotationally invariant magneto-elastic effects can also occur for itinerant electrons in metallic systems with deformation potential coupling. This question can be answered by investigating magneto-acoustic quantum oscillations (MAQO) as discussed thoroughly in Sect. 8.3.2. The same geometry as discussed above for rotational invariance with localised $4f$ electrons can be used. For example, in cubic LaB_6 the c_{44} mode showed MAQO for the geometries $k \parallel B \parallel [001], u \parallel [100]$ and $k \parallel [100], u \parallel B \parallel [001]$. This is the geometry to observe rotational invariance as shown in Fig. 13.2. But, as seen in Fig. 8.8a, no clear sign for a different field dependence of the two modes can be observed. On the other hand, for the $c_{11} - c_{12}$ mode such an effect was observed (not shown here) (Suzuki et al. [13.34]). In TmSb the c_{44} mode also shows MAQO (Nimori et al. [13.35]). These apparently show some effects arising from the antisymmetric part of the strain tensor, ω_{ij}, but they are rather small $\Delta c_{44}/c_{44} < 10^{-4}$ compared to the rotationally invariant effect due to the magneto-elastic $4f$-strain coupling of $\Delta c_{44}/c_{44} \sim 10^{-2}$ (see Fig. 13.4) in the same substance and field region.

13.3 Magneto-acoustic Birefringence Effects

Magneto-optic effects, like Faraday and Cotton–Mouton or Voigt effects, are well-known phenomena. The magneto-acoustic analoga exist also. Compounds with an n-fold rotation axis ($n > 2$) have doubly degenerate sound modes with propagation vector k along the symmetry axis. For example, in cubic materials with $k \parallel [001]$, there are two degenerate c_{44} modes with the sound velocity $v = (c_{44}/\rho)^{1/2}$ whose polarization vector e may be chosen perpendicularly, e.g. $e_x = [100]$ and $e_y = [010]$. Due to the magneto-elastic coupling, an applied field may lift this degeneracy. Experimentally, this is observed as a change of sound wave polarization along the propagation direction. Depending on whether $B \perp k$ or $B \parallel k$, either magneto-acoustic birefringence (Cotton Mouton–Voigt effects) or Faraday rotation (circular birefringence) of sound waves is observed (see Fig. 13.5). The former is simply due to a different effect of B on the sound velocities $v_{\parallel}(B)$ and $v_{\perp}(B)$, corresponding to polarization parallel and perpendicular to the field. The latter is a real finite frequency effect.

The acoustic Faraday effect for ferromagnetic materials was proposed by Kittel [13.37] and Vlasov and Ishmukhametov [13.38]. This was predicted by Kochalaev [13.39] for paramagnetic spin $\frac{1}{2}$ systems, see also Tucker [13.40]. The theory of this effect for CEF -split rare earth systems was given by Thalmeier and Fulde [13.41]. The related effect of dichroism (circular and linear absorption coefficients) is neglected in the following.

13.3 Magneto-acoustic Birefringence Effects

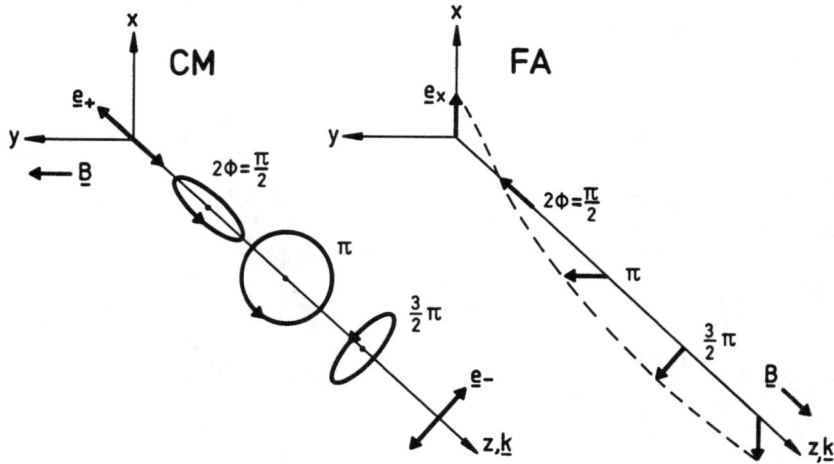

Fig. 13.5. The geometry of the propagation (k), polarization (e) and magnetic field (B) vectors for the Voigt or Cotton–Mouton (CM) and Faraday (FA) effects. The change of polarization state of transverse c_{44} modes is indicated. For clarity the polarization is not shown in perspective. Φ is the CM or FA rotation angle (from Thalmeier, Lüthi [13.36])

13.3.1 Voigt–Cotton–Mouton Geometry

The unequal sound velocities $v_\|(B)$ and $v_\perp(B)$ lead to different wave numbers $k_\| = \omega/v_\|$ and $k_\perp = \omega/v_\perp$ (with ω the frequency) for the two eigenmodes with polarization parallel and perpendicular to B. If we specify $B \parallel [010]$ and $k \parallel [001]$, the displacement field of a sound wave initially polarized along [110] for $z = 0$ is then given by

$$u(z,t) = \left[\left(\frac{\cos\varphi}{2}\right) e_+ + i \left(\frac{\sin\varphi}{2}\right) e_-\right] \exp[i(\omega t - kz)]$$

$$e_\pm = \frac{e_x \pm e_y}{\sqrt{2}} \qquad k = \frac{1}{2}(k_\| + k_\perp) \qquad \varphi = (k_\| - k_\perp)z. \tag{13.7}$$

These equations describe an elliptically polarized sound wave with major axis $e_\pm$ where the phase shift φ determines the ellipticity $u_y/u_x = \tan(\varphi/2)$. The phase shift per unit length can be determined experimentally and is given by

$$\frac{\varphi(B)}{z} = \omega \left(\frac{1}{v_\|} - \frac{1}{v_\perp}\right) = (2\pi/\lambda v(0)) \left[v_\perp(B) - v_\|(B)\right]. \tag{13.8}$$

$v_\|$ and $v_\perp$ can be taken either from a separate experiment or can be calculated from the coupled phonon–spin wave spectra in the case of a ferromagnet or from the strain susceptibilities (5.16c) for paramagnets.

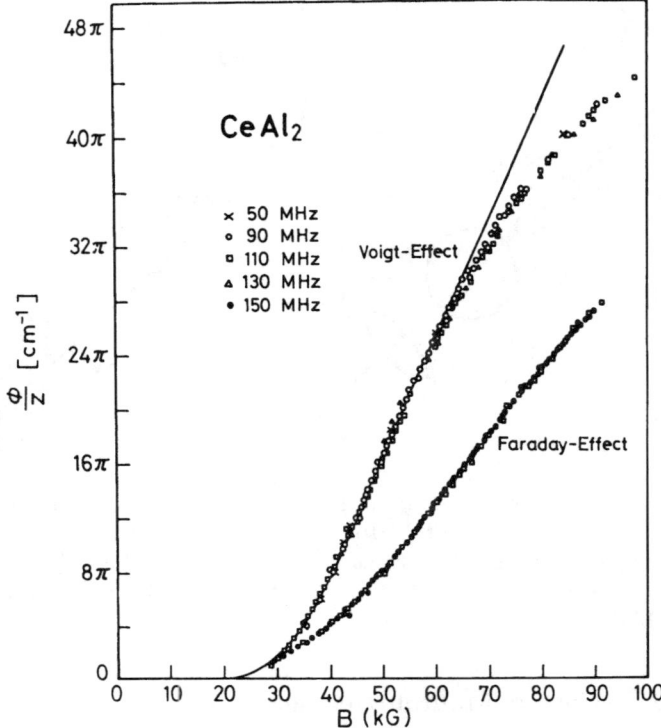

Fig. 13.6. CeAl$_2$: Phase difference Φ/cm for Cotton–Mouton and Faraday geometry for different Frequencies (scaled with (13.8)) and different echo numbers as a function of magnetic field. Solid line is calculated Φ/l for the Cotton–Mouton geometry for 90 MHz using velocity data of figure (13.1) for $T = 4.3$ K (Lüthi and Lingner [13.5])

As a first case let us consider the case of paramagnetic CeAl$_2$. This substance was introduced in Sect. 5.4 and further discussed in Sects. 13.1, 13.2. Figure 13.1 shows the velocities as a function of magnetic field. For our geometry we have the two modes which split for $B > 3$T. This leads to the amplitude modulated signal (see Fig. 4 of Lüthi and Lingner [13.5]). From the minima and maxima of this signal one can determine the phase $\varphi(B)/z$ for different sound wave frequencies. This is shown in Fig. 13.6 for CeAl$_2$ (Lüthi and Lingner [13.5]). It is seen that the phase is strictly proportional to ω. In the figure the different frequencies are scaled according to (13.8) to the 90 MHz frequency. The deviation for fields larger than 7T originate from the splitting of the echoes due to the large velocity changes. The absence of a phase change for $B < 2.5T$ is due to the same velocities for the two modes (Fig. 13.1).

Another Cotton–Mouton effect was observed in ferrimagnetic magnetite Fe$_3$O$_4$ above the Verwey transition $T_V = 120$ K (Moran and Lüthi [13.12],

Lüthi [13.42]). The properties of Fe_3O_4 are discussed in Sect. 7.1. In this case we have a coupling of one of the sound wave modes ($e \parallel B$) to the spin waves of the form (11.18a).

$$(\omega^2 - v^2 k^2)(\omega^2 - \omega_m^2) = \sigma \gamma B v^2 k^2 \qquad (13.9)$$

and for $e \perp B$ there is no coupling (see Sect. 11.2.2). The former mode exhibits a distinct field dependence for $B > 0.4\text{T}$ i.e. above the saturation field where there is a mono-domain magnet. The latter mode has no field dependence in the same field region. Therefore there is a strict proportionality between the phase change and the velocity change as expected (Moran and Lüthi [13.12]).

Other cases of magnetic substances investigated are $Gd_3Fe_5O_{12}$ (see Lüthi [13.43]), $Y_3Ga_{5-x}Fe_xO_{12}$ (Lüthi [13.44]), $RbNiF_3$ (Grishmanovsky et al. [13.45]). In metallic, non-magnetic materials, the Cotton–Mouton–Voigt effect has not, apparently, been studied so far (Gudkov and Gavenda [13.2]). But this effect has been observed for the giant magneto-strictive material $Tb_{0.3}Dy_{0.7}Fe_2$ (Cullen et al. [13.11]). The agreement with the coupled mode theory discussed above is satisfactory.

13.3.2 Faraday Geometry

In this configuration ($B \parallel k$), the rotational symmetry around the propagation axis is preserved contrary to the case of the Cotton–Mouton–Voigt effect. The lifting of the degeneracy is now directly due to the breaking of time-reversal invariance. The new eigenmodes are left and right handed circularly polarized. They are described by polarization vectors

$$e_L = \frac{e_x + i e_y}{\sqrt{2}} \qquad e_R = \frac{e_x - e_y}{\sqrt{2}}.$$

The phase velocity of the L and R–modes will be different, therefore the difference of wave numbers $k_L - k_R$ for a sound frequency ω will itself be proportional to ω. This follows because it must change sign under time reversal, i.e. $\omega \to -\omega$ and $R \leftrightarrow L$. A sound wave originally linearly polarized along e_x can be written as the superposition of L and R–polarized modes. If their wave numbers k_L and k_R differ slightly the displacement field will be given by

$$u(z,t) = (\cos\varphi\, e_x + \sin\varphi\, e_y)\exp[i(\omega t - kz)]$$

$$k = \frac{k_L + k_R}{2} \qquad \varphi = \frac{(k_L - k_R)z}{2}. \qquad (13.10)$$

This describes a linearly polarized sound wave whose polarization rotates during propagation (Fig. 13.5). The Faraday rotational angle is φ/z. Various expressions can be obtained for the different cases: ferromagnets, antiferromagnets, paramagnets, non-magnetic metals. Various cases are discussed below.

Fig. 13.7. Oscillograms of ultrasonic echo pulses detected by a piezoelectric quartz transducer at 528 MHz in YIG with [100] ∥ B ∥ k (a) B=2kOe, (b) B=1.5kOe (Matthews and LeCraw [13.46])

The magneto-acoustic Faraday effect was first shown experimentally in the ferrimagnet $Y_3Fe_5O_{12}$ (Matthews and Le Craw [13.46]). The structure of YIG was explained in Sect. 11.2.3 with Fig. 11.8. Figure 13.7 shows the original oscillograms for ultrasonic echoes at 528 MHz in YIG for two different magnetic field values (Matthews and Le Craw [13.46]). The geometry was $B \parallel k \parallel [100]$. Unlike acoustical activity, the Faraday angle – as a non-reciprocal effect – increases if the ultrasonic pulse propagates forward and backwards.

Instead of (13.9), this geometry (see (11.17)) results in the dispersion equation

$$(\omega - \omega_m)(\omega^2 - v^2 k^2) = \frac{\gamma b^2 k^2}{\rho M_0} \qquad (13.11)$$

which together with (13.10) and (13.11) gives

$$\frac{\Phi}{z} = \frac{k_L - k_R}{2} = \frac{\omega^2 b^2}{2} \rho v^3 \gamma M_0 \frac{1}{\left(\frac{\omega}{\gamma}\right)^2 - (B - NM_0)^2} . \qquad (13.12)$$

Using this formula, the experimental data could be fitted with the magneto-elastic coupling constant b and the demagnetizing factor N as fit parameters. Both parameters agreed with other determinations.

A similar experiment in YIG at 9.375GHz was performed by Guermeur et al. [13.47]. This is shown in Fig. 13.8a,b. Because of the high frequency, the spinwave–phonon intersection is at about 3.5kG (Fig. 11.6). With the saturation field of a few hundred Gauss, Faraday rotation can be observed for fields lower and higher than the intersection. Figure 13.8a gives the amplitude modulation and Fig. 13.8b the phase angle of the rotation. It is seen that the whole rotation is huge $\sim 20\pi$. Further experiments on YIG were made by Pavlenko et al. [13.48] and Lemanov et al. [13.49] and for YGaIG by Lüthi [13.44].

The magneto-acoustic Faraday effect has also been observed for paramagnetic Ni^{2+}-ions in MgO (Guermeur et al. [13.50]). The Ni ion concentration

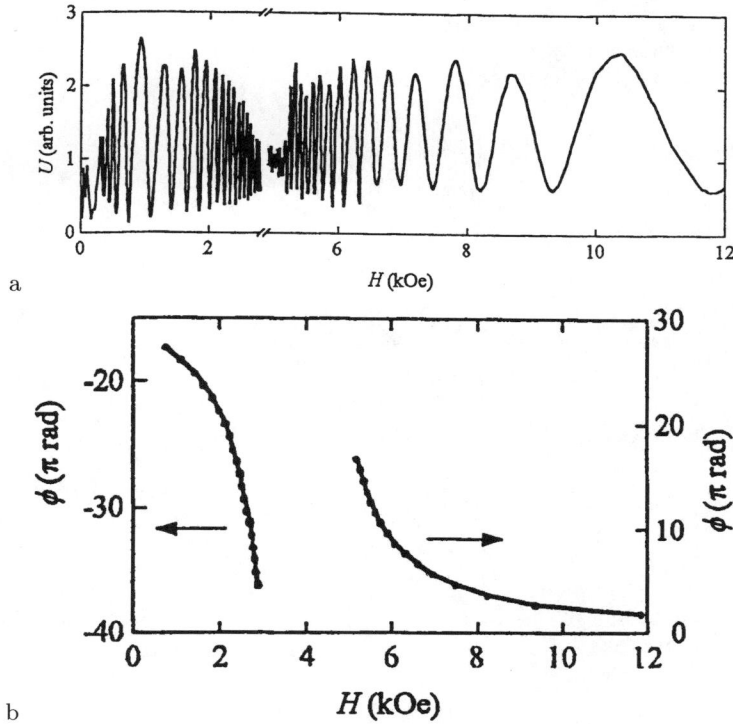

Fig. 13.8. (a) $Y_3Fe_5O_{12}$: Amplitude modulation versus magnetic field in Faraday geometry for 9.375GHz, (b) $Y_3Fe_5O_{12}$: Phase angle of the rotation (Guermeur et al. [13.47])

was 10^{20}/cm^3. The Ni^{2+} ion has an orbital singlet ground state in the octahedral environment. This is, therefore, a pure spin–phonon coupling. This system was mentioned before in Sect. 5.5 in conjunction with the ultrasonic paramagnetic resonance.

As shown in Fig. 13.6, the acoustic Faraday effect was also observed in the paramagnetic phase of CeAl$_2$. It is seen that the phase angle Φ/l is proportional to the frequency ω. Theoretically, one expects a ω^2 dependence (Thalmeier and Fulde [13.41]). This discrepancy cannot be explained with the fluctuations caused by the vicinity to the magnetic phase transition and is, as yet, unexplained.

Turning now to anti-ferromagnets: The magneto-acoustic Faraday effect was found in Cr$_2$O$_3$ with sound wave frequency of 8.89GHz at 4.2K (Boiteux et al. [13.51]). Cr$_2$O$_3$ is a uniaxial two-sublattice anti-ferromagnet with $T_N = 307$K. The Faraday effect was observed both in the anti-ferromagnetic and in the spin–flopped region with propagation direction along the 3-fold c-axis. In Fig. 13.9a, the microwave ultrasonic signal is shown as a function of applied field and in Fig. 13.9b the Faraday rotation per unit length ξ/π is given

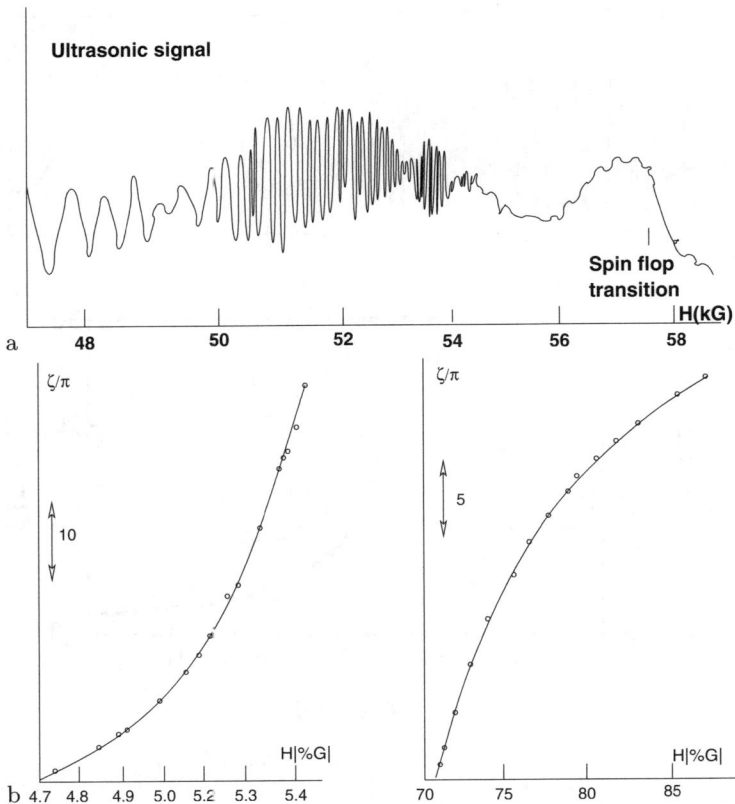

Fig. 13.9. (a) Ultrasonic signal as a function of applied field for 8.89GHz at $T = 4.2$ K for the anti-ferromagnet Cr_2O_3, (b) field dependence of the rotation ξ/π both for the anti-ferromagnetic and the spin flop phase (adapted from Boiteux et al. [13.51])

for the anti-ferromagnetic and the spin flop phase. If ξ/π versus $\Omega^2 - \omega^2$ is plotted, with Ω the anti-ferromagnetic resonance frequency, a straight line for the low field phase is obtained (Boiteux et al. [13.51]).

13.4 Acoustical Activity

In analogy to the optical activity (Landau–Lifshitz [13.52] and Pauli [13.53]), a similar effect should be present for acoustic waves, the so-called acoustical activity. In the optical activity the relation between displacement field $\boldsymbol{D}(\boldsymbol{r},t)$ and electrical field $\boldsymbol{E}$: $\boldsymbol{D} = \varepsilon \boldsymbol{E}$ has to be generalised to $D_i = \varepsilon_{ik} E_k + \gamma_{ikl}\frac{\partial E_k}{\partial x_l}$ because higher order terms in a/λ have to be considered with a the typical lattice constant and λ the wavelength of the light. In acoustics analogously, the Hooke relation (3.7) has to be generalised to include gradients of the deformation tensor (ε_{ik}) leading to spatial dispersion:

$$T_{ik} = c_{iklm}\varepsilon_{lm} + \gamma_{iklmn}\frac{\partial \varepsilon_{lm}}{\partial x_n} \ . \qquad (13.13)$$

Note that, in the case of acoustical activity, the tensor has a rank of 5 compared to a rank of 3 for optical activity. Equation (13.13) leads to effective elastic constants which depend on k, $c_{iklm}(\omega, k)$. The rotation of the plane of polarization of a transverse acoustic wave can be calculated with this relation. For isotropic materials, the wave vectors of the two circularly polarized waves 1, 2 (Kluge and Scholz [13.54]) is given by:

$$k_{1,2} = \frac{\omega}{v}\left(1 \pm \frac{1}{2}\frac{\gamma\omega}{\mu v}\right) \qquad (13.13a)$$

with μ the Lamé shear elastic constant and the sound velocity $v^2 = \mu/\rho$ (see Sect. 3.3). Unlike the Faraday effect, this acoustic activity is a reciprocal effect. These relations indicate that the components of the stress tensor depend not only on the strain tensor but also on the spatial change of the strain. The parameters γ_{ijklmn} change sign under mirror reflection, therefore they are $\neq 0$ only in materials without a spatial center of symmetry. A detailed discussion has been given by Portigal and Burstein [13.55].

Acoustical activity has been observed in α quartz by two groups (see Pine [13.56], Joffrin and Levelut [13.57]). They propagate shear waves along the trigonal c-axis around 1GHz and 9Ghz respectively using re-entrant cavities for transmission and analyser. Care has to be taken to avoid normal acoustic birefringence from misalignment and to avoid refraction effects. The rotation due to acoustical activity gives from (13.13a) $\varphi/L = \omega(1/v_1 - 1/v_2) = \gamma\omega^2/\rho v^4$. Experiments produce values $\sim 127°$ at 1GHz for a crystal length of $L = 1$cm and 8.5×10^3deg/cm for 9GHz.

13.5 Non-reciprocal Surface Acoustic Wave Effects

The non-reciprocal effects of SAW in Voigt geometry are discussed here (Field perpendicular to the propagation direction and in the surface plane). The symmetry effect can be explained directly using the equivalent of Fig. 3.3 but with a magnetic field B in the plane of the surface perpendicular to the propagation direction k. Figure 13.10 presents the geometry for the symmetry argument with $B = (0, B, 0)$ and $k = (k, 0, 0)$. The following symmetry operations are considered:

1. reflection on mirror plane $(y-z)\sigma_{yz}$: $k \to -k$, $B \to -B$ (axial vector)
2. reflection on mirror plane $(x-y)\sigma_{xy}$: $k \to k$, $B \to -B$ (axial vector)

For bulk waves, σ_{yz} : $\omega_b(k, B) = \omega_b(-k, -B)$ and σ_{xy} : $\omega_b(k, B) = \omega_b(k, -B)$ still apply as well as with time reversal symmetry T : $\omega_b(k, B) =$

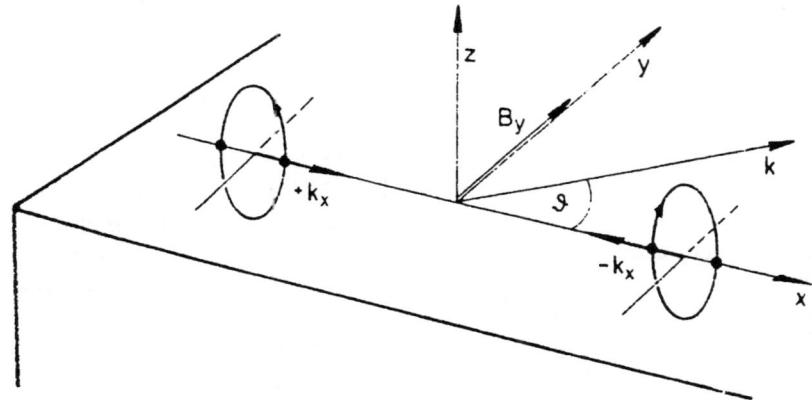

Fig. 13.10. Geometry for non-reciprocal surface waves (Heil et al. [13.60])

$\omega_b(-\boldsymbol{k}, -\boldsymbol{B})$. These relations express the well-known fact that bulk excitation frequencies are even functions of k_x and B_y. For surface waves, in general there is a non-reciprocity because the second symmetry operation σ_{xy} is no longer valid on the surface of the crystal (loss of mirror symmetry) i.e. $\omega(\boldsymbol{k}, \boldsymbol{B}) \neq \omega(\boldsymbol{k}, -\boldsymbol{B})$ and also $\omega(\boldsymbol{k}, \boldsymbol{B}) \neq \omega(-\boldsymbol{k}, \boldsymbol{B})$. Because of the still valid time reversal symmetry T and reflection symmetry σ_{yz}, $\omega(\boldsymbol{k}, \boldsymbol{B}) = \omega(-\boldsymbol{k}, -\boldsymbol{B})$ is also true. These considerations are valid for different surface waves such as ferromagnetic spinwaves, surface polaritons, dipolar anti-ferromagnetic spinwaves, surface acoustic waves SAW, etc. For a survey of various non-reciprocal effects with surface wave excitations, but not the effect discussed here (see e.g. Cottam and Tilley [13.58], Camley [13.59]).

The clearest most spectacular non-reciprocal effect for surface waves has been proposed for magneto-static modes (Damon and Eshbach [13.61]) and observed with Brillouin scattering (Gruenberg and Metawe [13.62], Sandercock and Wettling [13.63]). For a discussion of magneto-static modes, see Sect. 11.2.2. Here in the geometry of Fig. 13.10 a surface spinwave propagates in the $+\boldsymbol{k}$ direction but does not exist for the $-\boldsymbol{k}$ direction. The frequency (ω/γ) reads (Damon and Eshbach [13.61])

$$\frac{\omega}{\gamma} = \frac{1}{2}\left[\frac{B}{\cos\theta} + (B + 4\pi M)\cos\theta\right] . \tag{13.14}$$

Here θ is the angle between $\boldsymbol{k}$ and the x-direction (Fig. 13.10) and M the magnetization. For $\theta = 0$, $\cos\theta = 1$ and $\omega/\gamma > 0$, but for $\theta = \pi$ it follows that $\cos\theta = -1$ and $\omega/\gamma < 0$, i.e. the wave is non-propagating. Stable solutions exist only for $\theta < \theta_c$ with $\cos\theta_c = \left(\frac{B}{B+4\pi M}\right)^{1/2}$. Physically, this comes about because surface wave excitations in the continuum limit (SAW, surface polaritons etc), have particle, field or spin polarizations which are not linear

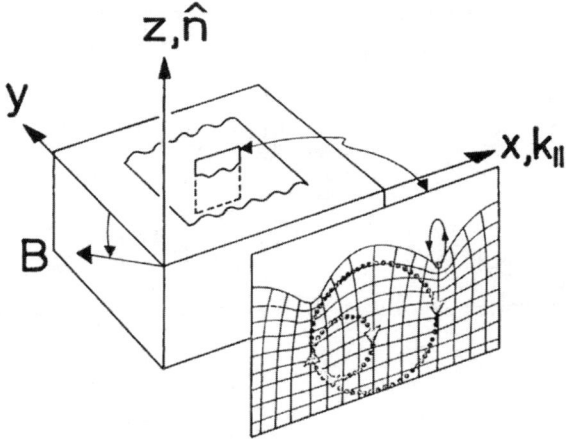

Fig. 13.11. Picture of a SAW displacement field with the particle rotational sense is given. Electronic cyclotron orbits for B_y configuration are indicated by dotted lines (Heil et al. [13.60])

but elliptical (see e.g. Sect. 3.4 and Fig. 3.4 and Fig. 13.10). Therefore, the rotation sense is different for $+k_x$ and $-k_x$ propagation. A magnetization vector has a rotation sense given by the applied field but not by the wave vector $\mathbf{k}$. Therefore for propagation in the $-\mathbf{k}$ direction the magnetization rotation is not reversed.

Turning now to SAW, non-reciprocal effects could occur with surface magneto-elastic waves. This has been suggested by Scott and Mills [13.64] using the symmetry argument given above. If this idea is applied to a well-characterised CEF system like CeAl$_2$, non-reciprocal behaviour for static strain susceptibilities are not found, but only when finite frequency effects are taken into account (Camley and Fulde [13.65]). This non-reciprocal effect could be resonantly enhanced when the energy of the SAW is close to the difference in energy between two CEF levels. Such experiments have not yet been performed.

A typical non-reciprocal SAW experiment was carried out in Al (Heil et al. [13.60, 13.66]). Here, it is the coupling of the SAW to conduction electrons on cyclotron orbits which is responsible for the non-reciprocal effect. A schematic drawing of the effect is shown in Fig. 13.11. Again the non-reciprocal effect arises because the rotational sense of the cyclotron orbits depends on the direction of the magnetic field and the sense of the particle motion depends on the sign of k_x.

The result of such an experiment is shown in Fig. 13.12. Here relative velocity changes for SAW and bulk waves are shown for different magnetic field directions. In the low field region, geometric resonances are observed (see Sect. 8.3.2 for a discussion of this effect). For fields above 0.5T (not shown in Fig. 13.12), magneto-acoustic quantum oscillations – discussed in

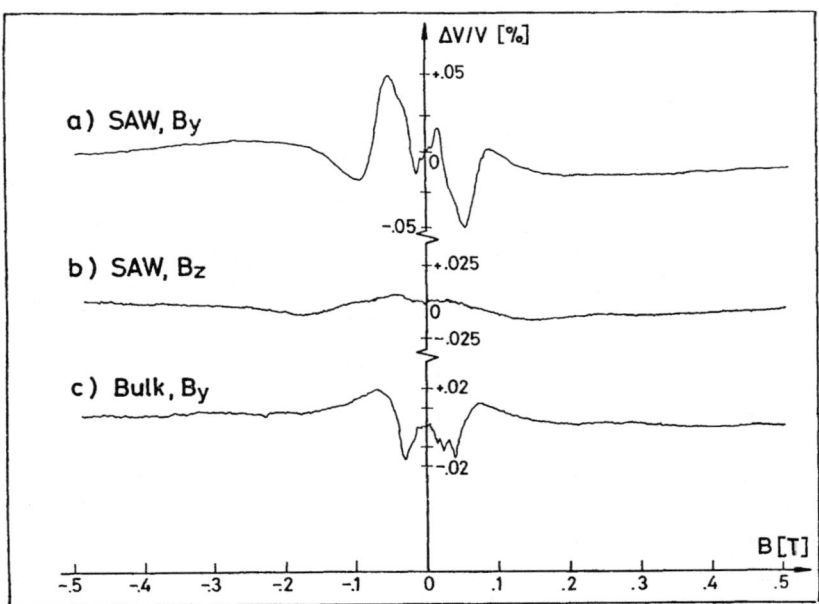

Fig. 13.12. Relative velocity changes for SAW and bulk waves on Al at 4.2 K for different magnetic field directions (Heil et al. [13.66]). Note the non-reciprocal effect for SAW in geometry (a) and the absence for other geometries (b) and for bulk waves (c)

Sect. 8.4.2 – are observed. For the case of B_y one observes a non-reciprocal effect in the region of geometric oscillations which is absent for the case of B_z. If the total velocity change is split into a symmetric, reciprocal and an antisymmetric, non-reciprocal part with $\Delta v(B) = \Delta_s v(B) + \Delta_a v(B)$ and with $\Delta_{s,a} v(B) = \frac{1}{2}[\Delta v(B) \pm \Delta v(-B)]$, then theory gives (Heil et al. [13.66]) for the non-reciprocal part

$$\frac{\Delta_a v(k, B)}{v} = AkR \frac{\omega_c \tau}{1 + 4(\omega_c \tau)^2} \qquad (13.15)$$

which has the necessary property $\Delta_a v(-k, B) = \Delta_a v(k, -B) = -\Delta_a v(k, B)$. This is only valid for

$$kR = \frac{kl_e}{\omega_c} \tau < 1,$$

with R the cyclotron radius (see Fig. 13.11), l_e the electronic mean free path, $\omega_c = eB/mc$ the cyclotron frequency and τ the electronic relaxation time.

Note that the kinetic stress tensor is due to impurity scattering and the effect would disappear for $\tau \to \infty$. This occurrence of dissipation factors is a general feature, also found in non-reciprocal reflectivity (Remer et al. [13.67, 13.68]).

In addition to the case of Al, such a non-reciprocal effect for SAW was also observed for Ga (Heil et al. [13.66]). It should also be possible to investigate semiconductors in this way. According to (13.15) higher frequencies should enhance this effect.

13.6 Surface Acoustic Wave Effects in the Integral and Fractional Quantum Hall Effect

A particularly nice example of SAW effects in solid state physics is the investigation of high frequency SAW attenuation and velocity shifts in the heterostructure material GaAs/AlGaAs exhibiting the integral and fractional quantum Hall effect (QHE). The conductivity within the two-dimensional electron gas can be determined with SAW (Wixforth et al. [13.69]). By increasing the SAW frequency, one probes – with corresponding small wavelengths – the frequency–wave-vector dependent conductivity $\sigma(q,\omega)$ which can be different from the dc conductivity. For a review of these experiments, especially the fractional quantum Hall effect, see Willett [13.70].

The actual experiment is performed with a device as discussed in Sect. 2.5 and shown in Fig. 2.5. Inter-digital transducers are put on a piezoelectric GaAs. The heterostructure GaAs/Al$_{0.3}$Ga$_{0.7}$As lies between the transmitter and the receiver with a Si dopant layer, donating the electrons which form the two-dimensional layer between GaAs and GaAlAs. The inter-digital SAW transducers have a fundamental frequency and odd harmonics in the range of 19 to 100 MHz for the integral QHE (Wixforth et al. [13.69], Schenstrom et al. [13.71], Drichko et al. [13.72]) and a fundamental frequency of 230 MHz and odd harmonics up to 1600 MHz for the fractional QHE (see Willett [13.70] and references therein).

In the two-dimensional electron gas with magnetic field perpendicular to the layer, the single particle states are quantized, forming Landau levels separated by $\hbar\omega_c$, as discussed (for the three-dimensional case) in Sect. 8.3.2. In contrast to the case shown in Fig. 8.6, in the quasi two-dimensional case, k_z is also quantized and the density of states is constant versus the energy for $B = 0$. The degeneracy of the Landau levels increases with magnetic field. The population of these levels is given by the filling factor $\nu = nhc/eB = n\Phi_0/B$, with n the electron density and $\Phi_0 = hc/e = 4.13 \times 10^{-7}$ Gauss/cm^2 the magnetic flux quantum. The integral quantum Hall effect gives plateaus of the Hall resistance of (Klitzing et al. [13.73])

$$R_H = \frac{1}{\nu}\left(\frac{h}{e^2}\right)(\nu = 1, 2, 3, \cdots). \tag{13.16}$$

The electronic states are localised between the adjacent Landau levels. For the integral QHE, the above-mentioned authors found the SAW analogue of Shubnikov–de Haas oscillations for the velocity and attenuation.

The first SAW experiment in such heterostructure materials was performed at 70 MHz and at 4 K (Wixforth et al. [13.69]). Magneto-acoustic quantum oscillations were observed and evaluated with a relaxation type interaction of the polarization field from the SAW with the conductivity $\sigma_{xx}(\omega)$ of the two-dimensional sheet of charge at the hetero-junction interface. The attenuation Γ and the relative velocity change leads to (Wixforth et al. [13.69])

$$\Gamma = \frac{\alpha q \frac{\sigma_{xx}}{\sigma_m}}{1 + \left(\frac{\sigma_{xx}}{\sigma_m}\right)^2}$$
$$\frac{\Delta v}{v_0} = \frac{\alpha}{1 + \left(\frac{\sigma_{xx}}{\sigma_m}\right)^2}.$$
(13.17)

Here α is an effective piezoelectric coupling coefficient ($\alpha = 3.2 \times 10^{-4}$) and $\sigma_m = 3.5 \times 10^{-7} (\Omega/\text{cm}^2)$. Equation (13.17) predicts that with increasing conductivity, σ_{xx} (or ρ_{xx}) – the SAW attenuation – increases until a maximum is reached for $\sigma_{xx} = \sigma_m$. They were able to analyse their results with these formulas, relating damping and velocity change with the frequency dependent longitudinal conductivity $\sigma_{xx}(\omega)$.

In heterostructure samples of high quality and at very low temperature, characteristic plateaus at fractional values of the filling factor ν are also found (Tsui et al. [13.74]). High quality means that the boundary layer between GaAs and GaAlAs is perfectly crystalline and of high purity. A typical example is shown in Fig. 13.13. Here $\rho_{xy} = R_H$ is plotted in units of h/e^2 versus magnetic field, as determined from Hall experiments. Notice the plateaus for $\nu = 1, 2/3, 3/5, 4/7$ etc. In addition ρ_{xx} is plotted in the lower part. At filling factors ν, giving rise to plateaus in R_H, it is noticed that ρ_{xx} becomes very small.

The SAW experiment is now introduced. As pointed out above, the SAW velocity change and attenuation is proportional to $\sigma_{xx}(q,\omega)$. It is apparently difficult to measure this quantity for higher frequency (which can be different from the dc value) otherwise. With this relaxation model, SAW results exhibiting fractional quantum Hall effect, can also be described. An example is given in Fig. 13.14 for 235 MHz and 160 mK (Willett et al. [13.76]). Dotted lines give the fit with (13.17) and the parameters quoted above. In the region of the fractional quantum Hall effect the conductivity becomes very small and the SAW amplitude very large (attenuation small).

The good agreement with the relaxational model, shown in Fig. 13.14. is severely disturbed for the filling factor $\nu = \frac{1}{2}$ near 5.3T for a high mobility specimen as shown in Fig. 13.15. It is found that, with increasing frequency and low temperatures, a resonance-like anomalous SAW attenuation increase and velocity change occurs for fields around $\nu = \frac{1}{2}$ (Willett et al. [13.76], Willett [13.70]). This points to an enlarged conductivity. These experiments have given rise to new developments in the field of fractional quantum Hall effect.

13.6 Surface Acoustic Wave Effects in the Quantum Hall Effect 353

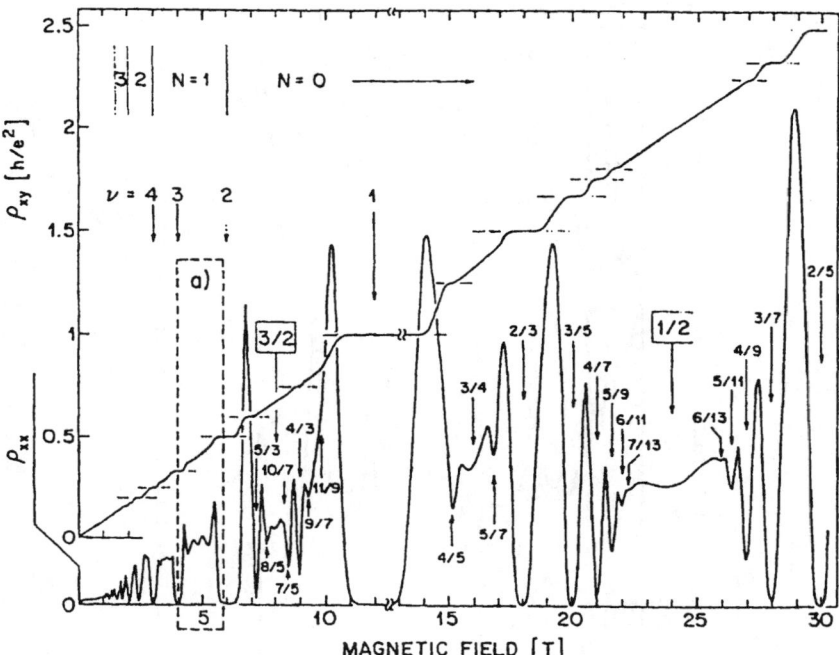

Fig. 13.13. Diagonal resistivity ρ_{xx} and Hall resistance ρ_{xy} of a high mobility heterostructure (Willett et al. [13.75])

While the excited states have energy gaps at the fractional quantum Hall plateaus, this is apparently not the case for $\nu = \frac{1}{2}$ (see Figs. 13.13, 13.15). This fact is explained with the concept of "composite fermions".

The composite fermion model was proposed for the fractional quantum Hall effect (Jain [13.77]). In the presence of an external magnetic field, n interacting electrons in a two-dimensional degenerate Landau level trap $2p$ flux quanta, with p integer but retaining otherwise their charge, spin and statistics. The resulting quasi particles move in the magnetic field that remains after $2pn$ flux quanta are removed by the formation of a new ensemble. In this way at a filling for $\nu = \frac{1}{2}$ the natural state would be a Fermi liquid of $p = 1$ composite fermions in the absence of an external magnetic field, since the total flux per unit surface for the electrons exactly equals, and thus cancels out, the externally applied field. For $\nu = \frac{1}{2}$ there are many such quasi particles which have a Fermi surface. Similar to a good metal (see Sect. 8.3), geometric resonances due to these composite fermions using high frequency SAW up to 5 GHz are found (Willett et al. [13.78]). For more details on composite fermions see the various articles in Heinonen [13.79].

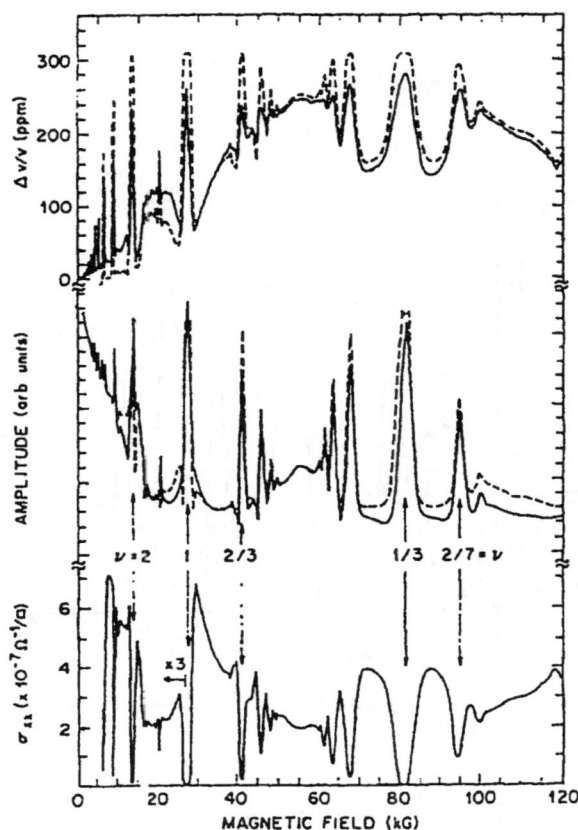

Fig. 13.14. DC conductivity, SAW amplitude and SAW velocity shifts versus magnetic field at 160 mK and 235 MHz. Solid lines measurements, dashed lines model calculation using σ_{xx} values from the figure (Willett et al. [13.76])

13.6 Surface Acoustic Wave Effects in the Quantum Hall Effect   355

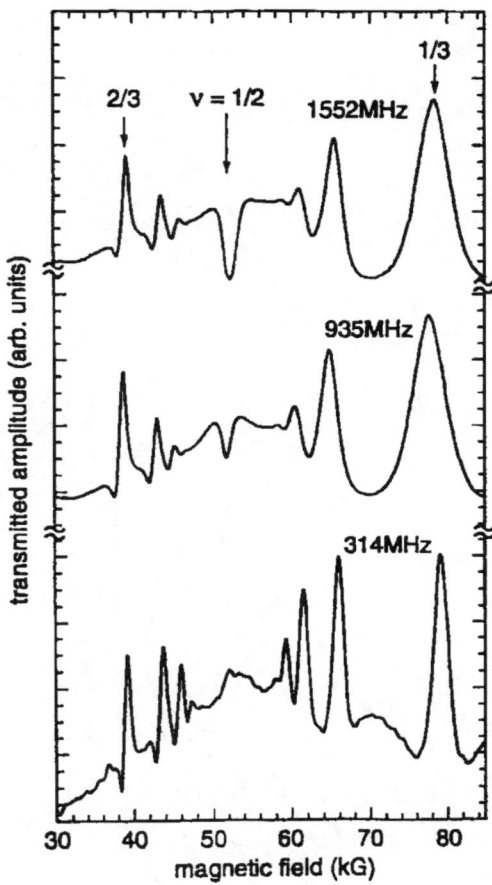

Fig. 13.15. Transmitted SAW amplitude near $\nu = 1/2$ for 3 SAW frequencies (Willett [13.70])

14 Ultrasonic Propagation in Tunneling Systems

Ultrasonic experiments in systems which have tunneling states are discussed briefly in this chapter. We can distinguish between systems which have individual isolated tunneling centers on the one hand, like KCl:Li or NaCl:OH or H in Nb and, on the other hand, amorphous materials with tunneling states which can interact and where a model of two-level systems can describe many phenomena. A similar case for quasi-crystals is also mentioned.

14.1 Crystalline Systems

For the first case of isolated tunneling centers, such a system has already been encountered in Sect. 5.5 where the transition metal Jahn–Teller ion Ni^{3+} in Al_2O_3 was discussed. For KCl:Li, an ultrasonic investigation has been carried out which showed that only the c_{44} mode (T_{2g}) couples to the Li^+ ion (Byer and Sack [14.1]). This fact indicates that the Li^+ ion is displaced from the lattice site along the various (111) directions. For Li^+ in KBr, no effect on the elastic modes is noticeable. It can be concluded, therefore, that in this case the Li^+ ion is not displaced and sits on a lattice site. Figure 14.1 shows the temperature dependence of the relative sound velocity (or compliance data) for the c_{44} mode and for different Li concentrations in KCl (Byer and Sack [14.1]). It is seen that the temperature dependence follows a $1/T$ law except for the highest concentration where one observes deviations from this law. For the temperature region of Fig. 14.1, $k_B T \gg \Delta$ (with Δ the tunnel splitting discussed below) and $tgh\Delta/k_B T \approx \Delta/k_B T$.

Another system, NaCl with OH impurities (NaCl:OH) in the ~ 100ppm concentration region, was investigated ultrasonically over a wide temperature range from room temperature down to 50 mK (Kanda et al. [14.2]). The positions of the potential minima for the OH ion deviate from the normal anion site toward the six equivalent (100) directions in contrast to the case KCl:Li discussed above. The quantum mechanical tunneling motion between the equivalent off-center positions of an impurity ion results in closely-spaced tunneling levels. These were investigated by millimeter-wave spectroscopy and also by phonon spectroscopy (Windheim and Kinder [14.3]). In the latter experiment, crystals with the low concentration of 1ppm OH ions were used with phonon frequencies in the 40–275 GHz range. In addition, the uniaxial

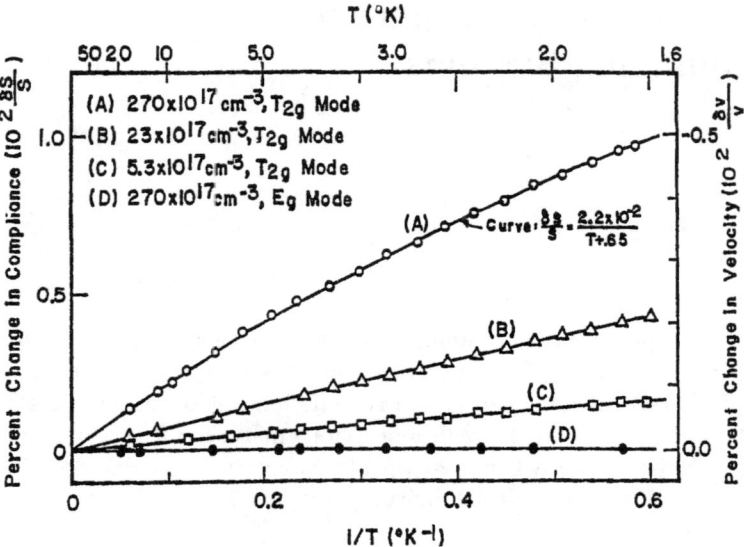

Fig. 14.1. Temperature dependence of the c_{44} mode for KCl:Li for different Li concentrations as indicated in the figure. The fractional compliance is plotted versus $1/T$ (Byer and Sack [14.1])

pressure dependence of the A_{1g}-E_g transition between the lowest tunneling energy states could be determined. These low energy levels govern the various low temperature thermodynamic quantities of the alkali halides with off-center impurities. Figure 14.2 shows the temperature dependence of the elastic modes for different OH concentrations together with theoretical fits (Kanda et al. [14.2]).

For the theoretical description, notice first that the OH concentrations are small enough to warrant a single impurity model without interaction. Since the strains couple to the quadrupole moment of the tunneling motion of the impurity OH ion, the formalism of strain–quadrupole coupling developed for rare earth ions in Sect. 5.2 can be used and describe the elastic constant with the strain susceptibility of (5.16c). Actually, the calculations (Kanda et al. [14.2], Yamada et al. [14.4]) take the whole energy level spectrum with Curie and Van Vleck terms into account and, because of the short relaxation times of the excited states, they do not include relaxation denominators into the strain susceptibilities, i.e. they can take the static strain susceptibility as in the case of the rare earth ions. As seen in Fig. 14.2, this description gives a good quantitative fit for the various elastic constants. The effect is proportional to the impurity concentration as would be expected.

More recently, another system of clathrate compounds $Eu_8Ga_{16}Ge_{30}$ and $Sr_8Ga_{16}Ge_{30}$, were investigated analogously with ultrasonic (RUS) techniques (Zerec et al. [14.5]). Clathrate compounds are composed of polyhedral cages

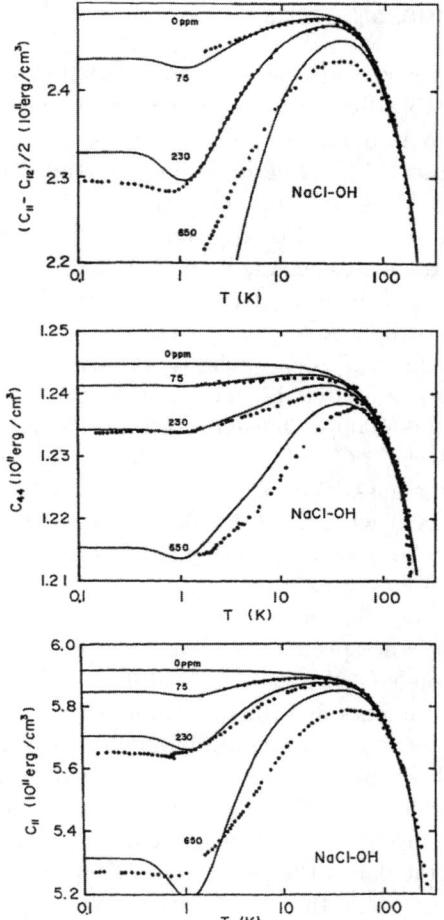

Fig. 14.2. Temperature dependence of $(c_{11} - c_{12})/2$, c_{44} and c_{11} for the OH concentrations 0ppm, 75ppm, 230ppm, 650ppm. Solid lines are strain susceptibility fits (Kanda et al. [14.2])

with 12–16 faces formed by Si, Ge or Ga atoms. The oversized cages accommodate guest atoms like Ba, Sr or Eu which exhibit large amplitude an-harmonic motion. This leads to strong scattering of acoustic phonons and a small thermal conductivity. The Eu nuclear density map shows four maxima, away from the cage center, leading to a fourfold split site model. The elastic response is due to the quadrupolar coupling with the elastic strain as in the case of NaCl:OH discussed above. The c_{44} and the $(c_{11} - c_{12})/2$ modes show characteristic temperature dependencies which can be described quantitatively with the strain susceptibilities (Van Vleck and Curie term), as discussed before (Chaps. 5 and 7).

14.2 Amorphous Systems

Many more sound experiments have been performed in amorphous materials. These materials exhibit distinct temperature dependences in various thermodynamic and transport quantities. There exist a number of review articles on this topic: Phillips [14.6], Hunklinger and Arnold [14.7], Hunklinger and Raychaudhuri [14.8]. We concentrate essentially on low temperatures because the effects are most pronounced there. Non-crystalline, amorphous solids can be considered as a homogeneous medium for sound wavelengths $\lambda > 1000$Å. A longitudinal and a degenerate transverse sound wave mode is therefore expected as in an isotropic solid (Sect. 3.4).

At low temperatures ($T < 1$K), the thermodynamic and thermal transport quantities are dominated by long wavelength phonons with average sound wavelengths $\lambda \sim 1000$Å. One expects a specific heat C obeying a Debye law $C \sim T^3$ but instead a $C \sim T$ law is found. For the thermal conductivity, one expects from the gas-kinetic expression (Sect. 4.5) $\kappa = C v_s l_{ph}/3 \sim T^3$ due to the specific heat with constant sound velocity v_s and constant mean free path l_{ph} due to boundary scattering. But at low temperatures $\kappa \sim T^n$, the exponent $n \sim 1.8$ is very close to 2. In fact the experimental investigation of vitreous SiO_2, Se, silica and germania based glasses, by means of thermal conductivity and specific heat, was the starting point for the physics of non-crystalline solids (Zeller and Pohl [14.9]).

Sound wave experiments show further anomalies at low temperatures. The temperature dependence of the longitudinal and transverse sound velocity is shown in Fig. 14.3 for borosilicate glass. It clearly exhibits a logarithmic temperature dependence for $T \leq 1$ K for both polarizations. Furthermore, the sound velocity is frequency independent in this region.

Even more spectacular is the power dependence of the ultrasonic attenuation at low temperatures. In Fig. 14.4, the power dependence for 940MHz longitudinal sound waves at 0.48 K is shown. With increasing acoustic intensity, the absorption decreases and finally saturates.

All these experimental facts at low temperatures can be explained satisfactorily by the model of the two-level system (Anderson et al. [14.12], Phillips [14.6]). The linear specific heat C, thermal conductivity and acoustic experiments could be accounted for by this model. One assumes a broad density of states of low energy excitations with a distribution of the double well potentials. Such a tunneling two-level system can be approximately treated as a particle in a double well potential. Especially for the acoustic experiments, a formal analogy between the two-level systems and the spin $S = \frac{1}{2}$ system is helpful in calculating velocity and attenuation changes and saturation effects. Considering the two-level Hamiltonian H_0 with energy splitting E and the strain-perturbed Hamiltonian H'.

$$H_0 = \frac{1}{2}\begin{pmatrix} E & 0 \\ 0 & -E \end{pmatrix} \qquad H' = \frac{1}{2}\begin{pmatrix} D & 2M \\ 2M & -D \end{pmatrix}\varepsilon$$

14.2 Amorphous Systems   361

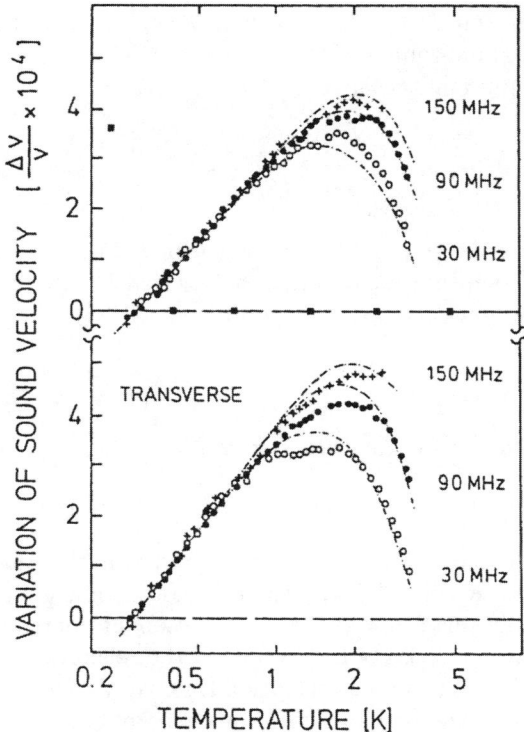

Fig. 14.3. Sound velocity changes versus temperature (ln scale) and for different frequencies in borosilicate glasses, longitudinal and transverse waves (Piché et al. [14.10])

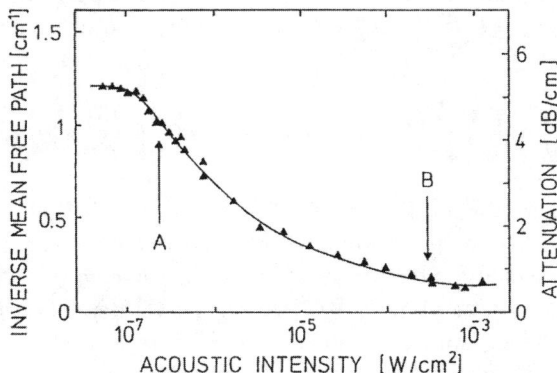

Fig. 14.4. Power dependence of the ultrasonic attenuation in borosilicate glass for 940 MHz longitudinal waves at 0.48 K (Arnold et al. [14.11])

with ε the strain and D, M the deformation potentials, then $H_0 + H'$ is equivalent to the Hamiltonian of the spin system $H_S = -\hbar\gamma(\boldsymbol{B} \cdot \boldsymbol{S} + \boldsymbol{b} \cdot \boldsymbol{S})$. Using the Pauli spin matrices gives

$$H_S = -\frac{\hbar\gamma}{2}\begin{pmatrix} B & 0 \\ 0 & B \end{pmatrix} - \frac{\hbar\gamma}{2}\begin{pmatrix} 0 & b \\ b & 0 \end{pmatrix}.$$

The strain terms correspond to the oscillating field b of the spin. Calculating the thermodynamical potential like the internal energy U, the frequency-dependent elastic constants $c(\omega)$ are obtained with the help of the strain susceptibilities

$$\chi_x = \frac{\partial \langle S_x \rangle}{\partial \varepsilon}, \quad \chi_z = \frac{\partial \langle S_z \rangle}{\partial \varepsilon}$$

and with n the number of molecules per volume.

$$c(\omega) = c_0 - n[4M^2\chi_x(\omega) + D^2\chi_z(\omega)] \tag{14.1}$$

where the strain susceptibilities can be calculated from the Bloch equations of the equivalent spin system. $\langle S_i \rangle$ are the thermodynamic expectation values. Two main contributions are important: the resonant contribution involving χ_x with the transverse relaxation time τ_2 and the relaxation contribution involving χ_z with the longitudinal relaxation time τ_1. Velocity and attenuation are then obtained by the real and imaginary part of $c(\omega)$:

$$v(\omega) = \frac{1}{2}\rho v Re[c(\omega)] \quad \text{and} \quad \alpha(\omega) = -\frac{\omega}{\rho v^2}Im[c(\omega)]. \tag{14.2}$$

The detailed calculations of these quantities are given in the reviews cited above and in Jäckle [14.13], Piché et al. [14.10]. For the resonant contribution, attenuation and elastic constants are related via the Kramers–Kronig relation. For $\hbar\omega < kT$ the relative sound velocity and attenuation are given by

$$\frac{\Delta v}{v} = \left(\frac{nM^2}{2\rho v^2}\right)\ln\left(\frac{k_B T}{\hbar\omega}\right) \quad \text{and} \quad \alpha(\omega, T) = \pi\omega\left(\frac{nM^2}{\rho v^2}\right)\tanh\left(\frac{\hbar\omega}{2k_B T}\right). \tag{14.3}$$

The experimental results (Figs. 14.3, 14.4), for which $\hbar\omega \ll kT$, show this behaviour. From fitting the formula to the experiment, the coupling constant M can be determined for longitudinal and transverse waves. We are dealing here with a resonant one-phonon process. The power dependence of the sound absorption can be explained by noting that with increased power, the upper level becomes more and more populated. Thus the number of effective two-level systems decreases and hence also the sound attenuation. This can be described with (14.3) by including a factor $(1 + I/I_c)^{-1/2}$ in the formula for the attenuation. For $I \ll I_c$, (14.3) is valid with the thermal distribution of the population difference between the two-levels. For $I \gg I_c$ the population difference disappears and the attenuation decreases. The same considerations as for saturation effects in NMR hold. One needs the energy gap $\Delta > kT$, a long relaxation time $\omega\tau > 1$ and the matrix element $M \neq 0$.

A stationary state in the presence of ultrasound, is described by the rate equation $n_1 W_{12} = n_2 W_{21} + n_2 A_{21}$ with n_1, n_2 the population of the lower and upper levels of the two level system. W_{12} is the probability of a transition from $1 \to 2$ and W_{21} from $2 \to 1$ respectively. For large ultrasound intensity, the spontaneous emission A_{12} can be neglected and one has with $W_{12} = W_{21}$, also $n_1 = n_2$, i.e. the ultrasound receives the same energy through induced emission as it loses through absorption. The wave then no longer experiences attenuation from the system. This effect is called in NMR resonance absorption. It cannot occur in a system of harmonic oscillators because of the availability of a large number of equidistant energy levels $E_n = (n+\frac{1}{2})\hbar\omega$. For an an-harmonic oscillator, the eigenfrequencies change with the amplitude.

This saturation effect leads to a number of interesting effects, such as hole burning, self-induced transparency and phonon echoes. The analogous effects are known from laser physics. An intense ultrasonic pulse leads to the equipartition of the tunneling system levels. In this case, a hole of a width $\Delta\omega$ is burnt into the spectral distribution at the energy corresponding to the applied frequency. With a second weaker pulse of different, variable frequency one can probe the population difference of the tunneling system. If τ_1 and τ_2 become long enough at low temperature, coherent phonon echoes can be generated. Such an experiment was performed in fused silica glass at a temperature of ~ 20 mK with ultrasound of 0.68 GHz (Graebner and Golding [14.14]). Two longitudinal pulses, a time interval τ_s apart, generate normal echoes at nT, $nT + \tau_s$, but also phonon echoes at $nT + (m+1)\tau_s$ with T the time between successive reflections. From such experiments, values for τ_1, τ_2 and the strain-two level coupling constant M can be obtained.

14.3 Recent Developments

In addition to the amorphous insulators, there are also metallic glasses. They have also been characterised with ultrasonic techniques. Apart from the normal conducting metallic glasses, there are superconducting metallic glasses. By applying high magnetic fields, sufficient to destroy the superconducting state, the influence of conduction electrons for the relaxation of the tunneling system can be studied. Many sound velocity and attenuation experiments in these systems have been performed, amongst them also phonon echo experiments in superconducting amorphous $Pd_{30}Zr_{70}$ (Weiss and Golding [14.15]). For a detailed discussion of glassy superconductors, see the reviews listed above and other more recent papers (Esquinazi et al. [14.16], Bezuglyi et al. [14.17]).

Recently, very low temperature experiments below 100 mK reveal interaction effects between two-level tunneling systems and a very sensitive dependence of the dielectric constant to small magnetic fields. In the presence of magnetic fields, flux periodicity in the energy levels of the tunneling system (Aharonov-Bohm effect) is predicted and observed. Since the relevant experimental tool in these experiments is the measurement of the dielectric constant we do not enter into a detailed discussion. Experimental and theoretical papers on these issues are by Kettemann et al. [14.18] and Strehlow et al. [14.19]; a recent review on these quantum phenomena in ultra-cold glasses is given by Enss [14.20]. Apparently nuclear spin interaction is of importance in this case.

14.4 Spin-Glass

Unlike the glass substances discussed above, to our knowledge, only one acoustic experiment has been performed for spin glasses (Moran et al. [14.21]). The cobalt and manganese aluminosilicate were investigated with 10MHz longitudinal and shear waves as a function of temperature. They exhibit a magnetic susceptibility maximum at low temperatures and also a broad specific heat peak. The acoustic waves showed spin glass effects of the order of 1 part in 10^4 for the relative velocity change $\Delta v/v$. If a background velocity is subtracted (as discussed in Sect. 4.2.1), broad minima in both longitudinal and transverse modes are observed. They can be related to the specific heat anomaly in a similar way as the velocity anomaly discussed for magnetic phase transitions in Sect. 6.1.3. With a free energy coupling $F = -Tf(T/T_c)$, a proportionality of the elastic constant with the specific heat C_V is obtained. The idea in the case of spin-glasses is to assume a mono-domain of predominantly magnetic ions (Co, Mn) with local anisotropy axis separated by Co, Mn poor regions. At low temperatures, the alignment of the anisotropy directions leads to maxima and minima in the various thermodynamic derivatives as observed.

For a general introduction to spin–glasses see Mydosh [14.22].

14.5 Quasi-crystals

Quasi-crystals are long-ranged ordered but have no periodic structure. Only a few of them are thermodynamically stable. One of them is the face-centered icosahedral structure (fci) Zn-Mg-Y. The actual composition of the specimen is $Zn_{62}Mg_{29}Y_9$. Using this material, it was possible to grow large single crystals (Langsdorf and Assmus [14.23]). Subsequently ultrasonic measurements were performed (Sterzel et al. [14.24]). The reason why this topic is discussed in this chapter on tunneling phenomena is the experimental fact that the longitudinal sound velocity has a logarithmic temperature dependence for temperatures below $\sim 1\,\text{K}$. In fact, it follows closely a temperature dependence given in (14.3). This behaviour can be again attributed to the occurrence of tunneling in a two-level system. Transverse sound waves exhibit a normal temperature dependence, given by the background elastic constant (Fig. 4.1 of Chap. 4). Similar effects were found in another quasi-crystalline system Al-Mn-Pd (Vernier et al. [14.25]).

An ultrasonic comparison of the Zn-Mg-Y quasi-crystal with the hexagonal parent compound Zn_2Mg was made. For the quasi-crystal, the sound propagation was along the fivefold symmetry axis. In addition, the polarization of the transverse wave was chosen in three different directions, but with the same value of the absolute sound velocity within the accuracy of the measurement. Within the error limits, the absolute values agreed also with the values of the parent crystal Zn_2Mg.

Thermal conductivity measurements of another quasi-crystal $Al_{70}Mn_9Pd_{21}$ have led to contributions both from conduction electrons as well as phonons (see Chernikov et al. [14.26]). Subtracting the electronic part using the Wiedemann-Franz law for this metallic sample resulted in a phonon thermal conductivity exhibiting a power law temperature dependence for $T \leq 1.6\,\mathrm{K}$ with a dependence $\sim T^2$ apparently in agreement with the tunneling model (with constant density of states and using the Debye model). The overall features of the temperature dependence of the thermal conductivity in quasi-crystals seems to be similar to the case of amorphous materials.

15 Conclusion and Outlook

Many applications of sound wave experiments to various solid state phenomena have been given in the foregoing chapters. Practically all main fields of solid state physics have been covered, such as metals and semiconductors, superconductors, magnetic materials, ferroelectric, low dimensional materials and many more. Sound wave propagation near different kinds of phase transitions have been studied, the interaction with collective modes have been investigated and various types of ground states and excited states have been analysed.

Sound waves are limited in k–space to the $k = 0$ point, the Γ point – a negligible range of phase space. This is, compared to neutron scattering, a distinct disadvantage. The high resolution for sound attenuation and especially sound velocity experiments, however, make this technique indispensable for many purposes. In addition, the small budget apparatus and installations make it an important and versatile instrument for condensed matter physics.

In the future, ultrasonics will be of importance in material testing. But also for solid state physics research, various modern techniques will be further applied. We have shown in previous chapters that many different effects occur for sound wave propagation in solids. With the new materials being produced in the future, sound waves can help to characterise these materials. The various effects for the different materials explained in this book can then hopefully be applied.

The parameter space is continuously extended in solid state physics. The temperature gets into the milli-Kelvin region, pulsed magnetic fields slowly approach the 100 Tesla mark and pressures of 10 GPa and higher are dealt with. Ultrasonics can also be applied in these extended regimes as shown in several places in this book. Novel techniques will also be applied in these and other cases.

A Mass Systems and Units

Table A.1. Conversion table for some physical quantities

		SI-units	Gaussian units
Magnetic field	B	1 tesla	10^4 Gauss
Magnetization	M	1 amp/m	10^{-3} emu/cm^3
Current	I	1 amp	3×10^9 statamp
Resistance	R	1 ohm	$1/9 \times 10^{-11}$ s cm^{-1}
Potential	V	1 volt	$1/300$ statvolt
Charge	q	1 coulomb	3×10^9 statcoulomb
Pressure	p	1 Pa	10^{-5} bar = 10 dyn/cm^2
		1 bar	10^5 Pa = 10^6 dyn/cm^2 = 1 atm
			1 GPa = 10^{10} dyn/cm^2 = 10^{10} erg/cm^3

Table A.2. Physical constants

Light velocity	c	2.9979×10^{10} cm s^{-1}
Planck constant	$\hbar = h/2\pi$	1.0547×10^{-27} erg s
Boltzmann constant	k	1.38066×10^{-16} erg K^{-1}
Avogadro constant	N_A	6.0221×10^{23} mol^{-1}
Bohr-magneton	μ_B	0.9274×10^{-20} erg gauss^{-1}
Electronic charge	e	4.8022×10^{-10} esu
Electron mass	m	9.1072×10^{-28} g
Magnetomechanical ratio $(\gamma = g\frac{e}{2mc} = g\frac{\mu_B}{\hbar})$	γ	1.76×10^7 (Gs)$^{-1}$

Table A.3. Conversion table for energy units

	f [GHz]	T [K]	E [meV]	k [cm^{-1}]
1 GHz	1	0.048	$4.136 \ 10^{-3}$	0.033
1 K	20.83	1	0.086	0.695
1 meV	242	11.606	1	8.072
1 cm^{-1}	29.98	1.439	0.124	1

B Wave Equation for Sound Waves

Section 3.1 introduced the stress and strain tensors. Hooke's law (3.7), in Voigt notation reads (3.7a)

$$T_i = \sum_n c_{in}\varepsilon_n .$$

In order to obtain the equation of motion (3.6), it is necessary to consider the variation of the stress across an infinitesimal parallelepiped with its sides parallel to the rectangular axes (Fig. B.1). The stress components will vary across the side faces as shown in the figure. The force acting on each face is obtained by taking the stress on each face times the area of the face.

As can be seen from the Fig. B.1, six separate forces are acting parallel to each axis. Considering the resultant force acting in the x-direction gives

$$\left(T_{xx} + \frac{\partial T_{xx}}{\partial x}dx\right)dydz - T_{xx}dydz + \left(T_{xy} + \frac{\partial T_{xy}}{\partial y}dy\right)dxdz - T_{xy}dxdz$$
$$+ \left(T_{xz} + \frac{\partial T_{xz}}{\partial z}dz\right)dxdy - T_{xz}dxdy = \left(\frac{\partial T_{xx}}{\partial x} + \frac{\partial T_{xy}}{\partial y} + \frac{\partial T_{xz}}{\partial z}\right)dxdydz .$$

According to Newton's second law, this force is equal to the $\rho\,dx\,dy\,dz\frac{\partial^2 u_x}{\partial t^2}$. Neglecting body forces such as gravity, i.e. we get $\rho\frac{\partial^2 u_x}{\partial t^2} = \frac{\partial T_{xx}}{\partial x} + \frac{\partial T_{xy}}{\partial y} + \frac{\partial T_{xz}}{\partial z}$ or more generally

$$\rho\frac{\partial^2 u_i}{\partial t^2} = \sum_k \frac{\partial T_{ik}}{\partial x_k} ,$$

which is (3.6) from the text. Finally Hooke's law (3.7) for plane waves $u_i = Ue_i \exp(i(\boldsymbol{kr}-\omega t))$ with U the amplitude, e_i the component of the polarization vector, results in

$$-\rho\omega^2 e_i = \sum_{klm} c_{iklm}k_k k_m e_l$$

which is (3.8) from the text.

As an example in calculating the sound velocities for different directions of a cubic crystal: We have $c_{11} = c_{22} = c_{33}$ and $c_{12} = c_{21} = c_{13} = c_{31} = c_{23} =$

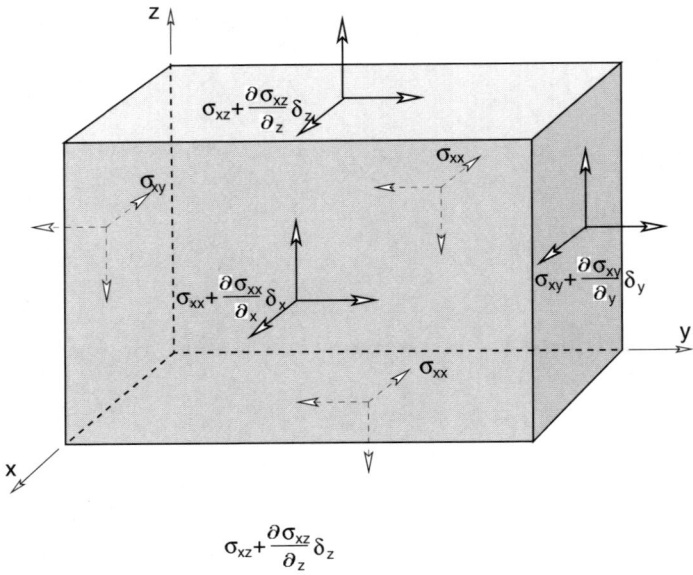

Fig. B.1. Stresses acting on a small rectangular parallelepiped. The σ_{ij} correspond to T_{ij} of the text

c_{32} and $c_{44} = c_{55} = c_{66}$. The plane wave Ansatz, with the equations above and (3.6) and (3.7) results in

$$(-\rho\omega^2 + c_{11}k_x^2 + c_{44}k_z^2 + c_{44}k_y^2)e_x$$
$$+(c_{12} + c_{44})k_x k_y e_y + (c_{12} + c_{44})k_x k_z e_z = 0$$

$$(-\rho\omega^2 + c_{11}k_y^2 + c_{44}k_z^2 + c_{44}k_x^2)e_y$$
$$+(c_{12} + c_{44})k_x k_y e_x + (c_{12} + c_{44})k_y k_z e_z = 0$$

$$(-\rho\omega^2 + c_{11}k_z^2 + c_{44}k_y^2 + c_{44}k_x^2)e_z$$
$$+(c_{12} + c_{44})k_z k_y e_y + (c_{12} + c_{44})k_z k_x e_x = 0. \quad \text{(B.1)}$$

The determinant of this linear system of equations for e_x, e_y, e_z gives a cubic root for $\omega^2(\boldsymbol{k})$ or three solutions for the sound waves, namely one longitudinal and two transverse for symmetry directions. This set of equations results in the phase velocities in the different directions as given in Table 3.1.

C Elastic Constants and Symmetry Strains for the Crystal Classes

Table C.1. Elastic constants for the different crystal systems (in Voigt notation and $c_{ij,kl} = c_{kl,ij}$)

System	c_{ij}	# of el. const.	Basis functions	Representation
Triclinic (C_i)	11 12 13 16 15 16 22 23 24 25 26 33 34 35 36 44 45 46 55 56 66	21	$x^2, y^2, z^2, xy, yz, xz$	$\Gamma = A_g$
monoclinic (C_{2h})	11 12 13 0 0 16 22 23 0 0 26 33 0 0 36 44 45 0 55 0 66	13	x^2, y^2, z^2, xy xz, yz	$\Gamma = 4A_g + 2B_g$

orthorhombic (D_{2h})	11 12 13
	22 23
	33
	          0
	    44
	      55
	        66
		9	$x^2 + y^2 + z^2$
			$2z^2 - x^2 - y^2, x^2 - y^2$
			xy
			xz
			yz
				$\Gamma = 3A_{1g} + B_{1g} + B_{2g} + B_{3g}$

tetragonal (D_{4h})	11 12 13
	   11 13
	      33
	          0
	    44
	      44
	        66
		6	$x^2 + y^2 + z^2$
			$2z^2 - x^2 - y^2$
			$x^2 - y^2$
			xy
			xz, yz
				$\Gamma = 2A_{1g} + B_{1g} + B_{2g} + E_g$

trigonal (D_{3d})	11 12 13 14
	   11 13 14
	      33 0
	         44
	           44
	             $66 = (11 - 12)/2$
		6	$x^2 + y^2 + z^2$
			$2z^2 - x^2 - y^2$
			$x^2 - y^2, xy$
			xz, yz
				$\Gamma = 2A_{1g} + 2E_g$

hexagonal (D_{6h})	11 12 13
	   11 13
	      33
	          0
	    44
	      44
	        $66 = (11 - 12)/2$
		5	$x^2 + y^2 + z^2$
			$2z^2 - x^2 - y^2$
			$x^2 - y^2, xy$
			xz, yz
				$\Gamma = 2A_{1g} + E_{1g} + E_{2g}$

cubic (O_h)	11 12 12
	   11 12
	      11
	         0
	    44
	      44
	        44
		3	$x^2 + y^2 + z^2$
			$2z^2 - x^2 - y^2, x^2 - y^2$
			xy, yz, zx
				$\Gamma = A_{1g} + E_g + T_{2g}$

D g-Factor, Steven's Factors, CEF Operators

D.1 3d and 4f Ions: Landé g_J Factor, Steven's Factors

Table D.1. Landé g_J factor, Steven's factors α, β and γ

Rare earth	ground state $^{2S+1}L_J$	Landé factor g_J	Steven's factor α	Steven's factor $\beta \times 10^4$	Steven's factor $\gamma \times 10^6$
Ce^{3+}	$4f^1\ {}^2F_{5/2}$	6/7	$-2/35$	63.492	0.0
Pr^{3+}	$4f^2\ {}^3H_4$	4/5	$-52/2475$	-7.3462	61.0
Nd^{3+}	$4f^3\ {}^4I_{9/2}$	8/11	$-7/1089$	-2.9111	-38.0
Sm^{3+}	$4f^5\ {}^6H_{5/2}$	2/7	13/315	25.012	0.0
$Gd^{3+}\ Eu^{2+}$	$4f^7\ {}^8S_{7/2}$	2/1			
Tb^{3+}	$4f^8\ {}^7F_6$	3/2	$-1/99$	1.2244	-1.12
Dy^{3+}	$4f^9\ {}^6H_{15/2}$	4/3	$-2/315$	-0.59200	1.03
Ho^{3+}	$4f^{10}\ {}^5I_8$	5/4	$-1/450$	-0.33300	-1.30
Er^{3+}	$4f^{11}\ {}^4I_{15/2}$	6/5	4/1575	0.44400	2.07
Tm^{3+}	$4f^{12}\ {}^3H_6$	7/6	1/99	1.6325	-5.60
Yb^{3+}	$4f^{13}\ {}^2F_{7/2}$	8/7	2/63	17.316	148.0

Table D.2. 3d Electrons

			examples used in this book	
for $3d^1\ {}^2D$ and $3d^6\ {}^5D$	$\alpha = -2/21$	$\beta = 2/63$	$3d^1$: Ti^{3+}, V^{4+}	$3d^6$: Fe^{2+}
for $3d^2\ {}^3F$ and $3d^7\ {}^4F$	$\alpha = -2/105$	$\beta = -2/315$		$3d^7$: Ni^{3+}
for $3d^3\ {}^4F$ and $3d^8\ {}^3F$	$\alpha = 2/105$	$\beta = 2/315$	$3d^3$: Cr^{3+}	$3d^8$: Ni^{2+}
for $3d^4\ {}^5D$ and $3d^9\ {}^2D$	$\alpha = 2/21$	$\beta = -2/63$		$3d^9$: Cu^{2+}

D.2 Cubic CEF and Quadrupolar Operators

Crystal Electric Field Hamiltonian

$$H_{CEF} = B_4 \left(O_4^0 + 5O_4^4\right) + B_6 \left(O_6^0 - 21O_6^4\right)$$
$$= A_4 \langle r^4 \rangle \beta_J \left(O_4^0 + 5O_4^4\right) + A_6 \langle r^6 \rangle \gamma_J \left(O_6^0 - 21O_6^4\right)$$

$$O_4^0 = 35 J_z^4 - 30 J(J+1) J_z^2 + 25 J_z^2 - 6 J(J+1) + 3 J^2 (J+1)^2$$
$$O_4^4 = \frac{1}{2} \left[J_+^4 + J_-^4\right].$$

For O_6^n operators see Hutchings [5.8].

Single Ion Magneto-elastic Hamiltonian

$$H_{me} = g_{\Gamma 11} \varepsilon_V (O_4^0 + 5 O_4^4) + g_{\Gamma 12} \varepsilon_V (O_6^0 - 21 O_6^4)$$
$$+ g_{\Gamma 3}(\varepsilon_3 O_2^0 + 3\sqrt{2} \varepsilon_2 O_2^2) + g_{\Gamma 5}(\varepsilon_{xy} O_{xy} + \varepsilon_{xz} O_{xz} + \varepsilon_{yz} O_{yz})$$

$$O_2^0 = 2 J_z^2 - J_x^2 - J_y^2 = 3 J_z^2 - J(J+1)$$

$$O_2^2 = J_x^2 - J_y^2$$

$$O_{xy} = \frac{1}{2}[J_x J_y + J_y J_x].$$

For O_2^4 operators see (7.4).

E Ultrasonic Attenuation in Metals

As mentioned in Sect. 8.2, many calculations have been performed for ultrasonic attenuation in metals for both limits $ql_e < 1$ and $ql_e > 1$ with l_e the mean free path of the electrons. Brief derivations are given here for the $ql_e < 1$ limit. The $ql_e > 1$ limit has to be solved using, e.g., the Boltzmann equation. The Pippard result for arbitrary ql_e will be quoted without a derivation.

A calculation for $ql_e < 1$ is outlined here using the concept of the viscosity of an electron gas. Such a calculation has been performed by Mason [8.9]. We use the formula of sound attenuation in a viscous medium (Landau–Lifshitz [4.27]):

$$\alpha = \frac{\mathrm{d}E}{\mathrm{d}t(2vE_0)} \quad \text{with} \quad E_0 = \frac{1}{2}\rho v^2 V_0$$

and

$$\frac{\mathrm{d}E}{\mathrm{d}t} = -\frac{1}{2}k^2 v^2 V_0 \left[\frac{4}{3}\eta + \zeta + \kappa\left(\frac{1}{C_V} - \frac{1}{C_p}\right)\right].$$

In a solid we can set for the specific heats $C_V \sim C_p$ and the compressional viscosity ζ can be neglected compared to the shear viscosity η. Therefore the attenuation α for compressional waves reads

$$\alpha = \frac{4}{3}\frac{\omega^2}{\rho v^3}\eta. \tag{E.1}$$

For the viscosity, the gaskinetic expression $\eta = Al_e n m v_F$ is taken with A a numerical constant, which is often taken as $A \approx \frac{1}{3}$, and v_F the Fermi velocity. This gives

$$\alpha = \frac{2}{9}nmv_F q^2 \frac{l_e}{\rho v_L}, \tag{E.2}$$

where $\omega = v_L q$ is used. Notice the dependence $q^2 l_e$ as in (8.5) of the text. The prefactor in this equation is slightly different (taken from Morse [8.12]).

ql_e Arbitrary

The formulas by Pippard for longitudinal and shear waves are quoted here for completeness without deriving it. For a derivation of these results see the reviews given in Sect. 8.2.1.

E Ultrasonic Attenuation in Metals

Ultrasonic attenuation for $\omega\tau \ll 1$ with $a = ql_e$:

Longitudinal waves $\quad \alpha_L = \dfrac{Nm}{\rho v_L \tau} \left\{ \dfrac{a^2 tg^{-1} a}{3(a - tg^{-1} a)} - 1 \right\}$

Transverse waves $\quad \alpha_T = \dfrac{Nm}{\rho v_T \tau} \left\{ \dfrac{2}{3} \dfrac{a^3}{(1+a^2)tg^{-1}a - a} - 1 \right\} .$

(E.3)

These formulas lead to (8.5) and (8.10) for $ql_e < 1$ and $ql > 1$ respectively. These formulas are further discussed in Sects. 8.2.1 and 8.3.1.

F Free Energy of Electrongas

Taking the Helmholtz free energy (Sect. 4.1) $F = U - TS$, the internal energy

$$U = \sum_j N_j E_j$$

and the entropy results in

$$S = k_B \ln W = k_B \ln \Pi_j g_j!/N_j!(g_j - N_j)!$$
$$= k_B \sum_j [g_j \ln g_j - N_j \ln N_j - (g_j - N_j)\ln(g_j - N_j)] .$$

Here, g_j is the degeneracy of the state j and $N_j/g_j = (1 + \exp((E_j - E_F)/kT))^{-1} = f_j$, the Fermi–Dirac distribution function. In addition, the Stirling formula for large numbers ($x \gg 1$)

$$\ln x! = x \ln x - x$$

was used. Inserting all these expressions into the free energy F gives

$$F = \sum g_j f_j E_j - k_B T \sum_j g_j \ln \left\{1 + e^{\frac{E_F - E_j}{k_B T}}\right\} - \sum_j g_j (E_j - E_F) f_j$$

or for k-states

$$F(T, \varepsilon, n) = n E_F - k_B T \sum_k \ln \left\{1 + e^{\frac{E_F - E_k}{k_B T}}\right\} , \tag{F.1}$$

where $n = \sum_j g_j f_j = \sum_k f_k$ is the number of electrons. This is the free energy expression for an electrongas with Fermi–Dirac statistics and it is the formula used in the text of Chap. 8, (8.6). Its generalisation for the presence of a magnetic field is mentioned in Sect. 8.3.2. The free energy expression (F.1) or (8.6) will be used to calculate the elastic constants for a metal in Sect. 8.2.1. For doing this, the strain dependence of E_F, E_k and n will be introduced.

G Order–Disorder Phase Transition

A discussion on simple order–disorder model of a phase transition as it occurs, e.g. in KCN (Sect. 7.4.2) or in MAMC (Chap. 12) follows. To be specific, the model of KCN is used where the (CN)$^-$ anions orient with their long axis along the different (110) directions. This is followed with some modifications, viz the outline of Rehwald et al. [7.177]. Denoting the occupation probability for the six directions with $n_i = N_i/N$ and $\sum_{i=1}^{6} n_i = 1$, we get for the internal energy $U = a\sum_i n_i^2 + b(n_1 n_2 + n_3 n_4 + n_5 n_6)$ and for the configurational entropy $S = -k_B \sum_i n_i \ln n_i$. The free energy $F = U - TS$ with the Lagrange parameter λ reads $F^\star = F - \lambda(\sum_i n_i - 1)$. Minimisation of $\frac{\partial F^\star}{\partial n_i} = 0$ gives the equilibrium conditions. With the three-dimensional order parameter

$$\eta_1 = n_2 - n_1, \quad \eta_2 = n_4 - n_3, \quad \eta_3 = n_6 - n_5 \tag{G.1}$$

all $\eta_i = 0$ for complete disorder and $\eta = 1$ for complete order along say (110) and $\eta = -1$ for order along (1$\bar{1}$0). This order parameter G.1 belongs to the three-fold degenerate representation T_{2g} and lead to the orthorhombic phase D_{2h}^{25}. The stability conditions $\frac{\partial^2 F^\star}{\partial n_i^2} > 0$, $\frac{\partial^2 F^\star}{\partial n_1^2}\frac{\partial^2 F^\star}{\partial n_2^2} - \frac{\partial^2 F^\star}{\partial n_1 \partial n_2} > 0$ leads to the transition temperature $kT_c^0 = (b-a)/6$. The order parameter–strain coupling is of the bilinear type

$$F_c = g(\eta_1 \varepsilon_{xz} + \eta_2 \varepsilon_{yz} + \eta_3 \varepsilon_{xy}). \tag{G.2}$$

From this follows (Sect. 4.3.4) for the elastic constant c_{44} (4.13) as discussed in the text on KCN (Sect. 7.4.2). For the second order phase transition, $T_c = T_c^0 + g^2/a'c_{44}^0$ with a' the Landau expansion parameter and T_c^0 the entropy driven phase transition. T_c, therefore, is, in addition, driven by the strain–order parameter coupling.

The six occupation probabilities above form a basis for a reducible representation of the cubic point group O_h. Its decomposition gives the irreducible representations $A_{1g} + E_g + T_{2g}$ of which the T_{2g} representation has been treated. The E_g representation gives another order parameter coupling to the E_g type strains with the $c_{11} - c_{12}$ elastic constant which also exhibits some softening (Rehwald et al. [7.177]). The order–disorder treatment above does not distinguish between head and tail of the CN molecules. Consideration of this phenomenon leads to the second phase transition for KCN at 83 K (see Table 7.4). An analogous description with three order parameters can be given for MAMC (see Blinc et al. [12.111] and Goto et al. [12.105]).

References

[1.1] W.P. Mason et al.: editor, *Physical Acoustics, Principles and Methods* (Academic Press, 1964)
[1.2] R. Truell, C. Elbaum, B. Chick: *Ultrasonic Methods in Solid State Physics* (Academic Press, 1969)
[1.3] R.T. Beyer, S.V. Letcher: *Physical Ultrasonics* (Academic Press, 1969)
[1.4] J.W. Tucker, V.W. Rampton: *Microwave Ultrasonics in Solid State Physics* (North Holland, 1972)
[1.5] A.E.H. Love: *Theory of Elasticity* (Cambridge University Press, 1934)
[1.6] H. Kolsky: *Stress waves in solids* (Dover Publ., 1953)
[1.7] Landau–Lifshitz: *Theory of Elasticity* (Addison-Wesley, 1959)
[1.8] D. Vollhardt and P. Wölfle: *The Superfluid Phases of Helium 3* (Taylor and Francis, 1990)
[1.9] M.J. Stephen and J.P. Straley: Rev. Mod. Phys. **46**, 617 (1974)
[1.10] P.G. de Gennes and J. Prost: *The Physics of Liquid Crystals* (Oxford Science Publ., 1995)
[1.11] J.P. Wolfe: *Imaging Phonons* (Cambridge University Press, 1998)
[1.12] W.E. Bron: Rep. Progr. Phys. **43**, 301 (1980)
[1.13] A.S. Nowick and B.S. Berry: *Anelastic Relaxation in Crystalline Solids* (Academic Press, 1972)
[1.14] N.W. Ashcroft and N.D. Mermin: *Solid State Physics* (Holt, Reinhart and Winston, 1976)
[1.15] H. Ibach and H. Lüth: *Festkörperphysik* (Springer, 1988)
[1.16] C. Kittel: *Introduction to Solid State Physics* (Wiley, 1971)
[2.1] A. Read, C.A. Wert, M. Metzger: Methods of experimental physics Vol. **6a** (1974)
[2.2] R. Truell, C. Elbaum, B. Chick: *Ultrasonic Methods in Solid State Physics* (Academic Press, 1969)
[2.3] D.I. Bolef and J.G. Miller: in *Physical Acoustics* Vol. IVA (Acad. Press, 1971)
[2.4] E. R. Fuller Jr., A.V. Granato, J. Holder, E.R. Naimon: in *Methods of experimental physics Vol. 11* (1974)
[2.5] A. Migliori and J.L. Sarrao: *Resonant Ultrasound Spectroscopy* (J. Wiley, 1997)
[2.6] J.W. Tucker, V.W. Rampton: *Microwave Ultrasonics in Solid State Physics* (North Holland, 1972)
[2.7] H. Ogi, Y. Kawasaki, M. Hirao, H. Leadbetter: J. Appl. Phys. **92**, 2451 (2002)
[2.8] W.P. Mason: *Piezo-electric Crystals* (Van Nostrand, 1959)

[2.9] N.F. Foster in: *Handbook of Thin Film Technology*, ed. L.I. Maissel, R. Glang, chapter 15 (McGraw-Hill, 1970)
[2.10] E.R. Dobbs: *Physical Acoustics* Vol. X, (Acad. Press, 1974)
[2.11] V.D. Buchelńikov and A.N. Vasil'ev: Sov. Phys. Usp. **35**, 192 (1992)
[2.12] G. Gorodetsky, B. Lüthi, T.J. Moran and M.E. Mullen: J. Appl. Phys. **43**, 1234 (1972)
[2.13] J. Heil, I. Kouroudis, B. Lüthi and P. Thalmeier: J. Phys. C **17**, 2433 (1984)
[2.14] B. Lüthi, G. Bruls, P. Thalmeier, B. Wolf, D. Finsterbusch, I. Kouroudis: J. Low Temp. Phys. **95**, 257 (1994)
[2.15] P.W. Wallace and C.W. Garland: Rev. Sci. Instr. **57**, 3085 (1986)
[2.16] F. Pobell: *Matter and Methods at low temperatures* (Springer, 1996)
[2.17] B. Wolf: Thesis, Universität Frankfurt (1993)
[2.18] T.J. Kim and W. Grill: Ultrasonics **36**, 233 (1998)
[2.19] M. Niksch and W. Grill: Acoustica **64**, 26 (1987)
[2.20] J.P. Wolfe: *Imaging Phonons* (Cambridge University Press, 1998)
[2.21] A. Migliori, W.M. Visscher, S.E. Brown, Z. Fisk, S.W. Cheong, B. Alten, E.T. Ahren, K.A. Kubat-Martin, J.D. Maynard, Y. Huang, D.R. Kirk, K.A. Gillis, H.K. Kim, M.H.W. Chan: Phys. Rev. B **41**, 2098 (1990)
[2.22] A. Migliori, W.M. Visscher, S. Wong, S.E. Brown, I. Tanaka, H. Kojima, P.B. Allen: Phys. Rev. Lett. **64**, 2458 (1990)
[2.23] M. Lei, J.L. Sarrao, W.M. Visscher, T.M. Bell, J.D. Thompson, A. Migliori, U.W. Welp, B.W. Veal: Phys. Rev. B **47**, 6154 (1990)
[2.24] J. Paglione et al.: Phys. Rev. B **65**, 220506(R) (2002)
[2.25] R.G. Leisure and F.A. Willis: J. Phys. Cond. Matter **9**, 6001 (1997)
[2.26] J. Schreuer: IEEE Transact. Ultrasound **49**, 1474 (2002)
[2.27] A. Migliori, J.L. Sarrao, W.M. Visscher, T.M. Bell, M. Lei, Z. Fisk, R.G. Leisure: Physica B **183**, 1 (1993)
[2.28] A.S. Novick and B.S. Berry: *Anelastic relaxation in Crystalline Solids* (Academic Press, 1972)
[2.29] M. Barmatz and B. Golding: Phys. Rev. B **9**, 3064 (1974)
[2.30] B.S. Berry and W.C. Pritchet, IBM J. Res. Dev., **19**, 334 (1975)
[2.31] P. Esquinazi, J. Low: Temp. Phys.**85**, 139 (1991)
[2.32] E.L. Hahn: Phys. Rev.**80**, 580 (1959)
[2.33] N.S. Shiren and T.G. Kazyaka: Phys. Rev. Lett. **20**, 1304 (1972)
[2.34] J. E. Graebner and B. Golding: Phys. Rev. B **19**, 964 (1979)
[2.35] S.N. Popov and N.N. Krainik: Sov. Phys. Solid State **12**, 2440 (1970)
[2.36] R.B. Thompson and C.F. Quate: Appl. Phys. Lett. **16**, 295 (1970)
[2.37] A. Billmann, Ch. Frenois, J. Joffrin, A. Levelut, S. Ziolliewicz: J. de Physique **34**, 453 (1973)
[2.38] K. Kajimura in: *Physical Acoustics* Vol. XVI, 295 (Acad. Press, 1982)
[2.39] R.L. Melcher and N.S. Shiren: in: *Physical Acoustics Vol. XVI*, 341 (Acad. Press, 1982)
[2.40] K. Fossheim and R. M. Holt: in *Physical Acoustics Vol. XVI*, 217 (Acad. Press, 1982)
[2.41] F. Herlach: Rep. Prog. Phys.**62**, 859 (1999)
[2.42] F. Herlach and N. Miura: editors, *High Magnetic Fields*, Vol. **1-3** (World Scientific, 2003)
[2.43] B. Wolf, B. Lüthi, S. Schmidt, H. Schwenk, M. Sieling, S. Zherlitsyn, I. Kouroudis: Physica B **294–295**, 612 (2001)

[2.44]   A.M. Tishin, K.A. Gschneidner, Jr., V.K. Pecharsky: Phys. Rev. B **59**, 503 (1999)
[2.45]   R.M. White: Proc.IEEE **58**, 1238 (1970)
[2.46]   K. Dransfeld and E. Salzmann in: *Physical Acoustics Vol. VII* 219 (Acad. Press, 1970)
[2.47]   A.A. Oliner editor: *Acoustic Surface Waves*, Topics in Appl.Physics Vol. 24 (Springer, 1978)
[2.48]   W. Kress and F.W. Wette: editors, *Surface Phonons*, Springer Series in Surface Science 21 (1991)
[2.49]   E.H. Jacobsen: J. Acoust. Soc. Am. **32**, 949 (1960)
[2.50]   K.N. Baranskii, Doklady: Akad. Nauk S.S.S.R **114**, 517 (1957)
[2.51]   E.H. Jacobsen: Phys. Rev. Lett. **2**, 249 (1959)
[2.52]   H. Bömmel and K. Dransfeld: Phys. Rev. Lett. (1959)
[2.53]   H. Bömmel and K. Dransfeld: Phys. Rev. **117**, 1245 (1960)
[2.54]   M. Pomerantz: Phys. Rev. **139A**, 501 (1965)
[2.55]   J.N. Lange: Phys. Rev.**176**, 1030 (1968)
[2.56]   P.J. King and H.M. Rosenberg: Proc. Roy. Soc. A **315**, 369 (1970)
[2.57]   J. Iluker and E.H. Jacobsen: *Phys. Acoustics*, Vol. V, p. 221 (Acad. Press, 1968)
[2.58]   R.T. Beyer, S.V. Letcher: *Physical Ultrasonics* (Academic Press, 1969)
[2.59]   W. Eisenmenger and A.H. Dayem: Phys. Rev. Lett. **18**, 125 (1967)
[2.60]   W. Eisenmenger: in *Physical Acoustics*, Vol. XII (Acad. Press, 1976)
[2.61]   H. Kinder: Int. Conf. *Phonon Scattering Solids*, Paris (1972)
[2.62]   J.R. Sandercock: in Light Scattering in Solids III, 173, ed. M. Cardona and G. Güntherodt (Springer, 1982)
[2.63]   A.S. Pine in: *Light Scattering in Solids*, Topics in Applied Physics Vol. 8, ed. M. Cardona (Springer, 1975)
[2.64]   R. Mock and G. Güntherodt: J. Phys. C **17**, 5635 (1984)
[2.65]   G.K. White: Cryogenics **1**, 151 (1961)
[2.66]   G. Brändli and R. Griessen: Cryogenics **13**, 299 (1973)
[2.67]   H.R. Ott and B. Lüthi: Z. Physik B **28**, 141 (1977)
[2.68]   M. Lang: Thesis TU Darmstadt (1991)
[2.69]   E. du Tremolet de Lacheisserie: *Magnetostriction* (CRC Press, 1993)
[2.70]   P. Brüesch, Phonons: *Theory and Experiment III* chapter 4, Springer Series in Solid State Science 66 (1987)
[2.71]   R. Berman: *Thermal Conductivity in Solids* (Clarendon Press Oxford, 1976)
[3.1]   Landau-Lifshitz: *Theory of Elasticity* (Addison-Wesley, 1959)
[3.2]   R. Truell, C. Elbaum, B. Chick: *Ultrasonic Methods in Solid State Physics* (Academic Press, 1969)
[3.3]   R.T. Beyer, S.V. Letcher: *Physical Ultrasonics* (Academic Press, 1969)
[3.4]   D.C. Wallace: *Thermodynamics of crystals* (John Wiley & Sons, 1972)
[3.5]   W. Ludwig: *Festkörperphysik* (Akad. Verlagsgesellschaft, 1978)
[3.6]   R.N. Thurston: *Physical Acoustics*, Vol. 1b (1964)
[3.7]   L. Bonsall and R.L. Melcher: Phys. Rev. B *14*, 1128 (1976)
[3.8]   P. Blanchfield and G.A. Saunders: J. Phys. C *12*, 4673 (1979)
[3.9]   M.E. Lines: Physics Reports **55**, 133 (1979)
[3.10]   M. Tinkham: *Group Theory and Quantum Mechanics* (McGraw-Hill, 1964)

References

[3.11]   G. Burns: *Introduction to Group Theory with Applications* (Academic Press, 1977)
[3.12]   T. Inui, Y. Tanabe, Y. Onodeara: *Group Theory and its application in Physics* (Springer, 1990)
[3.13]   M. Wagner: *Gruppentheoretische Methoden in der Physik* (Vieweg, 1998)
[3.14]   J.H. Van Vleck: J. Chem. Phys. **7**, 72 (1939)
[3.15]   J. Rouchy and E. du Tremolet de Lacheisserie: Z. Phys. B **36**, 67 (1979)
[3.16]   R.N. Thurston and K. Brugger: Phys. Rev. **133**, A1604 (1964)
[3.17]   K. Brugger: J. Appl. Phys. **36**, 759 (1965)
[3.18]   D.E. Eastman: Phys. Rev. **148**, 530 (1966)
[3.19]   M. Born and K. Huang: *Dynamical Theory of Crystal Lattices* (Oxford University Press, 1954)
[3.20]   H. Kolsky: *Stress waves in solids* (Dover Publ., 1953)
[3.21]   R. Stoneley: Proc. Roy. Soc. *A232*, 447 (1955)
[3.22]   D.C. Gazis, R. Herman, R. Wallis: Phys. Rev. **119**, 533 (1960)
[3.23]   C. Lingner and B. Lüthi: Phys. Rev. B **23**, 256 (1981)
[3.24]   C. Lingner, J. Heil, B. Lüthi: J. Appl. Phys. **52**, 2270 (1981b)
[3.25]   R.M. White: Proc. IEEE **58**, 1238 (1970)
[3.26]   A.A. Maradudin: *Festkörperprobleme XXI*, 25 (1981)
[3.27]   W. Kress and F.W. Wette, editors: *Surface Phonons*, Springer Series in Surface Science 21 (1991)
[3.28]   E.C. Svensson, B.N. Brockhouse, J.M. Rowe: Phys. Rev. **155**, 619 (1967)
[3.29]   P. Brüesch, Phonons: *Theory and Experiment III* (Springer, 1993)
[4.1]   P.C. Hohenberg and B.I. Halperin: Rev. Mod. Phys. **49**, 435 (1977)
[4.2]   R. Becker: *Theorie der Wärme* (Springer, 1966)
[4.3]   D.C. Wallace: *Thermodynamics of Crystals* (John Wiley& Sons, 1972)
[4.4]   H.B. Callen: *Thermodynamics and an Introduction to Thermostatistics* (John Wiley & Sons, 1985)
[4.5]   W. Ludwig: *Festkörperphysik* (Akad. Verlagsgesellschaft, 1978)
[4.6]   G. Leibfried and W. Ludwig: *Solid State Physics*, Vol. 12, 276 (Academic Press, 1961)
[4.7]   R.J. Schiltz and J.F. Smith: J. Appl. Phys. **45**, 4681 (1974)
[4.8]   Y.P. Varshny: Phys. Rev. B **2**, 3952 (1970)
[4.9]   G.W. Shannette and J.F. Smith: J. Appl. Phys. **40**, 79 (1969)
[4.10]   O.L. Anderson: J. Phys. Chem. Solids **24**, 909 (1963)
[4.11]   G.A. Alers, *Physical Acoustics*, Vol. 3B (Academic Press, 1965)
[4.12]   O.L. Anderson: *Physical Acoustics*, Vol. 3B (Academic Press, 1965)
[4.13]   R.F.S. Hearmon: Rev. Mod. Phys. **18**, 409 (1946)
[4.14]   R.F.S. Hearmon: Phil. Mag. Suppl. **5**, 323 (1956)
[4.15]   G. Simmons and H. Wang: *Single Crystal Elastic Constants and Calculated Aggregated Properties* (MIT Press, 1971)
[4.16]   M. Levy, H.E. Bass, R.R. Stern, editors: *Handbook of elastic properties of Solids, Liquids and Gases*, Vol. I-4 (Acad. Press, 2001)
[4.17]   H.J. Maris: *Physical Acoustics*, Vol. 8, p. 279 (Academic Press, 1971)
[4.18]   J.-C. Tolédano and P. Tolédano: *The Landau Theory of Phase Transitions* (World Scientific, 1987)
[4.19]   W. Rehwald: Adv. Phys. **22**, 721 (1973)
[4.20]   B. Lüthi and W. Rehwald: in *Structural phase transitions I, Topics in Current Physics*, Vol. 23, ed. K.A. Müller and H. Thomas (Springer, 1981)

[4.21] E.K.H. Salje: *Phase transitions in ferroelastic and co-elastic crystals* (Cambridge University, 1990)
[4.22] L.P. Kadanoff, W. Götze, D. Hamblen, R. Hecht, E.A.S. Lewis, V.V. Palciauskas, M. Rayl, J. Swift, D. Aspnes, J. Kane: Rev. Mod. Phys. **39**, 395 (1967)
[4.23] R.A. Cowley: Phys. Rev. B **13**, 4577 (1976)
[4.24] R.N. Thurston and K. Brugger: Phys. Rev. 133, A1604 (1964)
[4.25] J. Slonczewski and H. Thomas: Phys. Rev. B**1**, 3599 (1970)
[4.26] K. Kawasaki: in *Phase transitions and critical phenomena*, Vol. 5a, 166 (Acad. Press, 1976)
[4.27] Landau–Lifshitz: *Fluid mechanics*, p. 304 (Addison-Wesley, 1959b)
[4.28] W. Henkel, J. Pelzl, K.H. Höck, H. Thomas: Z. Phys. B**37**, 321 (1980)
[4.29] F. Schwabl: Phys. Rev. B **7**, 2038 (1973)
[4.30] K.K. Murata: Phys. Rev.B **13**, 4015 (1976)
[4.31] J. Als-Nielsen and R.J. Birgeneau: Am. J. Phys. **45**, 554 (1977)
[4.32] R. Folk, H. Iro, F. Schwabl: Phys. Rev. B **20**, 1229 (1979)
[4.33] H.J. Maris: Phil. Mag. **16**, 331 (1967)
[5.1] J.H. Van Vleck: J. Chem. Phys. **7**, 72 (1939)
[5.2] I. Waller: Z. Phys. **79**, 370 (1932)
[5.3] M. Tinkham: *Group Theory and Quantum Mechanics* (McGraw-Hill, 1964)
[5.4] A. Abragam and B. Bleaney: *Electron Paramagnetic Resonance of Transition Ions* (Clarendon Press, 1970)
[5.5] R.M. White: *Quantum Theory of Magnetism* (Springer, 1983)
[5.6] G. Williams and L.L. Hirst: Phys. Rev. **185**, 407 (1969)
[5.7] P. Fulde: in *Handbook on the Physics and Chemistry of Rare Earths*, Vol. 2 (North-Holland, 1979)
[5.8] M.T. Hutchings: in *Solid State Physics*, 16, 227 (Academic Press, 1964)
[5.9] W.E. Wallace: *Rare Earth Intermetallics* (Academic Press, 1973)
[5.10] K.W.H. Stevens: Proc. Phys. Soc. A**65**, 209 (1952)
[5.11] R.J. Elliott and K.W.H. Stevens: Proc. Roy. Soc. A**218**, 553 (1953)
[5.12] K.R. Lea, J.M. Leask, W.P. Wolf: J. Phys. Chem. Solids **23**, 1381 (1962)
[5.13] P. Morin and D. Schmitt: in *Ferromagnetic materials*, Vol 5, 1 (Elsevier Science Publ., 1990)
[5.14] M. Born and K. Huang: *Dynamical Theory of Crystal Lattices* (Oxford University Press, 1954)
[5.15] H. Stern: J. Phys. Chem. Solids **26**, 153 (1965)
[5.16] R.J. Schiltz and J.F. Smith: J. Appl. Phys. **45**, 4681 (1974)
[5.17] J. Rouchy, P. Morin, E. du Tremolet de Lacheisserie: J. Magn. Magn. Mat. **23**, 59 (1981)
[5.18] I.K. Kamilov and Kh.K. Aliev: Uspekhi **41**, 865 (1998)
[5.19] M. Lang, R. Modler, U. Ahlheim, R. Helfrich, P.H.P. Reinders, F. Steglich, W. Assmus, W. Sun, G. Bruls, D. Weber, B. Lüthi: Physica Scripta Vol. T**39**, 135 (1991)
[5.20] R.J. Birgeneau, E. Bucher, L. Passell, K.C. Turberfield: Phys. Rev. B**4**, 718 (1971)
[5.21] H.R. Ott and B. Lüthi: Z. Physik: B **28**, 141 (1977)
[5.22] K. Andres, S. Darack and H.R. Ott: Phys. Rev. B **19**, 5475 (1979)
[5.23] V.M.T.S. Barthem, D. Gignoux, A. Nait-Saada, D. Schmitt, G. Creuzet: Phys. Rev. B **37**, 1733 (1988)

[5.24] B. Lüthi and H.R. Ott: Solid State Comm. **33**, 717 (1980)
[5.25] N.P. Kolmakova, R.Z. Levitin, V.N. Orlov, N.F. Vedernikov: J. Magn. Magn. Mat. **87**, 218 (1990)
[5.26] Z.A. Kazei, N.P. Kolmakova, O.A. Shichkina: Phys. Solid State **39**, 91 (1997)
[5.27] J.H. Van Vleck: *The Theory of Electric and Magnetic Susceptibility* (Oxford University Press, 1932)
[5.28] B.R. Cooper: in *Magnetic Properties of Rare Earth Metals*, ed. R.J. Elliott, chapter 2 (Plenum Press, 1972)
[5.29] M.E. Mullen, B. Lüthi, P.S. Wang, E. Bucher, L.D. Longinotti, J.P. Maita, H.R. Ott: Phys. Rev. B **10**, 186 (1974)
[5.30] C. Lingner and B. Lüthi: Phys. Rev. B **23**, 256 (1981)
[5.31] I.A. Campbell and G. Creuzet: J. Phys. F **15**, 2559 (1985)
[5.32] J.D. Greiner, R.J. Schiltz, J.J. Toennies, F.H. Spedding, J.F. Smith: J. Appl. Phys. **44**, 3862 (1973)
[5.33] B. Lüthi, M.E. Mullen, E. Bucher: Phys. Rev. Lett. **31**, 95 (1973)
[5.34] S.B. Palmer and J. Jensen: J. Phys. C **11**, 2465 (1978)
[5.35] J. Coqblin: *The Electronic Structure of Rare Earth Metals and Alloys* (Acad. Press, 1977)
[5.36] J.J. Rhyne: in *Magn. Prop. Of Rare Earth Metals*, chapter 4, ed. R.J. Elliott (Plenum Press, 1972)
[5.37] S.J. Allen: Phys. Rev. 167, 492 (1968)
[5.38] M. Kataoka and J. Kanamori: J. Phys. Soc. Jpn.**32**, 113 (1972)
[5.39] P.M. Levy: J. Phys. C **6**, 3545 (1973)
[5.40] G.A. Gehring and K.A. Gehring: Rep. Progr. Phys. **38**, 1 (1975)
[5.41] R. L. Melcher: in *Phys. Acoustics*, Vol. XII (Academic Press, 1976)
[5.42] H. Thomas: in *Electron-Phonon Interactions and Phase Transitions* (Plenum Publishing, 1977)
[5.43] H. Freese and W. Döring: Z. Phys. B **34**, 135 (1979)
[5.44] B. Lüthi and C. Lingner: Z. Phys. B **34**, 157 (1979)
[5.45] J. R. Cullen and A. E. Clark: Phys. Rev. B **15**, 4510 (1977)
[5.46] C. Lingner and B. Lüthi: J.M.M.M. **36**, 86 (1983)
[5.47] M. Loewenhaupt, B.D. Rainford, F. Steglich: Phys. Rev. Lett.**42**, 1709 (1979)
[5.48] W. Reichert, N. Nöcker: J. Phys. R. 14, L135 (1984)
[5.49] P. Thalmeier and P. Fulde: Phys. Rev. Lett. **49**, 1588 (1982)
[5.50] P. Thalmeier: J. Phys C. **17**, 4153 (1984)
[5.51] G. Güntherodt, A. Jayaraman, B. Batlogg, M. Croft, E. Melczer: Phys. Rev. Lett. **51**, 2330 (1983)
[5.52] G. Hampel and R. Blick: J. Low Temp. Phys. **99**, 71 (1995)
[5.53] A.E. Clark: AIP Conf. Proc. No. 34, **13** (1976)
[5.54] A.E. Clark: in *Ferromagnetic Materials*, Vol. 1, 531 (North-Holland, 1980)
[5.55] B.R. Cooper: Phys. Rev. B **17**, 293 (1978)
[5.56] B. Lüthi, P.S. Wang, E. Bucher: Proc. 1st CEF Conference, ed. R.A.B. Devine (Montreal. 1974)
[5.57] J.W. Tucker and V.W. Rampton: *Microwave Ultrasonics in Solid State Physics* (North Holland, 1972)
[5.58] W.E. Bron: Rep. Progr. Phys. **43**, 301 (1980)
[5.59] G. Gorodetsky, G.A. Shaulov, V. Volterra, J. Makovsky: Phys. Rev. B. **13**, 1205 (1976)

[5.60] N.S. Shiren: Phys. Rev. **128**, 2103 (1962)
[5.61] E.H. Jacobsen and K.W.H. Stevens: Phys. Rev. **129**, 2036 (1963)
[5.62] W.I. Dobrov: Phys. Rev. 134, A **734** (1964)
[5.63] M.D. Sturge, J.T. Krause, E.M. Gyorgy, R.C. LeCraw, F.R. Merritt: Phys. Rev. **155**, 218 (1967)
[5.64] E.M. Gyorgy, R.C. Le Craw, M.D. Sturge: J. Appl. Phys. **37**, 1303 (1966)
[5.65] S. Sugano and Y. Tanabe: J. Phys. Soc. Jpn. **13**, 880 (1958)
[5.66] E.B. Tucker: Phys. Rev. Lett. **6**, 183 (1961)
[5.67] E.B. Tucker: Phys. Rev. Lett. **6**, 547 (1961)
[5.68] W. Grill and B. Lüthi: Phys. Rev. B **24**, 3571 (1981); Err. 25, 2933 (1982)
[5.69] E. Feher and M.D. Sturge: Phys. Rev. **172**, 244 (1968)
[5.70] E.I. Golovenchits, V.A. Sanina, T.A. Shaplygina: Sov. Phys. JETP **53**, 992 (1981)
[5.71] Th. Woike, W. Krasser, P.S. Bechthold, S. Haussühl: Phys. Rev. Lett. **53**, 1767 (1984)
[5.72] A.V. Kimel, R.V. Pisarev, J. Hohlfeld, Th. Rasing: Phys. Rev. Lett. **89**, 287401 (2002)
[5.73] W. G. Proctor and W.H. Tantilla, Phys. Rev. **101**, 1757 (1956)
[5.74] D.I. Bolef: in *Physical Acoustics*, IVA (Academic Press, 1966)
[5.75] E.H. Gregory and H.E. Bömmel: Phys. Rev. Lett. **15**, 404 (1965)
[5.76] P. G. De Gennes, P.A. Pincus, F. Hartmann-Boutron, J.W. Winter: Phys. Rev. **129**, 1105 (1963)
[5.77] Kh.G. Bogdanova, V.A. Golenishchev, A.A. Monakhov: JETP Lett. **25**, 268 (1977)
[6.1] L.P. Kadanoff: in *Phase transitions and critical phenomena*, Vol. 5a, 1 (Academic Press, 1976)
[6.2] L.M. Levinson, M. Luban, S. Shtrikman: Phys. Rev. **187**, 715 (1969)
[6.3] D.H. Martin: *Magnetism in Solids* (MIT Press, 1967)
[6.4] R.M. White: *Quantum Theory of Magnetism* (Springer, 1983)
[6.5] A. Ikushima and R. Feigelson: J. Phys. Chem. Solids **32**, 417 (1971)
[6.6] B. Lüthi and R.J. Pollina: Phys. Rev.**167**, 488 (1968)
[6.7] B. Lüthi, R.J. Pollina, T.J. Moran: J. Phys. Chem. Solids **31**, 1741 (1970)
[6.8] A. Ikushima: Phys. Lett. 29A, 417 (1969)
[6.9] R.J. Pollina and B. Lüthi: Phys. Rev. **177**, 841 (1969)
[6.10] M. Tachiki, S. Maekawa, R. Treder, M. Levy: Phys. Rev. Lett. **34**, 1579 (1975)
[6.11] C.W. Garland: in *Physical Acoustics, Vol. VII*, 52 (Academic Press, 1970)
[6.12] K. Kawasaki: in *Phase transitions and critical phenomena*, Vol. 5a, 166, ed. C. Domb and M.S. Green (Academic Press, 1976)
[6.13] B. Lüthi: in *Dynamical Properties of Solids*, Vol. 3, ed. G.K. Horten, A.A. Maradudin (North-Holland, 1980)
[6.14] I.K. Kamilov and Kh.K. Aliev: Uspekhi **41**, 865 (1998)
[6.15] F. Reif: *Fundamentals of Statistical Mechanics*, Chapt. 15 (McGraw-Hill, 1965)
[6.16] H. Stern: J. Phys. Chem. Solids **26**, 153 (1965)
[6.17] K. Tani and H. Mori: Progr. theor. Phys. **39**, 876 (1968); Phys. Lett. 19, 627 (1966)
[6.18] H.S. Bennett and E. Pytte: Phys. Rev. 155, 553 (1967); ibid 164, 712 (1967)

[6.19]   H. Okamoto: Progr. Theor. Phys. **37**, 1348 (1967)
[6.20]   K. Kawasaki: Solid State Comm. **6**, 57 (1968)
[6.21]   P.G. de Gennes: Magnetism Vol. 3 (Academic Press, 1963)
[6.22]   B.I. Halperin and P.C. Hohenberg: Phys. Rev.**177**, 952 (1969)
[6.23]   G.E. Laramore and L.P. Kadanoff: Phys. Rev. **187**, 619 (1969)
[6.24]   A. Pawlak: Eur. Phys. J. B4, 179 (1998)
[6.25]   H. S. Bennett: Phys.Rev. 181, 978(1969)
[6.26]   B. Lüthi and R.J. Pollina: Phys. Rev. Lett. **22**, 717 (1969)
[6.27]   K. Kawasaki: Phys. Lett. 26A, 543 (1969)
[6.28]   D.L. Huber: Phys. Rev. B **3**, 836 (1971)
[6.29]   Y. Itoh: J. Phys. Soc. Jpn. **38**, 336 (1975)
[6.30]   A. Bachellerie and Ch. Frenois: J. de Physique **35**, 437 (1974)
[6.31]   E.J. Samuelsen, M.T. Hutchings, G. Shirane: Physica **48**, 13 (1970)
[6.32]   G. Gorodetsky, B. Lüthi, T.J. Moran: Intern. J. Magn. **1**, 295 (1971)
[6.33]   B. Golding, M. Barmatz, E. Buehler, M.B. Salomon: Phys. Rev. Lett. **30**, 968 (1973)
[6.34]   A. Bachellerie, J. Joffrin, A. Levelut: Phys. Rev. Lett. **30**, 617 (1973)
[6.35]   B. Golding: Phys. Rev. Lett.**34**, 1102 (1975)
[6.36]   T.J. Moran and B. Lüthi: Phys. Rev. B **4**, 122 (1971)
[6.37]   T. Jimbo and C. Elbaum: Phys. Rev. B10, 213 1 (1974)
[6.38]   T.J. Moran and B. Lüthi: J. Phys. Chem. Solids **31**, 1735 (1970)
[6.39]   M. Tachiki, M.C. Lee, M. Levy: Phys. Rev. Lett. **29**, 488 (1972)
[6.40]   S. Maekawa, R.A. Treder, M. Tachiki, M.C. Lee, M. Levy: Phys. Rev. B **13**, 1284 (1976)
[6.41]   R. Treder and M. Levy: J. Magn. Magn. Mat. **5**, 9 (1977)
[6.42]   T. Komatsubara, A. Ishizaki, S. Kusaka, E. Hirahara: Solid State Comm. **14**, 741 (1974)
[6.43]   R. Erdem: Phys. Lett. A **312**, 238 (2003) and ref. therein
[6.44]   Kh.K. Aliev et al.: JETP 68, 1096 (1989)
[6.45]   M. Suzuki: J. Phys. Soc. Jpn. **50**, 1149 (1981)
[6.46]   A. Pawlak: Eur. Phys. J. Acta Phys. Pol. A **97**, 909 (2000)
[6.47]   K. Kawasaki and A. Ikushima: Phys. Rev. B **1**, 3143 (1970)
[6.48]   V.V. Kashcheev: Phys. Lett. 25A, 71 (1967)
[6.49]   L.P. Kadanoff, W. Götze, D. Hamblen, R. Hecht, E.A.S. Lewis, V.V. Palciauskas, M. Rayl, J. Swift, D. Aspnes, J. Kane: Rev. Mod. Phys. **39**, 395 (1967)
[6.50]   B. Golding: Phys. Rev. Lett. **20**, 5 (1968)
[6.51]   T. J. Moran and B. Lüthi: Phys. Lett. 29A, 665 (1969b)
[6.52]   B. Golding and M. Barmatz: Phys. Rev. Lett. **23**, 223 (1969)
[6.53]   M. Tachiki and S. Maekawa: Progr. Theor. Phys. **51**, 1 (1974)
[6.54]   R.J. Elliott, editor, *Magnetic Properties of Rare Earth Metals* (Plenum Press, 1972)
[6.55]   V.D. Buchelnikov, N. K. Dan'shin, L.T. Tsymbal, V.G. Shavrov, Uspekhi **39**, 547 (1996)
[6.56]   H. Horner and C.M. Varma: Phys. Rev. Lett. **20**, 845 (1968)
[6.57]   R.M. Hornreich and S. Shtrikman: J. Phys. C9, L683 (1976)
[6.58]   J.R. Shane: Phys. Rev. Lett. **20**, 728 (1968)
[6.59]   G. Gorodetsky, B. Lüthi: Phys. Rev. B2, 3688 (1970)
[6.60]   G. Gorodetsky, S. Shaft and B.M. Wanklyn: Phys. Rev. B **14**, 2051 (1976)

[6.61]	G. Gorodetsky and S. Shaft: Phys. Rev. B **23**, 6755 (1981)
[6.62]	N.K. Dan'shin et al.: JETP **66**, 1227 (1987)
[6.63]	N.K. Dan'shin et al.: Sov. Phys. Solid State **31**, 832 (1989)
[6.64]	T.J. Moran and B. Lüthi: Phys. Rev. **187**, 710 (1969)
[6.65]	H. Schwenk, S. Bareiter, C. Hinkel, B. Lüthi, Z. Kakol, A. Koslowski, J.M. Honig: Eur. Phys. J. B **13**, 491 (2000)
[6.66]	M. Long, A.R. Wazzan, R. Stern: Phys. Rev. **178**, 775 (1969)
[6.67]	C.D. Graham: J. Phys. Soc. Japan **17**, 1310 (1962)
[6.68]	W. Henggeler, T. Chattopadhyay, B. Roessli, P. Vorderwisch, P. Thalmeier, D.I. Zhigunov, S.N. Barilo, A. Furrer: Phys. Rev. B**55**, 1269 (1997)
[6.69]	S. Zherlitsyn, G.A. Zvyagina, V.D. Fil, I.M. Vitebskii, S.N. Barilo, D.I. Zhigunov: Low Temp. Phys. **19**, 934 (1993)
[6.70]	S. Zherlitsyn, V.D. Fil, D. Finsterbusch, J. Molter, B. Wolf, G. Bruls, B. Lüthi, S.N. Barilo, D.I. Zhigunov: Physica B **211**, 168 (1995)
[6.71]	E. Fawcett: Rev. Mod. Physics **60**, 209 (1988)
[6.72]	M.B. Walker: Phys. Rev. B **22**, 1338 (1980)
[6.73]	M.O. Steinitz, L.H. Schwartz, J.A. Marcus, E. Fawcett, W.A. Reed: Phys. Rev. Lett. **23**, 979 (1969)
[6.74]	W.C. Muir, E. Fawcett, J.M. Perz: J. Magn. Magn. Mat. **69**, 113 (1987)
[6.75]	G. Grüner: *Density Waves in Solids* (Addison-Wesley Publ. Company, 1994)
[6.76]	S. Zherlitsyn, G. Bruls, A. Goltsev, B. Alavi, M. Dressel: Phys. Rev. B **59**, 13861 (1999)
[7.1]	P. Papon, J. Leblond, P.H.E. Meijer: *The Physics of Phase Transitions* (Springer, 2002)
[7.2]	H.E. Stanley: *Introduction to Phase Transitions and Critical Phenomena* (Oxford University Press, 1971)
[7.3]	R. Blinc and B. Zeks: *Soft modes in ferroelectrics and antiferroelectrics* (North-Holland, 1974)
[7.4]	T. Goto and B. Lüthi: Adv. Phys. **52**, 67 (2003)
[7.5]	T. Goto, Y. Nemoto, A. Ochiai, T. Suzuki: Phys. Rev. B **59**, 269 (1999)
[7.6]	A. Ochiai, T. Suzuki, T. Kasuya: J. Phys. Soc. Jpn. **59**, 4129 (1990)
[7.7]	M. Kohgi, K. Iwasa, J.M. Mignot, A. Ochiai, T. Suzuki: Phys. Rev. B 56, R11388 (1997)
[7.8]	B. Schmidt, H. Aoki, T. Cichorek, J. Custers, P. Gegenwart, M. Kohgi, M. Lang, C. Langhammer, A. Ochiai, S. Paschen, F. Steglich, T. Suzuki, P. Thalmeier, B. Wand, A. Yaresko: Physica B **300**, 121 (2001)
[7.9]	E.J. Verwey: Nature **144**, 327 (1939)
[7.10]	J.M. Zuo, J.C.H. Spence, W. Petuskey: Phys. Rev. B **42**, 8451(1990)
[7.11]	S. Chikazumi: AIP Conf. Proc. **29**, 382 (1975)
[7.12]	H. Schwenk, S. Bareiter, C. Hinkel, B. Lüthi, Z. Kakol, A. Koslowski, J.M. Honig: Eur. Phys. J. B **13**, 491 (2000)
[7.13]	T.J. Moran, B. Lüthi: Phys. Rev. 187, 710 (1969)
[7.14]	J. Garcia, G. Subias, M.G. Proletti, J. Blasco, H. Renevier, J.L. Hodeau, Y. Joly: Phys. Rev. B **63**, 054110 (2001)
[7.15]	J.P. Wright, J.P. Attfield, P.G. Radaelli: Phys. Rev. Lett. **87**, 266401 (2001)
[7.16]	M. Isobe and Y. Ueda: J. Phys. Soc. Jpn. **65**, 1178 (1996)

References

[7.17] H. G. v. Schnering, Yu. Grin, M. Kaupp, M. Somer, R.K. Kremer, O. Jepsen: Z. Kristallogr. **213**, 246 (1998)

[7.18] H. Smolinski, C. Gross, W. Weber, U. Peuckert, G. Roth, M. Weiden, C. Geibel: Phys. Rev. Lett. **80**, 5164 (1998)

[7.19] T. Ohama, A. Goto, T. Shimizu, E. Ninomiya, H. Sawa, M. Isobe, Y. Ueda: Phys. Rev. B**59**, 3299 (1999)

[7.20] T. Ohama, A. Goto, T. Shimizu, E. Ninomiya, H. Sawa, M. Isobe, Y. Ueda: J. Phys. Soc. Jpn. **69**, 2751 (2000)

[7.21] A. N. Vasil'ev, A.I. Smirnov, M. Isobe, Y. Ueda: Phys. Rev. B **56**, 5065 (1997)

[7.22] S. Schmidt, W. Palme, B. Lüthi, M. Weiden, R. Hauptmann, C. Geibel: Phys. Rev. B **57**, 2687 (1998)

[7.23] J. Hemberger, M. Lohmann, M. Nicklas, A. Loidl, M. Klemm, G. Obermeier, S. Horn: Europhys. Lett. **42**, 661 (1998)

[7.24] S. Luther, H. Nojiri, M. Motokawa, M. Isobe, Y. Ueda: J. Phys. Soc. Jpn. **67**, 3715 (1998)

[7.25] J. Lüdecke, A. Jobst, S. van Smaalen, E. Morré, C. Geibel, H.G. Krane: Phys.Rev.Lett. **82**, 3633 (1999)

[7.26] J.L. de Boer, A Meetsma, J. Baas, T.T.M. Palstra: Phys. Rev. Lett. **84**, 3962 (2000)

[7.27] H. Nakao, K. Ohwada, N. Takesue, Y. Fujii, M. Isobe, Y. Ueda, M. v. Zimmermann, J.P. Hill, D. Gibbs, J.C. Woicik, I. Koyama, Y. Murakami: Phys. Rev. Lett. **85**, 4349 (2000)

[7.28] H. Sawa, E. Ninomiya, T. Ohama, H. Nakano, K. Ohwada, Y. Murakami, Y. Fujii, Y. Noda, M. Isobe, Y. Ueda: J. Phys. Soc. Jpn. **71**, 385 (2002)

[7.29] D. C. Johnston, R. K. Kremer, M. Troyer, X. Wang, A. Klümper, S. Bud'ko, A.F. Panchula, P.C. Canfield: Phys. Rev. B **61**, 9558 (2000)

[7.30] J. Riera and D. Poilblanc: Phys. Rev. B **59**, 2667 (1999)

[7.31] H. Schwenk, S. Zherlitsyn, B. Lüthi, E. Morré, C. Geibel: Phys. Rev. B **60**, 9194 (1999)

[7.32] M. Köppen, D. Pankert, R. Hauptmann, M. Lang, M. Weiden, C. Geibel, F. Steglich: Phys. Rev. B **57**, 8466 (1998)

[7.33] B. Lüthi, S. Zherlitsyn, C. Geibel: J. Phys. Soc. Jpn. **72**, 195 (2003)

[7.34] P. Fertey, M. Poirier, M. Castonguay, J. Jegoudez, A. Revcolevschi: Phys. Rev. B **57**, 13698 (1998)

[7.35] W. Schnelle, Yu. Grin, R.K. Kramer: Phys. Rev. B **59**, 73 (1999)

[7.36] Y. Tokura, A. Urushibara, Y. Moritama, T. Arima, A. Asamitsu, G. Kido, N. Furukawa: J. Phys. Soc.Jpn. **63**, 3931 (1994)

[7.37] P.G. de Gennes: Phys. Rev. **118**, 141 (1960)

[7.38] A.J. Millis, R. Mueller, B.J. Shraiman et al.: Phys. Rev.B **54**, 5404 (1996)

[7.39] H. Hazama, T. Goto, Y. Nemoto, Y. Tomioka, A. Asamitsu, Y. Tokura: Phys. Rev. B **62**, 15012 (2000)

[7.40] H. Hazama et al.: J. Phys. Soc. Jpn. **71** 139 suppl. (2002)

[7.41] A. Tamaki, T. Goto, S. Kunii, T. Suzuki, T. Fujimori, T. Kasuya: J. Phys. C **18**, 5849 (1985)

[7.42] H. Schwenk: Thesis Universität Frankfurt (2000)

[7.43] M.D. Kaplan and B.G. Vekhter: *Cooperative Phenomena in Jahn-Teller Crystals* (Plenum Press, 1995)

[7.44] R. Orbach and M. Tachiki: Phys. Rev. **158**, 524 (1967)

[7.45]   P. Fulde and I. Peschel: Adv. Phys. **21**, 1 (1972)
[7.46]   P.M. Levy, P. Morin, D. Schmitt: Phys. Rev. Lett. **42**, 1417 (1979)
[7.47]   G.A. Gehring and K.A. Gehring: Rep. Progr. Phys. **38**, 1 (1975)
[7.48]   R.L. Melcher: in *Phys. Acoustics*, Vol. XII (Acad. Press, 1976)
[7.49]   H. Thomas: in: *Electron-Phonon Interactions and Phase Transitions* (Plenum Publishing, 1977)
[7.50]   K.I. Kugel and D.I. Khomskii: Sov. Phys. Usp. **25**, 231 (1982)
[7.51]   P. Morin and D. Schmitt: in *Ferromagnetic materials*, Vol. 5, 1, ed. K.H.J. Buschow and E.P. Wohlfarth (North-Holland, 1990)
[7.52]   J.D. Dunitz and L.E. Orgel: J. Phys. Chem. Solids **3**, 20 (1957)
[7.53]   J.B. Goodenough: *Magnetism and the chemical bond* (Interscience Publishers, 1966)
[7.54]   E. Pytte: Phys. Rev. B**3**, 3503 (1971)
[7.55]   M. Kataoka and J. Kanamori: J. Phys. Soc. Jpn.**32**, 113 (1972)
[7.56]   Y. Kino, B. Lüthi, M.E. Mullen: J. Phys. Soc. Jpn. **33**, 687 (1972)
[7.57]   Y. Kino and B. Lüthi, Solid State Comm. **9**, 805 (1971)
[7.58]   Y. Kino: private communication (2002)
[7.59]   Y. Kino and S. Miyahara: J. Phys. Soc. Jpn.**20**, 1522 (1965)
[7.60]   Y. Kino, B. Lüthi, M.E. Mullen: Solid State Comm.**12**, 275 (1973)
[7.61]   H. Martinho, N.D. Moreno, A. Sanjurjo, C. Rettori, A.J. Garcia-Adera, D.L. Huber, S.B. Oseroff, W. Ratcliff II, S.W. Cheong, P.G. Pagliuso, J.L. Sarrao, G.B. Martins: J. Appl. Phys. **89**, 7050 (2001)
[7.62]   S.H. Lee, C. Broholm, T.H. Kim, W. Ratcliff II, S.W. Cheong: Phys. Rev. Lett. **84**, 3718 (2000)
[7.63]   C.J. Kroese and W.J.A. Maaskant: Chem. Phys. **5**, 224 (1974)
[7.64]   H.A. Graf, G. Shirane, U. Schotte, H. Dachs, N. Pyka, M. Iizumi: J. Phys. CM **1**, 3743 (1989)
[7.65]   D.I. Khomskii: JETP Lett. **25**, 544 (1978)
[7.66]   S. Hirotsu: J. Phys. C **10**, 967 (1977)
[7.67]   U. Förster, H.A. Graf, U. Schotte, U. Stuhr: J. Phys. CM **9**, 1067 (1997)
[7.68]   Y. Ishikawa et al.: Proc. Int. Conf. on Ferrites, Kyoto (1970)
[7.69]   Y. Ishikawa and Y. Syono: Phys. Rev. Lett. **26**, 1335 (1971)
[7.70]   R.L. Melcher and B.A. Scott: Phys. Rev. Lett. **28**, 607 (1972)
[7.71]   G. Gorodetsky, B. Lüthi, B.M. Wanklyn: Solid State Comm. **9**, 2157 (1971)
[7.72]   R.L. Melcher, E. Pytte and B.A. Scott: Phys. Rev. Lett. **31**, 307 (1973)
[7.73]   A.H. Cooke, S.J. Swithenby, M.R. Wells: Solid State Comm. **10**, 256 (1972)
[7.74]   J.R. Sandercock, S.B. Palmer, R.J. Elliott, W. Hayes, S.R.P. Smith, A.P. Young: J. Phys. C **5**, 3126 (1972)
[7.75]   R.T. Harley and D.I. Manning: J. Phys. C **11**, L633 (1978)
[7.76]   R.J. Birgeneau, J.K. Kjems, G. Shirane, L.G. Van Uitert: Phys. Rev. B **10**, 2512 (1974)
[7.77]   P.A. Fleury, P.D. Lazay, L.G. Van Uitert: Phys. Rev. Lett. **33**, 492 (1974)
[7.78]   P.W. Anderson and E.I. Blount: Phys. Rev. Lett. **14**, 217 (1965)
[7.79]   B. Lüthi, R. Sommer and P. Morin: J. Magn. Magn. Mat. **13**, 198 (1979)
[7.80]   R. Takke, N. Dolezal, W. Assmus, B. Lüthi: J. Magn. Magn. Mat. 23, 247 (1981b)
[7.81]   K. Knorr, B. Renker, W. Assmus, B. Lüthi, R. Takke, H.J. Lauter: Z. Phys. B **39**, 151 (1980)

[7.82] Y. Nemoto, T. Goto, S. Nakamura, H. Kitazawa: J. Phys. Soc. Jpn. **65**, 2571 (1996)

[7.83] O. Suzuki, T. Horino, Y. Nemoto, T. Goto, A. Dönni, T. Komatsubara, M. Ishikawa: Physica B **259–261**, 334 (1999)

[7.84] Y. Nemoto, T. Yagaguchi, T. Horino, M. Akatsu, T. Yanagisawa, T. Goto, O. Suzuki, A. Dönni, T. Komatsubara: Phys. Rev. B **68**, 184109 (2003)

[7.85] Y. Shimizu and O. Sakai: J. Phys. Soc. Jpn. **65**, 2632 (1996)

[7.86] E. Zirngiebl, B. Hillebrands, S. Blumenröder, G. Güntherodt, K. Winzer, Z. Fisk: Phys. Rev. B **30**, 4052 (1984)

[7.87] B. Lüthi, S. Blumenröder, B. Hillebrands, E. Zirngiebl, G. Güntherodt, K. Winzer: Z. Physik B 58, 31 (1984)

[7.88] O. Sakai, R. Shiina, H. Shiba, P. Thalmeier: J. Phys. Soc. Jpn. **66**, 3005 (1997)

[7.89] M. Sera and S. Kobayashi: J. Phys. Soc. Jpn. **68**, 1664 (1999)

[7.90] R. Shiina, H. Shiba, P. Thalmeier: J. Phys. Soc. Jpn. **66**, 1741 (1997)

[7.91] K. Hanzawa: J. Phys. Soc. Jpn. **70**, 468 (2001)

[7.92] S. Nakamura, T. Goto, O. Suzuki, S. Kunii, S. Sakatsume: Phys. Rev. B **61**, 15203 (2000)

[7.93] M. Akatsu, T. Goto, Y. Nemoto, O. Suzuki, S. Nakamura, S. Kunii: J. Phys. Soc. Jpn. **71**, 205 (2003)

[7.94] K. Kubo and Y. Kuramoto: J. Phys. Soc. Jpn **72**, 1859 (2003)

[7.95] S. Nakamura, T. Goto, S. Kunii, K. Iwashita, A. Tamaki: J. Phys. Soc. Jpn. **63**, 623 (1994)

[7.96] S. Zherlitsyn, B. Wolf, B. Lüthi, M. Lang, P. Hinze, E. Uhrig, W. Assmus, H.R. Ott, D.P. Young, Z. Fisk: Eur. Phys. J. B**22**, 327 (2001)

[7.97] T. Matsumura, Y. Haga, Y. Nemoto, S. Nakamura, T. Goto, T. Suzuki: Physica B **206&207**, 380 (1995)

[7.98] T. Matsumura, S. Nakamura, T. Goto, H. Amitsuka, K. Matsuhira, T. Sakakibara, T. Suzuki: J. Phys. Soc. Jpn. **67**, 612 (1998)

[7.99] H.R. Ott, B. Lüthi, P.S. Wang: in *Valence Instabilities* 289 (Plenum Press, 1976)

[7.100] E. Clementeyev, R. Köhler, M. Braden, J.M. Mignot, C. Vettier, T. Matsumara, T. Suzuki: Physica B **230-232**, 735 (1997)

[7.101] J.M. Mignot, P. Link, A. Gusakov, T. Matsumura, T. Suzuki: Physica B **281&282**, 470 (2000)

[7.102] R. Shiina, H. Shiba, O. Sakai: J. Phys. Soc. Jpn. **68**, 2105 (1999)

[7.103] A.V. Nikolaev and K.H. Michel: Phys. Rev. B **63**, 104105 (2001)

[7.104] H.R. Ott, K. Andres, P.S. Wang, Y.H. Wong, B. Lüthi: in *Crystal Field Effects in Metals and Alloys*, ed. A. Furrer, p. 84 (Plenum Press, 1977)

[7.105] R. Settai, S. Araki, P. Ahmet, M. Abliz, K. Sugiyama, Y. Onuki, T. Goto, H. Mitamura, T. Goto, S. Takayanagi: J. Phys. Soc. Jpn. **67**, 636 (1998)

[7.106] E. Bucher, J.P. Maita, G.W. Hull, L.D. Longinotti, B. Lüthi, P.S. Wang: Z. Phys. B **25**, 41 (1976)

[7.107] H. Suzuki, M. Kasaya, T. Miyazaki, Y. Nemoto, T. Goto: J. Phys. Soc. Jpn. **66**, 2566 (1997)

[7.108] F. Levy: Phys. Kondens. Mat. **10**, 85 (1969)

[7.109] E. Bucher, R.J. Birgeneau, J.P. Maita, G.P. Felcher, T.O. Brun: Phys. Rev. Lett. **28**, 746 (1972)

[7.110] K.W.H. Stevens and E. Pytte: Solid State Comm. **13**, 101 (1973)

[7.111] T.J. Moran, R.L. Thomas, P.M. Levy, H.H. Chen: Phys. Rev. B **7**, 3238 (1973)
[7.112] M.E. Mullen, B. Lüthi, P.S. Wang, E. Bucher, L.D. Longinotti, J.P. Maita, H.R. Ott: Phys. Rev. B **10**, 186 (1974)
[7.113] Y. Nakanishi, T. Sakon, M. Motokawa, M. Ozawa, T. Suzuki, M. Yoshizawa: Phys. Rev. B **68**, 144427 (2003)
[7.114] F. Hulliger: *Handbook on the Physics and Chemistry of Rare Earths*, Vol. 4 (North-Holland, 1980)
[7.115] J. Kötzler and G. Raffius: Z. Physik B **38**, 139 (1980)
[7.116] D. Mukamel and S. Krinsky: Phys. Rev. B13, **5065**, 5078 (1976)
[7.117] S. A. Brazovskii, I.E. Dzyaloshinskii, B.G. Kukharenko: JETP **43**, 1178 (1976)
[7.118] D. Endoh, T. Goto, A. Tamaki, B. Liu, M. Kasaya, T. Fujimura, T. Kasuya: J. Phys. Soc. Jpn. **58**, 940 (1989)
[7.119] E. V. Sampathkumaran, I. Das, R. Vijayaraghavan, A. Hayashi, Y. Ueda, M. Ishikawa: Z. Phys. B **92**, 191 (1993)
[7.120] T. Yanagisawa, T. Goto, Y. Nemoto, S. Miyata, R. Watanuki, K. Suzuki: Phys. Rev. B **67**, 115129 (2003)
[7.121] A. Kiss and P. Fazekas: Phys. Rev. B **68**, 174425 (2003)
[7.122] E.R. Callen and H.B. Callen: Phys. Rev. **129**, 578 (1963)
[7.123] M. Niksch, W. Assmus, B. Lüthi, H.R. Ott, J.K. Kjems: Helv. Phys. Acta **55**, 588 (1982)
[7.124] B. Lüthi, M. Niksch, R. Takke, W. Assmus, W. Grill: in: *Crystalline Electric Field Effects in f-electron Magnetism*, ed. by Guertin, Suski, Zolnierek (Plenum Publ., 1982)
[7.125] T. Tayama, T. Sakakibara, K. Kitami, M. Yokoyama, K. Tenya, H. Amitsuka, D. Aoki, Y. Onuki, Z. Kletowski: J. Phys. Soc. Jpn. **70**, 248 (2001)
[7.126] T. Udagawa, K. Morita, O. Suzuki, T. Takamasu, S. Kato, H. Kitazawa, G. Kido, A. Tamaki: J. Phys. Soc. Jpn. **71** (2002) Suppl. p. 133
[7.127] O G. Brandt and C.T. Walker: Phys. Rev. Lett. **18**, 11 (1967)
[7.128] S J. Allen: Phys. Rev. **167**, 492 (1968)
[7.129] J. Faber, G.H. Lander, B.R. Cooper: Phys. Rev. Lett. **35**, 1770 (1975)
[7.130] G. Solt and P. Erdös: Phys. Rev. B **22**, 4718 (1980)
[7.131] W.J.L. Buyers, A.F. Murray, T.M. Holden, E.C. Svensson, P. de V. Du Plessis, G.H. Lander, O. Vogt: Physica B+C **102**, 291 (1980)
[7.132] K.A. McEwen, U. Steigenberger, K.N. Clausen, J. Kulda, J.G. Park, M.B. Walker: J. Magn. Magn. Mat. **177–181**, 37 (1998)
[7.133] K. Andres, D. Davidov, P. Dernier, F. Hsu, W. A. Reed, G.J. Nieuwenhuys: Solid State Comm. **28**, 405 (1978)
[7.134] H.R. Ott, K. Andres, P.H. Schmidt: Physica B&C **102**, 148 (1980)
[7.135] N. Lingg, D. Maurer, V. Müller, K.A. McEwen: Phys. Rev. B **60**, R8430 (1999)
[7.136] M. Yoshizawa, B. Lüthi, T. Goto, T. Suzuki, B. Renker, A. deVisser, P. Frings, J.J.M. Franse: JMMM **52**, 413 (1985)
[7.137] M. Amara, D. Finsterbusch, B. Luy, B. Lüthi, F. Hulliger, H.R. Ott: Phys. Rev. B **51**, 16407 (1995)
[7.138] R. J. Birgeneau: AIP Conf. Proc. N o. **10**, 1664 (1973)
[7.139] T. Suzuki, I. Ishii, N. Okuda, K. Katoh, T. Tabataka, T. Fujita, A. Tamaka et al.: Phys. Rev. B **62**, 49 (2000)

[7.140] T. Akazawa, T. Suzuki, H. Goshima, T. Tahara, T. Fujita, T. Takabatake, H. Fujii: J. Phys. Soc. Jpn. **67**, 3256 (1998)
[7.141] C.F. van Doorn and P. de V. du Plessis: J. Magn. Magn. Mat. **5**, 164 (1977)
[7.142] D. Kaczorowski, D. Finsterbusch: B. Lüthi, Int. J. Mod. Phys. **7**, 212 (1993)
[7.143] J. Wong et al.: Science **301**, 1078 (2003)
[7.144] H.M. Leadbetter and R.L. Moment: Acta Metall. **24**, 891 (1976)
[7.145] H.H. Teitelbaum and P.M. Levy: Phys. Rev. B **14**, 3058 (1976)
[7.146] L.L. Hirst: Adv. Phys. **27**, 231 (1978)
[7.147] R.M. White: *Quantum Theory of Magnetism*, Springer (1983)
[7.148] A. Fert, R. Asomoza, D.H. Sanchez, D. Spanjaard, A. Friederich: Phys. Rev. B **16**, 5040 (1977)
[7.149] K.M. Leung, D.L. Huber, B. Lüthi: J. Appl. Phys. **50**, 1831 (1979)
[7.150] K.M. Leung and D.L. Huber: Phys. Rev. B **19**, 5483 (1979)
[7.151] K.W. Becker, P. Fulde, J. Keller, P. Thalmeier: J. Phys. Colloq. C**5**, 35 (1979)
[7.152] K. Aizu: J. Phys. Soc. Jpn. **27**, 387, Err. 1374 (1969)
[7.153] G.R. Barsch and J.A. Krumhansl: Phys. Rev. Lett. **53**, 1069 (1984)
[7.154] D.J. Gunton and G.A. Saunders: Solid State Comm. **14**, 865 (1974)
[7.155] M.P. Brassington and G.A. Saunders: Phys. Rev. Lett. **48**, 159 (1982)
[7.156] J.K. Liakos and G.A. Saunders: Phil. Mag. A **46**, 217 (1982)
[7.157] N. Nakanishi: Progr. In Materials Science **24**, 143 (1980)
[7.158] A. Nagasawa, N. Nakanishi, K. Enami: Phil Mag. A **43**, 1345 (1981)
[7.159] E.K.H. Salje: *Phase transitions in ferroelastic and co-elastic crystals* (Cambridge University, 1990)
[7.160] W. Bührer, R. Gotthardt, A. Kulik, O. Mercier, F. Staub: J. Phys. F 13, L77 (1983)
[7.161] I. Müller: Phys. Blätter, **44**, Mai (1988)
[7.162] T.M. Brill, S. Mittelbach, W. Assmus, M. Müllner, B. Lüthi: J. Phys.: Condens.Matter **3**, 9621 (1991)
[7.163] C. Zener: Phys. Rev. **71**, 846 (1947)
[7.164] X. Ren, N. Miura, J. Zhang, K. Otsuka, K. Tanaka, M. Koiwa, T. Suzuki, Yu. I. Chuneyakov, M. Asai: Mat. Sci. Eng. A **312**, 196 (2001)
[7.165] F. Falk and P. Konopka: J. Phys. C. M. **2**, 61 (1990)
[7.166] Y. Ishibashi and M. Iwate: J. Phys. Soc. Jpn. **72**, 1675 (2003)
[7.167] J.M. Zhang and G.Y. Guo: Phys. Rev. Lett. **78**, 4789 (1997)
[7.168] A.N. Vasil'ev, V.D. Buchel'nikov, T. Takagi, V.V. Khovailo, E.I. Estrin: Physics-Uspekhi **46**, 559 (2003)
[7.169] E.F. Wassermann: in: *Handbook of Magnetic Materials*, 5, 237, ed. K.H.J. Buschow (North-Holland, 1986)
[7.170] G. Hausch and H. Warlimont: Acta Metallurgica **21**, 401 (1973)
[7.171] G. Hausch: Phys. Stat. Sol. (a) **15**, 501 (1973)
[7.172] G. Hausch: J. Phys. Soc. Jpn. **37**, 819 (1975)
[7.173] C.W. Garland: in: *Physical Acoustics*, Vol. VII, 52 (Acad. Press, 1970)
[7.174] W. Rehwald: Adv. Phys. **22**, 721 (1973)
[7.175] B. Lüthi and W. Rehwald: in *Structural phase transitions*, I, Topics in Current Physics Vol. 23 (1981)
[7.176] S. Haussühl: Solid State Comm. **13**, 147 (1973)

[7.177] W. Rehwald, J.R. Sandercock and M. Rossinelli: Phys. Stat. Sol. A **42**, 699 (1977)
[7.178] W. Krasser, U. Buchenau and S. Haussühl: Solid State Comm. **18**, 287 (1976)
[7.179] K.H. Michels and J. Naudts: J. Chem. Phys. **67**, 547 (1977)
[7.180] U.T. Höchli, K. Knorr, A. Loidl: Adv. Phys. **39**, 405 (1990)
[7.181] P.S. Peercy, I.J. Fritz, G.A. Samara: J. Phys. Chem. Sol. **36**, 1105 (1975)
[7.182] D.B. McWhan, R.J. Birgeneau, W.A. Bonner, H. Taub, J.D. Axe: J. Phys. C8, L81 (1975)
[7.183] K. Fossheim and R.M. Holt: Phys. Rev. Lett. **45**, 730 (1980)
[7.184] K. Fossheim and R.M. Holt: *Physical Acoustics*, Vol. XVI, 218 (1982)
[7.185] K. Fossheim and B. Berre: Phys. Rev. B **5**, 3292 (1972)
[7.186] F. Schwabl and Iro: Ferroelectrics **35**, 215 (1981)
[7.187] P. Selgert, C. Lingner, B. Lüthi: Z. Phys. B **55**, 219 (1984)
[7.188] W. Henkel, J. Pelzl, K.H. Höck, H. Thomas: Z. Phys. B **37**, 321 (1980)
[7.189] H. Kameyama, Y. Ishibashi, Y. Takagi: J. Phys. Soc. Jpn. **46**, 566 (1979)
[8.1] V.V. Gudkov and J.D. Gavenda: *Magnetoacoustic Polarization Phenomena in Solids* (Springer, 2000)
[8.2] J. Ziman: *The Physics of Electrons and Phonons* (Oxford University Press, 1960)
[8.3] A.B. Pippard: Proc. Roy. Soc. A **257**, 165 (1960)
[8.4] E. Fawcett, R. Griessen, W. Joss, M.J.G. Lee, J.M. Perz: in *Electrons at the Fermi Surface*, ed. M. Springford (Cambridge University Press, 1980)
[8.5] J. Labbe and J. Friedel: J. Phys. 27, 153, **303**, 708 (1966)
[8.6] D.M. Gray and A.M. Gray: Phys. Rev. B **14**, 669 (1976)
[8.7] M. Niksch, B. Lüthi, J. Kübler: Z. Phys. B **68**, 291 (1987)
[8.8] A.I. Akhiezer, M.I. Kaganov, G.I. Liubarskii: Sov. Phys. JETP **5**, 685 (1957)
[8.9] W.P. Mason: Phys. Rev. **97**, 557 (1955)
[8.10] M.S. Steinberg: Phys. Rev. **111**, 425 (1958)
[8.11] T. Holstein: Phys. Rev. **113**, 479 (1959)
[8.12] R.W. Morse: in *Progress in Cryogenics*, **1**, 219 (1959)
[8.13] R.G. Chambers: *Proc. VIIIth Internat. Conf. Low Temp. Phys.* (Univ. Toronto Press, 1961)
[8.14] J.A. Rayne and C.K. Jones: in *Physical Acoustics*, Vol. VII, 149 (Acad. Press, 1970)
[8.15] A.A. Abrikosov: *Theory of Metals* (North Holland, 1988)
[8.16] H. Ihrig, D.T. Vigren, J. Kübler, S. Methfessel: Phys. Rev. B **8**, 4525 (1973)
[8.17] K. Knorr, B. Renker, W. Assmus, B. Lüthi, R. Takke, H.J. Lauter: Z. Phys. B **39**, 151 (1980)
[8.18] R.A. Alpher and R.J. Rubin: J. Acoust. Soc. **26**, 452 (1954)
[8.19] Y. Shapira: in *Physical Acoustics*, Vol. V, 1 (Acad. Press, 1969)
[8.20] S. Rodriguez: Phys. Rev. **130**, 1778 (1963)
[8.21] H.N. Spector: Phys. Rev. **120**, 1261 (1960)
[8.22] V.M. Kontorovich: Sov. Phys. Usp. **27**, 134 (1984)
[8.23] A.B. Pippard: Phil. Mag. **2**, 1147 (1957)
[8.24] M.H. Cohen, M.J. Harrison, W.A. Harrison: Phys. Rev. **117**, 937 (1960)
[8.25] J.B. Ketterson and R.W. Stark: Phys. Rev. **156**, 748 (1967)

[8.26]   D. Shoenberg: *Magnetic oscillations in metals* (Cambridge University Press, 1984)
[8.27]   B.W. Roberts: in *Physical Acoustics*, IV Part B (Acad. Press, 1968)
[8.28]   L.R. Testardi and J.H. Condon: *Physical Acoustics*, Vol. VIII, 59 (Acad. Press, 1971)
[8.29]   Y. Onuki, T. Goto, T. Kasuya: in *Materials Science and Technology* Vol. **3**, 547 (VCH Verlagsgesellschaft, 1991)
[8.30]   L. Onsager: Phil. Mag. **43**, 1006 (1952)
[8.31]   I.M. Lifshitz and A.M. Kosevich: JETP **2**, 636 (1956)
[8.32]   M. Kataoka and T. Goto: J. Phys. Soc. Jpn. **62**, 4352 (1993)
[8.33]   L. Taillefer and G.G. Lonzarich: Phys. Rev. Lett. **60**, 1570 (1988)
[8.34]   G. Zwicknagl: Adv. Phys. **41**, 203 (1992)
[8.35]   T. Suzuki, T. Goto, A. Tamaki, T. Fujimura, Y. Onuki, T. Komatsubara: J. Phys. Soc. Jpn. **54**, 2367 (1985)
[8.36]   H. Matsui, T. Goto, M. Kataoka, T. Suzuki, H. Harima, S. Kunii, R. Takayama, O. Sakai: J. Phys. Soc. Jpn. **64**, 3315 (1995)
[8.37]   T. Suzuki, T. Goto, T. Fujimura, S. Kunii, T. Suzuki, T. Kasuya: J. Magn. Magn. Mat. 52, 261 (1985)
[8.38]   B. Wolf, G. Bruls, D. Finsterbusch, I. Kouroudis, B. Lüthi: Physica B **211**, 233 (1995)
[8.39]   M. Hunt, P. Meeson, P.A. Probst, P. Reinders, M. Springford: Physica B **165&166**, 337 (1990)
[8.40]   R. Settai, T. Goto, S. Sakatsume, Y.S. Kwon, T. Suzuki, Y. Kaneta, O. Sakai: J. Phys. Soc. Jpn.**63**, 3026 (1994)
[8.41]   R. Settai, T. Goto, Y. Onuki: J. Phys. Soc.Jpn. **61**, 609 (1992)
[8.42]   T. Goto, S. Sakatsume, A. Sawada, H. Matsui, R. Settai, S. Nakamura, Y. Ohtani, H. Goto, Y. Ohe, A. Tamaki, Y. Fuda, K. Abe, Y. Yamamoto, T. Fujimura: Cryogenics **32**, 902 (1992)
[8.43]   T. Goto, T. Suzuki, Y. Ohe, T. Fujimura, A. Tamaki: J.Magn. Magn. Mat. **76&77**, 305 (1988)
[8.44]   S. Nimori, M. Kataoka, T. Goto, G. Kido: Phys. Rev. B **67**, 224103 (2003)
[8.45]   B. Wolf, R. Blick, G. Bruls, B. Lüthi, Z. Fisk, J.L. Smith, H.R. Ott: Z. Phys. B**85**, 159 (1991)
[8.46]   R. Corcoran, P. Meeson, P.A. Probst, M. Springford, B. Wolf, R. Blick, G. Bruls, B. Lüthi, Z. Fisk, J.L. Smith, H.R. Ott: Z. Phys. B **91**,135 (1993)
[8.47]   J.R. Feller, J.B. Ketterson, D.G. Hinks, D. Dasgupta, B.K. Sarma: Phys. Rev. B **62**, 11538 (2000)
[8.48]   Y. Yoshida, A. Mukai, R. Settai, U. Miyake, Y. Inada, Y. Onuki, K. Betsuyaku, H. Harima, T.D. Matsuda, Y. Aoki, H. Sato: J. Phys. Soc. Jpn. **68**, 3041 (1999)
[8.49]   A.P. Mackenzie, S.R. Julien, A.J. Diver, G.J. McMullan, M.P. Ray, G.G. Lonzarich, Y. Maeno, S. Nishizaki, T. Fujita: Phys. Rev. Lett.**70**, 3786 (1996)
[8.50]   H. Matsui, Y. Yoshida, A. Mukai, R. Settai, Y. Onuki, H. Takei, N. Kimura, H. Aoki, N. Toyota: J. Phys. Soc. Jpn. **69**, 3769 (2000)
[8.51]   H. Matsui, M. Yamaguchi, Y. Yosida, A. Mukai, R. Settai, Y. Onuki, H. Takei, N. Toyota: J. Phys. Soc. Jpn. **67**, 3687 (1998)
[8.52]   J. Paglione, C. Lupien, W.A. MacFarlane, J.M. Perz, L. Taillefer: Phys. Rev. B 65, 220506(R) (2002)

[8.53]   N.G. Einspruch: in: *Solid State Physics*, Vol. 17, 217 (Acad. Press, 1965)
[8.54]   R.W. Keyes: in *Solid State Physics*, Vol. 20, 37 (Acad. Press, 1967)
[8.55]   W.P. Mason and T.B. Bateman: Phys. Rev. **134**, A1387 (1964)
[8.56]   M. Pomerantz, R.W. Keyes, P.E. Seiden: Phys. Rev. Lett. **9**, 312 (1962)
[8.57]   G. Weinreich: Phys. Rev. **107**, 317 (1957)
[8.58]   A.R. Hutson and D.L. White: J. Appl. Phys. **33**, 40 (1962)
[8.59]   A.R. Hutson, J.H. Mc Fee, D.L. White: Phys. Rev. Lett. **7**, 237 (1961)
[8.60]   A.M. Toxen and S. Tansal: Phys. Rev. Lett. **10**, 481 (1963)
[8.61]   K. Walther: Phys. Rev. Lett. **16**, 642 (1966)
[8.62]   L. Esaki: Phys. Rev. Lett. **8**, 4 (1962)
[9.1]   P.W. Anderson: Phys. Rev. **124**, 41 (1961)
[9.2]   G. Zwicknagl, A.N. Yaresko, P. Fulde: Phys. Rev. B **65**, 081103 (2002)
[9.3]   P. Thalmeier and B. Lüthi: in: *Handbook on the Physics and Chemistry of Rare Earths*, Vol. 14 (North-Holland, 1991)
[9.4]   N.B. Brandt and V.V. Moshalkov: Adv. Phys. **33**, 373 (1984)
[9.5]   G. Stewart: Rev. Mod. Physics **56**, 755 (1984)
[9.6]   P. Lee, T.M. Rice, J.W. Serene, L.J. Sham, J.W. Wilkins: Comments Cond. Mat. Phys. **17**, 361 (1986)
[9.7]   P. Fulde, J. Keller, G. Zwicknagl: in: Solid State Physics, 41, 1 (Acad. Press, 1988)
[9.8]   H.R. Ott: *Progress Low Temperature Physics*, Vol. XI (1987)
[9.9]   N. Grewe and F. Steglich: *Handbook on the Physics and Chem. of Rare Earths*, Vol. 14 (North-Holland, 1991)
[9.10]   D.W. Hess, P.S. Riseborough, J.L. Smith: in: Encyclopedia Appl. Phys. **7**, 435 (1993)
[9.11]   A.C. Hewson: *The Kondo Problem to Heavy Fermions* (Cambridge University Press, 1993)
[9.12]   C. Varma: Rev. Mod. Phys. **48**, 219 (1976)
[9.13]   A. Jayaraman: in *Handbook on the Physics and Chemistry of Rare Earths*, Vol. 2 (North-Holland, 1979)
[9.14]   J.M. Lawrence, P.S. Riseborough, R.D. Parks: Reports Progress Phys. 44,1 (1981)
[9.15]   P. Wachter: *Handbook on the Physics and Chemistry of Rare Earths*, Vol. 19 (North-Holland, 1994)
[9.16]   J.S. Schilling: Adv. Phys. **28**, 657 (1979)
[9.17]   F. Thomas, C. Ayache, I.A. Fomine, J. Thomasson, C. Geibel: J. Phys. C. M. 8, L51 (1996)
[9.18]   E. Bucher, K. Andres, F.J. di Salvo, J.P. Maita, A.C. Gossard, A.S. Cooper, G.W. Hull: Phys. Rev. B **11**, 500 (1975)
[9.19]   T. Penney: in: *Moment Formation in Solids*, NATO ASI Series B: Physics, Vol. **117**, 47 (1984)
[9.20]   I. Nowik: in *Valence Instability and related narrow band phenomena* 261 (Plenum Press, 1977)
[9.21]   M. Loewenhaupt and K.H. Fischer: in *Handbook Magnetic Materials* **7**, 503 (1993)
[9.22]   A.P. Murani: in *Concepts in electron correlation*, NATO science series Vol. 110, ed. A.C. Hewson, V. Zlatic, p. 297 (Kluwer Acad. Publ. 2003)
[9.23]   J. Kondo: Progr. Theor. Phys. **28**, 846 (1962)
[9.24]   Y. Onuki and T. Komatsubara: J. Magn. Magn.Mat. **63&64**, 281 (1987)

[9.25] K. Kadowaki and S.B. Woods: Solid State Comm. **58**, 507 (1986)
[9.26] A. Jayaraman, V. Narayanamurti, E. Bucher, R.G. Maines: Phys. Rev.Lett. **25**, 1430 (1970)
[9.27] L. L. Hirst: J. Phys. Chem. Solids **35**, 1285 (1974)
[9.28] T. Penney, R. Melcher, F. Holtzberg, G. Güntherodt: AIP Conf. Proc. **29**, 392 (1975)
[9.29] T. Hailing, G.A. Saunders, H. Bach: Phys. Rev. B **29**, 1848 (1984)
[9.30] H. A. Mook, R.M. Nicklow, T. Penney, F. Holtzberg, M.W. Shafer, Phys. Rev. B **18**, 2925 (1978)
[9.31] H. Bilz, G. Güntherodt, W. Kleppmann and W. Kress: Phys. Rev. Lett. **43**, 1979 (1979)
[9.32] N. Grewe, H.J. Leder, P. Entel: *Festkörperprobleme*, XX (Vieweg, 1980)
[9.33] N. Wakabayashi: Phys. Rev.B **22**, 5833 (1980)
[9.34] G. Pastor, A. Caro, B. Alascio: Phys. Rev. B **36**, 1673 (1987)
[9.35] B. Lüthi: J. Magn.Magn. Mat. **52**, 70 (1985)
[9.36] H. Boppart: J. Magn. Magn.Mat. **47&48**, 436 (1985)
[9.37] R.D. Parks and J.M. Lawrence: AIP Conf.Proc. (1977)
[9.38] H. Wehr, K. Knorr, R. Feile: Solid State Comm. **40**, 507 (1981)
[9.39] R. Mock, B. Hillebrands, H. Schmidt, G. Güntherodt, Z. Fisk, A. Meyer: J. Magn. Magn. Mat. **47&48**, 312 (1985)
[9.40] T. Penney, B. Barbara, T.S. Plaskett, H.E. King, S.J. LaPlaca: Solid State Comm. **44**, 1199 (1982)
[9.41] J.L. Sarrao: Physica B **259-261**, 128 (1999)
[9.42] I. Felner and I. Nowik: Phys. Rev. B **33**, 617 (1986)
[9.43] B. Kindler, D. Finsterbusch, R. Graf, F. Ritter, W. Assmus, B. Lüthi: Phys. Rev. B **50**, 704 (1994)
[9.44] S. Zherlitsyn, B. Lüthi, B. Wolf, J.L. Sarrao, Z. Fisk, V. Zlatic: Phys. Rev. B **60**, 3148 (1999)
[9.45] A.L. Cornelius, J.M. Lawrence, J.L. Sarrao, Z. Fisk, M.F. Hundley, G.H. Kwei, J.D. Thompson, C.H. Booth, F. Bridges: Phys. Rev. B **56**, 7993 (1997)
[9.46] I.V. Svechkarev, A.S. Panfilov, S.N. Dolya, H. Nakamura, M. Shiga, P. Schlottmann: Phys. Rev. B **64**, 214414 (2001)
[9.47] E. Müller-Hartmann: Solid State Comm. **31**, 113 (1979)
[9.48] C.D. Immer, J.L. Sarrao, Z. Fisk, A. Lacerda, C. Mielke, J.D. Thompson: Phys. Rev. B **56**, 71 (1997)
[9.49] M.O. Dzero, L.P. Gorkov, A.K. Zvezdin: J. Phys. CM 12, l711 (2000)
[9.50] A.V. Goltsev and G. Bruls: Phys. Rev. B **63**, 155109 (2001)
[9.51] H. Wada, A. Nakamura, A. Mitsuda, M. Shiga, T. Tanaka, H. Mitamura, T. Goto: J. Phys. CM **9**, 7913 (1997)
[9.52] V.V. Platonov, O.M. Tatsenko, V.D. Selemir, M. Shiga: Phys. of the Solid State **44**, 315 (2002)
[9.53] R. Takke, M. Niksch, W. Assmus, B. Lüthi, R. Pott, R. Schefzyk, D. K. Wohlleben: Z. Phys. B44, 33 (1981)
[9.54] D. Lenz, H. Schmidt, S. Ewert, W. Boksch, R. Pott, D. Wohlleben: Solid State Comm. **52**, 759 (1984)
[9.55] B. Lüthi, G. Bruls, P. Thalmeier, B. Wolf, D. Finsterbusch, I. Kouroudis: J. Low Temp. Phys. **95**, 257 (1994)
[9.56] J.D. Thompson and J.M. Lawrence: in *Handbook on the Physics and Chem. of Rare Earths*, Vol. 19 (North-Holland, 1994)

[9.57]   K. Bömken, D. Weber, M. Yoshizawa, W. Assmus, B. Lüthi, E. Walker: J. Magn. Magn. Mat. **63&64**, 315 (1987)
[9.58]   E.S. Clementyev, P.A. Alekseev, M. Braden, J.M. Mignot, G. Lapertot, V.N. Lazukov, I.P. Sadikov: Phys. Rev. B **57**, R8099 (1998)
[9.59]   G. Creuzet and D. Gignoux: Phys. Rev. B **33**, 515 (1986)
[9.60]   E.V. Nefedova, P.A. Alekseev, V.N. Lazukov, I.P. Sadikov: JETP **96**, 1113 (2003)
[9.61]   B. Butler, D. Givord, F. Givord, S.B. Palmer: J. Phys. C 13, L743 (1980)
[9.62]   F.F. Voronov, V.A. Goncharova, O. Stal'gorova: Sov. Phys. JETP **49**, 687 (1979)
[9.63]   Y. Kuramoto: Z. Phys. B **37**, 299 (1980)
[9.64]   E. Müller-Hartmann: in *Electron Correlation and Magnetism in narrow band systems* p. 178 (Springer, 1981)
[9.65]   H.J. Schmidt and E. Müller-Hartmann: Z. Phys. B **60**, 363 (1985)
[9.66]   R. Mock, E. Zirngiebl, B. Hillebrands, G. Güntherodt, F. Holtzberg: Phys. Rev. Lett. **57**, 1040 (1986)
[9.67]   E. Zirngiebl, S. Blumenröder, R. Mock, G. Güntherodt: J. Magn. Magn. Mat. **54–57**, 359 (1986)
[9.68]   S. Nakamura, T. Goto, S. Kunii, K. Iwashita, A. Tamaki: J. Phys. Soc. Jpn. **63**, 623 (1994)
[9.69]   D. Sherrington and S. von Molnar: Solid State Comm. **16**, 1347 (1975)
[9.70]   D.I. Khomskii: JETP Lett. **27**, 331 (1978)
[9.71]   H. Capellmann and S. Lipinski: Z. Phys. B **83**, 199 (1991)
[9.72]   F. Milstein and K. Huang: Phys. Rev. B **19**, 2030 (1979)
[9.73]   R. Lakes: Science **235**, 1038 (1987)
[9.74]   A. W. Lipsett and A.I. Beltzer: J. Acoust. Soc. Am.**84**, 2179 (1988)
[9.75]   P. Thalmeier: J. Phys. C **20**, 4449 (1987)
[9.76]   J. Keller, B. Bulla, Th. Höhn, K.W. Becker: Phys. Rev. B **41**, 1878 (1990)
[9.77]   D. Weber, M. Yoshizawa, I. Kouroudis, B. Lüthi, E. Walker: Europhys. Lett.**3**, 827 (1987)
[9.78]   T. Suzuki, T. Goto, A. Tamaki, T. Fujimura, Y. Onuki, T. Komatsubara: J. Phys. Soc. Jpn. **54**, 2367 (1985)
[9.79]   M. Yoshizawa, B. Lüthi, K.D. Schotte: Z. Phys. B **64**, 169 (1986)
[9.80]   J.W. Allen and R.M. Martin: Phys. Rev. Lett. **49**, 1106 (1982)
[9.81]   M. Lavagna, C. Lacroix and M. Cyrot: Phys. Lett. 90A, 210 (1982)
[9.82]   H. Razafimandimby, P. Fulde and J. Keller: Z. Phys. B **54**, 111 (1984)
[9.83]   G. Lapertot, R. Calemczuk, C. Marcenat, J.Y. Henry, J.X. Boucherle, J. Flouquet, J. Hausmann, R. Cibin, J. Cors, D. Jaccard, J. Sierro: Physica B **186-.188**, 454 (1993)
[9.84]   S. Raymond, J.P. Rueff, S.M. Shapiro, P. Wochner, F. Sette, P. Lejay: Solid Sate Comm. **118**, 473 (2001)
[9.85]   K.W. Becker and P. Fulde: Z. Phys. B 65, 313 (1987); ibid. 67, 35 (1987)
[9.86]   K.W. Becker and P. Fulde: Europhys. Lett. 1, 669 (1986)
[9.87]   R. Mock and G. Güntherodt: Z. Phys. B **74**, 315 (1989)
[9.88]   K.W. Becker and P. Fulde: Phys. Lett. A 125, 68 (1987b)
[9.89]   C.J. Pethick, D. Pines, K.F. Quader: Phys. Lett. A **125**, 485 (1987)
[9.90]   K.H. Bennemann, M.L. Kulic, V. Müller: Phys. Lett. A **120**, 413 (1987)
[9.91]   K.D. Schotte, D. Förster, U. Schotte: Z. Phys. B **64**, 165 (1986)
[9.92]   V. Müller, D. Maurer, K. de Groot, E. Bucher, H.E. Bömmel: Phys. Rev. Lett. **56**, 248 (1986)

[9.93] Landau–Lifshitz: *Fluid mechanics*, p. 304 (Addison-Wesley, 1959b)
[9.94] P. Coleman: Phys. Rev. B **35**, 5072 (1987)
[9.95] G. Bruls, B.Lüthi, D. Weber, B. Wolf, P. Thalmeier: Phys. Scripta T **35**, 82 (1991)
[9.96] T. Goto, K. Morita, H. Matsui, S. Nakamura, R. Settai, Y. Haga, T. Suzuki, M. Kataoka, S. Saketsume: Physica B **199&200**, 517 (1994)
[9.97] B. Lüthi and C. Lingner: Z. Phys. B **34**, 157 (1979)
[9.98] D. Nikl, I. Kouroudis, W. Assmus, B. Lüthi, G. Bruls, U. Welp: Phys. Rev. B **35**, 6864 (1987)
[9.99] B. Lüthi, S. Blumenröder, B. Hillebrands, E. Zirngiebl, G. Güntherodt, K. Winzer: Z. Physik B **58**,31(1984)
[9.100] K. Andres, J.E. Graebner and H.R. Ott: Phys. Rev. Lett. **35**, 1779 (1975)
[9.101] M. Niksch, B. Lüthi and K. Andres: Phys. Rev. B **22**, 5774 (1980)
[9.102] S. Nakamura, T. Goto, Y. Isikawa, S. Sakatsume, M. Kasaya: J. Phys. Soc. Jpn. **60**, 2305 (1991)
[9.103] S. Holtmeier, C. Hinkel, M. Aigner, G. Bruls, D. Finsterbusch, B. Wolf, W. Assmus, B. Lüthi: Physica B **230-232**, 658 (1997)
[9.104] S. Paschen, B. Wand, G. Sparn, F. Steglich, Y. Echizen, T. Takabatake: Phys. Rev. B **62**, 14912 (2000)
[9.105] B. Lüthi and H.R. Ott: Solid State Comm. **33**, 717 (1980)
[9.106] D. Endoh, T. Goto, A. Tamaki, B. Liu, M. Kasaya, T. Fujimura, T. Kasuya: J. Phys. Soc. Jpn. **58**, 940 (1989)
[9.107] A. de Visser, J.J.M. Franse, A. Menovsky: J. Phys. F 15, L53 (1985)
[9.108] J. Paglione, C. Lupien, W.A. MacFarlane, J.M. Perz, L. Taillefer: Phys. Rev. B **65**, 220506(R) (2002)
[9.109] M. Yoshizawa, B. Lüthi, T. Goto, T. Suzuki, B. Renker, A. deVisser, P. Frings, J.J.M. Franse, JMMM **52**, 413 (1985)
[9.110] B. Wolf, W. Sixl, R. Graf, D. Finsterbusch, G. Bruls, B. Lüthi, E.A. Knetsch, A.A. Menovsky, J.A. Mydosh: J. Low Temp. Phys. **94**, 307 (1994)
[9.111] B. Wolf, S. Zherlitsyn, M. Lang, B. Lüthi: Acta Phys. Polonica B **34**, 269 (2003)
[9.112] B. Lüthi, B. Wolf, P. Thalmeier, M. Günther, W. Sixl, G. Bruls: Phys. Lettt. A **175**, 237 (1993)
[9.113] P. Santini and G. Amoretti: Phys. Rev. Lett. **73**, 1027 (1994)
[9.114] R. Modler, K. Gloos, H. Schimanski, C. Geibel, M. Günther, G. Bruls, B. Lüthi, T. Komatsubara, N. Sato, C. Schank, F. Steglich: Physica B **186-188**, 294 (1993)
[9.115] F.P. Milliken. T. Penney, F. Holtzberg, Z. Fisk: J. Magn. Magn. Mat. **76&77**, 201 (1988)
[9.116] E. Stryjewski and N. Giordano: Adv. Phys. **26**, 487 (1977)
[9.117] B. Lüthi, P. Thalmeier, G. Bruls, D. Weber: J. Magn. Magn. Mat. **90&91**, 37 (1990)
[9.118] J. Flouquet, S. Kambe, L.P. Regnault, P. Haen, J.P. Brison, F. Lapierre, P. Lejay: Physica B **215**, 77 (1995)
[9.119] I. Kouroudis, D. Weber, M. Yoshizawa, B. Lüthi, L. Puech, P. Haen, J. Flouquet, G. Bruls, U. Welp, J.J.M. Franse, A. Menovsky, E. Bucher, J. Hufnagl: Phys. Rev. Lett.**58**, 820 (1987)
[9.120] G. Bruls, D. Weber, B. Lüthi, J. Flouquet, P. Lejay: Phys. Rev. B 42, 4329 (1990)

[9.121] P. Thalmeier and P. Fulde: Europhys. Lett. **1**, 367 (1986)
[9.122] A.B. Kaiser and P. Fulde: Phys. Rev. B **37**, 5357 (1988)
[9.123] K. Held, M. Ulmke, N. Blümer, D. Vollhardt: Phys. Rev. B **56**, 14469 (1997)
[9.124] Y. Kuramoto and Y. Kitaoka: *Dynamics of Heavy Electrons* (Oxford Science Publ., 2000)
[9.125] D. Weber, I. Kouroudis, B. Lüthi, G. Bruls, M. Yoshizawa, P. Haen, J. Flouquet, E. Bucher, J. Hufnagl: J. Magn. Magn. Mat. **76&77**, 315 (1988)
[9.126] J. M. Mignot, J. Flouquet, P. Haen, F. Lapierre, L. Puech, J. Voiron: J. Magn. Magn. Mat. **76&77**, 97 (1988)
[9.127] K. Matsuhira, T. Sakakibara, A. Nomachi, T. Tayama, K. Tenya, H. Amitsuka, K. Maezawa, Y. Onuki: J. Phys. Soc.Jpn. **68**, 3402 (1999)
[9.128] G. Hampel, G. Bruls, D. Weber, I. Kouroudis, B. Lüthi: Physica B **161**, 333 (1989)
[9.129] B. Wolf, S. Zherlitsyn, H. Schwenk, S. Schmidt, B. Lüthi: J. Magn. Magn. Mat. **226–230**, 107 (2001)
[9.130] A. de Visser, F.R. de Boer, A.A. Menovsky, J.J.M. Franse: Solid State Comm. **64**, 527 (1987)
[9.131] K. Sugiyama, H. Fuke, K. Kindo, K. Shimohata, A.A. Menovsky, J.A. Mydosh, M. Date: J. Phys. Soc. Jpn. **59**, 3331 (1990)
[9.132] C. Broholm, H. Lin, P.T. Matthews, T.E. Mason, W.J.L. Buyers, M.F. Collins, A.A. Menovsky, J.A. Mydosh, J.K. Kjems: Phys. Rev. B **43**, 12809 (1991)
[9.133] M. Jaime, K.H. Kim, G. Jorge, S. McCall, J.A. Mydosh: Phys. Rev. Lett. **89**, 287201 (2002)
[9.134] A. Suslov, J.B. Ketterson, D.G. Hinks, D.F. Agterberg, B.K. Sarma: Phys. Rev. B **68**, 020406 (R) (2003)
[9.135] T. Goto, T. Suzuki, Y. Ohe, T. Fujimura, S. Sakatsume, Y. Onuki, T. Komatsubara: J. Phys. Soc. Jpn. **57**, 2612 (1988)
[9.136] H.V. Löhneysen, H.G. Schlager, A. Schröder: Physica B **186–188**, 590 (1993)
[9.137] D.L. Cox and A. Zawadowski: Exotic Kondo effects in Metals (Taylor and Francis, 1999)
[9.138] H. v.Löhneysen: J. Phys. Cond. Matter **8**, 9689 (1996)
[9.139] P. Fulde and M. Loewenhaupt: Adv. Phys. **34**, 589 (1985)
[9.140] D.L. Cox: Phys. Rev. Lett. **59**, 1240 (1987)
[9.141] C.L. Seaman, M.B. Maple, B.W. Lee, S. Ghamaty, M.S. Torikachvili, J.S. Kang, L.Z. Liu, J.W. Allen, D.L. Cox: Phys. Rev. Lett. **67**, 2882 (1991)
[9.142] B. Andraka and A.M. Tsvelick: Phys. Rev. Lett. **67**, 2886 (1991)
[9.143] D. Finsterbusch, H. Willig, B. Wolf, M. Amara, G. Bruls, B. Lüthi, M. Waffenschmidt, O. Stockert, H. v.Löhneysen: Physica B **223&224**, 329 (1996a)
[9.144] D. Finsterbusch, H. Willig, B. Wolf, G. Bruls, B. Lüthi, M. Waffenschmidt, O. Stockert, A. Schröder, H. v. Löhneysen: Ann. Phys. **5**, 184 (1996b)
[9.145] M. Amara, D. Finsterbusch, B. Luy, B. Lüthi, F. Hulliger, H.R. Ott: Phys. Rev. B **51**, 16407 (1995)
[9.146] Y. Nakanishi, M. Yoshizawa, T. Yamaguchi, H. Hazama, Y. Nemoto, T. Goto, T.D. Matsuda, H. Sugiwara, H. Sato: J. Phys. C. M. **14**, L715 (2002)

References

[9.147] S. Doniach: in *Valence Instabilities* 169 (Plenum Press, 1977)
[9.148] L. Zhu, M. Garst, A. Rosch, Q. Si: Phys. Rev. Lett. **91**, 066404 (2003)
[9.149] R. Küchler, N. Oeschler, P. Gegenwart, T. Cichorek, K. Neumaier, O. Tegus, C. Geibel, J.A. Mydosh, F. Steglich, L. Zhu, Q. Si: Phys. Rev. Lett. **91**, 066405 (2003)
[10.1] J.R. Schrieffer: *Theory of Superconductivity* (Benjamin, 1964)
[10.2] W. Buckel: Supraleitung (VCH, 1990)
[10.3] M. Tinkham: *Introduction to Superconductivity* (Mc Graw Hill, 1975)
[10.4] D.R. Tilley and J. Tilley: *Superfluidity and Superconductivity* (Adam Hilger, 1990)
[10.5] T. Tsuneto: *Superconductivity and Superfluidity* (Cambridge, 1998)
[10.6] V.P. Mineev and K.V. Samokhin: Introduction to Unconventional Superconductivity (Gordon and Breach, 1999)
[10.7] W.H. Keesom and P.H. van Laer: Physica **5**, 193 (1938)
[10.8] L.R. Testardi: Phys. Rev. B **3**, 95 (1971)
[10.9] L.R. Testardi: *Phys. Acoustics*, Vol. X, 193 (Acad. Press, 1974)
[10.10] A.B. Pippard: Phil. Mag. **2**, 1147 (1957)
[10.11] S. Zherlitsyn, B. Lüthi, B. Wolf: to be publ. (2004)
[10.12] H. Bömmel: Phys. Rev. **96**, 220 (1954)
[10.13] H. Bömmel: Phys. Rev. **100**, 758 (1955)
[10.14] J. Bardeen, L.N. Cooper, J.R. Schrieffer, Phys. Rev. **108**, 1175 (1957)
[10.15] R.W. Morse: in: Progress in Cryogenics **1**, 219 (1959)
[10.16] M. Levy in: *Physical Acoustics*, Vol. XX, 1 (Acad. Press, 1992)
[10.17] M. Levy: Phys. Rev. **131**, 1497 (1963)
[10.18] M. Gottlieb, M. Garbuny, C.K. Jones: in *Physical Acoustics*, Vol. VII, 1 (Acad. Press, 1970)
[10.19] J. A. Rayne and C. K. Jones: in *Physical Acoustics*, Vol. VII, 149 (Acad. Press, 1970)
[10.20] B.W. Batterman and C.S. Barrett: Phys. Rev. **149**, 296 (1966)
[10.21] M. Weger and I.B. Goldberg: Solid State Physics **28**, 1 (1973)
[10.22] J. Müller: Rep. Progr. Phys. **43**, 641 (1980)
[10.23] W. Rehwald: Phys. Lett. **27A**, 287 (1968)
[10.24] J.D. Axe and G. Shirane: Phys. Rev. B **8**, 1965 (1973)
[10.25] J. Labbé and J. Friedel: J. Phys. 27, 153, **303**, 708 (1966)
[10.26] L.P. Gor'kov and O.N. Dorokhov: J. Low Temp. Phys. **22**, 1 (1976)
[10.27] R.N. Bhatt: Phys. Rev. B **16**, 1915 (1977)
[10.28] W. Dieterich and P. Fulde: Z. Phys. **248**, 154 (1971)
[10.29] J.P. Maita and E. Bucher: Phys. Rev. Lett. **29**, 931 (1972)
[10.30] W. Dieterich: Phys. Lett. 37A, 409 (1971)
[10.31] O. Fischer: Appl. Phys. **16**, 1 (1978)
[10.32] O. Pena and M. Sergent: Progr. Solid State Chem. **19**, 165 (1989)
[10.33] B. Wolf, J. Molter, G. Bruls, B. Lüthi, L. Jansen: Phys. Rev. B **54**, 348 (1996)
[10.34] H. Keiber, C. Geibel, B. Renker, H. Rietschel, H. Schmidt, H. Wühl, G. Stewart: Phys. Rev. B **30**, 2542 (1984)
[10.35] B. Lüthi, M. Herrmann, W. Assmus, H. Schmidt, H. Rietschel, H. Wühl, U. Gottwick, G. Sparn, F. Steglich: Z. Phys. B **60**, 387 (1985)
[10.36] D. Rainer and G. Bergmann: J. Low Temp. Phys. **14**, 501 (1974)
[10.37] H. Goshima, T. Suzuki, T. Fujita, R. Settai, H. Sugawara, Y. Onuki: Physica B **223&224**, 172 (1996)

[10.38] B. Wolf, C. Hinkel, S. Holtmeier, D. Wichert, I. Kouroudis, G. Bruls, B. Lüthi, M. Hedo, Y. Inada, E. Yamamoto, Y. Haga, Y. Onuki: J. Low Temp. Phys. **107**, 421 (1997)
[10.39] M. Yoshizawa, M. Tamura, M. Ozawa, D.H. Yoon, H. Sugawara, H. Sato, Y. Onuki: J. Phys. Soc. Jpn. **66**, 2355 (1997)
[10.40] R.N. Shelton, A.C. Lawson, K. Baberschke: Solid State Comm. **24**, 465 (1977)
[10.41] T. Nakama et al.: unpublished (1997)
[10.42] M. Ozawa, M. Yoshizawa, H. Sugawara, Y. Onuki: Physica B **206&207**, 267 (1995)
[10.43] M. Levy, M.F. Xu, B.K. Sarma, K.J. Sun: in *Physical Acoustics*, Vol. XX, 237 (Acad. Press, 1992)
[10.44] S. Zherlitsyn, B. Lüthi, V. Gusakov, B. Wolf, F. Ritter, D. Wichert, S. Barilo, S. Shiryaev, C. Escribe-Filippini, J.L. Tholence: Eur. Phys. J. B**16**, 59 (2000)
[10.45] A. Haas, D. Wichert, G. Bruls, B. Lüthi, G. Balakrishnan, D. McK. Paul: J. Low Temp. Phys. **114**, 285 (1999)
[10.46] K.J. Sun and M. Levy: in: *Physical Acoustics*, Vol. XX, 191 (1992)
[10.47] K. H. Müller and V.N. Narozhnyi: Rep. Progr. Phys. **64**, 943 (2001)
[10.48] T. Ichitsubo, H. Ogi, S. Nishimura, T. Seto, M. Hirao, H. Inui: Phys. Rev. B **66**, 052514 (2002)
[10.49] T. Ichitsubo, H. Ogi, S. Nishimura, T. Seto, M. Hirao, H. Inui: Phys. Rev. B **69**, 099901(E) (2004)
[10.50] J.G. Bednorz and K.A. Müller: Z. Phys. B **64**, 189 (1986)
[10.51] T. Goto, B. Lüthi, R. Geick, K. Strobel: Phys. Rev. B **22**, 3452 (1980)
[10.52] Y. Maeno: in: *Materials and Crystallogr. Aspects of HT_c-Supercond.* 203 (Kluwer Acad.Publ., 1994)
[10.53] A. Migliori, W.M. Visscher, S.E. Brown, Z. Fisk, S.W. Cheong, B. Alten, E.T. Ahrens, K.A. Kubat-Martin, J.D. Maynard, Y. Huang, D.K. Kirk, K.A. Gillis, H.K. Kim, M.H.W. Chan: Phys. Rev. B **41**, 2098 (1990)
[10.54] T. Suzuki, M. Nohara, Y. Maeno, T. Fujita, I. Tanaka, H. Kojima: J. Supercond. **7**, 419 (1994)
[10.55] R.J. Birgeneau, C.Y. Chen, D.R. Gabbe, H.P. Jenssen, M.A. Kastner, C.J. Peters, P.J. Picone, T. Thio, T.R. Thurston, H.L. Tuller, J.D. Axe, P. Böni, G. Shirane: Phys. Rev. Lett. **59**, 1329 (1987)
[10.56] A. Migliori, W.M. Visscher, S. Wong, S.E. Brown, I. Tanaka, H. Kojima, P.B. Allen: Phys. Rev. Lett. **64**, 2458 (1990)
[10.57] M. Nohara, T. Suzuki, Y. Maeno, T. Fujita, I. Tanaka, H. Kojima: Springer Proc. in Phys., Vol. **60**, 213 (1992)
[10.58] B. Golding: in: *Physical Acoustics*, XX, 349 (Acad. Press, 1992)
[10.59] C. Hucho and M. Levy: Phys. Rev. Lett. **77**, 1370 (1996)
[10.60] M. Boekholt, J. V. Harzer, B. Hillebrands, G. Güntherodt: Physica C**179**, 101 (1991)
[10.61] R. Mock and G. Güntherodt: J. Phys. C **17**, 5635 (1984)
[10.62] M. Saint-Paul, J.L. Tholence, H. Noel, J.C. Levet, M. Potel, P. Gougeon: Physica C**166**, 405 (1990)
[10.63] E.H. Brandt: Rep. Progr. Phys. **58**, 1465 (1995)
[10.64] G. Blatter, M.V. Feigel'man, V.B. Geshkenbein, A.I. Larkin, V.M. Vinokur: Rev. Mod. Phys. **66**, 1125 (1994)

[10.65] A. Ikushima, T. Suzuki, N. Tanaka, S. Nakajima: J. Phys. Soc. Jpn. **19**, 2235 (1964)
[10.66] K. Kudo: Physica C **385**, 501 (2003)
[10.67] P.W. Anderson and Y.B. Kim: Rev. Mod. Phys. **36**, 39 (1964)
[10.68] P.H. Kes, J. Aarts, J. van den Berg, C.J. van der Beck, J.A. Mydosh: Supercond. Sci. Technol. **1**, 242 (1989)
[10.69] J. Pankert, G. Marbach, A. Comberg, P. Lemmens, P. Fröning, S. Ewert: Phys. Rev. Lett. **65**, 3052 (1990)
[10.70] J. Pankert: Physica C **168**, 335 (1990)
[10.71] P. Lemmens, P. Fröning, S. Ewert, J. Pankert, G. Marbach, A. Comberg: Physica C **174**, 289 (1991)
[10.72] P. Lemmens, P. Fröning, S. Ewert, J. Pankert, H. Passing, A. Comberg: Physica B **165&166**, 1275 (1990)
[10.73] S.B. Roy and P. Chaddah: Pramana **53**, 659 (1999)
[10.74] A.A. Abrikosov and L.P. Gorkov: JETP **12**, 1243 (1961)
[10.75] K. Maki: Progr. Theor. Phys. **31**, 731 (1964)
[10.76] K. Maki and P. Fulde: Phys. Rev. **140**, A1586 (1965)
[10.77] E. Krätzig: Phys. Rev. B **7**, 119 (1973)
[10.78] D. Saint James and P.G. de Gennes: Phys. Lett. **7**, 306 (1963)
[10.79] H.P. Fredericksen, M. Levy, M. Tachiki, M. Ashkin, J.R. Gavaler: Solid State Comm. **48**, 883 (1983)
[10.80] F. Akao: Phys. Lett. **30A**, 409 (1969)
[10.81] H.P. Fredericksen, H.L. Salvo Jr., M. Levy, R.H. Hammond, T.H. Geballe: Phys. Lett. **75A**, 389 (1980)
[10.82] M.C. Jain and D.R. Tilley: J. Low Temp. Phys. **9**, 499 (1972)
[10.83] D.A. Robinson, K. Maki, M. Levy: Phys. Rev. Lett. **32**, 709 (1974)
[10.84] D. Vollhardt and P. Wölfle: *The Superfluid Phases of He$_3$* (Taylor and Francis, 1990)
[10.85] P. Fulde, J. Keller, G. Zwicknagl: in: Solid State Physics, **41**, 1 (1988)
[10.86] N.K. Sato, N. Aso, K. Miyake, R. Shiina, P. Thalmeier, G. Varelogiannis, C. Geibel, F. Steglich, P. Fulde, T. Komatsubara: Nature **410**, 340 (2001)
[10.87] P. Thalmeier, M. Jourdan, M. Huth: Physik Journal 1, 51 (2002)
[10.88] C. Petrovic, P.G. Pagliuso, M.F. Hundley, R. Moshovich, J.L. Sarrao, J.D. Thompson, Z. Fisk, P. Monthoux: J. Phys. CM **13**, L337 (2001a)
[10.89] C. Petrovic, R. Moshovich, M. Jaime, P.G. Pagliuso, M.F. Hundley, J.L. Sarrao, Z. Fisk, J.D. Thompson: Europhys. Lett. **53**, 354 (2001b)
[10.90] P.A. Lee, T.M. Rice, J.W. Serene, L.J. Sham, J.W. Wilkins: Comments Cond. Matter Phys. **12**, 99 (1986)
[10.91] L.P. Gor'kov: Sov. Sci. Rev. A. Phys. Vol. **9**, 1 (1987)
[10.92] H.R. Ott: Progr. Low Temperature Physics, Vol. XI (1987)
[10.93] M. Sigrist and K. Ueda: Rev. Mod. Phys. **63**, 239 (1991)
[10.94] N. Grewe and F. Steglich: *Handbook on the Physics and Chem. of Solids*, Vol. 14, (1991)
[10.95] B.K. Sarma, M. Levy, S. Adenwalla, J.B. Ketterson: *Phys. Acoustics*, Vol. XX, 108 (Acad. Press, 1992)
[10.96] H.R. Ott, H. Rudigier, T.M. Rice, K. Ueda, Z. Fisk, J.L. Smith: Phys. Rev. Lett. **52**, 1915 (1984)
[10.97] D.J. Bishop, C.M. Varma, B. Batlogg, E. Bucher, Z. Fisk, J.L. Smith, Phys. Rev. Lett. **53**, 1009 (1984)

[10.98]   V. Müller, D. Maurer, E.W. Scheidt, Ch. Roth, K. Lüders, E. Bucher, H.E. Bömmel: Solid State Comm. 57, 319 (1986)
[10.99]   P. Thalmeier, B. Wolf, D. Weber, G. Bruls, B. Lüthi, A.A. Menovsky: Physica C 175, 61 (1991)
[10.100] J. Moreno and P. Coleman: Phys. Rev. B 53 R2995, (1996)
[10.101] G. Bruls, D. Weber, B. Wolf, P. Thalmeier, B. Lüthi, A. de Visser, A. Menovsky: Phys. Rev. Lett. 65, 2294 (1990)
[10.102] R. Fisher, S. Kim, B.F. Woodfield, N.E. Phillips, L. Taillefer, K. Hasselbach, J. Flouquet, A.L. Giorgi, J.L. Smith: Phys. Rev. Lett. 62, 1411 (1989)
[10.103] S. Addenwalla, S.W. Lin, Q.Z. Ran, Z. Zhao, J.B. Ketterson, J.A. Sauls, L. Taillefer, D.G. Hinks, M. Levy, B.K. Sarma: Phys. Rev. Lett. 65, 2298 (1990)
[10.104] V. Müller, Ch. Roth, D. Maurer, E.W. Scheidt, K. Lüders, E. Bucher, H.E. Bömmel: Phys. Rev. Lett. 58, 1224 (1987)
[10.105] N.H. van Dijk, A. de Visser, J.J.M. Franse, S. Holtmeier, L. Taillefer, J. Flouquet: Phys. Rev. 48, 1299 (1993)
[10.106] B. Bogenberger, H. v. Löhneysen, T. Trappmann, L. Taillefer: Physica B 186-188, 248 (1993)
[10.107] G.E. Volovik and L.P. Gor'kov: JETP 61, 843 (1985)
[10.108] J.A. Sauls: Adv. Phys.43, 113 (1994)
[10.109] G. Aeppli, E. Bucher, C. Broholm, J.K. Kjems, J. Baumann, J. Hufnagl: Phys. Rev. Lett. 60, 615 (1988)
[10.110] P. Frings, B. Renker, C. Vettier: Physica B+C 151, 499 (1988)
[10.111] R. Joynt: Sup. Sci. Technol. 1, 210 (1988)
[10.112] R. Joynt and L. Taillefer: Rev. Mod. Phys. 74, 235 (2002)
[10.113] K. Machida, M. Ozaki and T. Ohmi: J. Phys. Soc. Jpn. 58, 4116 (1989)
[10.114] P. Thalmeier, B. Wolf, D. Weber, R. Blick, G. Bruls, B. Lüthi: J. Magn. Magn. Mat. 108, 109 (1992)
[10.115] H. Monien, K. Scharnberg, L. Tewordt, D. Walker: Solid State Comm. 61, 581 (1987)
[10.116] A. Schenstrom, M.F. Xu, Y. Hong, D. Bein, M. Levy, B.K. Sarma, S. Addenwalla, Z. Zhao, T. Tokuyasu, D.W. Hess, J.B. Ketterson, J.A. Sauls, D.G. Hinks: Phys. Rev. Lett. 62, 332 (1989)
[10.117] B.S. Shivaram, Y.H. Jeong, T.F. Rosenbaum, D.G. Hinks: Phys. Rev. Lett. 56, 1078 (1986)
[10.118] B. Ellman, L. Taillefer, M. Poirier: Phys. Rev. B 54, 9043 (1996)
[10.119] M. J. Graf, S.K. Yip, J.A. Sauls: Phys. Rev. B 62, 14393 (2000)
[10.120] M. Boukhny, G.L. Bullock, B.S. Shivaram, D.G. Hinks: Phys. Rev. Lett. 73, 1707 (1994)
[10.121] F. Steglich, J. Aarts, C.D. Bredl, W. Lieke, D. Meschede, W. Franz, H.Schäfer: Phys. Rev. Lett. 43, 1982 (1979)
[10.122] W. Sun, M. Brand, G. Bruls, W. Assmus: Z. Phys. B 80, 249 (1990)
[10.123] S. Nüttgens, W. Assmus, D. Finsterbusch, B. Wolf, G. Bruls, B. Lüthi: Physica B 259-261, 681 (1999)
[10.124] M. Lang, R. Modler, U. Ahlheim, R. Helfrich, P.H.P. Reinders, F. Steglich, W. Assmus, W. Sun, G. Bruls, D. Weber, B. Lüthi: Physica Scripta Vol. T39, 135 (1991)
[10.125] G. Bruls, B. Lüthi, D. Weber, B. Wolf, P. Thalmeier: Physica Scripta T 35, 82 (1991)

[10.126] B. Wolf, G. Bruls, D. Finsterbusch, I. Kouroudis, B. Lüthi: Physica B **211**, 233 (1995)
[10.127] G. Bruls, B. Wolf, D. Finsterbusch, P. Thalmeier, I. Kouroudis, W. Sun, W. Assmus, B. Lüthi, M. Lang, K. Gloos, F. Steglich, R. Modler: Phys. Rev. Lett. **72**, 1754 (1994)
[10.128] D. Finsterbusch: Thesis Universität Frankfurt (2003)
[10.129] O. Stockert, E. Faulhaber, G. Zwicknagl, N. Stüsser, H.S. Jeevan, M. Deppe, R. Borth, R. Küchler, M. Loewenhaupt, C. Geibel, F. Steglich: Phys. Rev. Lett. **92**, 136401 (2004)
[10.130] F. Thomas, P. Ayache, I.A. Famine, J. Thomasson, C. Geibel: J. Phys. CM **8**, L51 (1996)
[10.131] B. Wolf, W. Sixl, R. Graf, D. Finsterbusch, G. Bruls, B. Lüthi, E.A. Knetsch, A.A. Menovsky, J.A. Mydosh: J. Low Temp. Phys. **94**, 307 (1994)
[10.132] B. Golding, D.J. Bishop, B. Batlogg, W.M. Haemmerle, Z. Fisk, J.L. Smith: Phys. Rev. Lett. **55**, 2479 (1985)
[10.133] B. Batlogg et al.: Phys. Rev. Lett. **55**, 1319 (1985)
[10.134] T.M. Rice and M. Sigrist: J. Phys. CM **7**, L643 (1995)
[10.135] M. Sigrist, D. Agterberg, A. Furusaki, C. Honerkamp, K.K. Ng, T.M. Rice, M.E. Zhitomirsky: Physica C **317–318**, 134 (1999)
[10.136] A.P. Mackenzie, R.K.W. Haselwimmer, A.W. Tyler, G.G. Lonzarich, Y. Mori, S. Nishizuki, Y. Maeno: Phys. Rev. Lett. **80**, 161 (1998)
[10.137] Y. Maeno, T.M. Rice, M. Sigrist: Physics Today, January 2001, p. 42
[10.138] H. Matsui, Y. Yoshida, A. Mukai, R. Settai, Y. Onuki, H. Takei, N. Kimura, H. Aoki, N. Toyota: Phys. Rev. B **63**, 060505 (2001)
[10.139] J.D. Gavenda: Phys. Rev.B **66**, 216501 (2002)
[10.140] C. Lupien, W.A. MacFarlane, C. Proust, L. Taillefer, Z.Q. Mao, Y. Maeno: Phys. Rev. Lett.**86**, 5986 (2001)
[10.141] N. Okuda, T. Suzuki, Z. Mao, Y. Maeno, T. Fujita: J. Phys. Soc. Jpn. **71**, 1134 (2002)
[10.142] M. Sigrist: Progr. Theor. Phys. **107**, 917 (2002)
[10.143] D. Jerome and H.J. Schulz: Adv. Phys. **31**, 299 (1982)
[10.144] A.M. Kini, U. Geiser, H.H. Wang, K.D. Carlson, J.M. Williams, W.K. Kwok, K.G. Vandervoort, J.E. Thompson, D.L. Stupka, D. Jung, M.H. Whangbo: Inorg. Chem. **29**, 2555 (1990)
[10.145] M. Lang: Superconductivity Review **2**, 1 (1996)
[10.146] M. Lang and J. Müller: The Physics of Superconductivity II, **453**, ed. K.H. Bennemann and J.B. Ketterson (Springer, 2003)
[10.147] S. Lefebvre, P. Wzietek, S. Brown, C. Bourbonnais, D. Jérome, C. Mézière: Physica B **312–313**, 578 (2002)
[10.148] M. Yoshizawa, Y. Nakamura, T. Sasaki, N. Toyota: Solid State Comm. **89**, 701 (1994)
[10.149] T. Simizu, N. Yoshimoto, M. Nakamura, M. Yoshizawa: Physica B **281&282**, 896 (2000)
[10.150] D. Fournier, M. Poirier, M. Castonguay, K.D. Truong: Phys. Rev. Lett. **90**, 127002 (2003)
[10.151] B. Lussier, B. Ellmann, L. Taillefer: Phys. Rev. B **53**, 5145 (1996)
[10.152] M. A. Tanatar, M. Suzuki, S. Nagai, Z.Q. Mao, Y. Maeno, T. Ishiguro: Phys. Rev. Lett. **86**, 2649 (2001)

[10.153] K. Izawa, H. Takahashi, H. Yamaguchi, Y. Matsuda, M. Suzuki, T. Sasaki, T. Fukasi, Y. Yoshida, R. Settai, Y. Onuki: Phys. Rev. Lett. **86**, 2653 (2001a)
[10.154] M. Chiao, B. Lussier, B. Ellmann, L. Taillefer: Physica B 230.**232**, 370 (1997)
[10.155] K. Izawa, H. Yamaguchi, Y. Matsuda, H. Shishido, R. Settai, Y. Onuki: Phys. Rev. Lett. 87, 057002 (2001b)
[10.156] M. Chiao, R. W. Hill, C. Lupien, L. Taillefer, P. Lambert, R. Gagnon, P. Fournier: Phys. Rev. B **62**, 3554 (2000)
[10.157] I. Vekhter, P.J. Hirschfeld, J.P. Carbotte, E.J. Nicol: Phys. Rev. B **59**, R9023 (1999)
[10.158] P. Thalmeier and K. Maki: Euroohys. Lett. **58**, 119 (2002)
[10.159] E.D. Bauer, N.A. Frederick, P.C. Ho, V.S. Zapf, M.B. Maple: Phys. Rev. B **65**, 100506R (2002)
[10.160] M. Ichioka, N. Nakai, K. Machida: J. Phys. Soc. Jpn. **72**, 1322 (2003)
[10.161] J. Goryo: Phys. Rev. B **67**, 184511 (2003)
[10.162] K. Miyake, H. Kohno, H. Harima: J. Phys. CM **15**, L275 (2003)
[10.163] K. Izawa et al.: to be publ. (2003)
[10.164] T. Goto, Y. Nemoto, K. Sakai, T. Yamaguchi, M. Akatsu, T. Yanagisawa, H.Hazama, K. Onuki, H. Sugawara, H. Sato, Phys. Rev. B **69**, 180511(R) (2004)
[11.1]   H.-Y. Hao and H.J. Maris: Phys. Rev. B **64**, 064302 (2001)
[11.2]   S.O. Demokritov, B. Hillebrands, A.N. Slavin: Phys. Reports **348**, 441 (2001)
[11.3]   M. Schöbinger and R.J. Jelitto: Z. Phys. B **43**, 199 (1981)
[11.4]   A. Mooradian and A.L. Mc Whorter: Phys. Rev. Lett. **19**, 849 (1967)
[11.5]   R. Bowers, C. Legendy, F. Rose: Phys. Rev. Lett. **7**, 339 (1961)
[11.6]   P. Cotti, A. Quattropani, P. Wyder: Phys. kondens. Mat. **1**, 27 (1963)
[11.7]   R.S. Brazis, J.K. Furdyna and J.K. Pozela: Phys. stat. sol. (a) **53**, 11 (1979)
[11.8]   S.J. Buchsbaum and J.K. Galt: Phys. Fluids **4**, 1514 (1961)
[11.9]   E.V. Bezuglyi, N.G. Burma, E.Yu. Deineka, A.M. Stepanenko, V.D. Fil': Low Temp. Phys. **20**, 752 (1994)
[11.10] C.C. Grimes and S.J. Buchsbaum: Phys. Rev. Lett. **12**, 357 (1964)
[11.11] I. Rosenman: Solid State Comm.**3**, 405 (1965)
[11.12] T.G. Blaney: Phil. Mag. **15**, 707(1967)
[11.13] J.J. Quinn, S. Rodriguez: Phys. Rev. Lett. **11**, 552 (1963)
[11.14] D.M. Langenberg and J. Bok: Phys. Rev. Lett. **11**, 549 (1963)
[11.15] A.A. Abrikosov: Theory of Metals (North Holland, 1988)
[11.16] J.J. Quinn: Arkiv för Fysik **26**, 93 (1964)
[11.17] V.V. Gudkov and J.D. Gavenda: *Magnetoacoustic Polarization Phenomena in Solids* (Springer, 2000)
[11.18] C. Herring and C. Kittel: Phys. Rev. **81**, 869 (1951)
[11.19] A.M. Clogston, H. Suhl, L.R. Walker, P.W. Anderson: J. Phys. Chem. Solids **1**, 129 (1956)
[11.20] M. Sparks: *Ferromagnetic Relaxation Theory* (McGraw-Hill, 1964)
[11.21] R.M. White: *Quantum Theory of Magnetism* (Springer, 1983)
[11.22] L.R. Walker: J. Appl.Phys. **29**, 318 (1958)
[11.23] C. Kittel: Phys. Rev. **110**, 836 (1958)

[11.24] K.B. Vlasov and B.Kh. Ishmukhametov: JETP **9**, 921 (1959)
[11.25] A.S. Borovik-Romanov, S.K. Sinha, editors: *Spin Waves and Magnetic Excitations*, in: Modern Problems in Condensed Matter Sciencs Vol. 22.1 and 22.2, North-Holland (1988)
[11.26] M. Nielsen, H. Bjerrum Moller, A.R. Mackintosh: J. Appl. Phys. **41**, 1174 (1970)
[11.27] H. Chow and F. Keffer: Phys. Rev. B **7**, 2028 (1973)
[11.28] J. Jensen and S.B. Palmer: J. Phys. C **12**, 4573 (1979)
[11.29] E.A. Turov and V.G. Shavrov: Sov. Phys. Usp. **26**, 593 (1984)
[11.30] J.R. Eshbach: Phys. Rev. Lett. **8**, 357 (1962)
[11.31] R. Pauthenet: Annales Phys. **3**, 424 (1958)
[11.32] E. Schlömann: Adv. in *Quantum Electronics*, p. 437 (Columbia Univ. Press, 1961)
[11.33] E. Schlömann: J. Appl. Phys. **35**, 159 (1964)
[11.34] J.R. Eshbach: J. Appl. Phys. **34**, 1298 (1963)
[11.35] R.W. Damon and H. Van De Vaart: Proc. IEEE, 348 (1965)
[11.36] J. Jorzick, E. Demokritov, B. Hillebrands: Phys. Rev. Lett. **88**, 047204 (2002)
[11.37] H. Suhl: J. Phys. Chem. Sol. **1**, 201 (1957)
[11.38] E. Schlömann, J.J. Green, U. Milano: J. Appl. Phys. **31**, 386S (1960)
[11.39] E. Schlömann: J.Appl.Phys. **31**, 1647 (1960)
[11.40] R.L. Melcher and D.I. Bolef: Phys. Rev. **178**, 864 (1969): Phys. Rev. **186**, 491 (1969)
[11.41] R.L. Melcher: Phys. Rev. **B2**, 733 (1970a)
[11.42] R.L. Melcher: J. Appl. Phys. **41**, 1412 (1970b)
[11.43] D.S. Rimai: Phys. Rev. B **16**, 4069 (1977)
[11.44] D.S. Rimai, M.A. Dunn, J.C. Jamieson, M.H. Manghnani: Phys. Rev. B **19**, 3215 (1979)
[11.45] Y. Shapira: J. Appl. Phys. **42**, 1588 (1971)
[11.46] K. Tani: J. Phys. C **3**, 1597 (1970)
[11.47] M.E. Fisher: AIP Conf. Proc. **24**, 273 (1974)
[11.48] H. Rohrer: AIP Conf. Proc. **24**, 268 (1974)
[11.49] Y. Shapira and C.C. Becerra: Phys. Rev. Lett. **38**, 358 (1977)
[11.50] L.E. Svistov, V.L. Safonov, J. Löw, H. Benner: J. Phys. C.M. **6**, 8051 (1994)
[12.1] E. Dagotto and T.M. Rice: Science **271**, 618 (1996)
[12.2] Y. Narumi, R. Sato, K. Kindo, M. Hagiwara: J. Magn. Magn. Mat. **177–181**, 685 (1998)
[12.3] L.J. de Jongh and A.R. Miedema: Adv. Phys. **23**, 1 (1974)
[12.4] H.J. Mikeska and M. Steiner: Adv. Phys. **40**, 191 (1991)
[12.5] H.A. Bethe: Z. Phys.71,205 (1931)
[12.6] S. Eggert, I. Affleck, M. Takahashi: Phys. Rev. Lett. **73**, 332 (1994)
[12.7] S. Eggert: Phys. Rev.B **53**, 5116 (1996)
[12.8] A. Klümper: Eur. Phys. J. B**5**, 677 (1998)
[12.9] S. Takagi, H. Daguchi, K. Takeda, M. Mito, M. Takahashi: J. Phys. Soc. Jpn. **65**, 1934 (1996)
[12.10] P.R. Hammar, M.B. Stone, D.H. Reich, C. Broholm, P.J. Gibson, M.M. Turnbull, C.P. Landee, M. Oshikawa: Phys. Rev. B **59**, 1008 (1999)
[12.11] K. Fabricius, M. Karbach, U. Löw, K.H. Mütter: Phys. Rev. B **45**, 5315 (1992)

[12.12] B. Wolf, S. Zherlitsyn, B. Lüthi, N. Harrison, U. Löw, V. Pashchenko, M. Lang, G. Margraf, H.W. Lerner, E. Dahlmann, F. Ritter, W. Assmus, M. Wagner: Phys. Rev. B **69**, 092403 (2004)
[12.13] L.N. Bulaevskii: Sov. Phys. Solid State **11**, 921 (1969)
[12.14] F.D.M. Haldane: Phys. Rev. B **25**, 4925 (1982)
[12.15] W. Shiramura, K. Takatsi, B. Kurniawan, H. Tanaka, H. Uekusa, Y. Ohashi, K. Takizawa, H. Mitamura, T. Goto: J. Phys. Soc. Jpn. **67**, 1548 (1998)
[12.16] M. Oshikawa, M. Yamanaka, I. Affleck: Phys. Rev. Lett. **78**, 1984 (1997)
[12.17] M. Oshikawa: Phys. Rev. Lett. **84**, 1535 (2000)
[12.18] K. Totsuka: Eur. Phys. J. B **5**, 705 (1998)
[12.19] F.D.M. Haldane: Phys. Lett. **A93**, 464 (1983)
[12.20] D. P. Almond and J.A. Rayne: Phys. Lett. **24A**, 295 (1975)
[12.21] Y. Trudeau, M. Poirier, A. Caillé: Phys. Rev B **46**, 169 (1992)
[12.22] K. Kakurai, M. Steiner, R. Pynn, J.K. Kjems: J. Phys. C. M. **3**, 715 (1991)
[12.23] I. Affleck, J. Phys. CM **1**, 3047 (1989)
[12.24] J. Fivez, H. De Raedt, B. De Raedt: Phys. Rev. B **21**, 5330 (1980)
[12.25] J. Fivez: Phys. Rev. B **26**, 6169 (1982)
[12.26] H.J. Mikeska: J. Phys. C **11**, L29 (1978)
[12.27] M. Barmatz, L.R. Testardi, M. Eibschütz, H.J. Guggenheim: Phys. Rev. B **15**, 4370 (1977)
[12.28] E. Käräjämäki, R. Laiho, T. Levola: Phys. Rev. B **25**, 6474 (1982)
[12.29] S. Schmidt, S. Zherlitsyn, B. Wolf, H. Schwenk, B. Lüthi, H. Tanaka: Europhys. Lett. **53**, 591 (2001) addendum: 54, 554 (2001)
[12.30] H. Tanaka, W. Shiramura, T. Takatsu, B. Kurniawan, M. Takahashi, K. Kamishima, K. Takizawa, H. Mitamura, T. Goto: Physica B **246-247**, 230 (1998)
[12.31] B. Kurniawan, M. Ishikawa, T. Kato, H. Tanaka, K. Takizawa, T. Goto: J. Phys. C. M **11**, 9073 (1999)
[12.32] M. Matsumoto: Phys. Rev. B **68**, 180403R (2003)
[12.33] B. Wolf, S. Zherlitsyn, S. Schmidt, B. Lüthi: Europhys. Lett. **48**, 182 (1999)
[12.34] B. Wolf, S. Schmidt, H. Schwenk, S. Zherlitsyn, B. Lüthi: J. Appl. Phys. **87**, 7055 (2000)
[12.35] M. N. Braud, M. Couzi, N.B. Chauh, C. Coursville, B. Gallois, C. Hauw, A. Meresse: J. Phys. CM **2**, 8229 (1990)
[12.36] D. Finsterbusch: Diploma thesis, Universität Frankfurt (1991)
[12.37] B.S. Shastri and B. Sutherland: Physica B **108**, 1069 (1981)
[12.38] S. Miyahara and K. Ueda: Phys. Rev.Lett. **82**, 3701 (1999)
[12.39] S. Miyahara and K. Ueda: Phys. Rev. B **61**, 3417 (2000)
[12.40] E. Müller-Hartmann, R.P. Singh, C. Knetter, G.S. Uhrig: Phys. Rev. Lett. **84**, 1808 (2000)
[12.41] K. Totsuka, S. Miyahara, K. Ueda: Phys. Rev. Lett. **86**, 520 (2001)
[12.42] K.Y. Choi, Yu.G. Pashkevich, K.V. Lamonova, H. Kageyama, Y. Ueda, P. Lemmens: Phys. Rev. B **68**, 104418 (2003)
[12.43] S. Zherlitsyn, S. Schmidt, B. Wolf, H. Schwenk, B. Lüthi, H. Kageyama, Y. Ueda, K. Ueda: Phys. Rev. B **62**, R6097 (2000)
[12.44] B. Wolf, S. Zherlitsyn, S. Schmidt, B. Lüthi, H. Kageyama, Y. Ueda: Phys. Rev. Lett. **86**, 4847 (2001)

References

[12.45] H. Kageyama, K. Yoshimura, R. Stern, N.V. Mushnikov, K. Onizuda, M. Kato, K. Kosuge, C.P. Slichter, T. Goto, Y. Ueda: Phys. Rev. Lett. **82**, 3168 (1999)

[12.46] Z. Weihong, C.J. Hamer, J. Oitmaa: Phys. Rev. B **60**, 6608 (1999)

[12.47] H. Kageyama, M. Nishi, N. Aso, K. Onizuka, T. Yosihama, K. Nukui, K. Kodama, K. Kakurai, Y. Ueda: Phys. Rev. Lett. **84**, 5876 (2000)

[12.48] H. Kageyama, K. Onizuka, Y. Ueda, M. Nohara, H. Suzuki, H. Takagi: JETP **90**, 129 (2000)

[12.49] M. Lang et al.: unpubl. (2003)

[12.50] G. Gorodetsky, B. Lüthi, M. Eibschütz, H.J. Guggenheim: Phys. Lett. **56A**, 479 (1976)

[12.51] B. Wolf, B. Lüthi, S. Schmidt, H. Schwenk, M. Sieling, S. Zherlitsyn, I. Kouroudis: Physica B**294–295**, 612 (2001)

[12.52] S. Zherlitsyn, B. Lüthi, B. Wolf, J.L. Sarrao, Z. Fisk, V. Zlatic: Phys. Rev. B**60**, 3148 (1999)

[12.53] B. Wolf, S. Zherlitsyn, B. Lüthi, A.A. Menovsky: Acta Phys. Pol. **34**, 3148 (2003)

[12.54] S. Zherlitsyn, B. Wolf, B. Lüthi, M. Lang, P. Hinze, E. Uhrig, W. Assmus, H. R. Ott, D. P. Young, Z. Fisk: Eur.Phys.J. B**22**, 327 (2001)

[12.55] T. Saito, T. Terashima, M. Azuma, M. Takano, T. Goto, H. Ohta, W. Utsumi, P. Bordet, D.C. Johnston: J. Solid State Chem. **153**, 124 (2000)

[12.56] D. C. Johnston, T. Saito, M. Azuma, M. Takano, T. Yamauchi, Y. Ueda: Phys. Rev. B **64**, 134403 (2001)

[12.57] A. W. Garrett, S. E. Nagler, D.A. Tennant, B.C. Sales, T. Barnes: Phys. Rev. Lett. **79**, 745 (1997)

[12.58] G. S. Uhrig and B. Normand: Phys. Rev. B **58**, R14705 (1998)

[12.59] G. S. Uhrig and B. Normand: Phys. Rev. B, **63**, 134418 (2001)

[12.60] A.V. Prokofiev, F. Büllesfeld, W. Assmus, H. Schwenk, D. Wichert, U. Löw, B. Lüthi: Eur. Phys. J. B **5**, 313 (1998)

[12.61] Y. Narumi, S. Kimura, S. Hirai, K. Kindo, H. Schwenk, S. Schmidt, B. Wolf, B. Lüthi, T. Saito, M. Azuma, M. Takano: Physica B **294-295**, 71 (2001)

[12.62] B. Lüthi, B. Wolf, S. Zherlitsyn: *Concepts in Electron Correlation*, Proc. of ARW NATO workshop Hvar, A.C. Hewson, V. Zlatic eds., 179 (Kluwer Acad. Publ., 2003)

[12.63] K. Onizuka, H. Kageyama, Y. Narumi, K. Kindo, Y. Ueda, T. Goto: J. Phys. Soc. Jpn. **69**, 1016 (2000)

[12.64] H. Nojiri, H. Kageyama, K. Onizuka, Y. Ueda, M. Motokawa: J. Phys. Soc. Jpn. **68**, 2906 (1999)

[12.65] B. Lüthi, B. Wolf, S. Zherlitsyn, S. Schmidt, H. Schwenk, M. Sieling: Physica B **294–295**, 20 (2001)

[12.66] K. Kodama, M. Takigawa, M. Horvatic, C. Berthier, H. Kageyama, Y. Ueda, S. Miyahara, F. Becca, F. Mila: Science **298**, 395 (2002)

[12.67] S. Miyahara, F. Becca, F. Mila: Phys. Rev. B **68**, 024401 (2003)

[12.68] B. Kurniawan, H. Tanaka, K. Takatsu, W. Shiramura, T. Fukuda, H. Nojiri, M. Motokawa: Phys. Rev. Lett. **82**, 1281 (1999)

[12.69] M. Matsumoto and M. Sigrist: (to be published) (2004)

[12.70] H. Nojiri, Y. Tokunaga, M. Motokawa: J. Phys. (Paris) **49**, C8-1459 (1988)

[12.71] N. Stüsser, U. Schotte, A. Hoser, M. Meschke, M. Meissner, J. Wosnitza: J. Phys. C. M. **14**, 5161 (2002)
[12.72] A. N. Vasil'ev, V. V. Pryadun, D. I. Khomskii, G. Dhalenne, A. Revcolevschi, M. Isobe, Y. Ueda: Phys. Rev. Lett. **81**, 1949 (1998)
[12.73] M. Hofmann, T. Lorenz, G. S. Uhrig, H. Kierspel, O. Zabara, A. Freimuth, H. Kageyama, Y. Ueda: Phys. Rev. Lett. **87**, 047202 (2001)
[12.74] K. Kudo, T. Noji, Y. Koike, T. Nishizaki, N. Kobayashi: J. Phys. Soc. Jpn. **70**, 1448 (2001)
[12.75] A. V. Sologubenko, K. Giannó, H.R. Ott, U. Ammerahl, A. Revcolevschi: Phys. Rev. Lett. **84**, 2714 (2000)
[12.76] C. Hess, C. Baumann, U. Ammerahl, B. Büchner, F. Heidrich-Meisner, W. Brenig, A. Revcolevschi: Phys. Rev. B**64**, 184305 (2001)
[12.77] H. Schwenk, S. Zherlitsyn, B. Lüthi, E. Morre, C. Geibel: Phys. Rev. B **60**, 9194 (1999)
[12.78] M. Matsuda and K. Katsumata: Phys. Rev. B **53**, 12201 (1996)
[12.79] D. Koenig, U. Löw, S. Schmidt, H. Schwenk, M. Sieling, W. Palme, B. Wolf, G. Bruls, B. Lüthi, M. Matsuda, K. Katsumata: Physica B **237–238**, 117 (1997)
[12.80] K. Kudo, S. Ishikawa, T. Noji, T. Adachi, Y. Koike, K. Maki, S. Tsuji, K. Kumagai: J. Phys. Soc. Jpn. **70**, 437 (2001)
[12.81] R.L. Douglass: Phys. Rev. **129**, 1132 (1963)
[12.82] B.Lüthi: J. Phys. Chem. Sol.**23**, 35 (1962)
[12.83] G. Slack and D.W. Oliver: Phys. Rev. B**4**, 592 (1971)
[12.84] D.J. Sanders and D. Walton: Phys. Rev.B **15**, 1489 (1977)
[12.85] J. Takeya, I. Tsukada, Y. Ando, T. Masuda, K. Uchinokura: Phys. Rev. B **62**, R9260 (2000)
[12.86] H. Miike and K. Hirakawa: J. Phys. Soc. Jpn. **38**, 1279 (1975)
[12.87] M. Köppen, M. Lang, R. Helfrich, F. Steglich, P. Thalmeier, B. Schmidt, B. Wunl, D. Pankert, H. Benner, H. Aoki, A. Ochiai: Phys. Rev. Lett. **82**, 4548 (1999)
[12.88] A.V. Sologubenko and H.R. Ott: in: *Strong Interactions in Low Dimensions*, ed. D. Baeriswyl, L. Degiorgi (Kluwer, 2004)
[12.89] R.E. Peierls: *Quantum Theory of Solids* (Oxford University Press, 1955)
[12.90] S. Roth: *One-Dimensional Metals* (VCH, 1995)
[12.91] W. Kohn: Phys. Rev. Lett. **2**, 393 (1959)
[12.92] B. Renker, H. Rietschel, L. Pintschovius, W. Gläser, P. Brüesch, D. Kuse, M.J. Rice: Phys. Rev. Lett. **30**, 1144 (1973)
[12.93] B. Renker, H. Rietschel, L. Pintschovius, W. Gläser, P. Brüesch, D. Kuse, M.J. Rice: Phys. Rev. Lett. **32**, 836 (1974)
[12.94] S. Kurihara, H. Fukuyama and S. Nakajima: J. Phys. Soc. Jpn. **47**, 1403 (1979)
[12.95] L.N. Bulaevskii: Sov. Phys. Uspekhi **18**, 131 (1975)
[12.96] H. Doi, H. Nagasawa, T. Ishiguro, S. Kagoshima: Solid State Comm. **24**, 729 (1977)
[12.97] M. Saint-Paul, P. Monceau, F. Levy: Solid State Comm. **67**, 581 (1988)
[12.98] M. Saint-Paul, S.Holtmeier, R. Britel, M. Monceau, R. Currat, F. Levy: J. Phys. C.M **8**, 2021 (1996)
[12.99] E. Pytte: Phys. Rev. B**10**, 4637 (1974)
[12.100] C. Ecolivet, M. Saint-Paul, G. Dhalenne, A. Revcolevschi: J. Phys. C. M. **11**, 4157 (1999)

[12.101] M. Saint-Paul, G. Reményi, N. Hegmann, P. Monceau, G. Dhalenne, A. Revcolevschi: Phys. Rev. B **52**, 15298 (1995)
[12.102] M. Poirier, M. Castouguay, A. Revcolevschi, G. Dhalenne: Phys. Rev. B **51**, 6147 (1995)
[12.103] M. Saint.Paul, N. Hegmann, G. Reményi, P. Monceau, G. Dhalenne, A. Revcolevschi: J. Phys. CM **9**, L231 (1997)
[12.104] T. Goto, B. Lüthi, R. Geick, K. Strobel: J. Phys. C **12**, L303 (1979)
[12.105] T. Goto, B. Lüthi, R. Geick, K. Strobel: Phys. Rev. B **22**, 3452 (1980)
[12.106] T. Goto, M. Yoshizawa, A. Tamaki, T. Fujimura: J. Phys. C **15**, 3041 (1982)
[12.107] S. Zherlitsyn, A.A. Stepanov, V.D. Fil', V.P. Popov: Sov. J. Low Temp. Phys. **15**, 712 (1989)
[12.108] T. Suzuki, Y. Masahito, T. Goto, T. Yamakami, M. Takahashi, T. Fujimura: J. Phys. Soc. Jpn. **52**, 1669 (1983)
[12.109] A. Yoshihara, J.C. Burr, S.M. Mudare, E.R. Bernstein, J.C. Raich, J. Chem. Phys. 803816 (1984)
[12.110] A. Yoshihara, T. Suzuki, T. Yamakami, T. Fujimuro: J.Phys.Soc.Jpn. **54**, 3376 (1985)
[12.111] R. Blinc, B. Zeks, R. Kind: Phys. Rev. B **17**, 3409 (1978)
[12.112] Y. Ishibashi and I. Suzuki: J. Phys. Soc.Jpn. **53**, 903 (1984)
[12.113] R. Geick and B. Lüthi: J.Phys. Soc. Jpn. **54**, 3199 (1985)
[12.114] M. Couzi, J. de Physique: I **1**, 743 (1991)
[12.115] T. Nikuni, M. Oshikawa, A. Oosawa, H. Tanaka: Phys. Rev. Lett. **84**, 5868 (2000)
[12.116] Ch. Rüegg, N. Cavadini, A. Furrer, H. U. Güdel, K. Krämer, H. Mutka, A. Wildes, K. Habicht, P. Vorderwisch: Nature **423**, 62 (2003)
[12.117] E.Ya. Sherman, P. Lemmens, B. Busse, A. Osawa, H. Tanaka: Phys.Rev. Lett. **91**, 057201 (2003)
[13.1] M. Fink: Physics Today, March 1997 p. 34
[13.2] V.V. Gudkov and J.D. Gavenda: *Magnetoacoustic Polarization Phenomena in Solids* (Springer, 2000)
[13.3] C.D. Bredl, F. Steglich, K.D. Schotte: Z. Phys. B **29**, 327 (1978)
[13.4] M. Loewenhaupt, B.D. Rainford, F. Steglich: Phys. Rev. Lett. **25**, 1709 (1979)
[13.5] B. Lüthi and C. Lingner: Z. Phys. B **34**, 157 (1979)
[13.6] E. Fawcett, V. Plazhnikov, H. Klimker: Phys. Rev. B **43**, 8531 (1991)
[13.7] S. Roth, A. Wurzinger, H.G. Purwins: Phys. Lett. 72A, 242 (1979)
[13.8] R. Schefzyk, W. Lieke, F. Steglich, T. Goto, B. Lüthi: J. Magn. Magn. Mat. **45**, 229 (1984)
[13.9] J. Rouchy, P. Morin and E. du Tremolet de Lacheisserie: J. Magn. Magn. Mat.**23**, 59 (1981)
[13.10] J. Rouchy and A Waintal: Solid State Comm. **17**, 1227 (1975)
[13.11] J.R. Cullen, S. Rinaldi, G.V. Blessing: J. Appl. Phys. **49**, 1960 (1978)
[13.12] T.J. Moran and B. Lüthi: Phys. Rev. **187**, 710 (1969)
[13.13] W.P. Mason: Phys. Rev. **82**, 715 (1951)
[13.14] E. du Tremolet de Lacheisserie: *Magnetostriction* (CRC Press, 1993)
[13.15] R. Becker and W. Döring: *Ferromagnetismus* (Springer, 1939)
[13.16] R.M. Bozorth: *Ferromagnetism* (Van Nostrand, Princeton, 1951)
[13.17] L. Néel: J. phys. radium **5**, 241 (1944)

References 415

[13.18] G. Simon: Annalen Phys. **7**, 23 (1958)
[13.19] G. Gorodetsky, B. Lüthi, B.M. Wanklyn: Solid State Comm. **9**, 2157 (1971)
[13.20] W.F. Brown Jr.: *Magnetoelastic Interactions* (Springer, 1966)
[13.21] H.F. Tiersten: J. Math. Phys. **5**, 1298 (1964)
[13.22] D.E. Eastman: Phys. Rev. **148**, 530 (1966)
[13.23] R.L. Melcher: Phys. Rev. Lett. 25, 1201 (1970)
[13.24] R. L. Melcher: Proc. Int. School of Physics "Enrico Fermi", Course LII (Acad. Press, 1972)
[13.25] V. Dohm and P. Fulde: Z. Phys. B **21**, 369 (1975)
[13.26] T. Goto, A. Tamaki, T. Fujimura, H. Unoki: J. Phys. Soc.Jpn. **55**, 1613 (1986)
[13.27] J. Jensen: Phys. Rev. B **37**, 9495 (1988)
[13.28] A. Einstein and W.J. de Haas: Verh. d. D. Phys. Ges. **17**, 152 (1915)
[13.29] P.S. Wang and B. Lüthi: Phys. Rev. B **15**, 2718 (1977)
[13.30] L. Bonsall and R.L. Melcher: Phys. Rev. B **14**, 1128 (1976)
[13.31] R.J. Birgeneau, E. Bucher, L. Passell, K.C. Turberfield: Phys. Rev. B**4**, 718 (1971)
[13.32] V. Dohm: Z. Phys. B **23**, 153 (1976)
[13.33] J. Rouchy and. E. du Tremolet de Lacheisserie: Z. Phys. B **36**, 67 (1979)
[13.34] T. Suzuki, T. Goto, T. Fujimura, S. Kunii, T. Suzuki, T. Kasuya: J. Magn. Magn. Mat. **52**, 261 (1985)
[13.35] S. Nimori, M. Kataoka, T. Goto, G. Kido: Phys. Rev. B **67**, 224103 (2003)
[13.36] P. Thalmeier and B. Lüthi: *Handbook on the Physics and Chemistry of Rare Earths*, Vol. 14, K.A. Gschneider, Jr, and L. Eyring editors (North Holland, 1991)
[13.37] C. Kittel: Phys. Rev. **110**, 836 (1958)
[13.38] K.B. Vlasov and B.Kh. Ishmukhametov: JETP **9**, 921 (1959)
[13.39] B.I. Kochalaev: Sov. Phys. Solid State **4**, 1145 (1962)
[13.40] J.W. Tucker: J. Phys. C **6**, 255 (1973)
[13.41] P. Thalmeier and P. Fulde: Z. Phys. B**29**, 299 (1978)
[13.42] B. Lüthi: Appl. Phys. Lett. **8**, 107 (1966)
[13.43] B. Lüthi: Appl. Phys. Lett. **6**, 234 (1965)
[13.44] B. Lüthi: J. Appl. Phys. **37**, 990 (1966)
[13.45] A.N. Grishmanovskii et al.: Sov. Phys. Solid State **14**, 2050 (1973)
[13.46] H. Matthews and R.C. Le Craw: Phys. Rev. Lett.**8**, 397 (1962)
[13.47] R. Guermeur, J. Joffrin, A. Levelut and J. Penne: Solid State Comm.**5**, 369 (1967)
[13.48] A.V. Pavlenko et al.: Sov. Phys. Solid State **11**, 2673 (1970)
[13.49] V.V. Lemanov, A.V. Pavlenko, A.N. Grishmanovsky: Sov. Phys. JETP **32**, 389 (1971)
[13.50] R. Guermeur, J. Joffrin, A. Levelut, J. Penne: Solid State Comm. **6**, 519 (1968)
[13.51] M. Boiteux, P. Doussineau, B. Ferry, J. Joffrin, A. Levelut: Phys. Rev. B **4**, 3077 (1971)
[13.52] Landau–Lifshitz: *Electrodynamics of Continuous Media* (Pergamon Press, 1960)
[13.53] W. Pauli: Optik und Elektronentheorie, ETH Zürich (1948)
[13.54] G. Kluge and G. Scholz: Acustica **16**, 60 (1966)

[13.55] D.L. Portigal and E. Burstein: Phys. Rev. **170**, 673 (1968)
[13.56] A.S. Pine: Phys. Rev. B **2**, 2049 (1970)
[13.57] J. Joffrin and A. Levelut: Solid State Comm. **8**, 1573 (1970)
[13.58] M.G. Cottam and D.R. Tilley: *Introduction to Surface and Superlattice Excitations* (Cambridge Univ. Press, 1989)
[13.59] R. Camley: Surf. Sci. Reports **7**, 103 (1987)
[13.60] J. Heil, B. Lüthi and P. Thalmeier: Phys. Rev. B **25**, 6515 (1982)
[13.61] R.W. Damon and J.R. Eshbach: J. Phys. Chem. Solids **19**, 308 (1961)
[13.62] P. Grünberg and F. Metawe: Phys. Rev. Lett. **39**, 1561 (1977)
[13.63] J.R. Sandercock and W. Wettling: J. Appl. Phys. **50**, 7784 (1979)
[13.64] R.Q. Scott and D.L. Mills: Phys. Rev.B **15**, 3545 (1977)
[13.65] R. Camley and P. Fulde: Phys. Rev. B **8**, 4137 (1984)
[13.66] J. Heil, I. Kouroudis, B. Lüthi and P. Thalmeier: J. Phys. C **17**, 2433 (1984)
[13.67] L. Remer, E. Mohler, W. Grill, B. Lüthi: Phys. Rev. B **30**, 3277 (1984)
[13.68] L. Remer, B. Lüthi, H. Sauer, R. Geick, R.E. Camley: Phys. Rev. Lett. **56**, 2752 (1986)
[13.69] A. Wixforth, J.P. Kotthaus, G. Weimann: Phys. Rev. Lett. **56**, 2104 (1986)
[13.70] R.L. Willett: in *Composite Fermions*, ed. O. Heinonen (World Scientific, 1998)
[13.71] A. Schenstrom, Y.J. Qian, M. F. Xu, H.P. Baum, M. Levy, B.K. Sarma: Solid State Comm. **65**, 739 (1988)
[13.72] I.L. Drichko, A.M. Dinkonov, I.Yu. Smirnov, Yu.M. Galperin, A.I. Toropov: Phys. Rev. B **62**, 7470 (2000)
[13.73] K. von Klitzing, G. Dorda, M. Pepper: Phys. Rev. Lett. **45**, 494 (1980)
[13.74] D.C. Tsui, H.L. Störmer, A.C. Gossard: Phys. Rev. Lett. **48**, 1559 (1982)
[13.75] R.L. Willett, J.P. Eisenstein, H.L. Störmer, D.C. Tsui, A.C. Gossard, J.H. English: Phys. Rev. Lett. **59**, 1776 (1987)
[13.76] R.L. Willett, M.A. Paalanen, R.R. Ruel, K.W. West, L.N. Pfeiffer, D.J. Bishop: Phys. Rev. Lett. **65**, 112 (1990)
[13.77] J.K. Jain: Phys. Rev. Lett. **63**, 199 (1989)
[13.78] R.L. Willett, R.R. Ruel, K.W. West, L.N. Pfeiffer: Phys. Rev. Lett. **71**, 3846 (1993)
[13.79] O. Heinonen: editor, *Composite Fermions* (World Scientific, 1998)
[14.1] N.E. Byer and H.S. Sack: J. Phys. Chem. Solids **29**, 677 (1968)
[14.2] E. Kanda, T. Goto, H. Yamada, S. Suto, S. Tanaka, T. Fujita, T. Fujimura: J. Phys.Soc. Jpn. **54**, 175 (1985)
[14.3] R. Windheim and H. Kinder: Phys. Lett. **51A**, 475 (1975)
[14.4] H. Yamada, S. Tanaka, Y. Kayanuma, T. Kojima: J. Phys. Soc. Jpn. **54**, 1180 (1985)
[14.5] I. Zerec, V. Keppens, M.A. McGuire, D. Mandrus, B.C. Sales, P. Thalmeier: Phys. Rev. Lett. **92**, 195502 (2004)
[14.6] W.A. Phillips: J. Low Temp. Phys. **7**, 351 (1972)
[14.7] S. Hunklinger and W. Arnold: Phys. Acoust. XII, 155 (1976)
[14.8] S. Hunklinger and A.K. Raychaudhuri: Progr. in Low Temp. Phys. **9**, 265 (1986)
[14.9] R.C. Zeller and R.O. Pohl: Phys. Rev. B **4**, 2029 (1971)
[14.10] L. Piche, R. Maynard, S. Hunklinger, J. Jäckle: Phys. Rev. Lett. **32**, 1426 (1974)

[14.11] W. Arnold, S. Hunklinger, S. Stein, K. Dransfeld: J. Non-Cryst. Solids **14**, 192 (1974)
[14.12] P.W. Anderson, B.I. Halperin, C.M. Varma: Philos. Mag. **25**, 1 (1972)
[14.13] J. Jäckle: Z. Phys. **257**, 212 (1972)
[14.14] J.E. Graebner and B. Golding: Phys. Rev. B **19**, 964 (1979)
[14.15] G. Weiss and B. Golding: Phys. Rev. Lett. **60**, 2547 (1988)
[14.16] P. Esquinazi, R. König, F. Pobell: Z. Phys. B **87**, 305 (1992)
[14.17] E. V. Bezuglyi, A.L. Gaiduk, V.D. Fil, S. Zherlitsyn, W.L. Johnson, G. Bruls, B. Lüthi, B. Wolf: Phys. Rev. B **62**, 6656 (2000)
[14.18] S. Kettemann, P. Fulde, P. Strehlow: Phys. Rev. Lett. **83**, 4325 (1999)
[14.19] P. Strehlow, M. Wohlfahrt, A.G.M. Jansen, R. Haueisen, G. Weiss, C. Enss, S. Hunklinger: Phys. Rev. Lett. **84**, 1938 (2000)
[14.20] C. Enss: Physica B **316–317**, 12 (2002)
[14.21] T.J. Moran, N.K. Batra, R.A. Verhelst, A.M. de Graaf: Phys. Rev. B **11**, 4436 (1975)
[14.22] J.A. Mydosh: Spin Gasses, Taylor and Francis (1993)
[14.23] A. Langsdorf and W. Assmus: J. Crystal Growth **192**, 152 (1998)
[14.24] R. Sterzel, C. Hinkel, A. Haas, A. Langsdorf, G. Bruls, W. Assmus: Europhys. Lett. **49**, 742 (2000)
[14.25] N. Vernier, G. Bellessa, B. Perrin, A. Zarembowitch, M. de Boissieu: Europhys. Lett. **22**, 187 (1993)
[14.26] M.A. Chernikov, A. Bianchi, H.R. Ott: Phys. Rev. B **51**, 153 (1995)

Index

absolute sound velocity   12, 14
acousto-electric effect   177
adiabatic elastic constant   63, 64, 90, 200, 202, 203, 212
Alpher–Rubin effect   164, 165, 169
Alven waves   275, 277
an-harmonic effects   36
anti-ferroquadrupolar   83, 129, 131–133, 137, 140, 141
aspherical Coulomb charge scattering   144, 145
attenuation of sound   7–9, 11, 52, 60, 61, 94–101, 146, 150, 153, 155, 156, 158–161, 164–167, 169, 177–179, 227–229, 252, 257, 258, 262, 264, 266, 328, 361

background elastic constant   49, 50, 57, 61, 161, 301, 365
BCS theory   211, 223, 226, 228, 229, 233, 242, 246, 247, 250, 259
bilinear coupling   57, 113
Bose–Einstein condensation   328
Brillouin scattering   5, 23, 24, 126–129, 151, 155, 191, 240, 286, 301, 326, 327, 348
Brillouin zone   21, 45, 144, 152, 162, 172, 174, 231, 239, 323

CEF crystalline electric field   68, 69, 74–80, 86–89, 121, 126, 127, 133–135, 137–139, 141, 196, 197, 204, 205, 211, 332, 336–338
charge order   110–119, 291, 296, 319, 320, 325

de Haas van Alphen effect   167, 169–175, 349

Debye temperature   45, 50, 51, 65, 133
deformation potential   157–163, 172, 196, 205, 231, 232, 234, 340, 362
dimerization   115, 291, 295–299, 325
dipolar interaction   70, 277, 278, 339
dispersion   44, 45, 61, 86, 102, 125, 127, 128, 200, 301

Einstein–de Haas experiment   336
elastic anisotropy   38, 133
elastic isotropy   27, 38, 242
elastic stability   27, 37, 43
energy gap   224–229, 248, 249, 265, 267, 293–298, 312, 323, 325, 363
ESR electron spin resonance   312, 317, 318
exchange striction   72, 73, 94–96, 101, 137, 150, 296, 300, 306–309, 317, 324, 325, 328

Faraday effect   277, 284, 289, 331, 340–347
Fermi liquid   86, 181, 185, 196, 219–221, 353
ferrimagnet   271, 277, 278, 280, 284, 285, 287, 323, 334, 342, 344
ferro-elastic phase transition   147–150
first order phase transition   55, 56, 58, 103, 106, 114, 121, 130, 136, 328
fluctuation   47, 53, 55, 59, 60, 62, 93, 94, 96–101, 110, 118, 155, 156, 163, 181–184, 186, 187, 195, 199–201, 212, 215, 219, 221, 222, 233, 239, 240, 247, 257, 293, 301, 303, 325, 345
flux line lattice   223, 241–243, 245

geometric resonance   167–170, 349, 350, 353

Index

Ginzburg criterion 62, 63, 93, 111, 127, 146, 225
Grüneisen parameter 49, 50, 65, 73, 76, 108, 186, 191, 193, 194, 196–209, 211–222, 250, 252, 256, 257, 259, 262

heavy fermion 173, 181–188, 196, 197, 199–206, 208, 211–227, 233, 234, 248–252, 257, 258, 268, 310
Heisenberg exchange 70, 80, 94, 103
helicon wave 274, 276, 277
Helmholtz free energy 48, 193, 202

Jahn–Teller, cooperative effect 83, 119–125, 129, 132, 135, 144, 146, 302

Kondo effect 182, 184, 185, 219, 220
Kondo lattice 181, 183–185, 188, 194

Landau theory of phase transition 52–55, 59, 62, 63, 111, 149, 155, 225, 227, 255, 310
Landau–Khalatnikov theory 60–62, 86, 99, 110, 265, 266
lattice dynamics 30, 44–46, 51, 64, 65, 188, 189, 324

magneto-acoustic effect 106, 157, 169, 241, 331, 333, 344, 345
magneto-caloric effect 20, 218, 222, 253, 295, 310, 311
magneto-elastic coupling 25, 71–73, 76, 79–86, 94, 120, 130, 133, 196, 197, 205, 280–283, 288, 306, 318, 332, 338, 340, 344
magneto-strictive 11, 25, 80, 86, 125, 172, 194, 215, 253, 283, 288, 334, 339, 343
magnon 104, 105, 277, 280, 292, 328
marginal dimensionality 62, 63, 93, 103
martensitic phase transition 147, 148, 150, 162, 231
meta-magnetic transition 174, 175, 201, 213–219
microwave ultrasonics 21, 24, 52, 87, 345

mixed valence compounds 110, 111, 182, 183, 188–195, 199, 206, 324
mode–mode coupling theories 59–61, 96–98
multipolar coupling 137, 138, 146

non-Fermi liquid 219–222
nuclear acoustic resonance 91, 92, 100
nuclear magnetic resonance 100

phonon dispersion 44, 45, 188
phonon echoes 16, 17, 153, 154, 363
piezo-distortive ferro-electrics 151

quadrupole moment 70, 71, 98, 119, 120, 137, 145
quantum critical point 219–222, 268, 304

RUS, resonant ultrasound spectroscopy 14–16, 176, 206, 208, 239, 358

SAW, surface acoustic waves 5, 20, 21, 24, 37, 39–43, 79, 80, 106, 196, 246, 247, 331, 347–355
scaling 47, 54, 55, 60, 93, 96, 97, 100, 101, 153, 193, 194, 202–204, 212, 214, 215, 222
SDW, spin density waves 106–108
spin ladders 296, 321–323
spin reorientation phase transition 93, 94, 103, 106, 114, 155, 283, 326
spinwave 94, 104, 105, 277–286, 288, 323, 328, 344, 348
strain–order parameter coupling 52–62, 103–106, 110, 117, 152, 197, 227, 232, 248, 249, 251, 255, 256, 259, 261, 267, 268, 303, 325, 326
superconductivity, conventional 223, 227–229
superconductivity, strong coupling 233, 234, 252
superconductivity, unconventional 221, 223, 247–268
susceptibility, magnetic 62, 77, 78, 82, 193, 214, 215, 217, 218, 302
susceptibility, strain 57, 61, 78, 79, 82, 86, 121, 131, 134, 138, 141, 170, 305, 307, 332, 337, 358, 359, 362

TAFF (thermally assisted flux flow) model 242–245
thermal conductivity 25, 26, 36, 64, 201, 206, 234, 251, 258, 267, 319–323, 359, 360, 365
thermal expansion 25, 36, 49, 50, 64, 73–77, 108, 117, 125, 135, 141, 146, 150, 183, 193, 194, 201, 204, 206–209, 212, 222, 253, 259, 289, 308, 309, 325, 334
thermodynamic potential 25, 47, 48, 202, 203, 214
third order elastic constant 37, 147

transducer 2, 5–9, 11–14, 16, 19–22, 153, 277, 344, 351

Verwey transition 114, 115, 342
vibrating reed technique 5, 8, 16, 240, 300
Voigt effect 284, 333, 340, 341, 343
vortices 223, 225, 241, 242

wave vector 59, 153, 158, 167–169, 273, 347

Zeeman term 53, 70

Index of Materials

AgCd   147
Al   349–351
Al_2O_3 22, 87, 88, 357
$AuAgCd_2$ 147
$AuCuZn_2$ 147

$Ba_{1-x}K_xBiO_3$ 230, 236
$BaMnF_4$ 153
BEDT-TTF   267
Bi   179
$Bi_{12}GeO_{20}$ 153
$Bi_2Sr_2CaCu_2O_{8+x}$Bi-22/2   240
Bi-2223   243

$Ca_2Sr(C_2H_5CO_2)_3$ 155
$Ca_3Rh_4Sn_{13}$ 230, 236
$Ca_9La_5Cu_{24}O_{41}$ 323
CaF_2 22
CdS   17, 22, 178
Ce   183, 186, 195, 199
$Ce_{1-x}La_xRu_2Si_2$ 222
$Ce_{1-x}Th_x$ 190, 191
$Ce_3Pd_{20}Ge_6$ 129, 131
$Ce_{.74}Th_{.26}$ 186
$Ce_xLa_{1-x}B_6$ 132
$Ce_xLa_{1-x}Cu_6$ 185
CeAg   129, 130, 131, 155, 186, 198
$CeAl_2$ 42, 43, 79, 84–86, 185, 186, 198, 205, 332–334, 337, 338, 342, 345, 349
$CeAl_3$ 185, 186, 197, 199, 205
CeAs   204
CeB_6 129, 131–134, 146, 173, 186, 198, 205
$CeBe_{13}$ 183, 186, 193, 195
CeBi   86, 204
$CeCoCu_5$ 249, 268
$CeCoIn_5$ 267

$CeCu_2$ 173
$CeCu_2Si_2$ 173, 185, 186, 197, 204, 205, 212, 213, 250, 251, 258–262, 268
$CeCu_{5.9}Au_{0.1}$ 220
$CeCu_6$ 173, 174, 185, 186, 197, 198, 203, 204, 212, 219, 220
$CeIn_3$ 198
$CeIr_2$ 235
$CeIrIn_5$ 249, 268
CeNi   183, 186, 193, 194
$CeNi_2Ge_2$ 222
$CeNi_5$ 194, 195
CeNiSn   205, 206, 219
CeP   204
$CePb_3$ 186, 198, 205
$CePd_2Al_3$ 131, 186
$CePd_3$ 183, 186, 193, 194, 195
$CeRu_2$ 51, 186, 194, 195, 230, 234, 235, 236, 244, 244, 245, 246
$CeRu_2Si_2$ 186, 197, 199, 203, 204, 208, 209, 212, 213, 214–219
CeSb   173, 204
$CeSn_3$ 185, 186, 193, 194, 195, 207
CeTe   76
CoF_2 289
CoO   94
CoPt   334
Cr   94, 106–108, 289
Cr_2O_3 98, 99, 289, 345, 346
$CsCuCl_3$ 123, 124, 297, 300, 302, 317
CsH_2AsO_4 152
$CsNiCl_3$ 300, 301, 302, 306
$CsNiF_3$ 300, 301
Cu   45, 200
Cu-Al-Zn   147
CuCCP   293, 294, 310, 311, 312
$CuGeO_3$ 117, 118, 292, 296, 323–325
CuPt   337

Index of Materials

CuPzN   293

Dy   94, 99, 100, 103
DyAsO$_4$ 126
DyB$_6$ 132, 133
DyP   135
DyPO$_4$ 129, 213
DySb   136
DyVO$_4$ 126, 127, 144, 331, 335

Er$_{1-x}$Ho$_x$Rh$_4$B$_4$ 236
ErFeO$_3$ 103, 106, 283
ErSb   136
Eu$_{0.6}$Sn$_{0.4}$Mo$_6$S$_8$Br$_{0.1}$ 232, 233, 243, 244
Eu$_3$S$_4$ 111
Eu$_8$Ga$_{16}$Ge$_{30}$ 358
EuB$_6$ 133, 310
EuCrO$_3$ 91
EuCu$_2$Si$_2$ 186, 191
EuNi$_2$(Si$_{1-x}$Ge$_x$)$_2$ 193
EuO   98–100
EuTe   289

Fe$_2$O$_3$ 103, 289
Fe$_2$TiO$_4$ 121, 125
Fe$_{3-x}$Zn$_x$O$_4$ 114, 115
Fe$_3$O$_4$ 102, 103, 106, 110, 111, 113, 114, 334, 342, 343
Fe$_{50}$Ni$_{35}$Cr$_9$ 150
Fe$_{51.5}$Co$_{23.4}$Ni$_{25.1}$ 150
Fe$_{64}$Ni$_{31}$Co$_5$ 150
Fe$_{72}$Pt$_{28}$ 150
FeBO$_3$ 289
FeCl$_2$ 87, 213
FeF$_2$ 94, 95, 289

Ga   275, 351
GaAs   22, 273, 274
GaAs/AlGaAs   39, 351–355
Gd   94, 95, 98–101, 103, 106
Gd$_3$Fe$_5$O$_{12}$ 99, 102, 343
GdAl$_2$ 51, 72
GdAlO$_4$ 289
GdB$_6$ 133
GdCu$_2$Ge$_2$ 137
GdZn   72, 334
Ge   22, 177

HfCo$_2$ 51

HfV$_2$ 230, 233, 234, 251, 252
Ho   94, 99, 100, 102, 103
HoB$_2$C$_2$ 137
HoB$_6$ 132, 133
HoFeO$_3$ 106
HoP   135
HoSb   136
HoVO$_4$ 129, 337

In-Tl   147
InSb   22

K   276
K$_2$NiF$_4$ 303, 308, 309, 326
K$_2$SnCl$_6$ 155
K-Na–tartrate   152
KCl:Li   357, 358
KCN   151, 152, 381
KCP   291, 292, 324
KCuBr$_3$ 328
KCuCl$_3$ 301, 302
KCuF$_3$ 121, 323
KH$_2$PO$_4$ (KDP)   151, 152
KH$_3$(SeO$_3$)$_2$ 152
KMnF$_3$ 92, 153, 154

La$_{1-x}$Ca$_x$MnO$_3$ 110
La$_{1-x}$Ce$_x$Ru$_2$ 235
La$_{1-x}$Sr$_x$MnO$_3$ 118, 119
La$_{2-x}$Ba$_x$CuO$_4$ 238
La$_{2-x}$Sr$_x$CuO$_4$ 155, 237, 239, 308
La$_2$CuO$_4$ 15, 237, 326
La$_3$Ga$_5$SiO$_{14}$ 15
LaAg   131, 160, 162, 163, 165, 172, 173
LaAg$_{0.78}$In$_{0.22}$ 155
LaAg$_x$In$_{1-x}$ 147, 162, 163
LaAl$_2$ 50, 51
LaB$_6$ 133, 173, 174, 340
LaNi   194
LaRh$_2$ 235
LaRu$_2$ 235
LaSb   76
LaSn$_3$ 193
LiKSO$_4$ 15
LiYF$_4$ 32

MACdC   326
MAFeC   326, 327
MAMC   237, 326, 327, 381

Mg   167, 168
MgB$_2$ 230, 237
MgF$_2$ 288
MgO   16, 22, 87, 344
MnF$_2$ 99, 101, 102, 287–289, 336, 337
MnF$_3$ 121
MnP   99–102

Na$_2$[Fe(CN)$_5$NO].2H$_2$O   91
NaCl:OH   357, 359
NaCN   152
NaNO$_2$ 153
NaV$_2$O$_5$ 110, 111, 115–118, 319, 320, 325
Nb$_3$Al   230
Nb$_3$Ge   230
Nb$_3$Sn   230, 231, 247
NbN   247
NbTi   242
Nd$_{2-x}$Ce$_x$CuO$_4$ 106
NdB$_6$ 132, 133
NdSb   136
NH$_4$CuCl$_3$ 296, 297, 301, 302, 313, 317, 318, 328
Ni   42, 43, 98, 99, 102, 334
Ni$_2$(Medpt)$_2$(μ-ox)(H$_2$O)$_2$ 297, 299
Ni$_2$MnGa   150
Ni$_{80}$Fe$_{20}$ 286
Ni$_x$Zn$_{1-x}$Cr$_2$O$_4$ 122, 123
NiCr$_2$O$_4$ 121–123
NiF$_2$ 288
NiTi   147, 148

Pb   161, 246, 247
PbMo$_6$S$_8$ 230, 232, 233
Pd$_{30}$Zr$_{70}$ 363
Pr   81, 82
Pr$_3$S$_4$ 135
Pr$_3$Se$_4$ 135
Pr$_3$Te$_4$ 135
PrAlO$_3$ 129
PrB$_6$ 132, 133
PrCu$_2$ 129, 135
PrFe$_4$P$_{12}$ 220
PrNi$_5$ 76, 77, 137, 138, 146, 194, 206, 308
PrOs$_4$Sb$_{12}$ 268, 269
PrPb$_3$ 129, 137, 138, 139, 146
PrPtBi   129, 135

PrSb   73, 79, 135, 137, 138, 146
PrSn$_3$ 135
PrVO$_4$ 129
Pu   144

Rb$_2$CoF$_4$ 99, 101
Rb$_2$NaHoF$_6$ 155
Rb$_2$NaTmF$_6$ 155
RbH$_2$PO$_4$ 152
RbMnF$_3$ 92, 98–100, 102, 286, 289
RbNiF$_3$ 343

SbSI   17
Si   22, 351
SiO$_2$ 153, 360
Sm$_{0.75}$Y$_{0.25}$S   186, 188, 189, 195
Sm$_{1-x}$Y$_x$S   187, 188, 189
SmB$_6$ 133, 183, 186, 195, 206
SmIn$_3$ 136, 186
SmPd$_3$ 136, 186
SmS   183, 184, 186, 187, 188, 195, 206
SmSb   76, 79
SmSn$_3$ 186
SmTl$_3$ 136, 186
Sn   161, 246
Sr$_{12}$Ca$_2$Cu$_{24}$O$_{41}$ 119
Sr$_{14-x}$Ca$_x$Cu$_{24}$O$_{41}$ 322
Sr$_{14}$Cu$_{24}$O$_{41}$ 119, 319, 321, 323
Sr$_2$RuO$_4$ 15, 172, 175, 176, 248, 249, 251, 252, 264–268
Sr$_8$Ga$_{16}$Ge$_{30}$ 358
SrCu$_2$(BO$_3$)$_2$ 77, 222, 296, 303–310, 313–316, 319, 320
SrTiO$_3$ 153, 155

TaSe$_{42}$I   324
Tb   94, 99, 100, 103, 283, 284
Tb$_{0.3}$Dy$_{0.7}$Fe$_2$ 86, 334, 343
TbAl$_2$ 86
TbAs   136
TbAsO$_4$ 126
TbB$_6$ 132, 133
TbBi   136
TbFe$_2$ 86, 283, 334
TbP   135, 136
TbPO$_4$ 126
TbSb   136
TbVO$_4$ 126, 127, 128
TbZn   86

TeO$_2$ 151, 152
TlCuCl$_3$ 222, 292, 301, 302, 328
TmAsO$_4$ 126
TmCd   129, 130
TmFeO$_3$ 103–105
TmGa$_3$ 129
TMMC   302, 303
TmPO$_4$ 91, 126, 128
TmSb   73–79, 91, 135, 174, 337–340
TmSe   133, 183, 184, 186, 190, 191, 195
TmSe$_{1-x}$Te$_x$ 190
TmTe   129, 133, 134, 137, 146, 191
TmVO$_4$ 126, 127
TmZn   129, 130, 146

U$_{0.2}$Y$_{0.8}$Pd$_3$ 220
U$_3$P$_4$ 139, 144
UAl$_2$ 209
UAs   86
UBe$_{13}$ 174, 186, 250, 251, 262, 264, 265, 268
UCu$_2$Sn   139, 141, 143
UCuAs$_2$ 139, 144
UCuP$_2$ 139, 144
UN   139, 142, 144
UNiSn   139, 141, 143
UO$_2$ 84, 86, 139, 139, 140, 144
UPd$_2$Al$_3$ 131, 186, 211, 248, 250, 251, 262, 267, 268
UPd$_3$ 139, 140, 141, 142, 187, 220
UPt$_3$ 15, 77, 174, 175, 186, 197, 200, 201, 206, 206, 207–210, 212–216, 218, 219, 234, 248, 250–258, 261, 262, 265, 268, 308

URu$_2$Si$_2$ 12, 186, 198, 209–212, 217, 218, 250, 251, 261, 262, 263, 268, 310

V$_3$Ga   230
V$_3$Ge   230, 231
V$_3$Si   230, 232
VO$_2$P$_2$O$_7$ 295, 310, 311, 312, 313, 317

Y$_3$Ga$_{5-x}$Fe$_x$O$_{12}$ 343
YAl$_2$ 51
Yb$_3$Rh$_4$Sn$_{13}$ 230, 236
Yb$_4$As$_3$ 110, 111, 112, 323, 325
YBa$_2$Cu$_3$O$_{7-x}$ 15, 237, 240, 243
YbAgCu$_4$ 186, 191
YbAl$_2$ 183, 186, 191
YbB$_6$ 133
YbCu$_2$Si$_2$ 191
YbFeO$_3$ 106
YbIn$_{0.75}$Ag$_{0.25}$Cu$_4$ 192
YbIn$_{1-x}$Ag$_x$Cu$_4$ 191, 192, 310
YbInCu$_4$ 183, 186, 187, 191, 192, 199
YbPO$_4$ 129
YbRh$_2$(Si$_{0.95}$Ge$_{0.05}$)$_2$ 222
YD$_{0.1}$ 15
YGaIG   344
YIG   278, 279, 282–284, 323, 344

Zn$_2$Mg   365
Zn$_{62}$Mg$_{29}$Y$_9$ 364
ZnCr$_2$O$_4$ 123
ZnF$_2$ 288
ZrCo$_2$ 51
ZrV$_2$ 233, 251

Springer Series in
SOLID-STATE SCIENCES

Series Editors:
M. Cardona   P. Fulde   K. von Klitzing   R. Merlin   H.-J. Queisser   H. Störmer

90 **Earlier and Recent Aspects of Superconductivity**
Editor: J.G. Bednorz and K.A. Müller

91 **Electronic Properties and Conjugated Polymers III**
Editors: H. Kuzmany, M. Mehring, and S. Roth

92 **Physics and Engineering Applications of Magnetism**
Editors: Y. Ishikawa and N. Miura

93 **Quasicrystals**
Editor: T. Fujiwara and T. Ogawa

94 **Electronic Conduction in Oxides**
2nd Edition By N. Tsuda, K. Nasu, A. Fujimori, and K. Siratori

95 **Electronic Materials**
A New Era in MaterialsScience
Editors: J.R. Chelikowsky and A. Franciosi

96 **Electron Liquids**
2nd Edition By A. Isihara

97 **Localization and Confinement of Electrons in Semiconductors**
Editors: F. Kuchar, H. Heinrich, and G. Bauer

98 **Magnetism and the Electronic Structure of Crystals**
By V.A. Gubanov, A.I. Liechtenstein, and A.V. Postnikov

99 **Electronic Properties of High-T_c Superconductors and Related Compounds**
Editors: H. Kuzmany, M. Mehring and J. Fink

100 **Electron Correlations in Molecules and Solids**
3rd Edition By P. Fulde

101 **High Magnetic Fields in Semiconductor Physics III**
Quantum Hall Effect, Transport and Optics By G. Landwehr

101 **High Magnetic Fields in Semiconductor Physics III**
Quantum Hall Effect, Transport and Optics By G. Landwehr

102 **Conjugated Conducting Polymers**
Editor: H. Kiess

103 **Molecular Dynamics Simulations**
Editor: F. Yonezawa

104 **Products of Random Matrices**
in Statistical Physics By. A. Crisanti, G. Paladin, and A. Vulpiani

105 **Self-Trapped Excitons**
2nd Edition By K.S. Song and R.T. Williams

106 **Physics of High-Temperature Superconductors**
Editors: S. Maekawa and M. Sato

107 **Electronic Properties of Polymers**
Orientation and Dimensionality of Conjugated Systems Editors: H. Kuzmany, M. Mehring, and S. Roth

108 **Site Symmetry in Crystals**
Theory and Applications
2nd Edition By R.A. Evarestov and V.P. Smirnov

109 **Transport Phenomena in Mesoscopic Systems**
Editors: H. Fukuyama and T. Ando

110 **Superlattices and Other Heterostructures**
Symmetry and Optical Phenomena 2nd Edition
By E.L. Ivchenko and G.E. Pikus

111 **Low-Dimensional Electronic Systems**
New Concepts
Editors: G. Bauer, F. Kuchar, and H. Heinrich

112 **Phonon Scattering in Condensed Matter VII**
Editors: M. Meissner and R.O. Pohl

Springer Series in
SOLID-STATE SCIENCES

Series Editors:
M. Cardona   P. Fulde   K. von Klitzing   R. Merlin   H.-J. Queisser   H. Störmer

113 **Electronic Properties of High-T_c Superconductors**
Editors: H. Kuzmany, M. Mehring, and J. Fink

114 **Interatomic Potential and Structural Stability**
Editors: K. Terakura and H. Akai

115 **Ultrafast Spectroscopy of Semiconductors and Semiconductor Nanostructures**
By J. Shah

116 **Electron Spectrum of Gapless Semiconductors**
By J.M. Tsidilkovski

117 **Electronic Properties of Fullerenes**
Editors: H. Kuzmany, J. Fink, M. Mehring, and S. Roth

118 **Correlation Effects in Low-Dimensional Electron Systems**
Editors: A. Okiji and N. Kawakami

119 **Spectroscopy of Mott Insulators and Correlated Metals**
Editors: A. Fujimori and Y. Tokura

120 **Optical Properties of III–V Semiconductors**
The Influence of Multi-Valley Band Structures   By H. Kalt

121 **Elementary Processes in Excitations and Reactions on Solid Surfaces**
Editors: A. Okiji, H. Kasai, and K. Makoshi

122 **Theory of Magnetism**
By K. Yosida

123 **Quantum Kinetics in Transport and Optics of Semiconductors**
By H. Haug and A.-P. Jauho

124 **Relaxations of Excited States and Photo-Induced Structural Phase Transitions**
Editor: K. Nasu

125 **Physics and Chemistry of Transition-Metal Oxides**
Editors: H. Fukuyama and N. Nagaosa

126 **Physical Properties of Quasicrystals**
Editor: Z.M. Stadnik

127 **Positron Annihilation in Semiconductors**
Defect Studies. By R. Krause-Rehberg and H.S. Leipner

128 **Magneto-Optics**
Editors: S. Sugano and N. Kojima

129 **Computational Materials Science**
From Ab Initio to Monte Carlo Methods.   By K. Ohno, K. Esfarjani, and Y. Kawazoe

130 **Contact, Adhesion and Rupture of Elastic Solids**
By D. Maugis

131 **Field Theories for Low-Dimensional Condensed Matter Systems**
Spin Systems and Strongly Correlated Electrons.   By G. Morandi, P. Sodano, A. Tagliacozzo, and V. Tognetti

132 **Vortices in Unconventional Superconductors and Superfluids**
Editors: R.P. Huebener, N. Schopohl, and G.E. Volovik

133 **The Quantum Hall Effect**
By D. Yoshioka

134 **Magnetism in the Solid State**
By P. Mohn

135 **Electrodynamics of Magnetoactive Media**
By I. Vagner, B.I. Lembrikov, and P. Wyder